Wound Healing

This book discusses the molecular mechanisms, signaling pathways, and emerging technologies in wound healing. The initial chapters present wound classification and examine the important role of signaling pathways and biomarkers in tissue repair. The relationship between the skin microbiome and wound healing is also explored, emphasizing the influence of microbial communities on acute and chronic wounds. The book also explores the use of animal models in wound healing research, outlining the advantages and limitations of various species in preclinical studies. Additionally, it examines the natural and bio-based approaches, including the use of herbal remedies and animal-derived products in wound healing. Further, the book reviews the revolutionary role of artificial intelligence in wound assessment and precision medicine. Further, the book presents conventional and novel therapeutic strategies, including bioengineered materials and nanotechnology for stem cell-based treatments and extracellular matrix-based therapies. The significance of nutritional support is examined, detailing the role of macronutrients, micronutrients, and hydration in optimizing wound repair.

Key Features

- Covers the molecular mechanisms, classification, and physiological processes involved in tissue repair
- Examines key biological regulators and their roles in wound assessment and healing progression
- Highlights the impact of microbial communities on acute and chronic wound healing
- Discusses the advantages, limitations, and biomedical relevance of various animal models in wound healing studies
- Investigates the role of herbal remedies and animal-derived products in promoting wound healing

This book is intended for researchers and students of regenerative medicine, biomedical sciences, and pharmaceutical sciences.

Wound Healing

Molecular and Therapeutic Synergy

Edited by
Aarif Ali

CRC Press
Taylor & Francis Group
Boca Raton London New York

CRC Press is an imprint of the
Taylor & Francis Group, an **informa** business

Designed cover image: Shutterstock

First edition published 2026
by CRC Press
2385 NW Executive Center Drive, Suite 320, Boca Raton FL 33431

and by CRC Press
4 Park Square, Milton Park, Abingdon, Oxon, OX14 4RN

CRC Press is an imprint of Taylor & Francis Group, LLC

ISBN: 978-1-041-08340-5 (hbk)
ISBN: 978-1-041-08341-2 (pbk)
ISBN: 978-1-003-64495-8 (ebk)

DOI: 10.1201/9781003644958

Typeset in Times
by KnowledgeWorks Global Ltd.

Contents

Preface

Wound healing is a field that occupies a significant position in fundamental biology, translational research, and clinical medicine, showcasing one of the most remarkable examples of the body's natural ability to repair and regenerate. At the same time, it represents one of the most enduring challenges faced by healthcare systems globally. Despite thousands of years of medical advancements, the reality remains that wound healing is an imperfect and often unpredictable process, heavily influenced by age, comorbidities such as diabetes, risks of infection, vascular insufficiency, and even environmental conditions. Acute wounds can transition into chronic, non-healing states that not only affect quality of life but also create significant economic burdens on healthcare systems. Chronic wounds alone impact millions of people worldwide, leading to amputations, disabilities, and a reduced lifespan if not managed properly. Over the past two decades, advancements in molecular biology, genomics, proteomics, and bioinformatics have significantly broadened our understanding of the intricate regulatory networks that drive tissue repair. However, the challenge lies in translating this molecular knowledge into reproducible and effective clinical therapies. Traditional treatments such as dressings, debridement, and topical antimicrobials have saved numerous lives, yet they primarily provide symptomatic relief rather than addressing the fundamental molecular dysfunction that underlies impaired healing. It is within this gap between fundamental knowledge and practical application that new technologies and therapeutic strategies are emerging, with the potential to transform our understanding, diagnosis, and treatment of wounds in the twenty-first century.

This book, titled *Wound Healing: Molecular and Therapeutic Synergy*, serves as both a reference guide and a roadmap for researchers, clinicians, and innovators who are dedicated to understanding the biology of repair and applying these insights to patient-centered solutions. The chapters compiled in this volume encompass a variety of perspectives, including basics of wound healing, molecular signaling pathways, biomarkers, skin microbiome, animal models, psychological aspects, herbal remedies, nutritional support, animal-based products, and therapeutic approaches. More importantly the addition of artificial intelligence (AI) has become a transformative force in the field of wound healing. AI-powered algorithms and machine learning models are currently being developed to assess wound images, forecast healing paths, and customize treatment strategies by incorporating genomic, proteomic, and clinical information.

This volume emphasizes the collaboration of biology, therapeutics, and AI, marking the beginning of a new era in wound management where precision medicine, data science, and regenerative technologies come together. My goal is to offer not only a thorough reference but also a visionary outlook, encouraging clinicians, scientists, and technologists to adopt innovation and teamwork. Wound healing should no longer be seen as a fixed or solely clinical outcome, but rather as a dynamic process enhanced by molecular science, driven by therapeutic ingenuity, and directed by AI. I hope this book will inspire curiosity, promote cross-disciplinary conversations, and ultimately lead to breakthroughs that alleviate suffering, restore health, and elevate the standard of care for patients worldwide.

About the Editor

Aarif Ali holds a Doctorate of Philosophy in Clinical Biochemistry from the University of Kashmir, Srinagar, Jammu, and Kashmir, India, and earned his Masters in Clinical Biochemistry from the University of Kashmir, Srinagar, Jammu, and Kashmir, India. Dr. Aarif Ali serves as a reviewer of several high impact national as well as international scientific journals. He has published the books *Pharmacogenomics: From Discovery to Clinical Implementation*, *Biological Insights of Multi-Omics Technologies in Human Diseases*, *Fundamentals of Biochemistry*, and *Gene Editing by CRISPR-Cas*.

Dr. Ali has collaboratively worked on and published more than 100 research publications in different reputed national as well as international journals. He has been awarded the Young Scientist Award at the AI-CIAS-2024, with his contributions to the scientific community extending to numerous research publications covering various aspects of human health. Additionally, he has pursued continuous professional development through a range of training and certifications, showcasing his commitment to staying at the forefront of his field. Dr. Ali's dedication to research and his passion for contributing to scientific knowledge make him a valuable asset to the academic and scientific community.

1 Overview of Wound Healing

Hafsa Ihsan[1], *Huma Hameed*[1*], *Aroosh Khalid*[1], *Mahtab Ahmad Khan*[1], *Muhammad Zaman*[1], *Muhmmad Jamshaid*[1], *Ana Claudia Paiva-Santos*[2,3], *and Ali Irfan*[4]

[1] Faculty of Pharmaceutical Sciences, University of Central Punjab (UCP), Lahore, Pakistan

[2] Department of Pharmaceutical Technology, Faculty of Pharmacy of the University of Coimbra, University of Coimbra, Coimbra, Portugal

[3] REQUIMTE/LAQV, Group of Pharmaceutical Technology, Faculty of Pharmacy of the University of Coimbra, University of Coimbra, Coimbra, Portugal

[4] The Green Institute of Chemical, Biomedical and Environmental Sciences (GICBES), Lahore, Pakistan

1.1 SKIN

Skin is considered as one of the most multifunctional and vital organs of the human body. It acts as a barrier against many environmental threats, and it also preserves and regulates the body's internal conditions. The sensory functions and layered structural framework make the skin important for overall homeostasis as well as an efficient defensive mechanism against the external environment. Structurally, the human skin is composed of three layers: the first one is the epidermis, the second one is the dermis, and the third one is the hypodermis.

- *Epidermis*: The topmost layer is 'epidermis' that is made up of keratinocytes. These keratinocytes are the main components that act as the protective barrier (Abdo et al., 2020). The epidermis is made up of five distinct layers (from deepest to most superficial):
  - *Stratum Basale (Germinativum)*: This is a single layer of proliferative basal cells that are attached to the basement membrane; it contains melanocytes (which produce pigment), Merkel cells (which act as mechanoreceptors), and basal keratinocytes (Proksch et al., 2008).
  - *Stratum Spinosum*: This layer consists of several layers of keratinocytes that are interconnected by desmosomes; it also houses Langerhans cells, which play a role in the immune response.
  - *Stratum Granulosum*: This layer contains keratinocytes that have keratohyalin granules, contributing to the formation of the skin barrier.
  - *Stratum Lucidum (Present Only in Thick Skin)*: A thin, transparent layer located on the palms and soles.
  - *Stratum Corneum*: The outermost layer, which is made up of anucleate, flattened, keratin-rich corneocytes, embedded in lipid bilayers, serving as the primary barrier function. The epidermis lacks blood vessels and obtains nutrients through diffusion from the underlying dermis.

* Corresponding author: huma.hameed@ucp.edu.pk

DOI: 10.1201/9781003644958-1

- *Dermis*: The dermis, which is situated below the epidermis, is the thickest layer. It is the most structurally significant layer with a depth range of 1.5–4 mm. Fibroblasts are a predominant component in the dermis, and they are the primary cells responsible for the production of collagen and elastin. Collagen and elastin are the proteins that give the skin its strength and flexibility.
- *Hypodermis*: The hypodermis lies directly below the dermis and is mainly composed of adipose tissues and connective tissues. This layer is responsible for contributing to the overall strength and stability of the skin (Rittie, 2016). The slight acidic pH of the skin creates an environment that acts as a natural defense mechanism against the pathogenic microorganisms. The presence of Langerhans cells, located in the epidermis, is important for identifying and responding to potential infectious threats, acting as immune surveillance for the skin (Clayton et al., 2017). Despite all of its protective features, the skin is vulnerable to injury and can break. A compromise of the anatomical and physiological integrity of the tissues, which results in a loss of the normal structure and function, is known as a wound (Sharma et al., 2021).

1.2 WOUND

A wound can be characterized as a disruption in the skin or the mucous membranes which leads to the loss of integrity of the underlying tissues. Wounds can result from any kind of trauma to the skin, like burns, incisions, abrasions, or a combination of these. For each type of wound, the selection of a proper dressing and maintenance is important for reducing the risks of infections and further tissue damage (Sung et al., 2023). The wounds that have a coordinated and sequential healing mechanism without any delay or interruption are referred to as 'acute wounds'. Minor cuts and injuries often heal effectively, but there are certain factors that can interfere with this process, like oxidative stress, underlying medical conditions, and microbial infections. These elements can lead to impaired healing and can develop into a non-healing or inert wound, also referred to as 'chronic wounds' (Xiang et al., 2020). These chronic wounds have specific features like prolonged inflammation, persistent infection, and impaired response of the epidermal cells to the reparative signals (Firlar et al., 2022).

1.3 CLASSIFICATION OF WOUNDS

Different kinds of wounds result in the destruction of organs. Organ destruction is fully recovered after proper care of the wound. For caring of wounds, it is affirmed that the wounds are completely sanitized, covered, and cured to inhibit the contamination (Herman and Bordoni, 2023; Kujath and Michelsen, 2008; Onyekwelu et al., 2017; Wilkins and Unverdorben, 2013). Identification of wound and kind of wound is too much necessary for proper healing. The cause of wound and knowledge of its existence is needed for proper control of wound (Herman and Bordoni, 2023; Levy et al., 2013; Mioton et al., 2013; Vu et al., 2009). This will be feasible when types of wounds are properly differentiated, and the wounds are divided into various categories. Categorization of wounds is also necessary for differentiation of healthy and unhealthy person wounds (diabetic and undernourished people) because their wounds show serious alterations in inflammation and contamination by microorganisms and also show different capacity for proper healing of wounds (Herman and Bordoni, 2023; Jones, 2015; Onyekwelu et al., 2017; Percival, 2002; Vu et al., 2009). When the chances of complexities after operation, such as infection at surgical site, are predicted, then categorization of wounds can be accomplished (Herman and Bordoni, 2023; Kujath and Michelsen, 2008; Levy et al., 2013; Mioton et al., 2013; Onyekwelu et al., 2017; Wilkins and Unverdorben, 2013). The categorization of wounds has been tried many times by various medical experts, but it is complicated due to variations in wounds (Belkaid and Tamoutounour, 2016; Choi and Di Nardo, 2018; Rodrigues et al., 2018). The less inter-scorer reliability is the major difficulty in categorization of wounds (Herman and Bordoni, 2023; Levy et al., 2013; Onyekwelu et al., 2017). Furthermore, the categorization of wounds is not achieved efficiently in surgical wounds of neonates and children that are persistently sick. Distinct categorization is needed for this group of people (Baharestani, 2007; Herman and Bordoni, 2023; Onyekwelu et al., 2017).

1.3.1 Classification Based on Skin Intactness

This category of the wound is centered on the intactness of the skin. Skin intactness is achieved by complete skin renewal after proper wound control and scheme for healing. This classification method is wide and simple. It does not involve further categorization. When the classification of wounds is done on the basis of deep and outer surface wounds, then an alternative variant of same category appears (Meara et al., 2015). There are two categories of this method that are explained in Table 1.1.

TABLE 1.1
Various Wound Types Classified According to Categories of the Wounds

Category	Types of Wounds	Description	References
Intactness of skin	Closed wound	Skin intactness is maintained in this. There is no disturbance in integrity of skin.	Meara et al. (2015)
	Exposed wound	Exposed wounds are caused by skin tearing and result in disappearance of tissue.	Meara et al. (2015)
Cause of wound	Blunt force trauma	This wound occurs when non-sharp thing links with body parts.	Simon et al. (2017)
	Wounds due to burns	These wounds can occur when part of body or any other tissues are damaged due to burning.	Spanholtz et al. (2009)
	Injuries that can penetrate	This results in wound that is categorized as open wound due to puncturing of skin when any external thing penetrates in body.	Kuhajda et al. (2014)
	Surgical wounds	Surgical wounds are the wounds that can occur due to any cut or incision and cause the tissue damage.	Gamelli and He (2003)
Wound healing duration	Non-chronic wounds	These include the minor cuts and injuries like insect bites and burns; these injuries include the damage of tissues that are soft.	Moreo (2005) and Morton and Phillips (2016)
	Abrasion	These wounds include the burns due to rope and scraped knees and occur due to friction between skin and any surface that is considered as rough surface.	Korting et al. (2011)
	High velocity injuries	Tissues are damaged deeply due to gunshot of high-speed.	Hamdan (2006) and Lone et al. (2009)
	Contusions	Bone breakage occurs due to forceful hit on the body in this type of injury. This includes the loss of teeth due to ball hit etc.	Romano et al. (2017)
	Cut injury	Body part cuts due to heavy object falling on the body. This also involves the injuries due to road accidents as dermis is disrupted in this.	Montella et al. (2014) and Bowler (2001)
	Ulcers	These are also called radiation wounds. These include the tissues that are soft due to exposure to ionizing radiations. Therapy by chemotherapeutics is an example of this type.	Jacobson et al. (2017)

(Continued)

TABLE 1.1 (Continued)
Various Wound Types Classified According to Categories of the Wounds

Category	Types of Wounds	Description	References
	Lacerations	Soft tissues of the body are torn in this type. The tearing can be external and internal. Childbirth is an example of laceration.	Pergialiotis et al. (2014)
	Persistent or chronic wounds	Chronic wounds are caused by metabolic disturbances and also termed slowly healed wounds.	Morton and Phillips (2016) and Powers et al. (2016)
	Venous ulcers	These are also known as venous ulcers, stasis ulcers, or dermatitis. These occur in lower ends of legs and occur majorly in old people.	Vasudevan (2014)
	Diabetic ulcers	In this wound the healing of the wounds is difficult for the body and small wound converts into persistent wound due to undermine immune system and dysfunction of peripheral nerves.	Alexiadou and Doupis (2012)
	Decubitus ulcers	These are also known as pressure ulcers and bedsores. These are present in paralysis. Occurs due to inhibited blood flow to tissues and cessation of movements of body parts.	Bhattacharya and Mishra (2015)
	Wounds due to low blood flow	It is called ischemic wounds. It results in less supply of oxygen and nutrients to tissues of body for proper metabolism.	Quirinia (2000)
Origin of wounds	Mechanical wounds	Due to mechanical force on the body, the integrity of skin is disturbed. These wounds are divided into the wounds due to cuts, shots, incisions, etc.	Sveen et al. (2020)
	Chemical wounds		
	Base wounds	These wounds cause more damage than acidic wounds. Breakdown of proteins and cells occurs.	VanHoy et al. (2025)
	Acid wounds	These wounds occur when body comes in contact with acids and results in chest pain and vomiting.	VanHoy et al. (2025)
	Wounds caused by radiation	When the tissues are exposed to radiations, they will be damaged.	Iyer and Balasubramanian (2012)
	Wounds caused by thermal stress		
	Burning wounds	Skin is injured due to contact with heat, hot surface, fire, etc.	Schaefer and Tannan (2021)
	Freezing wounds	This type of wound occurs due to cold temperature and results in embolism due to contracted blood vessels.	Sarabahi and Tiwari (2012)
	Wounds caused by diseases		
	Bone infection	Microorganism (bacteria) is responsible for infection of bones.	Ikpeme et al. (2010)
	Diabetes	The chances of infection increased when the immune system of the body becomes weakened, and dysfunction of peripheral nerves occurs. This will result in the conversion of acute wound into large wounds.	Boniakowski et al. (2017) and Okonkwo and DiPietro (2017)

(Continued)

TABLE 1.1 *(Continued)*
Various Wound Types Classified According to Categories of the Wounds

Category	Types of Wounds	Description	References
	Gangrene	Bacterial infection causes the death of the cells that will result in death of tissue of surrounding area.	Tan et al. (2018)
	Immunosuppressive disorder wounds	These wounds occurred due to weak immune system at the surgical sites.	Raje and Dinakar (2015)
	Surgical wounds	Wound due to cut during surgery.	Onyekwelu et al. (2017)
	Microbial infection	The infection occurs due to entry of microorganisms through incision on the body and results in disease.	Aly (1996)
	Wounds caused by animals	Animal bite is responsible for this type of wounds.	Abrahamian and Goldstein (2011)
Based on the contamination degree	Clean wounds	The operative wounds are not infected and there is no damage inside the body occurred. Contamination of microbial flora of skin is present.	Herman and Bordoni (2023) and Sarabahi and Tiwari (2012)
	Clean-contaminated wounds	When the operative wounds are contaminated with penetration of microbial flora in respiratory and GIT in uncontrolled manner, the clean contaminated wounds are produced.	Devaney and Rowell (2004) and Margenthaler et al. (2003)
	Contaminated wounds	These kinds of wounds are noticeable by the intimation of microbial flora in the already clean area of the body due to cut in sterile procedure from the intestine.	Devaney and Rowell (2004) and Gorvetzian et al. (2018)
	Heavily contaminated wounds	Generally, these are ancient shocking injuries which have dead tissues that include the clinical contagion just like that an infection that is involving the restoration of bowel which includes perforations.	Gorvetzian et al. (2018) and Onyekwelu et al. (2017)

1.3.2 Classification of the Wounds Based on the Cause

The intensity of the wound is based on the procedure of damage and occurrence of concurrent conditions of the patients. This method of classification is explained on the basis of their cause or etiology (Gamelli and He, 2003; Kuhajda et al., 2014; Spanholtz et al., 2009).

1.3.3 Classification of the Wounds Based on the Wound Healing Duration

Recovery of wounds majorly depends on the duration of healing. When wounds are classified on the basis of surgery, it has two types, namely acute and chronic. These two conditions are viewed as source of the etiology and duration of wound healing (Bowler, 2001; Irfan-Maqsood, 2018; Moreo, 2005; Morton and Phillips, 2016; Powers et al., 2016). For chronic wounds the duration for wound healing is approx. six weeks. The wound that can be recovered in less time period is termed acute wound (Bowler, 2001; Irfan-Maqsood, 2018; Jacobson et al., 2017; Korting et al., 2011; Moreo, 2005). The wounds that heal in longer period of time do not pass through the typical phases of healing of wound in proper duration. Wounds that are categorized as chronic wounds stand at inflammation stage of wound healing (Alexiadou and Doupis, 2012; Quirinia, 2000). Vascular assessment is done for chronic wounds because the major key factor for their healing is the installation of cells of immune system through blood. The characterization of chronic wounds is also done through imbalance between the fabrication and destruction of different cells during healing procedure of wounds (Alexiadou and Doupis, 2012; Morton and Phillips, 2016; Powers et al., 2016). Different types of wounds occur based on the time span of healing of wounds, as explained in Table 1.1.

1.3.4 Classification of Accidental Wounds Based on Their Origin

The physical examination of wound provides the cause of injury on the skin. The operations for healing of wounds are yet unknown. Medical experts have problems in finding the compatible order of wounds because every tissue has a different healing perspective (Abrahamian and Goldstein, 2011; Ikpeme et al., 2010; Iyer and Balasubramanian, 2012; Okonkwo and DiPietro, 2017; Schaefer and Tannan, 2021; Sveen et al., 2020; VanHoy et al., 2025). The category has different types based on the origin of wounds, which are explained in Table 1.1.

1.3.5 Classification of the Wounds Based on the Contamination Degree

The inhibition of damage proliferation that occurred due to wounds is necessary. The cleanliness and dressing of the wounds is majorly required. The post-operative wounds are imagined as the wounds that are contaminated, and when some other things come across in contact with the wound, there are major chances of infection. The controlling scheme for wounds depends upon the level of contamination (Table 1.1). The best procedure for controlling the contamination of wounds is to close the wound. The CDC had categorized the wounds based on their contamination degree into four categories that are explained in Table 1.1 (Devaney and Rowell, 2004; Gorvetzian et al., 2018; Herman and Bordoni, 2023; Onyekwelu et al., 2017; Sarabahi and Tiwari, 2012).

1.4 DIFFERENCES IN HEALING MECHANISMS IN VARIOUS WOUND CATEGORIES

Wound healing is not a consistent process; it varies greatly based on the type, cause, depth, and duration of the wound. These differences affect the order, length, and effectiveness of the healing stages. Grasping the unique healing processes associated with various wound types is crucial for customized clinical treatments and for forecasting results.

1.4.1 Post-operation Wound Healing

Wound healing after surgery consistently starts with the formation of a blood clot within the area maintained by the sutured flap, regardless of the type or purpose of surgery. During the inflammatory phase, the inflammatory cells gather around the clot which serves as a temporary matrix for cells from the nearby tissues to migrate and proliferate in the wound area. After this, the healing process progresses as it normally does i.e. inflammation, proliferation, matrix formation and repair, and tissue remodeling and maturation (Kaner et al., 2017). Post-operative morbidity can be a factor for impaired wound healing in the early stages, and this can negatively affect the treatment outcome (Zhao et al., 2021).

1.4.2 Chronic Wound Healing

Skin wounds are classified as either acute or chronic which is based on the underlying pathogenesis (Tottoli et al., 2020). In individuals with a normal wound healing mechanism, acute wounds – caused by trauma, surgery, or burns – usually heal within three weeks to a month. However, the chronic wounds fail to heal properly due to certain factors like infections or abnormalities in the wound microenvironment (Powers et al., 2016). Any disruption in the wound healing process can result in chronic wound development which leads to persistent inflammation, scars, and unresolved proliferation. These chronic wounds often remain resistant to healing despite the conventional treatments like dressings, topical agents, physical therapy, and even skin grafting (Li et al., 2025; Tottoli et al., 2020; Wu et al., 2019). The chronic wounds vary in their pathophysiology and can be a result of underlying medical conditions like diabetes mellitus (Holzer-Geissler et al., 2022). These underlying diseases can cause lack of nutrients and oxygen, hyperinflammation, and other complications like bacterial infections, which worsens the chances of wound healing (Ganesh et al., 2015).

The slow healing rate of chronic wounds is mainly due to the prolonged inflammation, specifically due to the overpopulation of pro-inflammatory macrophages. In contrast, the quantity of anti-inflammatory macrophages is quite low (Krzyszczyk et al., 2018; Saleh et al., 2019). The macrophages in chronic wounds have exhibited a reduced ability to clear out dead cells, which leads to a highly inflammatory environment. This environment results in an excess of inflammatory mediators like interleukin-1β (IL-1β) and tumor necrosis factor-α (TNF-α) (Krzyszczyk et al., 2018). These pro-inflammatory macrophages also release matrix metalloproteinases (MMPs) which are responsible for breaking down the extracellular matrix (ECM) and eventually preventing the progressive wound healing (Raziyeva et al., 2021).

1.4.3 Diabetic Wound Healing

The normal wound healing mechanism is complex, and it involves overlapping of the wound healing phases (Dasari et al., 2021). However, it has been observed that diabetes causes disruption in almost all of these phases (Gorecka et al., 2019).

1.4.3.1 Inflammation in Diabetic Wounds

The diabetic wounds, when compared to the nondiabetic wounds, have a prolonged inflammatory phase which slows the healing process (Zhao et al., 2016). This is because, during inflammation in normal and healthy wound healing, the macrophages that arrive are the M1 macrophages, also known as the phagocytic and pro-inflammatory macrophages. Later on, these M1 macrophages are replaced by the M2, or anti-inflammatory macrophages, which produce the ECM and also support angiogenesis (Davis et al., 2018). Inflammation in diabetic wounds is impaired due to the excessive production of the pro-inflammatory cytokines and the inability of the macrophages to transition from the pro-inflammatory to the anti-inflammatory macrophages (Boniakowski et al., 2017; Davis et al., 2018). The oxidative stress in diabetic wounds also plays a role in accelerating tissue damage and impaired healing (Rosique et al., 2015).

1.4.3.2 Angiogenesis in Diabetic Wounds

Restoring the blood flow to the wound site is an important element in wound healing, and this step takes place in the proliferation phase (Demidova-Rice et al., 2012). Inadequate angiogenesis significantly impacts the normal wound healing mechanism. This happens because of the lack of proangiogenic factors as the reduced number of the macrophages responsible for producing these factors is lower (Okonkwo and DiPietro, 2017). Other than that, the anti-angiogenic factors are increased, while the factors that promote capillary maturation are decreased in diabetic wounds (Patel et al., 2019).

The latter stages of diabetic wound healing have exhibited obstruction of growth and stabilization of the new capillaries due to the lack of mature vasculature. This impairment slows down the healing process and eventually leads to the formation of a chronic wound or the reopening of the wound (Okonkwo and DiPietro, 2017).

1.4.4 Burn Wound Healing

Burn wound healing includes the same wound healing process like inflammation, proliferation, and then remodeling (Wu et al., 2025). To manage a burn wound, the treatment approach should include antibacterial property, support the wound healing process, absorb the exudates efficiently, and also provide mechanical flexibility to the affected site (Cook et al., 2022). However, the conventional treatment options like gauzes, skin grafts, and gel dressings often fail to meet the specific requirements of burn wound care at different stages (Zhao et al., 2023). The dressings, which have a fixed shape, often fail to cover the irregular surfaces of the wounds. The skin transplants are also limited due to the availability of autologous skin and are always at a potential risk of allogenic rejection (Singh et al., 2022). To overcome the complexities that arise during the healing of burn wounds, certain bioactive agents like growth factors, hyaluronic acid, and antibacterial agents – can be used (Babaluei et al., 2023; Liu et al., 2022; Wen et al., 2020). These growth factors, hyaluronic acid, along with elastin, are important because they are the components of the ECM (Li et al., 2023). A promising approach for improving the tissue repair process includes the development of an artificial dermis. Elastin-like proteins (ELP) are utilized as they are excellent in providing mechanical compliance and cell motility (Sun et al., 2023; Wu et al., 2023; Zhang et al., 2023). Hyaluronic acid is also considered as an ideal option for treating burn wounds as it is highly hydrophilic, has great swelling properties, is not that immunogenic, and has pro-wound healing properties (Yang et al., 2020).

1.4.5 Fracture Wound Healing

The process for fracture wound healing as well as bone repair are the postnatal mechanisms that are similar to the developmental stage of skeleton formation during embryogenesis (Gerstenfeld et al., 2003).

These processes start with an anabolic stage, which involves a growth in the tissue volume. This growth is a result of the differentiation and recruitment of stem cells that will eventually develop into skeletal tissues and vascular tissues. Right after that, a cartilaginous callus is formed right next to the fractured area. On the outer edges of the callus, the periosteum expands and the initial bone formation begins (Phillips, 2005). As the cartilage develops, the cells that are responsible for the formation of new blood vessels are recruited to nourish the emerging bone, and they begin to differentiate within the muscle sheath (Einhorn and Gerstenfeld, 2015). This results in an increased blood flow into the wounded area. The cartilage ECM begins to mineralize as the chondrocytes differentiate which marks the end of the anabolic phase, concluding it further with the apoptosis of the chondrocytes (Einhorn and Gerstenfeld, 2015).

The next phase after the anabolic phase is the catabolic phase which is characterized by a reduction in the callus tissue volume. While the cartilage resorption occurs, secondary bone is formed. The process of the resorption of the mineralized matrix by the osteoclasts and the development of the new bone by osteoblasts are known as coupled remodeling. The bone marrow space is reestablished, and the bone tissue and structure are regenerated. In the end, vascular remodeling takes place, and blood flow is returned to a normal rate (Holstein et al., 2013). These processes overlap each other, even though they occur in a specific sequence, resulting in a continuous regeneration and healing process (Einhorn and Gerstenfeld, 2015).

1.5 WOUND HEALING

After the skin has been disrupted, a complex and highly coordinated regenerative process is initiated. This is known as wound healing, and its purpose is to restore the physical and structural integrity of the skin. The wound healing process is a series of well-regulated cellular and molecular mechanisms which are designed to repair or replace the damaged tissue (Gonzalez et al., 2016). An efficient wound healing process is important for the prevention of complications like microbial infections and maintenance of homeostasis (Wang et al., 2024). This process involves the coordination of many tissue types and cells. This process regulates the cell migration, proliferation, deposition of ECM, remodeling, as well as inflammation and angiogenesis. Minor injuries and wounds usually take a few days to heal, but the wounds resulting from severe illness, major surgical procedures, or trauma can take up to several weeks to resolve. Such types of wounds tend to form fibrotic scars that have the ability to impair tissue functions. A deeper understanding of the cellular and molecular events that take place during the wound healing process can help develop advanced therapeutic strategies that can minimize scarring and treat chronic wounds effectively (Pena and Martin, 2024). The tissue-specific signals provided by the ECM proteins, like collagen, are known to be bioscaffolds for the skin as they are able to mimic the natural cellular microenvironment (Mouw et al., 2014). These proteins are also important for certain other wound healing pathways involving cell migration, adhesion, tissue regeneration, and remodeling (Pena and Martin, 2024). In cell adhesion, the main role in re-epithelization is played by specialized anchoring structures that connect the epithelial cells to the basement cells, called hemidesmosomes (Liu et al., 2019). Wounds have a huge impact on global health, affecting millions of individuals every year. The associated costs have been estimated to exceed $28 billion (Sen, 2023). In developed countries, 4 million cutaneous wounds are reported every year. This number is slightly higher in underdeveloped countries (Chen et al., 2023). Wounds have affected patients for centuries, which puts the patient and the caregivers under a lot of burden. Due to this, wounds have been considered as a 'silent epidemic' (Lindholm and Searle, 2016). The evolution of wound care devices shows how they progressed from simple topical applications to advanced, multifunctional technologies.

1.6 HISTORICAL PERSPECTIVE ON WOUND HEALING

Management of wounds has transitioned from simple cell-targeting ointments applied topically to complex approaches involving the development of biosensors and portable diagnostic tools. These advancements focus on continuous improvements toward more precise, responsive, and patient-oriented wound care solutions. This notable transformation of wound care devices is illustrated below (Figure 1.1), and it illustrates the collective acceleration toward betterment in this field (Wang et al., 2024).

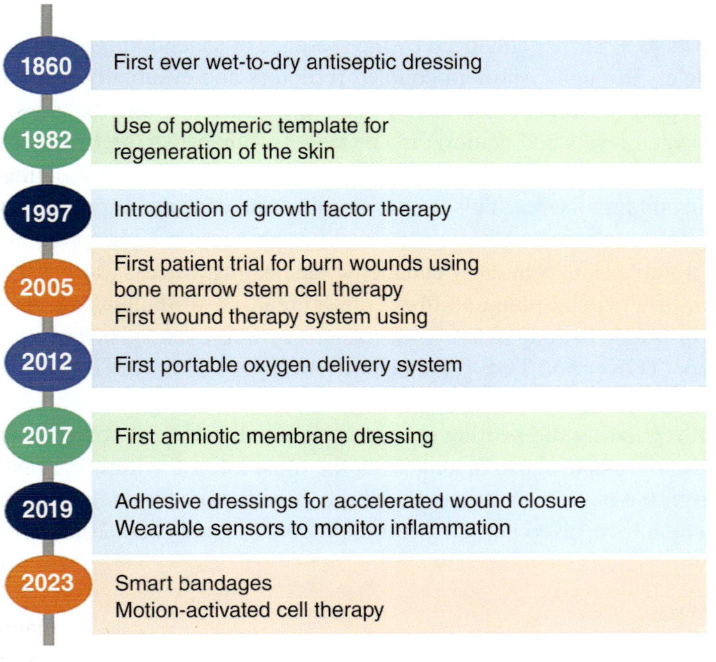

FIGURE 1.1 Timeline showing important breakthroughs in wound healing and management.

Rapid advancements are majorly attributed to the integration of multidisciplinary technologies – like biotechnology, nanotechnology, and digital health technology. This collaboration is the basis of the improvements made in wound care solutions (Sen, 2023).

1.7 WOUND HEALING PHASES

Wound healing is a complex and dynamic process that can progress or regress, depending upon various internal or external factors like the patient's overall health, the environmental conditions, and the presence or absence of any present infection or underlying health condition (Desai et al., 2023). The longer a wound takes to heal, the higher the chance of severe complications like amputations, organ damage, and even death (Hemphill and Nathanson, 2023). Wound healing is generally divided into four overlapping stages: hemostasis, inflammation, proliferation, and remodeling (Chen et al., 2023; Pang et al., 2023).

1.7.1 Hemostasis

The first step after there has been a wound is hemostasis. This phase occurs rapidly to form a blood clot, and its main purpose is to stop further bleeding from the site of injury (Zhou et al., 2023). Hemostasis can be further divided into three steps – coagulation, primary hemostasis, and secondary hemostasis. Right after trauma on the skin, the platelets stick to the exposed area of the wound and become activated. This activation results in the formation of fibrin. This activated fibrin creates a mesh-like scaffold, acting as a biological glue that enhances the adherence and aggregation of platelets and stops the blood from flowing out of the wound. The platelet–fibrin complex forms a clot and stabilizes the wound site. This is done in the coagulation phase (Tennent et al., 2007; Zhang and Zhao, 2020).

Simultaneously, the neuronal reflex mechanisms are triggered due to the injury of the vessel wall. As a result, rapid vasoconstriction takes place with the help of the vascular smooth muscles. The vasoconstriction is activated by the release of endothelin from the damaged endothelium, as well as from the catecholamines from the surrounding damaged cells. This is considered as primary hemostasis (Godo and Shimokawa, 2017). Vasoconstriction, along with the platelet plug formation, plays the key role in secondary hemostasis, making sure that a potentially life-threatening blood loss is prevented. Hemostasis is further enhanced by the presence of subendothelial collagen. This collagen activates the platelets through G-protein-coupled receptors and eventually facilitates the formation of a stable platelet plug (Pradhan et al., 2017). The vasoconstriction only happens for a few minutes before the low oxygen levels and acidosis in the vessel walls cause the blood vessels to relax and dilate again. This vasodilation, along with increased vessel wall permeability, leads to the leakage of blood into the surrounding tissues. This causes swelling and can sometimes restart the bleeding. To avoid this, the platelets release certain substances like sphingosine-1-phosphate which also helps in the formation of a stable clot, eventually enhancing the healing process (Sorg and Sorg, 2023).

The formation of a platelet plug and fibrin mesh creates a thrombus, and this starts the initial signaling events in wound healing by releasing various cytokines and growth factors like transforming growth factors (TGF-α and TGF-β), platelet-derived growth factor (PDGF), epidermal growth factor (EGF), insulin-like growth factor (IGF), and interleukin-1 (IL-1). These mediators play an important role in regulating the healing process, specifically when it comes to the recruitment of immune cells. The release of these mediators is the most intense within the first hour of platelet-activation; however, the platelets continue to secrete these factors for up to seven days, maintaining the cellular participants involved and the healing process (Senzel et al., 2009).

1.7.2 Inflammation

The second phase in wound healing after hemostasis is inflammation. In the inflammatory phase cellular debris, along with the invading microorganisms, are removed from the wound site. The inflammatory response recruits and activates the immune cells, specifically the neutrophils and the macrophages, which digest the bacteria, foreign particles, and nonviable tissues. Phagocytosis is essential for the wounds to progressively heal because wounds that contain unresolved bacterial load are unlikely to heal effectively (Sorg and Sorg, 2023). Thrombin also activates the release of certain chemical signals which call for more immune cells, deliver essential nutrients, and stimulate the production of additional growth factors. This will accelerate the healing process and also enhance the body's immune system against potential infections (Tuli et al., 2023). The intensity and the duration of the immune response is the main difference between acute and chronic inflammation. Acute inflammation is a short-term and low-intensity response, whereas chronic inflammation is a prolonged and heightened response. After the damaged tissue has been repaired and the infection is removed, the inflammatory process concludes, and this is known as the resolution of inflammation (Meizlish et al., 2021).

In between the acute and chronic states is a transitional state known as the OxInflammation which happens when oxidative stress interferes with the inflammation during wound healing. In such a case, the neutrophils produce certain immunological effectors, like interleukins (ILs) and interferon-γ (IFN-γ), which play an important part in recruiting the macrophages to the wound site (Funes et al., 2018; Kadomoto et al., 2021; Lopes et al., 2024; Murray, 2017). Once all the bacteria are cleared up from the wound area, the neutrophils gradually become less active over several days. The neutrophils must be removed in order to continue the healing process further. This removal can be done by either extrusion or apoptosis. Apoptosis is a form of programmed cell death, and it prevents excessive inflammation and tissue damage. Extrusion takes place at the wound surface and can be visibly seen as wound slough when cellular debris is being expelled (Sorg and Sorg, 2023). Pro-inflammatory macrophages exhibit the M1 phenotype, and they play an important role in recognizing and engulfing pathogens within the intracellular phagosomes. Apart from phagocytosis, the macrophages are responsible for secreting essential signaling molecules which activate keratinocytes, endothelial cells (ECs), and fibroblasts. This takes wound healing to the next phase, which is proliferation (Slauch, 2011).

1.7.3 Proliferation

This phase of the wound healing mechanism begins with the release of proangiogenic factors released from the activated platelets. These factors initiate the tissue repair mechanisms by forming a granulation tissue. This term is derived from the histological examination of the newly developed tissue as it has a granular appearance. During this phase, a new stroma is formed by fibroblasts, new blood vessels are also developed, which is known as angiogenesis, ECM is deposited, and collagen is also synthesized. Collectively, these processes restore the wound environment and support further tissue regeneration (Sorg and Sorg, 2023). Elastin is also formed in this phase, and fibroblasts develop into myofibroblasts which results in the constriction of the wound area (Lu et al., 2023). The blood clot that forms on the surface of the wound not only facilitates hemostasis and reduces blood loss but also acts as a reservoir for the growth factors. This wound microenvironment, which is now proliferated with signaling molecules, promotes the neovascularization as it is important for restoring blood perfusion for the provision of oxygen and nutrients that support the cellular metabolism throughout the healing process. The proliferation phase can be divided into two stages – the proangiogenic stage that is characterized by angiogenesis, and the anti-angiogenic stage. In the second stage, vascular pruning and maturation occur which lead to a stable, newly formed vascular network (Sorg et al., 2018).

Epithelialization begins just a few hours after any injury, which is marked by the transformation of stationary, cobblestone-shaped keratinocytes into migratory, flattened keratinocytes which cover the wound bed (Stone et al., 2016). To enable this migration, the keratinocytes must first loosen their intercellular and basal attachments, which are maintained by hemidesmosomes and desmosomes (Pastar et al., 2014). Once the keratinocytes are in the center of the wound, their migration stops and this signals the completion of wound coverage (Grambow et al., 2021).

1.7.4 Remodeling

In most clinical contexts, complete epithelialization is considered as the last stage of wound healing; however, the healing process actually continues for months or years. Many studies have shown that disruption during this phase can impair the healing process, leading to chronic wounds or abnormal scar formation known as keloids or hypertrophic scars (Wernig et al., 2017; Yates et al., 2010). In this final phase of wound healing, the neovasculature is regressed, granulation tissue is broken down, and the ECM undergoes remodeling. Collagen type III, which is majorly involved in the early granulation tissue, is replaced by a stronger collagen type I (Caley et al., 2015; Visse and Nagase, 2003). The collagen fibers help to close the wound fully, while the damaged cells and debris involved in the healing process are cleared through apoptosis. The skin in the wounded area becomes stronger as collagen cross-linking occurs (Sangnim et al., 2024).

Myofibroblasts also play an important role by producing MMPs as well as their inhibitors which are known as tissue inhibitors of metalloproteinases (TIMPs). This balance between MMPs and TIMPs is important for the proper remodeling of ECM and regeneration of stable tissues (Caley et al., 2015; Visse and Nagase, 2003).

1.8 WHY IS WOUND HEALING IMPORTANT?

Wound healing is important for preserving the skin's integrity as it acts as the body's first line of defense against various problems like infections, environmental threats, and blood loss. The body's immediate response to an injury is to stop the bleeding, which is done in the hemostasis phase by constriction of the blood vessels, activation of platelets and collagen at the injury site. The fibrin clot formed due to the activation of platelets acts as a barrier to pathogens. Efficient wound healing is necessary to ensure that the tissues restore their normal structure and function, and it also prevents impaired wound healing that can lead to complications like chronic wounds, infections, and scarring (Salvo and Sandoval, 2022). Wound healing is mainly impaired due to prolonged inflammation which disrupts the normal process and is a major factor in formation of chronic wounds. Instead of progressing through the natural healing processes, the wound becomes stuck in the inflammatory phase. This leads to continuous tissue damage, impaired cellular functions, and delayed repair of the injury site. Persistent inflammation also increases infection risks and prevents the formation of healthy tissues, eventually resulting in disrupted regeneration of wounds (Qian et al., 2016). The main task of the inflammatory phase is to clear the wound site of potential pathogens, cell debris, and dead tissues (Velnar et al., 2009). This involves macrophages and neutrophils which arrive at the wound site through chemotaxis and clear the pathogens. After the bacteria are eliminated, the neutrophils undergo apoptosis – a programmed and controlled cell death that is necessary to avoid additional tissue damage. The macrophages remain at the site of injury and release growth factors that help initiate tissue repair (Ramasastry, 2005; Velnar et al., 2009). If the leukocytes fail to clear the pathogens or the cell debris, then the transition of the inflammatory phase to the proliferation phase is delayed or completely stopped, leading to prolonged inflammation. The pro-inflammatory signals would keep the wound in a prolonged inflammatory state, and this creates an environment that is difficult to resolve, and it significantly impairs wound healing. This prolonged inflammation, combined with ischemia-induced oxygen and nutrient deprivation, as well as the presence of bacterial colonies are the key factors that contribute to the development of chronic wounds (Velnar et al., 2009).

1.9 FACTORS AFFECTING WOUND HEALING

Wound healing is a meticulously controlled biological process that encompasses inflammation, tissue regeneration, and remodeling. The effectiveness and speed of this process are greatly affected by numerous factors, which can be generally classified as local, systemic, and environmental.

Depending on the context of the wound and the overall health of the individual, these factors can either facilitate, hinder, or complicate the healing process.

1.9.1 Local Factors

Local factors operate directly at the site of the wound, affecting the efficiency, direction, and results of the healing process. Local elements like oxygen tension, microbial load, moisture equilibrium, tissue perfusion, and wound biochemistry are crucial and directly influence wound repair. To manage wounds effectively, it is essential to tackle these local determinants by enhancing hygiene, ensuring proper debridement, optimizing moisture and oxygen levels, and minimizing trauma.

1.9.1.1 Oxygenation

The production, epithelial formation, and development of collagen depend upon oxygen. Development of edema decreases due to constriction of vessels by increased oxygen, and it results in death of bacteria. When more hyperbaric oxygen (pure oxygen) is inhaled, it will lead to serious constriction of vessels and may obstruct the wound healing (Bilic et al., 2005; Kaufman et al., 1983; Sen, 2009). ATP is necessary for metabolism and living cells. Oxygen is necessary for production of ATP. Due to wound the vessels become damaged and the availability of oxygen reduces to injured area (Guo and DiPietro, 2010). For the release of substances that are necessary for blood vessel formation, epithelial formation, and production of growth factors, the low oxygen wound environment is the main part (Bishop, 2008; Darby and Hewitson, 2016; Guo and DiPietro, 2010). However, the production of reactive oxygen species and cells of immune system responsible for inflammation may hinder the wound healing procedure. In addition, the appropriate concentration of oxygen is important for prevention of infection of wounds and production of blood vessels (Darby and Hewitson, 2016).

1.9.1.2 Infection

Infection due to bacteria at injured site results in delaying the healing of wounds (Singh et al., 2017). Infection of wound occurs due to different species of bacteria. When infection occurs, the mechanism of inflammation starts with the arrival of different immune system cells (Guo and DiPietro, 2010). So, endotoxins release from the bacteria in response to phagocytosis by WBCs that results in cell death and inflammation due to increased activity of metalloproteinase, less release of factors responsible for growth, and higher release of cytokines responsible for inflammation (Edwards and Harding, 2004; Tottoli et al., 2020). Although rapid response of immune system for inflammation is initiator for wound healing, persistent inflammation is responsible for impairment of wound healing that affect the regeneration of epithelial and formation of tissues (Guo and DiPietro, 2010; Tottoli et al., 2020).

1.9.1.3 Edema

Edema is responsible for reduction of blood supply from arteries due to hindrance of lymphatic and venous return. The chances of infection and low oxygen supply are increased. Edema is also responsible for obstruction of respiratory movements due to burning of area around chest (Lund et al., 1992).

1.9.1.4 Moisture Balance

Keeping a damp wound environment enhances healing by facilitating cell movement, autolytic debridement, and the formation of new blood vessels. In contrast, a dry wound bed may result in drying out and scab development, hindering the migration of epithelial cells, whereas too much moisture can cause the surrounding tissue to become macerated and heighten the risk of infection (Winter, 1962).

1.9.1.5 Foreign Bodies

The existence of foreign substances, like sutures or debris, can trigger ongoing inflammation and act as a breeding ground for bacterial growth. Likewise, dead tissue needs to be removed to facilitate adequate granulation and epithelial growth. Nonviable tissue serves as both a physical and biochemical obstacle to the healing process.

1.9.1.6 Wound Dimensions, Depth, and Position

Larger and deeper wounds take longer to heal because of the significant tissue loss and the heightened need for cellular activity and matrix remodeling. Wounds located in regions with low blood supply (such as cartilage and tendons) or those subjected to high mechanical stress (like joints) may experience slower healing. Tension on the edges of a wound can hinder closure and result in dehiscence.

1.9.1.7 pH

Normal skin and healing wounds have an acidic pH of approximately 5.5, which facilitates keratinocyte proliferation and migration, regulates protease activity, and prevents bacterial colonization. Conversely, an alkaline wound environment encourages microbial growth and heightens the protease-driven breakdown of ECM and growth factors (Gethin, 2007).

1.9.2 Systemic Factors

Wound healing is influenced not just by the local conditions of the wound but also by systemic factors. These systemic factors play a crucial role in immune competence, cellular signaling, vascular supply, hormonal regulation, and metabolic balance. Often, systemic conditions create an unfavorable internal environment that hinders the smooth transition through the four traditional stages of wound healing – hemostasis, inflammation, proliferation, and remodeling resulting in delayed healing or the formation of chronic wounds.

1.9.2.1 Age and Immune Function

Healing of wounds is impaired due to aging. As the age of person increases, the thinning of layer of epidermis occurs due to changes in the systemic and metabolic activities (Beyene et al., 2020). The response of inflammation changes in aged people that results in the slow arrival of immune cells at the target site, less phagocytosis, and less release of factors responsible for growth (Beyene et al., 2020; Singh et al., 2017). In addition, the regeneration of epithelial activity of fibroblasts is also delayed in elderly people. The reconstruction of collagen also occurs in elderly people (Guo and DiPietro, 2010).

1.9.2.2 Gender

The response of immune system on inflammation and reconstruction phase is majorly affected by the different hormones present in males and females. In female patients less total body surface area, increased death rate, and increased complications commonly occur (Chang et al., 2005; Karimi et al., 2017; Ranganathan et al., 2015).

1.9.2.3 Nutritional Status (Role of Vitamins, Minerals, and Proteins)

Different vitamins, carbs, and proteins are necessary for appropriate regeneration of tissues and wound healing (Beyene et al., 2020). Delaying of inflammation process, less production of blood vessels, degradation of fibroblasts, and delayed formation of matrix occur due to low nutrient supply that will lead to impaired healing of wounds (Singh et al., 2017). Several nutrients are needed for proper wound healing, such as alpha linolenic acid, which is required for cell membrane production; vitamin A, required for differentiation of keratin cells; and carbs for production of collagen (Beyene et al., 2020; Singh et al., 2017). Proteins are also responsible for production of collagen. Zinc is responsible for production of deoxyribonucleic acid and ribonucleic acid that is necessary for differentiation of cells at the wound site (Wild et al., 2010). Low amount of iron is responsible for the impaired production of matrix of outside environment of cells. Iron is also necessary for transportation of oxygen because it is important part of hemoglobin (Beyene et al., 2020; Wild et al., 2010).

1.9.2.4 Medications

There are many drugs that affect the healing process of wounds by inhibiting the clotting factors, procedure of inflammation, and multiplication of cells (Guo and DiPietro, 2010). Although corticosteroids have best effect as reducing the inflammation and activating the response of the immune system, these have impact on the growth factors and inflammatory cytokines (responsible for activating the healing procedures) due to their effect in the systemic circulation (Franz, 2007).

Nonsteroidal anti-inflammatory drugs (NSAIDs) (most commonly used drugs) also have negative impact on healing procedure of wounds (Kaushal et al., 2006; Su et al., 2010). Studies on animals have shown that NSAIDs hinder the repairing of tissues by reducing the formation of prostaglandins by hindering cyclooxygenase-1 and cyclooxygenase-2. NSAIDs have negative impact on production of new blood vessels and lead to obstructing the rate of healing (Krischak et al., 2007). Remarkably, the use of NSAIDs for short period of time has positive impact on management of short-term pains. However, patients with diabetes are prone to obstruction of fibroblasts by using NSAIDs (Legendre et al., 2008). This hypothesis can also be applicable for corticosteroids. However, use of steroids in high doses for short period of time also has positive impact on managing the wound healing (Wang et al., 2013). Corticosteroids and vitamin A have opposed their effects on healing of wounds. Effect of corticosteroids on wound healing can be recovered by using vitamin A (Wicke et al., 2000).

Therapies by chemotherapeutics also have serious effects on healing of wounds. These medications have the potential to prolong the inflammatory phase of wound healing that results in less build-up of clotting protein and production of collagen (Gordon et al., 2009). The data of National Surgical Quality Improvement Program shows that after taking the primary chemotherapy, breast surgery has no serious complication on wounds (Decker et al., 2012).

1.9.2.5 Stress

Stress also has a worse effect on wound healing. The impact of stress occurs on the nervous system and causes the more release of some hormones from the hypothalamus (Guo and DiPietro, 2010; Takahashi et al., 2018). Release of adrenaline, noradrenaline, and glucocorticoids reduces the inflammatory cytokine release that results in the impaired wound healing due to prolonged procedure of inflammation (Takahashi et al., 2018).

1.9.3 Environmental Factors

Wound healing is affected not only by local wound conditions and internal systemic factors but is also significantly influenced by external environmental factors. These factors encompass the conditions in the wound care setting, the surrounding physical environment, exposure to contaminants, and external therapeutic interventions. Environmental influences can either promote or impede healing by changing microbial burden, moisture levels, temperature, and the physiological behavior of the cells involved in tissue repair.

1.9.3.1 Exposure to Radiation

The breakage of the strand of the DNA occurs due to exposure to the atomic particles of radiations (Vignard et al., 2013). The free radicals also produced due to exposure to ionizing rays ultimately destroy all the membranes of cells and proteins (Dormand et al., 2005). The wound healing (differentiation of cells) is adversely affected by damage of cells involved in recovery called fibroblast due to radiations (Tibbs, 1997). All these irregularities resulted in the impaired formation of epithelial membrane, increased chances of infection, and ultimately wound healing (Haubner et al., 2012; Tibbs, 1997).

1.9.3.2 Exposure to Smoking

There are many diseases that occur due to smoking like disruption in the healing process of wounds. Many studies proved that the components present in the cigarettes adversely interfere with the healing process of the wounds (Beahrs et al., 2019). Low oxygen supply is one of the crucial mechanisms that is linked with smoking and is responsible for less differentiation of the red blood cells and formation of new blood vessels in the damaged tissues of the body (Guo and DiPietro, 2010). There are also great chances of formation of blood clots due to smoking because it involves in the accumulation of thrombocytes and enhanced viscosity of blood. In addition, the components of cigarettes also disrupt the movement and differentiation of embryonic cells and reformation of the collagen that is involved in healing of wounds (Avishai et al., 2017; Guo and DiPietro, 2010). The chances of infections also increased due to smoking because it decreases the cells of immune system like WBCs and lymphocytes (Beahrs et al., 2019; Guo and DiPietro, 2010).

1.9.3.3 Exposure to Alcohol

Alcoholism also involves in the disruption of process of wound healing even use for shorter periods of time (Guo and DiPietro, 2010). The person's immune system is majorly affected by use of alcohol that will lead to increased chances of infections. Many studies showed that during initial phases the inflammatory response decreased due to less action of the WBCs, and ultimately, if alcohol is used for longer time period, then the inflammation can also be prolonged due to increased recruitment of inflammatory cells (Avishai et al., 2017). In addition, during initial phase of wound healing, the vascular endothelial growth factor receptors are also affected due to alcoholism that leads to decreased formation of new blood vessels at the injured area (Avishai et al., 2017; Guo and DiPietro, 2010). As a result, the low oxygen environment at the injured area due to alcohol use will lead to formation of oxidative species. Alcoholism also disturbs the reformation of matrix and less formation of collagen (Guo and DiPietro, 2010).

1.10 WOUND ASSESSMENT TECHNIQUES

The characterization of wounds is very important which is done by estimation of their severity, ability of its healing, and also assessment of proper strategy to treat the wounds according to their types. The evaluation by visual means is the conventional procedure that is adopted by practitioners. But now there are new methods available that provide the focused evaluation. The assessment of injured area and epithelial formation is done by currently available techniques.

1.10.1 Assessment of Cutaneous Wounds

The inspection of this kind of wound majorly relies on visual means. The evaluation of wounds of feet due to diabetes is also done by visual means. Initially, the evaluation of wounds on feet due to diabetes involves the area of wound, shape and size, neuronal assessment, and assessment of vessels. X-radiation and magnetic resonance imaging are preferred for severe wounds (Holt, 2013).

1.10.2 Assessment of Injuries Due to Burns

For accurate therapeutic outcomes, it is necessary to evaluate the depth of the wounds due to burn (Shupp et al., 2010). For evaluation of the depth of injured area, the practitioners use their experience and visual assessment. If the evaluation is exaggerated, then it will result in surgery that is not required that ultimately enhances the cost, chances of infection, and formation of scars on wound. If under-evaluation is done, then it will result in increase of the wound. The estimation shows that the evaluation by visual means is corrected in fewer cases (Jaskille et al., 2010). To evaluate the burn wounds accurately, the tissue analysis by biopsy is performed which is the best means of assessment. But this is very hurting situation for patient and also results in more traumatic and aggravate condition (Jaskille et al., 2010; Monstrey et al., 2008).

1.10.3 Wound Assessment by Spectrophotometry

As a non-invasive and objective optical technique, spectrophotometry has become a quantitative method for assessing the biochemical and physiological status of wounds, particularly focusing on tissue oxygenation, hemoglobin content, and perfusion. The principle behind spectrophotometry is that different chromophores, including oxyhemoglobin, deoxyhemoglobin, melanin, and bilirubin, exhibit unique light absorption and reflection characteristics across the visible and near-infrared spectrum (400–1000 nm). Through the analysis of light absorption patterns, clinicians are able to infer vital parameters concerning wounds. In wound care, different spectrophotometric devices are used, as mentioned in the following sections.

1.10.3.1 Imaging Spectrophotometry

Imaging of burn wounds is done by a device called spectrophotometric intracutaneous analysis. This involves the formation of computed charts of required colored components that are on the basis of evaluation of spectrum of light that is reflected back. This device has many advantages like it can be handled easily, provides fast imaging, and also has ability to give in-depth knowledge of burn wound (Tehrani et al., 2008). This device uses the light of 400-1000 nm for examination of skin. There is another procedure that can be used called HSI. This technique is useful for assessment of wounds due to diabetes (Khaodhiar et al., 2007) and burn (Calin et al., 2015).

1.10.3.2 Spectrophotometry by Using Near-Infrared (NIR) Imaging

Knowledge about the saturation of oxygen, quantity of components of Hb, and quantity of water (components of dermal tissues of the skin) is taken by using this NIR spectrophotometry. The devices used in this procedure are called near-infrared devices. This device brightens the tissues of the skin at the wavelength NIR. The reflected light is accumulated and examined by a thin glass fiber that transmits the light signals. Hemoglobin with and without oxygen and water are the components of skin that absorb light of wavelength of 700–1100 nm spectrum. Some amount of light is spread, and remaining light is reflected and detected via a device called detector. After this the collected light is evaluated that gives the quantity of different components of skin. The further evaluation is applicable for the evaluation of severity and healing progression of wounds. This technique is helpful in determining the depth of burn wounds and also quantifying the edema. If the burn is severe, then there will be changes in the depth of the burns due to less systemic circulation at injured area, and ultimately variations occur in water and Hb quantities and saturated oxygen. Different studies on animal models represent that if oxygen saturation value and Hb quantified, then the type of burn can be differentiated (Cross et al., 2007; Sowa et al., 2001). The amount of water determination will give further distinction and differentiate all types of burns in accordance with depth (Sowa et al., 2001). The power of near-infrared technique to quantify the water will be greatly helpful in determination of the presence of edema at the site (Cross et al., 2009). Near-infrared spectroscopy is greatly helpful in determining the wounds due to burns. It also gives the information about ulcers due to diabetes and differentiates whether the wound is due to diabetes or not. NIR also provides the prediction about healing process of wounds. A study in rat model describes that enhanced coefficients of near-infrared light represent that the wound is due to diabetes (Papazoglou et al., 2006; Weingarten et al., 2006). The value of increased coefficient is due to the major characteristics of ulcers due to diabetes that are the less perfusion and enhanced mismanagement of the tissues. Other studies that have done by the study on tissues with microscope confirm the correlation of coefficient of absorption with angiogenesis (Weingarten et al., 2008).

1.10.4 Wound Assessment by Imaging

Modern imaging technologies now allow for non-invasive, objective, and reproducible evaluations of wound features such as size, depth, perfusion, tissue composition, and healing progress. Wound imaging is crucial in contemporary wound care as it improves diagnostic precision, directs debridement, tracks treatment responses, and forecasts healing results. This technology is particularly beneficial for managing intricate or chronic wounds, including diabetic foot ulcers, pressure injuries, and venous leg ulcers.

1.10.4.1 Orthogonal Polarization Spectral Imaging (OPS)

OPS device is used to brighten the tissues of skin with using a light that is polarized. Then the light quanta that are spread by the tissues in response to light are collected by a polarizer that is placed in equilateral position against the light (Groner et al., 1999). The principle of this technique is the release of light quantum particles that emits from the surface structures. The Hb is visualized in small circulation by further evaluation (Goertz et al., 2010) and also allows the small micro-sized vessel quantification. This technique has less success in determining the types of wounds due to burns (Milner, 2002; Milner et al., 2005).

1.10.4.2 Spatial Frequency Domain Imaging

This technique includes the portion of curved design of disordered single-colored light of particular frequency. Then the reflected light is collected by using a camera that disperse the absorbed light and less spread degree of tissue under evaluation (Cuccia et al., 2009). The coloring compounds of tissues and water are assessed by using the calculated values of different wavelengths at specific degree of absorption. Then the results are corresponding to the tissues made with vessels and inflammation of dermal tissues (Nguyen et al., 2013a; Yafi et al., 2011). This is also suitable for wounds due to burns and able to determine the deep area of wounds due to burns (Nguyen et al., 2013a) and seriousness of damage of wounds (Ghassemi et al., 2014). This technique also determines the inflammation at the wound site by variation in the quantity of visual characteristics which corresponds to infection due to bacteria (Nguyen et al., 2013b).

1.10.4.3 Imaging with Thermal Radiations

Skin is involved in the emission of infrared radiations, and these emitted rays are collected by using specific recorders and correspond with the temperature. Then this correspondence permits the thermal imaging quantification that assessed for further production of colored images that show the temperature of skin tissues at specific area (Renkielska et al., 2005; Roback, 2010).

1.10.4.4 Active Dynamic Thermal Imaging

It includes the reporting of temperatures at equilibrium state by using a camera of infrared radiation that is followed by excitation of tissues thermally using a device which emit light. Further evaluation permits the diffusion of heat in tissues that will help in quantification of areas of wound (Prindeze et al., 2015; Renkielska et al., 2006).

1.10.4.5 High-Resolution Digital IR Imaging

Many studies have used this technique to classify the burn wounds and also give information about deep area of wounds by using the camera which takes images of great resolving power with IR thermal phenomenon (Hardwicke et al., 2013; Medina-Preciado et al., 2013; Prindeze et al., 2015).

1.10.4.6 Three-Dimensional Profilometry

The boundary, volume, and wound area can also be seen by using three-dimensional imaging through the intense beam of coherent monochromatic light (Kecelj-Leskovec et al., 2007; Marjanovic et al., 1998). In this system a scanner has been acknowledged for the implementation of technique that shows the surface morphology of wound (profilometry) of the legs (Marjanovic et al., 1998). The displacement measuring sensors are used in this system that quantify the distance to a substance by using the phenomenon of navigation. The two motion control motors for the scanning of wound by dot model and a colored map for outline of wound are produced in the system. A light detector is used to measure the boundary of the wounds. The estimation of wound volume is done using the particular method. The process of healing is also apprehended precisely by using different measurements after wound. This technique is applicable to investigate the varicose ulcers (Kecelj Leskovec et al., 2008; Kecelj-Leskovec et al., 2007).

1.10.5 Liquid Crystal Thermography

In this technique the temperature of the tissue is quantified by using a clear crystal which can change its color when exposed to heat. The heat that emit from the tissues is soaked by the clear crystal, and then the band of colors is shown by the plate that corresponds to the values of temperature (Roback, 2010).

1.10.6 Microscopy

Microscopy is crucial for enhancing our comprehension of the cellular, structural, and molecular processes that occur during wound healing. It allows for the direct observation of tissue structure, cellular behavior, ECM changes, and microbial presence, providing unique insights into the stages of wound repair: hemostasis, inflammation, proliferation, and remodeling. Utilizing a range of methods from traditional histology to cutting-edge high-resolution and live-imaging techniques, microscopy is applied in both clinical diagnostics and research environments to assess the progress of wound healing, identify pathological changes, and evaluate treatment results.

1.10.6.1 Light Microscopy

Light microscopy (LM) is a crucial and extensively utilized method for observing the cellular and structural characteristics of tissue with high resolution. In the fields of wound healing research and clinical pathology, LM is an indispensable instrument for analyzing histological alterations throughout all phases of tissue repair: hemostasis, inflammation, proliferation, and remodeling (Bancroft and Gamble, 2008). When paired with histochemical or immunohistochemical (IHC) staining techniques, LM offers significant insights into cellular makeup, tissue structure, angiogenesis, and the remodeling of the ECM.

1.10.6.2 Electron Microscopy

Electron microscopy (EM) is a highly effective imaging method that facilitates ultrastructural examination of biological tissues with nanometer-scale resolution (Bozzola and Russell, 1999). Unlike LM, which relies on visible light and optical lenses, EM utilizes a stream of electrons and electromagnetic lenses, enabling magnification of up to several million times and providing detailed views of cellular organelles, ECM components, and cell–cell interactions (Bozzola and Russell, 1999). In the field of wound healing research, EM provides valuable insights into the ultrastructural alterations that take place during different phases of tissue repair, such as cell migration, angiogenesis, fibrosis, and scar formation (Bozzola and Russell, 1999). EM is especially advantageous for investigating fine subcellular characteristics and interactions that conventional light or fluorescence microscopy cannot resolve. EM is still considered the gold standard for analyzing the ultrastructure of healing tissues. Whether investigating collagen arrangement, the morphology of ECs, or the colonization of pathogens, EM offers insights that are typically unattainable. Although its application is primarily limited to research environments due to technical and financial limitations, it continues to influence our molecular comprehension of wound healing and directs the advancement of regenerative therapies, biomaterials, and anti-biofilm approaches.

1.10.6.3 Fluorescence Microscopy

Fluorescence microscopy is a commonly utilized optical imaging method that facilitates the specific and high-resolution visualization of cellular and molecular activities during the process of wound healing. By tagging biological molecules with fluorophores, this method enables the detection, localization, and quantification of particular proteins, cells, or extracellular elements within damaged tissue. Its exceptional sensitivity and specificity render it a vital instrument for examining the intricate cellular events associated with inflammation, proliferation, angiogenesis, and tissue remodeling. Fluorescence microscopy has played a crucial role in uncovering the functions of growth factors, immune cells, matrix components, and signaling pathways that are pivotal in tissue repair, both *in vitro* and *in vivo*. By allowing the visualization of specific molecules, pathways, and cells, it provides essential insights into the cellular and molecular coordination of tissue repair.

1.10.6.4 Multi-Photon Microscopy

Multi-photon microscopy (MPM) is a sophisticated non-linear imaging method that facilitates high-resolution imaging of deep tissues with minimal phototoxic effects. Emerging in the 1990s, MPM has transformed live-cell and *in vivo* imaging by enabling the observation of dynamic biological activities within intact, living tissues, including those related to wound healing. In contrast to traditional fluorescence microscopy, which relies on single-photon excitation, MPM employs near-infrared (NIR) femtosecond pulsed lasers to excite fluorophores through the simultaneous absorption of two or more lower-energy photons. This multi-photon excitation is localized to the focal point, allowing for intrinsic optical sectioning and minimizing out-of-focus photodamage. MPM is a state-of-the-art technology that has greatly enhanced our comprehension of the spatio-temporal dynamics involved in wound healing (Yeh et al., 2004; Zhu et al., 2010). Its capability to visualize cellular behavior, collagen remodeling, and vascular alterations in living tissues renders it an essential instrument in wound biology research. Although it is presently limited to experimental environments, continuous improvements in imaging probes, miniaturized optics, and computational analysis may soon enable the application of MPM in clinical diagnostics and therapy monitoring.

1.10.6.5 Optoacoustic Microscopy

This is a technique that is used to explain the evaluation of dermal wounds due to burns and based on the brightness of skin tissues by using an intense beam of light (Hatanaka et al., 2010; Hirao et al., 2010; Sato et al., 2005; Yamazaki et al., 2005; Yamazaki et al., 2006; Zhang et al., 2006). Usually, this system uses the light source of the range that is suitable for blood to absorb it. So, this system quantifies the Hb levels in tissues. Light absorbing components of tissues partly assimilate the energy from the intense beam of light. This procedure originates the warm stretchy growth that results in the production of an optoacoustic signal that is further recorded by the photoelectric detector. This technique is applicable for the evaluation of wounds due to burns on the skins of rats (Sato et al., 2005; Zhang et al., 2006).

1.11 CONCLUSION

Wound healing is a multifaceted and meticulously controlled biological process that is crucial for restoring both the integrity and functionality of the skin, which is the largest organ of the body and serves as the primary defense mechanism. When an injury occurs, the body triggers a series of overlapping and interrelated healing phases: hemostasis, inflammation, proliferation, and remodeling. Each of these phases is governed by a dynamic interaction among resident skin cells, inflammatory cells, growth factors, cytokines, and the ECM. The inflammatory phase is responsible for eliminating pathogens and debris; the proliferative phase emphasizes angiogenesis, the formation of granulation tissue, and re-epithelialization; while the remodeling phase enhances the strength of the repaired tissue through collagen maturation and scar formation. Wound healing is affected by numerous factors, both local (such as infection, oxygen levels, moisture, and mechanical stress) and systemic (including diabetes, age, malnutrition, and immunosuppression). Moreover, environmental conditions like humidity, temperature, and exposure to pollutants can influence healing rates and results. These internal and external factors can hinder healing, encourage chronic conditions, or heighten the risk of infection and fibrosis. To assess the progress of wound healing and inform clinical decisions, a variety of wound assessment techniques have been established. Traditional approaches such as visual inspection and planimetry are now enhanced by advanced technologies like spectrophotometry, imaging techniques, and microscopy methods. Collectively, these instruments allow for more accurate, objective, and timely detection of healing issues, including infection, impaired angiogenesis, and inadequate re-epithelialization.

In conclusion, wound healing is a complex biological process influenced by various factors and scales, deeply connected to skin biology, systemic physiology, and environmental interactions. A comprehensive grasp of its phases, modulators, and evaluation methods is crucial not only for effective clinical management but also for creating innovative therapies, advanced dressings, and regenerative strategies designed to speed up healing and alleviate the burden of chronic wounds.

GLOSSARY

Accidental wounds: The injuries caused by unexpected and external trauma which damage the skin and the underlying tissues.

Anti-inflammatory macrophages: The macrophages responsible for reducing inflammation by releasing anti-inflammatory cytokines which eventually promote wound healing.

Apoptosis: A type of programmed cell death that removes unnecessary or damaged cells.

Blunt force trauma: Injuries that are caused by the impact of blunt or non-sharp objects.

Catecholamines: Group of hormones like adrenaline that are released in response to an injury.

Chemotaxis: This is the movement of immune cells in response to a chemical signal toward the wound site.

Collagen types I and III: Type III collagen is produced in the early phases of wound healing, and type I collagen, which is stronger, is produced in a later stage.

Contusions: The injuries that are caused by a blunt force trauma, leading to tissue damage, bruises, and swelling, without tearing or rupturing of the skin.

Diabetic ulcers: Chronic wounds that are often formed on the feet of diabetic patients and are difficult to heal due to impaired healing process.

Elastin: A protein that plays an important role in skin flexibility by allowing the skin to be stretched and come back to its original shape.

Endothelin: A substance that is released by damaged blood vessels causing vasoconstriction.

Epithelialization: A process in which new cells migrate toward the wound site and restore the skin.

Fibroblasts: The cells which are responsible to produce collagen and other components of the extracellular matrix.

Fibrotic scars: Thick scars that are formed due to excessive collagen production and deposition.

Hemidesmosomes: The supporting structures that attach the cells of the epithelium to the basement membrane.

Hypertrophic scars: These scars that are a result of excessive collagen production but do not grow outside the wound area.

Keloids: Scar tissues that grow beyond the wound boundary due to excessive collagen production.

Neovasculature: The new blood vessels that are formed during the healing process and provide oxygen and nutrition to the healed area.

Pro-inflammatory macrophages: The macrophages responsible for initiating and promoting inflammation.

Wound slough: A white or yellow tissue that is made up of dead cells and accumulates on the wound surface.

REFERENCES

Abdo, Joseph M, Nikolai A Sopko, and Stephen M Milner. 2020. "The applied anatomy of human skin: A model for regeneration." *Wound Medicine* 28: 100179.

Abrahamian, Fredrick M, and Ellie JC Goldstein. 2011. "Microbiology of animal bite wound infections." *Clinical Microbiology Reviews* 24 (2): 231–246.

Alexiadou, Kleopatra, and John Doupis. 2012. "Management of diabetic foot ulcers." *Diabetes Therapy* 3: 1–15.

Aly, Raza. 1996. "Microbial infections of skin and nails." *Medical Microbiology* 98.

Avishai, Eden, Kristina Yeghiazaryan, and Olga Golubnitschaja. 2017. "Impaired wound healing: Facts and hypotheses for multi-professional considerations in predictive, preventive and personalised medicine." *EPMA Journal* 8: 23–33.

Babaluei, Mercedeh, Fatemeh Mottaghitalab, Alexander Seifalian, and Mehdi Farokhi. 2023. "Injectable multifunctional hydrogel based on carboxymethylcellulose/polyacrylamide/polydopamine containing vitamin C and curcumin promoted full-thickness burn regeneration." *International Journal of Biological Macromolecules* 236: 124005.

Baharestani, Mona Mylene. 2007. "An overview of neonatal and pediatric wound care knowledge and considerations." *Ostomy/Wound Management* 53 (6): 34–36, 38, 40, passim.

Bancroft, John D., and Marilyn Gamble, eds. 2008. *Theory and Practice of Histological Techniques*. Elsevier Health Sciences.

Beahrs, Taylor R, James Reagan, Clayton C Bettin, Benjamin J Grear, G Andrew Murphy, and David R Richardson. 2019. "Smoking effects in foot and ankle surgery: An evidence-based review." *Foot & Ankle International* 40 (10): 1226–1232.

Belkaid, Yasmine, and Samira Tamoutounour. 2016. "The influence of skin microorganisms on cutaneous immunity." *Nature Reviews Immunology* 16 (6): 353–366.

Beyene, Robel T, Stephen Lentz Derryberry, and Adrian Barbul. 2020. "The effect of comorbidities on wound healing." *Surgical Clinics* 100 (4): 695–705.

Bhattacharya, S. and Mishra, R.K., 2015. "Pressure ulcers: current understanding and newer modalities of treatment." *Indian Journal of plastic surgery* 48(1): 004–016. https://doi.org/10.4103/0970-0358.155260

Bilic, I, NM Petri, J Bezic, D Alfirevic, Darko Modun, V Capkun, and Božo Bota. 2005. "Effects of hyperbaric oxygen therapy on experimental burn wound healing in rats: A randomized controlled study." *Undersea & Hyperbaric Medical* 32 (1): 1–9.

Bishop, Ah. 2008. "Role of oxygen in wound healing." *Journal of Wound Care* 17 (9): 399–402.

Boniakowski, Anna E, Andrew S Kimball, Benjamin N Jacobs, Steven L Kunkel, and Katherine A Gallagher. 2017. "Macrophage-mediated inflammation in normal and diabetic wound healing." *The Journal of Immunology* 199 (1): 17–24.

Bowler, Peter G., Brian I. Duerden, and David G. Armstrong. 2001. "Wound microbiology and associated approaches to wound management." *Clinical Microbiology Reviews* 14(2): 244–269.

Bozzola, John J., and Lonnie Dee Russell. 1999. *Electron Microscopy: Principles and Techniques for Biologists*. Jones & Bartlett Learning.

Caley, Matthew P, Vera LC Martins, and Edel A O'Toole. 2015. "Metalloproteinases and wound healing." *Advances in Wound Care* 4 (4): 225–234.

Calin, Mihaela Antonina, Sorin Viorel Parasca, Roxana Savastru, and Dragos Manea. 2015. "Characterization of burns using hyperspectral imaging technique—A preliminary study." *Burns* 41 (1): 118–124.

Chang, Eun J, Linda S Edelman, Stephen E Morris, and Jeffrey R Saffle. 2005. "Gender influences on burn outcomes in the elderly." *Burns* 31 (1): 31–35.

Chen, Yao, Yu Xiang, Hongmei Zhang, Tonghe Zhu, Sihao Chen, Jun Li, Juan Du, and Xiaoyu Yan. 2023. "A multifunctional chitosan composite aerogel based on high density amidation for chronic wound healing." *Carbohydrate Polymers* 321: 121248.

Choi, Jae Eun, and Anna Di Nardo. 2018. "Skin neurogenic inflammation." *Seminars in Immunopathology* 40(3): 249–259

Clayton, Kalum, Andres F Vallejo, James Davies, Sofia Sirvent, and Marta E Polak. 2017. "Langerhans cells—Programmed by the epidermis." *Frontiers in Immunology* 8: 1676.

Cook, Katherine A, Edith Martinez-Lozano, Robert Sheridan, Edward K Rodriguez, Ara Nazarian, and Mark W Grinstaff. 2022. "Hydrogels for the management of second-degree burns: Currently available options and future promise." *Burns & Trauma* 10: tkac047.

Cross, Karen M, Lorenzo Leonardi, Manuel Gomez, Jeri R Freisen, Michelle A Levasseur, Bernie J Schattka, Michael G Sowa, and Joel S Fish. 2009. "Noninvasive measurement of edema in partial thickness burn wounds." *Journal of Burn Care & Research* 30 (5): 807–817.

Cross, K.M., Leonardi, L., Payette, J.R., Gomez, M., Levasseur, M.A., Schattka, B.J., Sowa, M.G. and Fish, J.S., 2007. "Clinical utilization of near-infrared spectroscopy devices for burn depth assessment." *Wound Repair and Regeneration* 15 (3): 332–340. https://doi.org/10.1111/j.1524-475X.2007.00235.x

Cuccia, David J, Frederic Bevilacqua, Anthony J Durkin, Frederick R Ayers, and Bruce J Tromberg. 2009. "Quantitation and mapping of tissue optical properties using modulated imaging." *Journal of Biomedical Optics* 14 (2): 024012-024012-13.

Darby, Ian A, and Tim D Hewitson. 2016. "Hypoxia in tissue repair and fibrosis." *Cell and Tissue Research* 365: 553–562.

Dasari, Nina, Austin Jiang, Anna Skochdopole, Jayer Chung, Edward M Reece, Joshua Vorstenbosch, and Sebastian Winocour. 2021. "Updates in diabetic wound healing, inflammation, and scarring." *Seminars in Plastic Surgery 35* (3):153–158

Davis, Frank M, Andrew Kimball, Anna Boniakowski, and Katherine Gallagher. 2018. "Dysfunctional wound healing in diabetic foot ulcers: New crossroads." *Current Diabetes Reports* 18 (1): 8.

Decker, Marquita R, David Y Greenblatt, Jeff Havlena, Lee G Wilke, Caprice C Greenberg, and Heather B Neuman. 2012. "Impact of neoadjuvant chemotherapy on wound complications after breast surgery." *Surgery* 152 (3): 382–388.

Demidova-Rice, Tatiana N, Jennifer T Durham, and Ira M Herman. 2012. "Wound healing angiogenesis: Innovations and challenges in acute and chronic wound healing." *Advances in Wound Care* 1 (1): 17–22.

Desai, Nimeet, Dhwani Rana, Sagar Salave, Raghav Gupta, Pranav Patel, Bharathi Karunakaran, Amit Sharma, Jyotsnendu Giri, Derajram Benival, and Nagavendra Kommineni. 2023. "Chitosan: A potential biopolymer in drug delivery and biomedical applications." *Pharmaceutics* 15 (4): 1313.

Devaney, Lynn, and Katherine S Rowell. 2004. "Improving surgical wound classification—Why it matters." *AORN Journal* 80 (2): 208–223.

Dormand, Emma-Louise, Paul E Banwell, and Timothy EE Goodacre. 2005. "Radiotherapy and wound healing." *International Wound Journal* 2 (2): 112–127.

Edwards, Ruth, and Keith G Harding. 2004. "Bacteria and wound healing." *Current Opinion in Infectious Diseases* 17 (2): 91–96.

Einhorn, Thomas A, and Louis C Gerstenfeld. 2015. "Fracture healing: Mechanisms and interventions." *Nature Reviews Rheumatology* 11 (1): 45–54.

Firlar, Ilayda, Mine Altunbek, Colleen McCarthy, Murugan Ramalingam, and Gulden Camci-Unal. 2022. "Functional hydrogels for treatment of chronic wounds." *Gels* 8 (2): 127.

Franz, Michael G. 2007. "Optimizing healing of the acute wound by minimizing complications." *Current Problems in Surgery* 44: 679–766.

Funes, Samanta C, Mariana Rios, Jorge Escobar-Vera, and Alexis M Kalergis. 2018. "Implications of macrophage polarization in autoimmunity." *Immunology* 154 (2): 186–195.

Gamelli, Richard L, and Li-Ke He. 2003. "Incisional wound healing: Model and analysis of wound breaking strength." In Wound Healing: Methods and Protocols, edited by Luisa A. DiPietro and Aime L. Burns, (pp. 37–54). Totowa, NJ: Humana Press

Ganesh, Kasturi, Mithun Sinha, Shomita S Mathew-Steiner, Amitava Das, Sashwati Roy, and Chandan K Sen. 2015. "Chronic wound biofilm model." *Advances in Wound Care* 4 (7): 382–388.

Gerstenfeld, Louis C, Dennis M Cullinane, George L Barnes, Dana T Graves, and Thomas A Einhorn. 2003. "Fracture healing as a post-natal developmental process: Molecular, spatial, and temporal aspects of its regulation." *Journal of Cellular Biochemistry* 88 (5): 873–884.

Gethin, Georgina. 2007. "The significance of surface pH in chronic wounds." *Wounds UK* 3 (3): 52.

Ghassemi, Pejhman, Taryn E Travis, Lauren T Moffatt, Jeffrey W Shupp, and Jessica C Ramella-Roman. 2014. "A polarized multispectral imaging system for quantitative assessment of hypertrophic scars." *Biomedical Optics Express* 5 (10): 3337–3354.

Godo, Shigeo, and Hiroaki Shimokawa. 2017. "Endothelial functions." *Arteriosclerosis, Thrombosis, and Vascular Biology* 37 (9): e108–e114.

Goertz, Ole, Andrej Ring, Annette Köhlinger, Adrien Daigeler, Christoph Andree, Hans-Ulrich Steinau, and Stefan Langer. 2010. "Orthogonal polarization spectral imaging: A tool for assessing burn depths?" *Annals of Plastic Surgery* 64 (2): 217–221.

Gonzalez, Ana Cristina de Oliveira, Tila Fortuna Costa, Zilton de Araújo Andrade, and Alena Ribeiro Alves Peixoto Medrado. 2016. "Wound healing—A literature review." *Anais brasileiros de dermatologia* 91 (5): 614–620.

Gordon, Chad R, Yuri Rojavin, Mitul Patel, James E Zins, Generosa Grana, Brian Kann, Robert Simons, and Umar Atabek. 2009. "A review on bevacizumab and surgical wound healing: An important warning to all surgeons." *Annals of Plastic Surgery* 62 (6): 707–709.

Gorecka, Jolanta, Valentyna Kostiuk, Arash Fereydooni, Luis Gonzalez, Jiesi Luo, Biraja Dash, Toshihiko Isaji, Shun Ono, Shirley Liu, and Shin Rong Lee. 2019. "The potential and limitations of induced pluripotent stem cells to achieve wound healing." *Stem Cell Research & Therapy* 10: 1–10.

Gorvetzian, Joseph W, Katharine E Epler, Samuel Schrader, Joshua M Romero, Ronald Schrader, Alissa Greenbaum, and Rohini McKee. 2018. "Operating room staff and surgeon documentation curriculum improves wound classification accuracy." *Heliyon* 4 (8).

Grambow, Eberhard, Heiko Sorg, Christian GG Sorg, and Daniel Strüder. 2021. "Experimental models to study skin wound healing with a focus on angiogenesis." *Medical Sciences* 9 (3): 55.

Groner, Warren, James W Winkelman, Anthony G Harris, Can Ince, Gerrit J Bouma, Konrad Messmer, and Richard G Nadeau. 1999. "Orthogonal polarization spectral imaging: A new method for study of the microcirculation." *Nature Medicine* 5 (10): 1209–1212.

Guo, S, and Luisa A DiPietro. 2010. "Factors affecting wound healing." *Journal of Dental Research* 89 (3): 219–229.

Hamdan, TA. 2006. "Missile injuries of the limbs: An Iraqi perspective." *Journal of the American Academy of Orthopaedic Surgeons* 14 (10): S32–S36. https://doi.org/10.5435/00124635-200600001-00007

Hardwicke, Joseph, Richard Thomson, Amy Bamford, and Naiem Moiemen. 2013. "A pilot evaluation study of high resolution digital thermal imaging in the assessment of burn depth." *Burns* 39 (1): 76–81.

Hatanaka, Kousuke, Shunichi Sato, Daizoh Saitoh, Hiroshi Ashida, and Toshihisa Sakamoto. 2010. "Photoacoustic monitoring of granulation tissue grown in a grafted artificial dermis on rat skin." *Wound Repair and Regeneration* 18 (3): 284–290.

Haubner, Frank, Elisabeth Ohmann, Fabian Pohl, Jürgen Strutz, and Holger G Gassner. 2012. "Wound healing after radiation therapy: Review of the literature." *Radiation Oncology* 7: 1–9.

Hemphill, Meredith, and Abigail Nathanson. 2023. *The Practice of Clinical Social Work in Healthcare*. Springer.

Herman, Timothy F, and Bruno Bordoni. 2023. "Wound classification." StatPearls [Internet]. Treasure Island (FL): StatPearls Publishing; 2025 Jan.

Hirao, Akihiro, Shunichi Sato, Daizoh Saitoh, Nariyoshi Shinomiya, Hiroshi Ashida, and Minoru Obara. 2010. "In vivo photoacoustic monitoring of photosensitizer distribution in burned skin for antibacterial photodynamic therapy." *Photochemistry and Photobiology* 86 (2): 426–430.

Holstein, Joerg H, Bianca Karabin-Kehl, Claudia Scheuer, Patric Garcia, Tina Histing, Christoph Meier, Emanuel Benninger, Michael D Menger, and Tim Pohlemann. 2013. "Endostatin inhibits callus remodeling during fracture healing in mice." *Journal of Orthopaedic Research* 31 (10): 1579–1584.

Holt, Paula. 2013. "Assessment and management of patients with diabetic foot ulcers." *Nursing Standard (Through 2013)* 27 (27): 49.

Holzer-Geissler, Judith CJ, Simon Schwingenschuh, Martin Zacharias, Johanna Einsiedler, Sonja Kainz, Peter Reisenegger, Christian Holecek, Elisabeth Hofmann, Barbara Wolff-Winiski, and Hermann Fahrngruber. 2022. "The impact of prolonged inflammation on wound healing." *Biomedicines* 10 (4): 856.

Ikpeme, IA, NE Ngim, and AA Ikpeme. 2010. "Diagnosis and treatment of pyogenic bone infections." *African Health Sciences* 10 (1): 82.

Irfan-Maqsood, Muhammad. 2018. "Classification of wounds: Know before research and clinical practice." *Journal of Genes and Cells* 4 (1): 1–4.

Iyer, Subramania, and Deepak Balasubramanian. 2012. "Management of radiation wounds." *Indian Journal of Plastic Surgery* 45 (02): 325–331.

Jacobson, Lia K, Maxwell B Johnson, Raj D Dedhia, Solmaz Niknam-Bienia, and Alex K Wong. 2017. "Impaired wound healing after radiation therapy: A systematic review of pathogenesis and treatment." *JPRAS Open* 13: 92–105.

Jaskille, Amín D, Jessica C Ramella-Roman, Jeffrey W Shupp, Marion H Jordan, and James C Jeng. 2010. "Critical review of burn depth assessment techniques: Part II. Review of laser Doppler technology." *Journal of Burn Care & Research* 31 (1): 151–157.

Jones, Menna Lloyd. 2015. "A short history of the development of wound care dressings." *British Journal of Healthcare Assistants* 9 (10): 482–485.

Kadomoto, Suguru, Kouji Izumi, and Atsushi Mizokami. 2021. "Macrophage polarity and disease control." *International Journal of Molecular Sciences* 23 (1): 144.

Kaner, Doğan, Mouaz Soudan, Han Zhao, Georg Gaßmann, Anna Schönhauser, and Anton Friedmann. 2017. "Early healing events after periodontal surgery: Observations on soft tissue healing, microcirculation, and wound fluid cytokine levels." *International Journal of Molecular Sciences* 18 (2): 283.

Karimi, Karen, Iris Faraklas, Giavonni Lewis, Daniel Ha, Bridget Walker, Yan Zhai, Gareth Graves, and Sharmila Dissanaike. 2017. "Increased mortality in women: Sex differences in burn outcomes." *Burns & Trauma* 5.

Kaufman, Theodor, J Wesley Alexander, and Bruce G MacMillan. 1983. "Topical oxygen and burn wound healing: A review." *Burns* 9 (3): 169–173.

Kaushal, Mandeep, N Gopalan Kutty, and C Mallikarjuna Rao. 2006. "Nitrooxyethylation reverses the healing-suppressant effect of Ibuprofen." *Mediators of Inflammation* 2006 (1): 024396.

Kecelj Leskovec, Nada, Maja Pohar Perme, Matija Jezeršek, Janez Mozina, Miloš D Pavlović, and Tomaž Lunder. 2008. "Initial healing rates as predictive factors of venous ulcer healing: The use of a laser-based three-dimensional ulcer measurement." *Wound Repair and Regeneration* 16 (4): 507–512.

Kecelj-Leskovec, Nada, Matija Jezeršek, Janez Možina, Miloš D Pavlović, and Tomaž Lunder. 2007. "Measurement of venous leg ulcers with a laser-based three-dimensional method: Comparison to computer planimetry with photography." *Wound Repair and Regeneration* 15 (5): 767–771.

Khaodhiar, Lalita, Thanh Dinh, Kevin T Schomacker, Svetlana V Panasyuk, Jenny E Freeman, Robert Lew, Tiffany Vo, Alexander A Panasyuk, Christina Lima, and John M Giurini. 2007. "The use of medical hyperspectral technology to evaluate microcirculatory changes in diabetic foot ulcers and to predict clinical outcomes." *Diabetes Care* 30 (4): 903–910.

Korting, HC, C Schöllmann, and RJ White. 2011. "Management of minor acute cutaneous wounds: Importance of wound healing in a moist environment." *Journal of the European Academy of Dermatology and Venereology* 25 (2): 130–137.

Krischak, GD, Peter Augat, Lutz Claes, Lothar Kinzl, and Alexander Beck. 2007. "The effects of non-steroidal anti-inflammatory drug application on incisional wound healing in rats." *Journal of Wound Care* 16 (2): 76–78.

Krzyszczyk, Paulina, Rene Schloss, Andre Palmer, and François Berthiaume. 2018. "The role of macrophages in acute and chronic wound healing and interventions to promote pro-wound healing phenotypes." *Frontiers in Physiology* 9: 419.

Kuhajda, Ivan, Konstantinos Zarogoulidis, Ioanna Kougioumtzi, Haidong Huang, Qiang Li, Dryllis Georgios, Ioannis Kioumis, Georgia Pitsiou, Nikolaos Machairiotis, and Nikolaos Katsikogiannis. 2014. "Penetrating trauma." *Journal of Thoracic Disease* 6(Suppl 4): S461.

Kujath, Peter, and Angela Michelsen. 2008. "Wounds—From physiology to wound dressing." *Deutsches Ärzteblatt International* 105 (13): 239.

Legendre, Cécile, Clélia Debure, Sylvie Meaume, Catherine Lok, Jean Louis Golmard, and Patricia Senet. 2008. "Impact of protein deficiency on venous ulcer healing." *Journal of Vascular Surgery* 48 (3): 688–693.

Levy, Shauna M, Galit Holzmann-Pazgal, Kevin P Lally, Koya Davis, Lillian S Kao, and KuoJen Tsao. 2013. "Quality check of a quality measure: Surgical wound classification discrepancies impact risk-stratified surgical site infection rates in pediatric appendicitis." *Journal of the American College of Surgeons* 217 (6): 969–973.

Li, Jingyi, Yaqi Yang, Guofang Zhang, Jianfei Sun, Yang Li, and Bing Song. 2025. "Therapeutic advances of magnetic nanomaterials in chronic wound healing." *Nano Today* 60: 102554.

Li, Ming, Baimei Liu, Wei Xu, Lai Zhao, Zili Wang, Haonan He, Jingjing Li, Fan Wang, Chao Ma, and Kai Liu. 2023. "Engineering protein coacervates into a robust adhesive for real-time skin healing." *Engineering* 31: 76–85.

Lindholm, Christina, and Richard Searle. 2016. "Wound management for the 21st century: Combining effectiveness and efficiency." *International Wound Journal* 13: 5–15.

Liu, Nan, Hiroyuki Matsumura, Tomoki Kato, Shizuko Ichinose, Aki Takada, Takeshi Namiki, Kyosuke Asakawa, Hironobu Morinaga, Yasuaki Mohri, and Adèle De Arcangelis. 2019. "Stem cell competition orchestrates skin homeostasis and ageing." *Nature* 568 (7752): 344–350.

Liu, Wenshuai, Rui Gao, Chunfang Yang, Zujian Feng, Wenbin Ou-Yang, Xiangbin Pan, Pingsheng Huang, Chuangnian Zhang, Deling Kong, and Weiwei Wang. 2022. "ECM-mimetic immunomodulatory hydrogel for methicillin-resistant *Staphylococcus aureus*–infected chronic skin wound healing." *Science Advances* 8 (27): eabn7006.

Lone, Reyaz Ahmed, Mehmood Ahmad Wani, Abdul Majid Dar, Zahur Hussain, Mukhand Lal Sharma, Mohd Akbar Bhat, and Abdul Gani Ahangar. 2009. "Missile cardiac injuries: Review of 16 years' experience." *Turkish Journal of Trauma & Emergency Surgery* 15 (4): 353–356.

Lopes, Fernanda Barbosa, Mariáurea Matias Sarandy, Rômulo Dias Novaes, Giuseppe Valacchi, and Reggiani Vilela Gonçalves. 2024. "Oxinflammatory responses in the wound healing process: A systematic review." *Antioxidants* 13 (7): 823.

Lu, Zhuoting, Jingyu Li, Baohong Chen, Jiayu Feng, Qinglian Hu, Yuanxiang Jin, and Zhengwei Fu. 2023. "Mitochondria targeted nanoparticles potentiate tumor chemo-phototherapy by toxic oxidative stress mediated oxeiptosis." *Macromolecular Bioscience* 23 (11): 2300151.

Lund, Tjøstolv, Henning Onarheim, and Rolf K Reed. 1992. "Pathogenesis of edema formation in burn injuries." *World Journal of Surgery* 16 (1): 2–9.

Margenthaler, Julie A, Walter E Longo, Katherine S Virgo, Frank E Johnson, Charles A Oprian, William G Henderson, Jennifer Daley, and Shukri F Khuri. 2003. "Risk factors for adverse outcomes after the surgical treatment of appendicitis in adults." *Annals of Surgery* 238 (1): 59–66.

Marjanovic D, Dugdale RE, Vowden P, Vowden KR. 1998. "Measurement of the volume of a leg ulcer using a laser scanner." *Physiological Measurement19* (4): 535–542. https://doi.org/10.1088/0967-3334/19/4/008

Meara, John G., Craig D. McClain, Selwyn O. Rogers, Jr., and David P. Mooney 2015. *Global Surgery and Anesthesia Manual: Providing Care in Resource-Limited Settings*. Boca Raton, FL: Taylor & Francis.

Medina-Preciado, Jose David, Eleazar Samuel Kolosovas-Machuca, Ezequiel Velez-Gomez, Ariel Miranda-Altamirano, and Francisco Javier González. 2013. "Noninvasive determination of burn depth in children by digital infrared thermal imaging." *Journal of Biomedical Optics* 18 (6): 061204–061204.

Meizlish, Matthew L, Ruth A Franklin, Xu Zhou, and Ruslan Medzhitov. 2021. "Tissue homeostasis and inflammation." *Annual Review of Immunology* 39 (1): 557–581.

Milner, Stephen M. 2002. "Predicting early burn wound outcome using orthogonal polarization spectral imaging." *International Journal of Dermatology* 41 (10).

Milner, Stephen M, Satyanarayan Bhat, Salil Gulati, Giulio Gherardini, Christopher E Smith, and Roger J Bick. 2005. "Observations on the microcirculation of the human burn wound using orthogonal polarization spectral imaging." *Burns* 31 (3): 316–319.

Mioton, Lauren M, Sumanas W Jordan, Philip J Hanwright, Karl Y Bilimoria, and John YS Kim. 2013. "The relationship between preoperative wound classification and postoperative infection: A multi-institutional analysis of 15,289 patients." *Archives of Plastic Surgery* 40 (05): 522–529.

Monstrey, Stan, Henk Hoeksema, Jos Verbelen, Ali Pirayesh, and Phillip Blondeel. 2008. "Assessment of burn depth and burn wound healing potential." *Burns* 34 (6): 761–769.

Montella E, Schiavone D, Apicella L, Di Silverio P, Gaudiosi M, Ambrosone E, Moscaritolo E, Triassi M. 2014. "Cost-benefit evaluation of a preventive intervention on the biological risk in health: The accidental puncture during the administration of insulin in the University Hospital "Federico II" of Naples." *Annali di Igiene: Medicina Preventiva e di Comunita* 26 (3): 272–278. https://doi.org/10.7416/ai.2014.1985

Moreo, Kathleen. 2005. "Understanding and overcoming the challenges of effective case management for patients with chronic wounds." *The Case Manager* 16 (2): 62–67.

Morton, Laurel M, and Tania J Phillips. 2016. "Wound healing and treating wounds: Differential diagnosis and evaluation of chronic wounds." *Journal of the American Academy of Dermatology* 74 (4): 589–605.

Mouw, Janna K, Guanqing Ou, and Valerie M Weaver. 2014. "Extracellular matrix assembly: A multiscale deconstruction." *Nature Reviews Molecular Cell Biology* 15 (12): 771–785.

Murray, Peter J. 2017. "Macrophage polarization." *Annual Review of Physiology* 79 (1): 541–566.

Nguyen, John Quan, Christian Crouzet, Tuan Mai, Kathleen Riola, Daniel Uchitel, Lih-Huei Liaw, Nicole Bernal, Adrien Ponticorvo, Bernard Choi, and Anthony J Durkin. 2013a. "Spatial frequency domain imaging of burn wounds in a preclinical model of graded burn severity." *Journal of Biomedical Optics* 18 (6): 066010–066010.

Nguyen, Thu TA, Jessica C Ramella-Roman, Lauren T Moffatt, Rachel T Ortiz, Marion H Jordan, and Jeffrey W Shupp. 2013b. "Novel application of a spatial frequency domain imaging system to determine signature spectral differences between infected and noninfected burn wounds." *Journal of Burn Care & Research* 34 (1): 44–50.

Okonkwo, A Uzoagu, and Luisa A DiPietro. 2017. "Diabetes and wound angiogenesis." *International Journal of Molecular Sciences* 18 (7): 1419.

Onyekwelu, Ikemefuna, Ramakanth Yakkanti, Lauren Protzer, Christina M Pinkston, Cody Tucker, and David Seligson. 2017. "Surgical wound classification and surgical site infections in the orthopaedic patient." *JAAOS Global Research & Reviews* 1 (3): e022.

Pang, Qian, Zilian Jiang, Kaihao Wu, Ruixia Hou, and Yabin Zhu. 2023. "Nanomaterials-based wound dressing for advanced management of infected wound." *Antibiotics* 12 (2): 351.

Papazoglou, Elisabeth S, Leonid Zubkov, Linda Zhu, Michael S Weingarten, Sorn Tyagi, and Kambiz Pourrezaei. 2006. "Monitoring diabetic wound healing by NIR spectroscopy." 2005 IEEE Engineering in Medicine and Biology 27th Annual Conference.

Pastar, Irena, Olivera Stojadinovic, Natalie C Yin, Horacio Ramirez, Aron G Nusbaum, Andrew Sawaya, Shailee B Patel, Laiqua Khalid, Rivkah R Isseroff, and Marjana Tomic-Canic. 2014. "Epithelialization in wound healing: A comprehensive review." *Advances in Wound Care* 3 (7): 445–464.

Patel, Satish, Shikha Srivastava, Manju Rawat Singh, and Deependra Singh. 2019. "Mechanistic insight into diabetic wounds: Pathogenesis, molecular targets and treatment strategies to pace wound healing." *Biomedicine & Pharmacotherapy* 112: 108615.

Pena, Oscar A, and Paul Martin. 2024. "Cellular and molecular mechanisms of skin wound healing." *Nature Reviews Molecular Cell Biology* 25 (8): 599–616.

Percival, Nicholas J. 2002. "Classification of wounds and their management." *Surgery (Oxford)* 20 (5): 114–117.

Pergialiotis, Vasileios, Dimitrios Vlachos, Athanasios Protopapas, Kaliopi Pappa, and Georgios Vlachos. 2014. "Risk factors for severe perineal lacerations during childbirth." *International Journal of Gynecology & Obstetrics* 125 (1): 6–14.

Phillips Anthony M. 2005. "Overview of the fracture healing cascade." *Injury* 36 (3 Suppl): S5–S7. https://doi.org/10.1016/j.injury.2005.07.027

Powers, Jennifer G, Catherine Higham, Karen Broussard, and Tania J Phillips. 2016. "Wound healing and treating wounds: Chronic wound care and management." *Journal of the American Academy of Dermatology* 74 (4): 607–625.

Pradhan, Subhashree, Tanvir Khatlani, Angus C Nairn, and K Vinod Vijayan. 2017. "The heterotrimeric G protein Gβ1 interacts with the catalytic subunit of protein phosphatase 1 and modulates G protein–coupled receptor signaling in platelets." *Journal of Biological Chemistry* 292 (32): 13133–13142.

Prindeze, Nicholas J, Payam Fathi, Matthew J Mino, Neil A Mauskar, Taryn E Travis, Dereck W Paul, Lauren T Moffatt, and Jeffrey W Shupp. 2015. "Examination of the early diagnostic applicability of active dynamic thermography for burn wound depth assessment and concept analysis." *Journal of Burn Care & Research* 36 (6): 626–635.

Proksch, Ehrhardt, Johanna M. Brandner, and Jens-Michael Jensen. 2008. "The skin: An indispensable barrier." *Experimental Dermatology* 17 (12): 1063–1072.

Qian, Li-Wu, Andrea B Fourcaudot, Kazuyoshi Yamane, Tao You, Rodney K Chan, and Kai P Leung. 2016. "Exacerbated and prolonged inflammation impairs wound healing and increases scarring." *Wound Repair and Regeneration* 24 (1): 26–34.

Quirinia, Anne. 2000. "Ischemic wound healing and possible treatments." *Drugs of Today (Barcelona, Spain: 1998)* 36 (1): 41–53.

Raje, Nikita, and Chitra Dinakar. 2015. "Overview of immunodeficiency disorders." *Immunology and Allergy Clinics of North America* 35 (4): 599.

Ramasastry, Sai S. 2005. "Acute wounds." *Clinics in Plastic Surgery* 32 (2): 195–208.

Ranganathan, Kavitha, Jonathan Peterson, Shailesh Agarwal, Eboda Oluwatobi, Shawn Loder, Jonathan A Forsberg, Thomas A Davis, Steven R Buchman, Stewart C Wang, and Benjamin Levi. 2015. "Role of gender in burn-induced heterotopic ossification and mesenchymal cell osteogenic differentiation." *Plastic and Reconstructive Surgery* 135 (6): 1631–1641.

Raziyeva, Kamila, Yevgeniy Kim, Zharylkasyn Zharkinbekov, Kuat Kassymbek, Shiro Jimi, and Arman Saparov. 2021. "Immunology of acute and chronic wound healing." *Biomolecules* 11 (5): 700.

Renkielska, Alicja, Antoni Nowakowski, Mariusz Kaczmarek, Marek K Dobke, Jacek Grudziński, Andrzej Karmolinski, and Wojciech Stojek. 2005. "Static thermography revisited—An adjunct method for determining the depth of the burn injury." *Burns* 31 (6): 768–775.

Renkielska, Alicja, Antoni Nowakowski, Mariusz Kaczmarek, and Jacek Ruminski. 2006. "Burn depths evaluation based on active dynamic IR thermal imaging—A preliminary study." *Burns* 32 (7): 867–875.

Rittié, Laure. 2016. "Cellular mechanisms of skin repair in humans and other mammals." *Journal of Cell Communication and Signaling* 10: 103–120.

Roback, Kerstin. 2010. "An overview of temperature monitoring devices for early detection of diabetic foot disorders." *Expert Review of Medical Devices* 7 (5): 711–718.

Rodrigues, Melanie, Kosaric Nina, Bonham Clark A, Gurtner Geoffrey C. 2018. "Wound healing: a cellular perspective." *Physiological Reviews* 98(1): 1–23.

Romano, F., Iacobellis, F., Guida, F., Laccetti, E., Sorbo, A., Grassi, R. and Scaglione, M., 2017. Traumatic injuries: mechanisms of lesions. In *Diagnostic Imaging in Polytrauma Patients.* 35–55. Springer International Publishing.

Rosique, Rodrigo G, Marina J Rosique, and Jayme A Farina Junior. 2015. "Curbing inflammation in skin wound healing: A review." *International Journal of Inflammation* 2015 (1): 316235.

Saleh, Bahram, Harkiranpreet Kaur Dhaliwal, Roberto Portillo-Lara, Ehsan Shirzaei Sani, Reza Abdi, Mansoor M Amiji, and Nasim Annabi. 2019. "Local immunomodulation using an adhesive hydrogel loaded with miRNA-laden nanoparticles promotes wound healing." *Small* 15 (36): 1902232.

Salvo, Jessica, and Cristian Sandoval. 2022. "Role of copper nanoparticles in wound healing for chronic wounds: Literature review." *Burns & Trauma* 10: tkab047.

Sangnim, Tanikan, Vivek Puri, Divya Dheer, D Nagasamy Venkatesh, Kampanart Huanbutta, and Ameya Sharma. 2024. "Nanomaterials in the wound healing process: New insights and advancements." *Pharmaceutics* 16 (3): 300.

Sarabahi, Sujata, and VK Tiwari. 2012. *Principles and Practice of Wound Care*. JP Medical Ltd.

Sato, Shunichi, Mutsuo Yamazaki, Daizoh Saitoh, Hitoshi Tsuda, Yoshiaki Okada, Minoru Obara, and Hiroshi Ashida. 2005. "Photoacoustic diagnosis of burns in rats." *Journal of Trauma and Acute Care Surgery* 59 (6): 1450–1456.

Schaefer, TJ, and SC Tannan. 2021. "Thermal burns. [Updated 2020 Sep 18]." StatPearls (Internet) TreaSure Island (FL): StatPearls Publishing.

Sen, Chandan K. 2009. "Wound healing essentials: Let there be oxygen." *Wound Repair and Regeneration* 17 (1): 1–18.

Sen Chandan K. 2023. *Human wound and its burden: updated 2022 compendium of estimates*. New York, NY: Mary Ann Liebert, Inc.

Senzel, Lisa, Dmitri V Gnatenko, and Wadie F Bahou. 2009. "The platelet proteome." *Current Opinion in Hematology* 16 (5): 329–333.

Sharma, Akshay, Suryamani Khanna, Gaganjot Kaur, and Inderbir Singh. 2021. "Medicinal plants and their components for wound healing applications." *Future Journal of Pharmaceutical Sciences* 7 (1): 1–13.

Shupp, Jeffrey W, Teresa J Nasabzadeh, Dean S Rosenthal, Marion H Jordan, Philip Fidler, and James C Jeng. 2010. "A review of the local pathophysiologic bases of burn wound progression." *Journal of Burn Care & Research* 31 (6): 849–873.

Simon LV, Lopez RA, King KC. Blunt force trauma. In: StatPearls [Internet]. Treasure Island (FL): StatPearls Publishing; 2017. Available from: https://www.ncbi.nlm.nih.gov/books/NBK499884/

Singh, Ruby, Purandhi Roopmani, Meenakshi Chauhan, Suparna Mercy Basu, Waghela Deeksha, MD Kazem, Sarbani Hazra, Eerappa Rajakumara, and Jyotsnendu Giri. 2022. "Silver sulfadiazine loaded core-shell airbrushed nanofibers for burn wound healing application." *International Journal of Pharmaceutics* 613: 121358.

Singh, Shailendra, Alistair Young, and Clare-Ellen McNaught. 2017. "The physiology of wound healing." *Surgery (Oxford)* 35 (9): 473–477.

Slauch, James M. 2011. "How does the oxidative burst of macrophages kill bacteria? Still an open question." *Molecular Microbiology* 80 (3): 580–583.

Sorg, Heiko, and Christian GG Sorg. 2023. "Skin wound healing: Of players, patterns, and processes." *European Surgical Research* 64 (2): 141–157.

Sorg, Heiko, Daniel J Tilkorn, Ursula Mirastschijski, Joerg Hauser, and Robert Kraemer. 2018. "Panta Rhei: Neovascularization, angiogenesis and nutritive perfusion in wound healing." *European Surgical Research* 59 (3–4): 232–241.

Sowa, Michael G, Lorenzo Leonardi, Jeri R Payette, Joel S Fish, and Henry H Mantsch. 2001. "Near infrared spectroscopic assessment of hemodynamic changes in the early post-burn period." *Burns* 27 (3): 241–249.

Spanholtz, Timo A, Panagiotis Theodorou, Peymaneh Amini, and Gerald Spilker. 2009. "Severe burn injuries: Acute and long-term treatment." *Deutsches Ärzteblatt International* 106 (38): 607.

Stone, Rivka C, Irena Pastar, Nkemcho Ojeh, Vivien Chen, Sophia Liu, Karen I Garzon, and Marjana Tomic-Canic. 2016. "Epithelial-mesenchymal transition in tissue repair and fibrosis." *Cell and Tissue Research* 365: 495–506.

Su, Wen-Hsiang, Ming-Huei Cheng, Wen-Ling Lee, Tsung-Shan Tsou, Wen-Hsun Chang, Chien-Sheng Chen, and Peng-Hui Wang. 2010. "Nonsteroidal anti-inflammatory drugs for wounds: Pain relief or excessive scar formation?" *Mediators of Inflammation* 2010 (1): 413238.

Sun, Jing, Jinrui Zhang, Lai Zhao, Sikang Wan, Baiheng Wu, Chao Ma, Jingjing Li, Fan Wang, Xiwen Xing, and Dong Chen. 2023. "Contribution of hydrogen-bond nanoarchitectonics to switchable photothermal-mechanical properties of bioinorganic fibers." *CCS Chemistry* 5 (5): 1242–1250.

Sung, Samuel, Lindsay A Steele, Gregory E Risser, and Kara L Spiller. 2023. "Biomaterial-assisted macrophage cell therapy for regenerative medicine." *Advanced Drug Delivery Reviews* 199: 114979.

Sveen, Lene, Christian Karlsen, and Elisabeth Ytteborg. 2020. "Mechanical induced wounds in fish—A review on models and healing mechanisms." *Reviews in Aquaculture* 12 (4): 2446–2465.

Takahashi, Aki, Meghan E Flanigan, Bruce S McEwen, and Scott J Russo. 2018. "Aggression, social stress, and the immune system in humans and animal models." *Frontiers in Behavioral Neuroscience* 12: 56.

Tan, Jih Huei, Yuzaidi Mohamad, Chor Lip Henry Tan, Mahazir Kassim, and Theodore E Warkentin. 2018. "Concurrence of symmetrical peripheral gangrene and venous limb gangrene following polytrauma: A case report." *Journal of Medical Case Reports* 12: 1–5.

Tehrani, Hamid, Marc Moncrieff, Bruce Philp, and Peter Dziewulski. 2008. "Spectrophotometric intracutaneous analysis: A novel imaging technique in the assessment of acute burn depth." *Annals of Plastic Surgery* 61 (4): 437–440.

Tennent, Glenys A, Stephen O Brennan, Arie J Stangou, John O' Grady, Philip N Hawkins, and Mark B Pepys. 2007. "Human plasma fibrinogen is synthesized in the liver." *Blood* 109 (5): 1971–1974.

Tibbs, Martha K. 1997. "Wound healing following radiation therapy: A review." *Radiotherapy and Oncology* 42 (2): 99–106.

Tottoli, Erika Maria, Rossella Dorati, Ida Genta, Enrica Chiesa, Silvia Pisani, and Bice Conti. 2020. "Skin wound healing process and new emerging technologies for skin wound care and regeneration." *Pharmaceutics* 12 (8): 735.

Tuli, Hardeep Singh, Ruchira Joshi, Ginpreet Kaur, Vivek Kumar Garg, Katrin Sak, Mehmet Varol, Jagjit Kaur, Sulaiman Ali Alharbi, Tahani Awad Alahmadi, and Diwakar Aggarwal. 2023. "Metal nanoparticles in cancer: From synthesis and metabolism to cellular interactions." *Journal of Nanostructure in Chemistry* 13 (3): 321–348.

VanHoy, Tess B., Heidi Metheny, and Bhupendra C. Patel. Chemical burns. In: *StatPearls*. Treasure Island (FL): StatPearls Publishing; 2025. Available from: https://www.ncbi.nlm.nih.gov/books/NBK562202/

Vasudevan, Biju. 2014. "Venous leg ulcers: Pathophysiology and classification." *Indian Dermatology Online Journal* 5 (3): 366–370.

Velnar, Tomaz, Tracey Bailey, and Vladimir Smrkolj. 2009. "The wound healing process: An overview of the cellular and molecular mechanisms." *Journal of International Medical Research* 37 (5): 1528–1542.

Vignard, Julien, Gladys Mirey, and Bernard Salles. 2013. "Ionizing-radiation induced DNA double-strand breaks: A direct and indirect lighting up." *Radiotherapy and Oncology* 108 (3): 362–369.

Visse, Robert, and Hideaki Nagase. 2003. "Matrix metalloproteinases and tissue inhibitors of metalloproteinases: Structure, function, and biochemistry." *Circulation Research* 92 (8): 827–839.

Vu, Lan T, Kerilyn K Nobuhara, Hanmin Lee, and Diana L Farmer. 2009. "Conflicts in wound classification of neonatal operations." *Journal of Pediatric Surgery* 44 (6): 1206–1211.

Wang, Audrey S, Ehrin J Armstrong, and April W Armstrong. 2013. "Corticosteroids and wound healing: Clinical considerations in the perioperative period." *The American Journal of Surgery* 206 (3): 410–417.

Wang, Canran, Ehsan Shirzaei Sani, Chia-Ding Shih, Chwee Teck Lim, Joseph Wang, David G Armstrong, and Wei Gao. 2024. "Wound management materials and technologies from bench to bedside and beyond." *Nature Reviews Materials* 9 (8): 550–566.

Weingarten, Michael S, Elisabeth Papazoglou, Leonid Zubkov, Linda Zhu, Greg Vorona, and Adam Walchack. 2006. "Measurement of optical properties to quantify healing of chronic diabetic wounds." *Wound Repair and Regeneration* 14 (3): 364–370.

Weingarten, Michael S, Elisabeth S Papazoglou, Leonid Zubkov, Linda Zhu, Michael Neidrauer, Guy Savir, Kim Peace, John G Newby, and Kambiz Pourrezaei. 2008. "Correlation of near infrared absorption and diffuse reflectance spectroscopy scattering with tissue neovascularization and collagen concentration in a diabetic rat wound healing model." *Wound Repair and Regeneration* 16 (2): 234–242.

Wen, Qingyun, Suzanne M Mithieux, and Anthony S Weiss. 2020. "Elastin biomaterials in dermal repair." *Trends in Biotechnology* 38 (3): 280–291.

Wernig, Gerlinde, Shih-Yu Chen, Lu Cui, Camille Van Neste, Jonathan M Tsai, Neeraja Kambham, Hannes Vogel, Yaso Natkunam, D Gary Gilliland, and Garry Nolan. 2017. "Unifying mechanism for different fibrotic diseases." *Proceedings of the National Academy of Sciences* 114 (18): 4757–4762.

Wicke, Corinna, Betty Halliday, Daniel Allen, Nanette S Roche, Heinz Scheuenstuhl, Martin M Spencer, Anita B Roberts, and Thomas K Hunt. 2000. "Effects of steroids and retinoids on wound healing." *Archives of Surgery* 135 (11): 1265–1270.

Wild, Thomas, Arastoo Rahbarnia, Martina Kellner, Luboš Sobotka, and Thomas Eberlein. 2010. "Basics in nutrition and wound healing." *Nutrition* 26 (9): 862–866.

Wilkins, Robert G, and Martin Unverdorben. 2013. "Wound cleaning and wound healing: A concise review." *Advances in Skin & Wound Care* 26 (4): 160–163.

Winter, George D. 1962. "Formation of the scab and the rate of epithelization of superficial wounds in the skin of the young domestic pig." *Nature* 193 (4812): 293–294.

Wu, Yuan-Kun, Nai-Chen Cheng, and Chao-Min Cheng. 2019. "Biofilms in chronic wounds: Pathogenesis and diagnosis." *Trends in Biotechnology* 37 (5): 505–517.

Wu, Zhuoxia, Ming Li, Qianqian Liu, Jing Sun, Chen Tao, Bo Jia, Xiaoyan Sun, Hongjie Zhang, Kai Liu, and Fan Wang. 2025. "Engineered multifunctional artificial dermis for infected burn wound healing." *Advanced Functional Materials* 35 (8): 2415514.

Wu, Zhuoxia, Lianbo Zhang, Chen Tao, Shidong Wang, and Fan Wang. 2023. "Self-assembling NIR-responsive recombinant protein complex nanoparticles for synergistic treatment of malignant tumors." *ACS Applied Nano Materials* 6 (9): 7754–7762.

Xiang, Jinxi, Lan Shen, and Yanlong Hong. 2020. "Status and future scope of hydrogels in wound healing: Synthesis, materials and evaluation." *European Polymer Journal* 130: 109609.

Yafi, Amr, Thomas S Vetter, Thomas Scholz, Sarin Patel, Rolf B Saager, David J Cuccia, Gregory R Evans, Anthony J and Durkin. 2011. "Postoperative quantitative assessment of reconstructive tissue status in a cutaneous flap model using spatial frequency domain imaging." *Plastic and Reconstructive Surgery* 127 (1): 117–130.

Yamazaki, Mutsuo, Shunichi Sato, Hiroshi Ashida, Daizo Saito, Yoshiaki Okada, and Minoru Obara. 2005. "Measurement of burn depths in rats using multiwavelength photoacoustic depth profiling." *Journal of Biomedical Optics* 10 (6): 064011–064011.

Yamazaki, Mutsuo, Shunichi Sato, Daizoh Saitoh, Yoshiaki Okada, Hiroshi Ashida, and Minoru Obara. 2006. "Photoacoustic monitoring of neovascularities in grafted skin." *Lasers in Surgery and Medicine: The Official Journal of the American Society for Laser Medicine and Surgery* 38 (3): 235–239.

Yang, Bo, Jiliang Song, Yuhang Jiang, Ming Li, Jingjing Wei, Jiajun Qin, Wanjia Peng, Fernando López Lasaosa, Yiyan He, and Hongli Mao. 2020. "Injectable adhesive self-healing multicross-linked double-network hydrogel facilitates full-thickness skin wound healing." *ACS Applied Materials & Interfaces* 12 (52): 57782–57797.

Yates, Cecelia C, Priya Krishna, Diana Whaley, Richard Bodnar, Timothy Turner, and Alan Wells. 2010. "Lack of CXC chemokine receptor 3 signaling leads to hypertrophic and hypercellular scarring." *The American Journal of Pathology* 176 (4): 1743–1755.

Yeh, Alvin T, Bunsho Kao, Woong Gyu Jung, Zhongping Chen, J Stuart Nelson, and Bruce J Tromberg. 2004. "Imaging wound healing using optical coherence tomography and multiphoton microscopy in an in vitro skin-equivalent tissue model." *Journal of Biomedical Optics* 9 (2): 248–253.

Zhang, Hao F, Konstantin Maslov, George Stoica, and Lihong V Wang. 2006. "Imaging acute thermal burns by photoacoustic microscopy." *Journal of Biomedical Optics* 11 (5): 054033–054033-5.

Zhang, Miao, and Xia Zhao. 2020. "Alginate hydrogel dressings for advanced wound management." *International Journal of Biological Macromolecules* 162: 1414–1428.

Zhang, Peng, Jing Sun, Jingjing Li, Hongjie Zhang, and Kai Liu. 2023. "Biosynthesis, assembly, and biomedical applications of high-performance engineered proteins." *ACS Chemical Biology* 18 (7): 1460–1472.

Zhao, Han, Jingchao Hu, and Li Zhao. 2021. "The effect of low-level laser therapy as an adjunct to periodontal surgery in the management of postoperative pain and wound healing: A systematic review and meta-analysis." *Lasers in Medical Science* 36 (1): 175–187.

Zhao, Kelu, Bo Li, Yao Sun, Bo Jia, Jing Chen, Wenhao Cheng, Lai Zhao, Jingjing Li, Fan Wang, and Juanjuan Su. 2023. "Engineered bicomponent adhesives with instantaneous and superior adhesion performance for wound sealing and healing applications." *Advanced Functional Materials* 33 (38): 2303509.

Zhao, Ruilong, Helena Liang, Elizabeth Clarke, Christopher Jackson, and Meilang Xue. 2016. "Inflammation in chronic wounds." *International Journal of Molecular Sciences* 17 (12): 2085.

Zhou, Yuwei, Meiru Li, Weicheng Gao, Xueping Li, Lixia Long, Xin Hou, Jin Zhao, Sidi Li, and Xubo Yuan. 2023. "Microstructure-united heterogeneous sodium alginate doped injectable hydrogel for stable hemostasis in dynamic mechanical environments." *International Journal of Biological Macromolecules* 248: 125877.

Zhu, Xiaoqin, Shuangmu Zhuo, Liqin Zheng, Kecheng Lu, Xingshan Jiang, Jianxin Chen, and Bifang Lin. 2010. "Quantified characterization of human cutaneous normal scar using multiphoton microscopy." *Journal of Biophotonics* 3(1–2): 108–116.

2 Signaling Pathways in Wound Healing

Nahal Zahra[1], *Aarif Ali*[2*], *Javeed Iqbal Ahmad Bhat*[2], *Faisal Amin*[1], *Bashir Ahmad Malla*[3], *and Humayun Azad*[1]

[1] Division of Environmental Sciences, Sher-e-Kashmir University of Agricultural Sciences and Technology of Kashmir, Shalimar, Srinagar, Jammu & Kashmir, India

[2] Division of Basic Sciences and Humanities, Faculty of Horticulture, Sher-e-Kashmir University of Agricultural Sciences and Technology of Kashmir, Shalimar, Srinagar, Jammu & Kashmir, India

[3] Government Degree College Sumbal Sonawari Bandipora, Jammu & Kashmir, India

2.1 INTRODUCTION

Wound healing is a fundamental biological process essential for maintaining the integrity and functionality of the skin, the body's primary barrier against external harm. When injury occurs, the body initiates a complex and tightly regulated sequence of cellular and molecular events to repair the damaged tissue. This process typically unfolds through four overlapping and interdependent phases: haemostasis, inflammation, proliferation, and remodelling (Martin and Nunan, 2015). In acute wounds, this cascade progresses efficiently, leading to complete healing. However, in chronic wounds, this progression is disrupted, leading to a prolonged inflammatory phase, impaired angiogenesis, cellular senescence, and an inability to restore functional tissue (Demidova-Rice et al., 2012). Globally, chronic wounds, such as diabetic foot ulcers (DFUs), venous leg ulcers, and pressure sores, affect over 10 million individuals annually in the United States alone and cost an estimated USD 25 billion in treatment expenses (Bonnici et al., 2023; Leyane et al., 2021). These wounds are particularly prevalent among the elderly, immunocompromised patients, and individuals with metabolic diseases like diabetes mellitus. The persistent nature of chronic wounds leads to increased risk of infection, prolonged hospital stays, diminished quality of life, and, in severe cases, limb amputation or death.

At the molecular level, chronic wounds are characterised by sustained oxidative stress, dysregulated cytokine signaling, impaired keratinocyte and fibroblast migration, and defective extracellular matrix (ECM) remodelling (Basu and Martins-Green, 2022). Cellular signaling pathways that ordinarily drive wound resolution such as phosphoinositide 3-kinase/protein kinase B signaling pathway (PI3K/AKT) for survival and proliferation, wingless-related integration site/beta-catenin (Wnt/β-catenin) for granulation tissue formation, transforming growth factor beta (TGF-β) for fibroblast activation and collagen deposition, and nuclear factor erythroid 2-related factor 2 (Nrf2) for redox balance are often suppressed or aberrantly activated in chronic wound environments (Gumede et al., 2024; Krizanova et al., 2022). Furthermore, chronic wounds exhibit features reminiscent of tumour biology, including the upregulation of oncogenic markers such as erb-B2 receptor tyrosine

* Corresponding author: buttaarif31@gmail.com

DOI: 10.1201/9781003644958-2

kinase 2 (ErbB2) and syndecan-1 (SDC1) and the activation of gene networks linked to carcinogenesis (Basu and Martins-Green, 2022). Systems biology tools such as Weighted Gene Correlation Network Analysis (WGCNA) and transcriptome profiling have expanded our understanding of these complex networks. For example, WGCNA of chronic wound tissues has identified modules of co-expressed genes involved in inflammation, immune dysregulation, and metabolic collapse (Basu and Martins-Green, 2022). These insights suggest that chronic wounds are not simply delayed in healing but are reprogrammed at the molecular level. In light of these findings, therapeutic strategies targeting these signaling abnormalities have gained traction. Techniques like photobiomodulation (PBM) use specific light wavelengths to modulate intracellular pathways and promote tissue repair, while stem cell-based therapies, peptide delivery systems, and bioengineered scaffolds aim to restore lost regenerative potential (Bonnici et al., 2023; Leyane et al., 2021). Understanding how each signaling pathway contributes to wound pathology and how it can be therapeutically modulated is central to developing next-generation wound care strategies.

This chapter provides an in-depth overview of the major cellular signaling pathways implicated in chronic wound healing. It elaborates on their physiological roles, the consequences of their dysregulation, and how modern therapies seek to restore their function. Emphasis is placed on integrating mechanistic insights with translational therapeutic approaches to pave the way for more effective wound care.

2.2 PHASES OF WOUND HEALING

Wound healing is a sophisticated and finely coordinated biological process essential for the restoration of tissue integrity following physical injury. It involves an orchestrated interplay of various cell types, ECM components, signaling pathways, and bioactive molecules that collectively drive the transition from tissue damage to functional repair. This regenerative process is classically demarcated into four overlapping yet sequential phases: haemostasis, inflammation, proliferation, and remodelling (Krizanova et al., 2022). Each phase is characterised by a distinct but interdependent set of molecular events that are tightly regulated in both spatial and temporal dimensions.

2.2.1 Haemostasis: Immediate Defence Mechanism

The haemostatic phase constitutes the first physiological response to skin injury and begins within seconds of the insult (Martin and Nunan, 2015). Upon injury, vascular constriction is immediately triggered by the local release of vasoconstrictors such as serotonin and thromboxane A2. This is followed by the activation of platelets, which adhere to exposed collagen and release procoagulant factors including adenosine diphosphate (ADP, thrombin) and fibrinogen. These molecules activate the coagulation cascade, culminating in the conversion of fibrinogen to fibrin, which forms a mesh-like clot that not only prevents haemorrhage but also acts as a scaffold for immune cell recruitment and epithelial cell migration (Gumede et al., 2024). Moreover, activated platelets release a milieu of growth factors such as platelet-derived growth factor (PDGF) and TGF-β, which are pivotal for the recruitment of neutrophils and macrophages, thus seamlessly transitioning the wound milieu into the inflammatory phase (Bonnici et al., 2023).

2.2.2 Inflammation: Clearing the Wound Bed

The inflammatory phase typically initiates within hours of the injury and may persist for several days depending on the severity and biological context. This phase serves as the body's intrinsic clean-up mechanism, wherein immune cells remove cellular debris, necrotic tissue, and invading pathogens, thereby preparing the wound bed for tissue regeneration (Leyane et al., 2021). In the inflammatory phase of wound healing various cell signaling pathways work together to orchestrate the immune response aimed at eradicating pathogens, clearing away damaged tissue, and setting the stage for wound repair. The key signaling pathways during this phase are the following:

- *Nuclear Factor Kappa Beta (NF-κB) Pathway*: Triggered by toll-like receptors (TLRs) and cytokines such as tumour necrosis factor alpha (TNF-α) and interleukin-1 beta (IL-1β), NF-κB moves to the nucleus where it promotes the expression of pro-inflammatory genes like interleukins such as interleukin-6 (IL-6), interleukin-8 (IL-8), Cyclooxygenase-2 (COX-2), and inducible nitric oxide synthase (iNOS).
- *Mitogen-Activated Protein Kinase (MAPK) Pathway*: These kinases are activated by reactive oxygen species (ROS), cytokines, and stress.
- *Janus Kinase/Signal Transducer and Activator of Transcription Pathway (JAK/STAT)*: Cytokines such as IL-6 trigger signal transducer and activator of transcription 3 (STAT3) activation, which results in the transcription of genes that play a role in cell survival and immune regulation.
- *Nrf2 Pathway*: This pathway manages the cellular response to oxidative stress by controlling the expression of antioxidant genes, thereby reducing oxidative harm to adjacent tissues.

2.2.3 Proliferation: Rebuilding the Architecture

Following successful resolution of inflammation, the wound enters the proliferative phase, during which granulation tissue begins to form. This phase is marked by fibroblast proliferation, keratinocyte migration, angiogenesis, and ECM deposition, all of which contribute to restoring tissue architecture (Bonnici et al., 2023). A number of essential growth factors play a role, such as epidermal growth factor (EGF), fibroblast growth factor (FGF), vascular endothelial growth factor (VEGF), and TGF-β. A multitude of signaling pathways are active during this phase:

- *PI3K/AKT Pathway*: Stimulated by growth factors, it promotes cell survival, angiogenesis, and metabolic reprogramming. AKT also prevents apoptosis and facilitates fibroblast migration
- *Extracellular Signal-Regulated Kinase (ERK)/MAPK Pathway*: Triggered by EGF, FGF, and VEGF, ERK1/2 enhances cell cycle progression, encourages keratinocyte proliferation, activates fibroblasts, and fosters the growth of endothelial cells.
- *Wnt/β-Catenin Pathway*: Essential for the processes of re-epithelialisation and fibroblast function. Upon binding of Wnt ligands to their receptors, β-catenin avoids degradation, translocates to the nucleus, and promotes the expression of genes associated with proliferation and regeneration.
- *TGF-β*: TGF-β attaches to its receptor and activates SMAD family member 2/3 (SMAD2/3) through phosphorylation, resulting in the formation of complexes with SMAD4 that control transcription. This process results in: myofibroblast differentiation, collagen synthesis, and ECM production (Bonnici et al., 2023; Gumede et al., 2024).

2.2.4 Remodelling: From Granulation to Maturation

The ultimate phase of wound healing, termed remodelling, can continue for weeks or even months, depending on wound severity and various patient-related factors. It is focused on strengthening the newly formed tissue, remodelling the ECM, and restoring tissue tensile integrity. The ultimate phase consists of collagen remodelling, the apoptosis of myofibroblasts, and the creation of scar tissue (Martin and Nunan, 2015). The important pathways in this phase are as follows.

TGF-β/SMAD Pathway: Continues to support ECM production, but excessive activation may lead to fibrosis.

MMP/TIMP Balance

Matrix metalloproteinases (MMPs) and tissue inhibitors of metalloproteinases (TIMPs) are tightly regulated to ensure the controlled degradation of provisional ECM and facilitate the maturation of the dermal matrix. The regulation of their balance is closely managed by signaling pathways such as NF-κB, p38 MAPK, and TGF-β. Myofibroblasts continue to play a role in wound contraction, while endothelial cells undergo vascular pruning to refine the neovasculature (Krizanova et al., 2022).

Together, the four phases of wound healing reflect a biologically intricate yet elegantly structured programme that restores damaged tissue. A successful outcome hinges on the precise regulation of cellular functions via signaling pathways responsive to both biochemical cues and mechanical stress. In chronic wounds, disruption at any stage—especially inflammation and proliferation—can derail the entire cascade, leading to pathological healing outcomes. Understanding the molecular mechanisms governing each phase is therefore essential to devising therapeutic interventions that can overcome the intrinsic barriers of chronic wound repair (Figure 2.1).

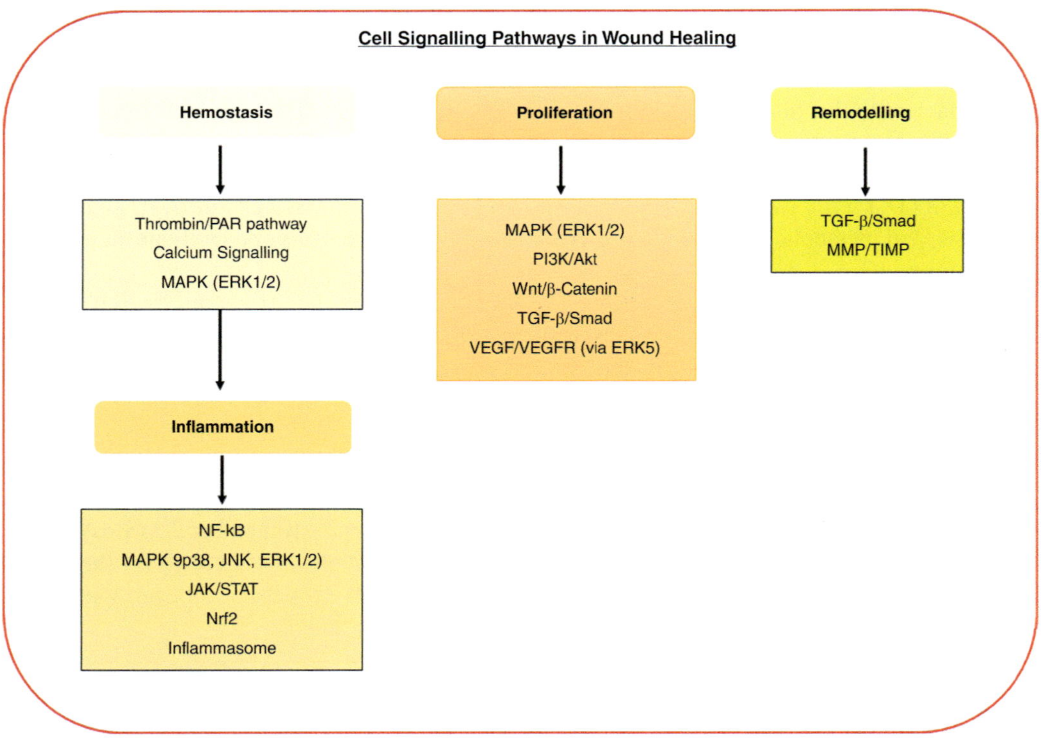

FIGURE 2.1 Key signaling pathways involved in wound healing.

2.3 KEY SIGNALING PATHWAYS IN WOUND HEALING

The restoration of tissue integrity following injury is dependent on the timely and accurate activation of several interrelated intracellular signaling pathways. These molecular circuits orchestrate a vast array of cellular responses ranging from proliferation, differentiation, migration, and apoptosis to ECM synthesis and immune modulation. In physiological wound healing, signaling pathways function in a spatially and temporally restricted manner, ensuring orderly progression through each healing phase. However, dysregulation of these signaling events can derail healing, contributing to chronicity.

This section outlines the seven most critical signaling pathways involved in wound repair, focusing on their biological functions, mechanisms of activation, and relevance to both acute and chronic wound states.

2.3.1 Phosphoinositide 3-Kinase (PI3K)/Protein Kinase B (AKT) Pathway: Master Regulator of Survival and Growth

The phosphoinositide 3-kinase (PI3K)/protein Kinase B (AKT) signaling axis is a pivotal intracellular pathway regulating cellular survival, metabolism, angiogenesis, proliferation, and migration. The PI3K/AKT pathway begins when cytokines and growth factors, including PDGF and EGF, attach to receptor tyrosine kinases (RTKs) (Jere et al., 2019). Following this, RTKs experience autophosphorylation of tyrosine residues within the cytosolic domain. Subsequently, PI3K associates with the activated RTKs through Src homology 2 domains. Upon binding, class I PI3Ks are recruited to the membrane where they catalyse the conversion of phosphatidylinositol-4,5-bisphosphate (PIP2) into phosphatidylinositol-3,4,5-triphosphate (PIP3) (Engelman et al., 2006). PIP3 provides a docking site at the plasma membrane for AKT, and PDK1 (3-phosphoinositide-dependent kinase-1). PDK1 is responsible for phosphorylating AKT at Thr308, and mTORC2 (mammalian target of rapamycin complex 2) phosphorylates it at Ser473, leading to AKT's complete activation. Phosphorylated AKT activates mTORC1, which subsequently facilitates the translation of various pro-migratory and proliferative RNA transcripts. In the context of normal wound healing, PI3K/AKT signaling boosts Epithelial–mesenchymal transition (EMT), cell proliferation, and angiogenesis while diminishing inflammation (Manning and Toker, 2017). Members of the PI3K-related kinase (PIKK) family, such as DNA-PK, are capable of phosphorylating AKT at serine 473. Additionally, AKT can undergo dephosphorylation by protein phosphatase 2A (PP2A) and PH-domain rich leucine-repeat-containing protein phosphatases (PHLPP1/2) (Bozulic and Hemmings, 2009; Manning and Cantley, 2007). Phosphatase and TENsin homolog deleted on chromosome 10 (PTEN) serves as a vital upstream regulator of the PI3K/AKT pathway, which can suppress cell proliferation and increase cell susceptibility to apoptosis. The primary substrate for PTEN is PIP3, which it dephosphorylates at site D3 to produce PIP2, thereby negatively influencing the PI3K/AKT pathway. When PTEN is inactivated, it leads to the persistent activation of PI3K/AKT. Moreover, PI3K/AKT binds to several downstream molecules to aid in cell motility. The Bcl-2 family members, including Bcl-2-associated death promoter (BAD), are key regulators of cell apoptosis. AKT downregulates BAD to suppress cell apoptosis and support cell survival (She et al., 2005). AKT is also capable of directly phosphorylating and inactivating Caspase-9 at the S196 site, which inhibits Caspase-9-mediated cell apoptosis. The mTORC1 oversees the initiation of translation and ribosome synthesis, subsequently promoting cell growth and proliferation (Skeen et al., 2006; Wullschleger et al., 2006). The AKT can phosphorylate mTOR2448 directly, leading to the activation of mTORC1. Additionally, it has the ability to phosphorylate TSC2 at S939 and T1462, which inhibits TSC2's function and subsequently activates mTORC1 indirectly via Rheb-GTP. In addition, AKT activates endothelial cell growth factor and phosphorylates endothelial NO synthase at S1177, leading to an increase in nitric oxide (NO) production within endothelial cells. This action fosters the growth and proliferation of these cells, boosts vascular permeability, and aids in angiogenesis. Upon activation, the downstream protein p70S6K of mTOR can enhance cellular movement. Moreover, the PI3K/AKT pathway can elevate the mRNA expression of matrix metalloproteinase-2 (MMP-2), which is essential for breaking down the ECM and encouraging cell invasion and metastasis.

The forkhead box O (FOXO) belongs to the transcription factor family and acts as an essential downstream component of the PI3K/AKT pathway (Calnan and Brunet, 2008). The mammalian FOXO family includes FOXO1, 3, 4, and 6, all of which exhibit a significant level of similarity in their structure, function, and regulation. Their primary differences lie in their tissue expression. Within the nucleus, FOXOs facilitate the transcription of various target genes linked to cellular physiological processes (Tia et al., 2018). As a result, FOXO proteins are involved in the progression of various skin disorders, such as acne and psoriasis, among others.

PI3K/AKT pathway is especially crucial in the proliferative phase, where endothelial cells form new capillaries and keratinocytes migrate to restore epithelial integrity (Gao et al., 2016). In chronic wounds, particularly DFUs, the PI3K/AKT pathway is often impaired due to oxidative stress, hyperglycaemia, or insulin resistance, resulting in poor angiogenesis and diminished keratinocyte migration (Jere et al., 2019). The PI3K/AKT pathway is well-established as being closely linked to the development of an epidermal barrier, which primarily relies on the proliferation and differentiation of keratinocytes. Research by Chang et al. (2019) showed that the binding of miR-126 to its target gene PLK2 enhances the keratinocytes migration and proliferation, thus significantly contributing to skin wound healing through the activation of the PI3K/AKT pathway. Jiang et al. (2020) also reported that miR-26a could lessen the migration of keratinocytes by influencing its target gene, ITGA5, thereby obstructing wound healing. EMT is shown to be involved in the process of skin wound healing, with the activation of PI3K believed to activate mTOR via AKT, thus accelerating EMT (Chen et al., 2017; Stone et al., 2016).

2.3.2 Wnt/β-Catenin Pathway: The Architect of Granulation Tissue

The Wnt/β-catenin pathway, commonly known as the canonical Wnt pathway, initiates when Wnt ligands bind to frizzled (Fz) receptors and low-density lipoprotein receptor-related protein 5/6 (LRP5/6) co-receptors on the cell surface. This binding action prevents the destruction complex, which includes adenomatous polyposis coli (APC), Axin, and glycogen synthase kinase 3β (GSK-3β), from functioning, resulting in the stabilisation and accumulation of β-catenin in the cytoplasm. Following this, β-catenin translocates to the nucleus, where it interacts with T-cell factor/lymphoid enhancer factor (TCF/LEF) transcription factors, thus promoting the expression of target genes involved in cell proliferation and differentiation (Clevers and Nusse, 2012). APC serves as a crucial suppressor of Wnt/β-catenin signaling, which is vital for maintaining cellular equilibrium during the wound healing process. It guarantees that β-catenin is broken down in non-stimulated states, thereby preventing abnormal cell growth and fibrosis. The regulation of APC is essential for controlled regeneration and epithelial repair throughout wound healing. GSK-3β, a serine/threonine kinase, serves a crucial inhibitory function in the canonical Wnt/β-catenin signaling pathway. It functions as a major negative regulator of Wnt/β-catenin signaling, essential for the degradation of β-catenin in the absence of Wnt ligands. In the context of wound healing, the precise regulation of GSK-3β is imperative: its inhibition can hasten healing, encourage cell proliferation, and enhance angiogenesis, whereas uncontrolled activity may hinder regeneration and worsen inflammation. In addition to its role in Wnt signaling, GSK-3β is engaged in various cellular functions, such as metabolism, regulation of the cell cycle, inflammation, apoptosis, and tissue repair. Therefore, GSK-3β is recognised as both a vital biomarker and a therapeutic target in wound healing studies.

The Wnt/β-catenin pathway is crucial for regulating fibroblast proliferation, keratinocyte migration, and granulation tissue formation, especially during the proliferative and remodelling phase. The activation of Wnt/β-catenin signaling promotes the transcription of genes associated with ECM deposition, angiogenesis, and fibroblast migration. Furthermore, this pathway is essential for activating keratinocyte stem cells, which are vital for re-epithelialisation and restoration of the epidermal barrier (Nusse et al., 2008). In the context of chronic wounds, Wnt signaling is frequently downregulated, contributing to fibroblast senescence, decreased ECM production, and reduced granulation tissue. Experimental models have demonstrated that PBM can reactivate Wnt/β-catenin signaling, enhancing regenerative outcomes in diabetic wound settings (Jere et al., 2022).

2.3.3 Transforming Growth Factor-β (TGF-β) Pathway: The Conductor of Tissue Repair

The TGF-β signaling pathway is widely acknowledged as a primary regulator, or "conductor," of tissue repair and wound healing. Its multifaceted actions on immune cells, epithelial cells, fibroblasts, and endothelial cells enable TGF-β to direct the inflammatory, proliferative, and remodelling phases of healing, coordinating cellular behavior, and ECM remodelling to re-establish tissue integrity (Barrientos et al., 2008). TGF-β represents a family of cytokines that consists of TGF-β1, TGF-β2, and TGF-β3, with TGF-β1 being the most prevalent and functionally significant during wound healing. When tissue is injured, TGF-β is released from platelets, macrophages, fibroblasts, and keratinocytes into the extracellular space. TGF-β interacts with the TGF-β type II receptor (TβRII), which is a serine/threonine kinase located on the cell membrane. TβRII then recruits and phosphorylates the TGF-β type I receptor (TβRI). Once activated, TβRI phosphorylates SMAD2 and SMAD3 (known as R-SMADs). These R-SMADs form a complex with SMAD4 (co-SMAD) and move into the nucleus (Derynck and Budi, 2019). This complex is responsible for regulating the transcription of genes that are involved in: synthesis of ECM, fibroblast-to-myofibroblast differentiation, cell motility, and differentiation (Derynck and Budi, 2019).

Following an injury, TGF-β1 is swiftly released from degranulated platelets and activated macrophages (Li et al., 2007; Werner and Grose, 2003). This factor draws neutrophils, monocytes, and macrophages to the wound area. TGF-β is instrumental in shifting macrophages from a pro-inflammatory (M1) phenotype to a pro-repair (M2) phenotype (Li et al., 2007; Werner and Grose, 2003). Moreover, it reduces the expression of pro-inflammatory cytokines (including TNF-α and IL-1β), which helps to prevent chronic inflammation (Li et al., 2007; Werner and Grose, 2003).

TGF-β encourages the growth of fibroblasts and triggers the differentiation of myofibroblasts via α-SMA expression, which is crucial for the process of wound contraction. It boosts angiogenesis by increasing VEGF levels, thereby facilitating the formation of blood vessels within the granulation tissue (Piek et al., 1999). Additionally, TGF-β aids in the migration of keratinocytes and the re-epithelialisation process, although it might restrict excessive cell proliferation (Piek et al., 1999). It also stimulates the production of ECM proteins such as collagen, elastin, and proteoglycans, thereby creating the wound matrix (Piek et al., 1999).

TGF-β plays an important role in preserving the balance between ECM deposition and degradation by influencing MMPs and TIMPs. Continuous activity of TGF-β1 results in heightened collagen cross-linking and the development of scar tissue. Conversely, TGF-β3 diminishes scarring and encourages a more structured process of tissue regeneration (Ferguson and O'Kane, 2004; Shah et al., 1995).

However, dysregulated or excessive TGF-β signaling can be detrimental, especially in chronic wounds. Persistent activation is associated with fibrosis, hypertrophic scarring, and healing arrest, particularly under hyperglycaemic conditions common in diabetic patients (Meng et al., 2016).

2.3.4 Notch Signaling Pathway: Cell Fate and Angiogenesis Modulator

The Notch signaling pathway is a highly conserved juxtacrine signaling mechanism that plays essential roles in cell fate determination, angiogenesis, keratinocyte differentiation, and immune regulation during wound healing. The Notch signaling pathway is characterised by the interaction of adjacent cells through Notch receptors and their corresponding ligands. In mammals, there exist four Notch receptors (Notch1–4) along with five primary ligands: Jagged1, Jagged2, Delta-like 1 (Dll1), Delta-like 3 (Dll3), and Delta-like 4 (Dll4). These ligands belong to the Delta/Serrate/LAG-2 (DSL) family. Activation is initiated when membrane-bound ligands such as Delta-like or Jagged proteins on one cell bind to Notch receptors (Notch1–4) on a neighbouring cell. This interaction initiates a series of proteolytic cleavages, ultimately leading to the release of the Notch intracellular domain (NICD). NICD translocates into the nucleus and modulates target gene transcription through RBP-Jκ-dependent transcriptional mechanisms (Chigurupati et al., 2007). In the context of cutaneous wound healing, Notch signaling has been shown to:

- Promote keratinocyte differentiation, aiding epidermal renewal and stratification (Blanpain et al., 2006).
- Enhance endothelial sprouting and vascular remodelling, largely through crosstalk with the VEGF signaling pathway (Liu et al., 2009).
- Regulate macrophage polarisation, driving the transition from pro-inflammatory M1 macrophages to reparative M2 macrophages, a key step in resolving inflammation and supporting granulation tissue formation (Monsalve et al., 2006).

However, in chronic wounds, Notch pathway components are frequently downregulated or dysregulated, contributing to:

- Persistent inflammation due to impaired M1-to-M2 macrophage transition;
- Insufficient angiogenesis, which limits oxygen and nutrient delivery to the wound bed;
- Keratinocyte dysfunction, resulting in defective re-epithelialisation and delayed closure (Chigurupati et al., 2007).

Given these dysfunctions, therapeutic modulation of the Notch pathway using small molecule modulators, biomaterials, or genetic tools is under active investigation as a promising strategy to reinstate normal healing dynamics in chronic wound environments.

2.3.5 JAK/STAT Pathway: Cytokine Signal Transducer

The Janus kinase (JAK)/signal transducer and activator of transcription (STAT) pathway is essential for wound healing, coordinating the responses to cytokines and growth factors that manage inflammation, cell proliferation, migration, and tissue remodelling. This pathway serves as a key signaling mechanism that connects extracellular signals to transcriptional responses throughout the intricate, multi-phase process of tissue repair. The JAK/STAT pathway functions as a signaling system within cells. It assists cells in responding to external signals—especially from molecules like cytokines and growth factors that are released when the body sustains an injury.

- *JAK* = Janus Kinase (an enzyme that starts the signal).
- *STAT* = Signal Transducer and Activator of Transcription (a protein that goes into the nucleus and turns genes on or off).

It acts like a messenger chain:

Signal from outside → JAK turns on → STAT moves into nucleus → cell responds (e.g., heals, grows, or resists damage).

The body responds to injury by mobilising immune cells. These cells produce cytokines such as IL-6 and interferons (interferon-gamma (IFN-γ) that bind to their specific surface receptors triggering the activation of JAKs. The activation of JAK kinases (JAK1, JAK2, JAK3, and TYK2) occurs in conjunction with the cytoplasmic domains of receptors. These JAKs phosphorylate the receptor, forming docking sites for STATs (STAT1–6). STAT1 and STAT3 are quickly activated in reaction to pro-inflammatory cytokines like IL-6, IFN-γ, and TNF-α. This stage brings immune cells (macrophages, neutrophils) to the site of injury. STAT3, specifically, facilitates the transition of macrophages to a reparative (M2) phenotype, which is essential for the resolution of inflammation. Once phosphorylated, STATs dimerise and move to the nucleus, where they play a role in regulating gene expression (Liu et al., 2023b). The activation of STAT3 promotes keratinocyte proliferation and migration, which are crucial for re-epithelialisation (Chen et al., 2016; Hu et al., 2023). Additionally, the activation of fibroblasts through JAK2/STAT3 increases the production of ECM components, including collagen and fibronectin (Chen et al., 2016; Hu et al., 2023).

Growth factors like EGF and IL-6 utilise JAK/STAT signaling to facilitate tissue regeneration. During the remodelling phase, the differentiation of myofibroblasts is partially controlled by JAK2/STAT3, which facilitates wound contraction. STAT3 also influences angiogenesis through the regulation of VEGF expression, aiding in the formation of new blood vessels (Figure 2.2).

Abnormal or prolonged activity of STAT3 may lead to fibrosis or chronic wounds (Pastar et al., 2014; Samad et al., 2025). The JAK/STAT pathway is essential for orchestrating wound healing, particularly through the action of STAT3, which governs inflammation, cell proliferation, ECM deposition, and angiogenesis. Modifying this pathway provides hopeful strategies for boosting healing in chronic wounds or decreasing fibrosis in hypertrophic scars.

2.3.6 Nuclear Factor Erythroid 2-Related Factor 2 (Nrf2): Guardian of Redox Homeostasis

Nuclear factor erythroid 2-related factor 2 (Nrf2) is a transcription factor essential for cellular antioxidant defence. Under unstressed conditions, Kelch-like ECH-associated protein 1 (Keapl) binds Nrf2 in the cytoplasm, targeting it for ubiquitination and proteasomal degradation. Upon oxidative stress, oxidative modifications of cysteine residues in Keapl disrupt this interaction, allowing Nrf2 to escape into the nucleus, where it binds antioxidant response elements (AREs) and stimulates the expression of enzymes like glutathione peroxidase (GPx), NAD(P)H quinone dehydrogenase 1 (NQO1), and heme-oxygenase-1 (HO-1) (Suntar et al., 2021). The early inflammatory stage of wound healing is marked by the production of ROS and the infiltration of neutrophils and macrophages. While moderate levels of ROS serve important signaling functions, an overabundance of ROS can harm proteins, lipids, and DNA, exacerbating tissue damage. Nrf2 serves a protective function by boosting the expression of antioxidant genes that counteract ROS, thereby averting excessive oxidative harm (Ahmed et al., 2017; Zhang et al., 2021). Additionally, Nrf2 has the ability to suppress the NF-κB pathway, an essential mediator of inflammation, thereby decreasing the levels of pro-inflammatory cytokines such as TNF-α, IL-6, and IL-1β (Ahmed et al., 2017). During the proliferative phase, keratinocytes, fibroblasts, and endothelial cells undergo proliferation and migration towards the wound bed. Nrf2 plays a significant role in this phase in various ways: it

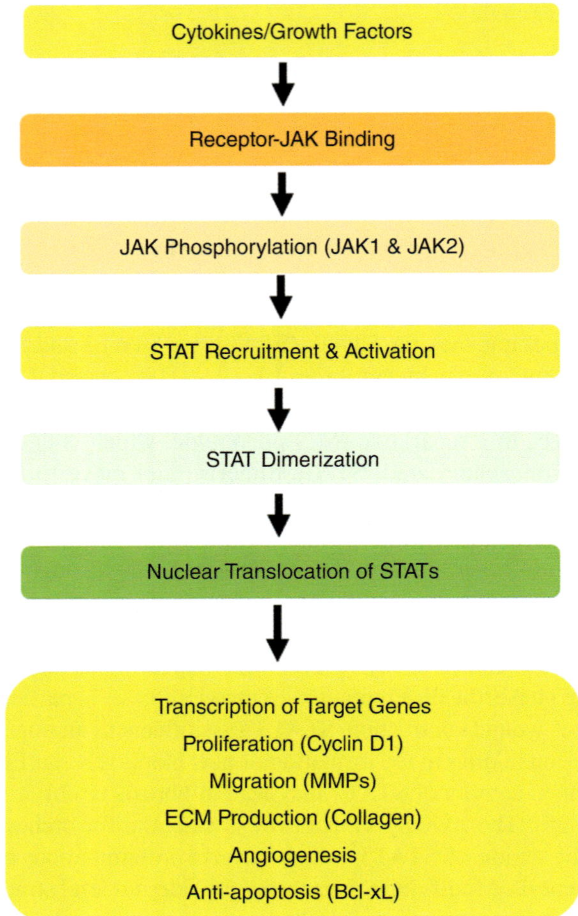

FIGURE 2.2 Overview of the activation process of JAK/STAT pathway.

encourages fibroblast migration and collagen production, which are essential for the formation of the ECM and wound contraction (Saha et al., 2020). Additionally, Nrf2 boosts the expression of VEGF, a powerful pro-angiogenic factor and this action promotes neovascularisation and guarantees sufficient oxygen and nutrient supply to the healing tissue (Saha et al., 2020). Furthermore, Nrf2 aids in re-epithelialisation by stimulating keratinocyte proliferation and migration, which is crucial for wound closure (Saha et al., 2020). During the remodelling phase, granulation tissue transforms into scar tissue via collagen remodelling, ECM turnover, and cell apoptosis (Schafer and Werner, 2008). Nrf2 plays a role in this process by:

- Regulating the expression of MMPs and TIMPs to ensure balanced ECM turnover.
- Enhancing antioxidant defences that protect against oxidative damage to newly formed tissue.
- Aiding mitochondrial function and bioenergetics essential for tissue regeneration.

In chronic wounds, especially those linked to diabetes mellitus, Nrf2 signaling frequently becomes dysregulated or suppressed (Long et al., 2016). High glucose concentrations and prolonged inflammation result in elevated ROS production, while the inadequate activation of Nrf2 hinders a proper antioxidant response (Long et al., 2016). This creates a continuous cycle of oxidative stress, inflammation, and slow healing. Research involving diabetic mouse models has demonstrated that mice lacking Nrf2 show slower wound closure, heightened inflammatory markers, and reduced collagen deposition (Long et al., 2016). Therapeutic activation of Nrf2 via compounds such as sulphoraphane and dimethyl fumarate or through light-based modalities has shown promise in enhancing antioxidant defences, reducing inflammation, and promoting wound closure in diabetic and other chronic wound models (Zhang et al., 2021).

2.3.7 Mitogen-Activated Protein Kinase (MAPK) Pathway

The MAPK pathway is among the most evolutionarily conserved signal transduction pathways found in eukaryotic cells. It is essential for regulating a variety of biological processes, such as cell proliferation, differentiation, migration, apoptosis, and stress responses all of which are critical for wound healing. The MAPK pathway consists of several subfamilies, which include the following:

- ERK1/2 (extracellular signal-regulated kinases)
- JNK (c-Jun N-terminal kinases)
- p38 MAPK
- ERK5 (Big MAP kinase 1)
- Each sub-pathway is triggered by specific stimuli and facilitates distinct yet overlapping functions throughout the wound healing process.

2.3.7.1 ERK1/2 MAPK

The ERK1/2 MAPK pathway is mainly triggered by growth factors like EGF, FGF, and VEGF, which are released at the onset of the wound healing process. ERK1/2 encourages mitogenesis, which is crucial for re-epithelialisation and the development of granulation tissue (Leyane et al., 2021; Pastore et al., 2008). It aids in the reorganisation of the cytoskeleton, allowing keratinocytes to migrate and close the wound (Leyane et al., 2021; Pastore et al., 2008). Additionally, through VEGF signaling, ERK1/2 promotes the creation of new blood vessels (Leyane et al., 2021; Pastore et al., 2008).

2.3.7.2 c-Jun N-Terminal Kinases (JNK)

JNK (c-Jun N-terminal kinase) is typically activated by pro-inflammatory cytokines such as TNF-α and IL-1β, along with environmental stressors like UV radiation or ROS, which are common during the early inflammatory stage of wound healing (Zenz et al., 2003). JNK is involved in controlling the generation of inflammatory cytokines and in drawing immune cells to the wound area (Zenz et al., 2003). Additionally, JNK affects the apoptosis of damaged or ageing cells, creating space for regeneration. It also aids in the cytoskeletal remodelling that is essential for the migration of keratinocytes (Zenz et al., 2003).

2.3.7.3 p38 MAPK

The p38 MAPK pathway exhibits a strong response to cellular stress and inflammatory cytokines. It is crucial during the inflammatory and initial proliferative stages. p38 enhances the transcription of pro-inflammatory cytokines (IL-6, TNF-α) that assist in the recruitment of immune cells (Ashcroft et al., 1999). Additionally, it regulates MMPs, which break down ECM components and facilitate tissue remodelling. p38 also plays a role in the expression of VEGF and other factors that promote angiogenesis (Ashcroft et al., 1999). Furthermore, it encourages the transformation of fibroblasts into myofibroblasts, which are essential for wound contraction (Ashcroft et al., 1999).

2.3.7.4 ERK5 (Big MAP Kinase 1)

ERK5, also referred to as Big MAP kinase 1 (BMK1), belongs to the MAPK family. It is both structurally and functionally different from ERK1/2, JNK, and p38 (Kondoh et al., 2006). ERK5 is mainly activated by oxidative stress, shear stress, growth factors, and hyperosmolarity. In contrast to other MAPKs, ERK5 possesses a distinctive C-terminal transactivation domain, which allows it to directly influence gene transcription. The main upstream activator of ERK5 is MEK5 (MAP kinase kinase 5) (Kondoh et al., 2006). Although it has been less researched compared to ERK1/2, JNK, or p38, ERK5 is gaining recognition for its role in maintaining vascular integrity, promoting angiogenesis, ensuring cell survival, and supporting endothelial function, all of which are essential for effective wound healing. ERK5 is predominantly found in endothelial cells and plays a vital role in their proliferation, migration, and survival. These activities are fundamental to angiogenesis, the process through which new blood vessels are formed, thereby restoring the supply of oxygen and nutrients to the wound area. ERK5 influences VEGF-induced angiogenesis by activating transcription factors such as myocyte enhancer factor 2C (MEF2C) (Hayashi and Lee, 2004). It enhances the expression of angiogenic genes, including VEGF receptor 2 (VEGFR2), (iNOS), and MMPs (Hayashi and Lee, 2004). ERK5 plays a significant role in the migration and differentiation of keratinocytes, which are essential processes in re-epithelialisation (Sun et al., 2001). It also supports fibroblast-driven matrix remodelling by modulating genes associated with the ECM (Sun et al., 2001). In keratinocytes, the signaling pathway of ERK5 has been associated with alterations in cell shape, movement, and the reorganisation of the cytoskeleton (Sun et al., 2001). In fibroblasts, ERK5 might affect collagen deposition and the activity of MMPs, thereby contributing to the formation and remodelling of granulation tissue (Sun et al., 2001). While ERK5 is not the main regulator of pro-inflammatory cytokines such as p38 or JNK, it plays a role in modulating the endothelial response to inflammatory signals: the activation of ERK5 safeguards endothelial cells from apoptosis induced by TNF-α (Pi et al., 2004). It helps maintain the integrity of the endothelial barrier during inflammatory stress, thereby minimising vascular leakage in wounds (Pi et al., 2004). Under oxidative stress conditions (for instance, during the inflammatory phase of wound healing), ERK5 becomes activated and aids in cell survival by enhancing the expression of antioxidant genes (Sohn et al., 2002). ERK5 interacts with transcription factors like Nrf2 and MEF2, which are crucial for regulating antioxidant and cytoprotective genes (Sohn et al., 2002). This interaction supports the survival of keratinocytes and fibroblasts in environments rich in ROS (Sohn et al., 2002).

2.3.8 Nuclear Factor Kappa Beta (NF-κB) Pathway

The nuclear factor kappa beta (NF-κB) pathway represents a vital signaling cascade that plays a key role in managing immune responses, inflammation, cell survival, and development. This pathway includes a group of transcription factors known as NF-κB, which, when activated, translocate to the nucleus to modulate gene expression. It is critically involved in all phases of wound healing, with its primary activity observed during the inflammatory phase. The NF-κB pathway governs the expression of numerous genes encoding pro-inflammatory cytokines, chemokines, adhesion molecules, growth factors, and matrix-degrading enzymes. Dysregulation of this pathway contributes to chronic wound pathogenesis by sustaining inflammation and impairing tissue repair mechanisms (Lawrence, 2009; Schafer and Werner, 2008; Verma, 2004). NF-κB comprises a group of transcription factors, including the key members RelA (p65), RelB, c-Rel, p50, and p52. In its inactive state, it is located in the cytoplasm, where it is associated with the inhibitor IκB (inhibitor of κB). Under basal conditions, NF-κB dimers, predominantly p65 (RelA)/p50 heterodimers, remain sequestered in the cytoplasm bound to inhibitors of κB (IκBs), preventing their nuclear translocation. Upon stimulation by pro-inflammatory cytokines such as TNF-α, IL-1β, pathogen-associated molecular patterns (PAMPs) via TLRs, and ROS, the IκB kinase (IKK) complex becomes activated (Cai et al., 2023; Lawrence, 2009). After phosphorylation, IκBs are subjected to ubiquitination and subsequent proteasomal degradation, which results in the release of NF-κB dimers. These dimers then move into the nucleus where they attach to κB enhancer elements located in the promoter regions of target genes, thereby initiating transcription. This canonical NF-κB pathway is responsible for regulating the expression of a variety of genes that manage immune responses, inflammation, cell proliferation, and apoptosis (Ahmad et al., 2006; Basu and Martins-Green, 2022).

The NF-κB pathway is rapidly activated during the early hours following tissue injury. NF-κB also regulates the expression of enzymes, including COX-2 and iNOS, further amplifying inflammatory signaling (Cai et al., 2023). Beyond inflammation, NF-κB supports cell proliferation and tissue regeneration. It regulates keratinocyte migration, fibroblast survival, and ECM remodelling. NF-κB modulates angiogenesis through upregulation of VEGF, interacting with hypoxia-inducible factors (HIFs) to promote neovascularisation (Xuan et al., 2016; Pastar et al., 2023). During the remodelling phase, NF-κB influences fibroblast-mediated collagen synthesis and myofibroblast differentiation. Tight regulation is essential, as excessive NF-κB activity may result in pathological fibrosis, whereas insufficient activity could impair wound contraction and closure (Piechota-Polanczyk and Fichna, 2014).

2.4 CROSSTALK AND INTEGRATION

Importantly, cellular signaling pathways involved in wound healing are not discrete, isolated modules but operate within a dynamic and tightly regulated network of interactions. These interactions commonly referred to as "crosstalk" allow the cell to integrate diverse environmental and intracellular cues, ensuring coordinated tissue repair. For example, TGF-β signaling can modulate both the Wnt/β-catenin and Notch pathways, thereby influencing fibroblast differentiation, ECM deposition, and keratinocyte proliferation (Gumede et al., 2024). Similarly, the PI3K/AKT pathway interacts closely with both Nrf2, which governs antioxidant defence, and mTOR, a master regulator of cell metabolism and growth (Jere et al., 2019). Such interplay ensures that redox balance and energy availability are tightly linked to cellular proliferation and migration processes critical to wound closure. In parallel, MAPK pathways, including ERK, JNK, and p38, exert regulatory control over JAK/STAT and TGF-β signaling cascades by phosphorylating shared transcription factors and upstream regulators (Rawlings et al., 2004). These intersecting nodes allow for feedback amplification or attenuation, depending on the wound's stage and local microenvironmental conditions.

This intricate crosstalk offers both redundancy and adaptability, which is biologically advantageous under conditions of stress or injury. However, it also presents therapeutic opportunities. Targeting multiple nodes within this network rather than single isolated pathways may yield superior regenerative outcomes, especially in complex chronic wounds where multiple repair mechanisms are impaired. Combination therapies that simultaneously activate PI3K/AKT and Nrf2 or inhibit hyperactive p38 while restoring ERK signaling are currently under investigation for their potential to accelerate tissue regeneration and re-epithelialisation (Gumede et al., 2024; Jere et al., 2019). A comprehensive understanding of the molecular signaling architecture that governs wound repair is essential for elucidating the mechanisms underlying impaired healing. The coordinated activity of pathways such as JAK/STAT, Nrf2, MAPKs and their interconnected networks forms the biochemical foundation of effective tissue regeneration. When these pathways are dysregulated as commonly observed in chronic wounds the result is a sustained inflammatory environment, oxidative stress, impaired cell migration, and delayed re-epithelialisation.

Recognising the complexity and interdependence of these molecular circuits not only deepens our understanding of wound pathophysiology but also highlights novel targets for therapeutic intervention. In the following sections, we will explore how these signaling pathways are altered in chronic wound environments and assess current and emerging strategies aimed at restoring signaling fidelity to promote complete and sustained healing.

2.5 DYSREGULATION OF SIGNALING PATHWAYS IN CHRONIC WOUNDS

Acute wound healing is a temporally orchestrated and self-limiting biological process that typically progresses through well-defined phases like haemostasis, inflammation, proliferation, and remodelling culminating in tissue repair within a standard timeframe of four to six weeks. In contrast, chronic wounds represent a pathological deviation from this tightly regulated programme, characterised by a failure to proceed through the sequential phases of healing, most notably a prolonged inflammatory phase (Cullen and Gefen, 2023; Demidova-Rice et al., 2012). Chronic wounds including DFUs, venous leg ulcers, and pressure ulcers are typified by persistent inflammatory cell infiltration, elevated levels of ROS, impaired angiogenesis, aberrant fibroblast function, and delayed keratinocyte migration (Huelsboemer et al., 2024). At the core of these dysfunctions lies the disruption of key intracellular signaling pathways that govern immune cell dynamics, redox homeostasis, ECM deposition, and vascular regeneration.

Among the major molecular drivers implicated in wound pathophysiology are the PI3K/AKT, MAPK, JAK/STAT, Nrf2, and TGF-β pathways, whose impaired or aberrant activation results in the chronicity of wounds (Bonnici et al., 2023). The sustained inflammatory environment further amplifies protease activity, degrades essential ECM components, and prevents appropriate cellular responses to growth factors and cytokines. A mechanistic understanding of how these pathways become dysregulated provides a critical foundation for the development of targeted therapies aimed at re-establishing cellular equilibrium and restoring effective tissue regeneration (Bonnici et al., 2023). In this section, we explore the mechanistic underpinnings of signaling dysfunction in chronic wounds, with an emphasis on how specific pathway alterations manifest clinically in different wound types and patient populations.

2.5.1 Persistent Inflammation: The Consequence of Aberrant JAK/STAT, NF-κB, and Notch Signaling

Chronic wounds are dominated by prolonged inflammation, often maintained by sustained activation of the JAK/STAT and NF-κB pathways. These pathways respond to cytokines such as IL-6, TNF-α, and IFN-γ and initiate transcription of further inflammatory mediators, leading to a self-perpetuating inflammatory loop (Krizanova et al., 2022). In normal wound healing, a switch from pro-inflammatory (M1) to anti-inflammatory (M2) macrophages is essential to transition from inflammation to proliferation. However, in chronic wounds, this transition is delayed or absent, with M1 macrophages remaining dominant. The Notch signaling pathway, known to regulate macrophage polarisation, is frequently underactive, thereby preventing resolution of inflammation (Gumede et al., 2024). This unresolved inflammation leads to:

- Increased neutrophil infiltration
- Excessive release of ROS and proteases
- Collateral tissue damage, impairing ECM stability and cellular function.

2.5.2 Oxidative Stress and Impaired Nrf2 Response

Uncontrolled oxidative stress is a defining characteristic of chronic wounds, especially among diabetic individuals. Factors such as hyperglycaemia, the accumulation of advanced glycation end-products (AGEs), and ischemia–reperfusion injury drive the overproduction of ROS. Under physiological conditions, Nrf2 regulates the transcription of antioxidant enzymes including glutathione S-transferase, superoxide dismutase, and HO-1 to maintain redox homeostasis (Baird and Dinkova-Kostova, 2011; Pasupuleti et al., 2020). However, in chronic wounds:

- Nrf2 activation is blunted—often due to sustained Keap1-mediated degradation (An et al., 2024; Li et al., 2019a).
- The antioxidant capacity is overwhelmed, leading to ROS accumulation that damages lipids, proteins, DNA, and ECM components (Pasupuleti et al., 2020).
- As a result, keratinocyte migration and fibroblast proliferation are significantly impaired, and cellular senescence is induced further stalling regeneration (Li et al., 2019a; Pasupuleti et al., 2020).

2.5.3 Fibroblast Senescence and Wnt/TGF-β Dysregulation

Fibroblasts are the principal producers of ECM and are essential for granulation tissue formation, wound contraction, and remodelling. However, in chronic wounds especially those associated with ageing or diabetes—fibroblasts often display a senescent phenotype. Senescent fibroblasts exhibit characteristic features including enlarged morphology, decreased proliferative capacity, elevated levels of senescence-associated β-galactosidase (SA-β-gal), and secretion of pro-inflammatory cytokines and proteases, collectively referred to as the senescence-associated secretory phenotype (SASP) (Bonnici et al., 2023; Wang and Shi, 2020).

Two key signaling pathways (Wnt/β-catenin and TGF-β) that regulate fibroblast proliferation, migration, and differentiation into myofibroblasts are often dysregulated in chronic wounds. Wnt/β-catenin signaling, crucial for cellular regeneration and ECM production, is frequently suppressed in diabetic ulcers, leading to defective tissue remodelling and impaired wound closure (Walczyńska et al., 2023). Similarly, disrupted TGF-β signaling contributes to reduced fibroblast responsiveness to mitogenic cues such as PDGF and TGF-β1 itself, resulting in insufficient collagen deposition and delayed wound contraction (Harding et al., 2005).

Paradoxically, overactivation of TGF-β in certain chronic wound environments may induce excessive ECM production and drive fibrotic scarring, indicating that precise modulation—not simply activation is required for therapeutic success (Liu et al., 2006). Thus, balanced signaling through Wnt and TGF-β is critical for appropriate fibroblast activity and effective wound resolution.

2.5.4 Angiogenesis Failure: The Role of PI3K/AKT and VEGF

Adequate vascularisation is indispensable for wound healing, as it delivers oxygen, nutrients, and immune cells to regenerating tissue. In chronic wounds, particularly venous ulcers and ischemic diabetic wounds, angiogenesis is markedly impaired. Central to this defect is suppression of the PI3K/AKT signaling pathway, a pivotal regulator of endothelial cell survival, migration, and tube formation (Bonnici et al., 2023). Under normal conditions, AKT activation stimulates VEGF expression and enhances endothelial cell stability (Bonnici et al., 2023; Shi et al., 2023). When this pathway is inactive, VEGF signaling diminishes, resulting in reduced capillary density within the wound bed. Consequently, hypoxia persists, exacerbating oxidative stress and tissue degradation.

Additionally, the expression of angiogenesis inhibitors such as thrombospondin-1 (TSP-1) is frequently upregulated in chronic wounds, further blocking neovascularisation. TSP-1 acts through interactions with endothelial receptors (e.g., CD36, CD47), directly triggering antiangiogenic signaling and sequestering pro-angiogenic growth factors (Rusnati et al., 2010; Streit et al., 2000). Elevated TSP-1 levels have been linked to impaired granulation and delayed vascularisation in ischemic and chronic wound environments. Together, PI3K/AKT pathway suppression and TSP-1 upregulation create a severely antiangiogenic environment, preventing the necessary vascular redevelopment required for tissue regeneration.

2.5.5 Re-epithelialisation Failure and Keratinocyte Dysfunction

Re-epithelialisation, driven by keratinocyte migration and proliferation, is essential for restoring the epidermal barrier. In chronic wounds, this critical process is severely compromised, with keratinocytes exhibiting reduced migratory capacity, diminished proliferation, and impaired responsiveness to growth factors (Pastar et al., 2023). This dysfunction is driven by multiple intersecting factors:

- Elevated pro-apoptotic signaling renders keratinocytes more susceptible to cell death.
- Suppression of ERK/MAPK and PI3K/AKT pathways, which normally drive keratinocyte motility and proliferation, further impedes epidermal restoration (Miles et al., 2021).
- Growth factor unresponsiveness, notably to EGF and insulin-like growth factor-1 (IGF-1), limits mitogenic stimulation (Barrientos et al., 2008).
- Excessive MMP activity degrades the basement membrane and ECM, disrupting keratinocyte adhesion and forward movement (Armstrong and Jude, 2002).

Histologically, chronic wounds frequently display an epithelial tongue that fails to extend across the wound bed, reflecting a halted re-epithelialisation process (Bartolo et al., 2022).

2.5.6 Gene Expression Parallels Between Chronic Wounds and Cancer

Emerging systems biology studies have shown that chronic wounds display striking gene expression similarities to tumours. Basu and Martins-Green (2022) used RNA sequencing (RNA-Seq) and WGCNA to demonstrate that chronic wound tissue:

- Overexpresses oncogenes such as ErbB2 (HER2) and SDC1.
- Shares modules with cancer transcriptomes associated with cell proliferation, matrix degradation, and immune escape.

This has led to the conceptualisation of chronic wounds as "tumours that fail to grow," where the environment is hostile to healing but rich in inflammatory and oncogenic signals. Such insights open avenues for repurposing cancer therapies (e.g., kinase inhibitors, immunomodulators) for chronic wound treatment.

2.6 SYSTEMS BIOLOGY AND OMICS APPROACHES TO UNDERSTANDING CHRONIC WOUNDS

As wound biology becomes increasingly defined at the molecular level, conventional histological and clinical observations are no longer sufficient to understand the complex, multi-layered dysregulation that defines chronic wounds. The emergence of systems biology, powered by high-throughput "omics" technologies such as transcriptomics, proteomics, epigenomics, and metabolomics has revolutionised the study of wound healing and chronicity. These tools allow researchers to decipher global gene expression profiles, identify dysregulated molecular networks, and uncover novel biomarkers and therapeutic targets in an unbiased, data-driven manner (Basu and Martins-Green, 2022). This section provides a comprehensive overview of how integrative systems biology has been applied to chronic wound research, with an emphasis on WGCNA, RNA-Seq, and the identification of cancer-like molecular features in non-healing wounds.

2.6.1 Transcriptomics: Mapping the Molecular Signature of Chronic Wounds

Transcriptomics, particularly RNA-Seq, enables the quantitative assessment of gene expression at the whole-genome level. By comparing tissue from acute wounds, chronic wounds, and healthy skin, researchers can generate differential expression profiles that highlight upregulated or downregulated genes associated with chronicity. In a landmark study by Basu and Martins-Green (2022) chronic wound samples from human patients were subjected to RNA-Seq analysis, revealing more than 1,800 genes differentially expressed in chronic vs. healing wounds. Among these:

- Oncogenes such as ErbB2, SDC1, and myelocytoma (Myc) were overexpressed.
- Pro-healing transcription factors like Nrf2, hypoxia-inducible factor-1α (HIF-1α), and SRY-box transcription factor 17 (SOX17) were under expressed.
- Immune-related genes, particularly those in the IL-6–JAK–STAT3 axis, showed persistent activation.

These transcriptomic profiles not only validated known players like VEGF, TNF-α, and TGF-β but also uncovered novel wound-specific expression patterns that could not be predicted from conventional assays.

2.6.2 Weighted Gene Co-expression Network Analysis (WGCNA): A Pathway-Centric Approach

While differential gene expression analysis provides a valuable gene-by-gene assessment of transcriptional changes, it often lacks insight into coordinated regulatory processes. WGCNA overcomes this limitation by grouping genes into modules based on shared expression patterns across multiple samples, enabling the identification of biologically relevant gene clusters and central regulatory nodes or "hub genes" (Langfelder and Horvath, 2008). In their comprehensive transcriptomic study on chronic wounds, Basu and Martins-Green (2022) applied WGCNA to RNA-Seq data obtained from patient-derived chronic wound tissue. This network-based approach revealed several co-expression modules significantly associated with hallmark features of chronic wounds, include the following:

- Persistent inflammation
- Impaired ECM remodelling
- Exacerbated oxidative stress
- Immune dysregulation.

Each module was interrogated for hub genes highly interconnected nodes likely serving as master regulators. Notably, the following hub genes were identified:

- *ErbB2 (HER2)*—A receptor tyrosine kinase implicated in carcinogenesis and cellular proliferation
- *SDC1*—A heparan sulphate proteoglycan involved in cell adhesion, inflammation, and tissue remodelling
- *Baculoviral Inhibitor of Apoptosis Repeat-Containing 5 (BIRC5) (Survivin)*—An anti-apoptotic protein frequently overexpressed in cancers
- *C-X-C Motif Chemokine Ligand 10 (CXCL10)*—A chemokine associated with T-cell recruitment and prolonged inflammation.

These findings reinforce the notion that chronic wounds may mirror cancer-like regulatory networks, exhibiting "tumour-like wound stasis," where inflammation, apoptosis resistance, and aberrant cell signaling converge to block healing progression (Basu and Martins-Green, 2022).

2.6.3 Proteomics and Post-Transcriptional Modifications

While transcriptomic analyses offer valuable insights into gene expression, they do not always correlate with protein abundance or activity due to post-transcriptional and post-translational regulation. Proteomic methodologies such as mass spectrometry (MS) and two-dimensional gel electrophoresis (2D-GE) provide direct measurement of proteins and their modifications, thereby offering a more functionally relevant perspective of wound biology (Rabilloud et al., 2010). Proteomic investigations in chronic wound contexts have revealed several key findings:

- Quantification of ECM proteins including collagens, fibronectin, and laminin has been achieved via MS and 2D-GE, elucidating deficiencies in granulation tissue composition (Dengjel et al., 2020).
- Elevated expression of proteolytic enzymes, such as MMP-9, cathepsins, and serine proteases, has been documented in chronic wound exudates and pressure ulcer biopsies (Edsberg et al., 2012; Taverna et al., 2015).
- Degradation of growth factor receptors and adhesion molecules has also been confirmed, underscoring the impact of proteolysis on cell signaling integrity (Kalogeropoulos, 2022).
- Advanced proteomic analyses have identified oxidative post-translational modifications (oxPTMs) notably protein carbonylation and nitrosylation, on key wound healing mediators such as integrins, collagenases, and antioxidant enzymes, correlating with compromised protein function and healing failure (Dalle-Donne et al., 2006).

Collectively, these proteomic findings highlight that post-translational modifications and protein-level dysregulation play vital roles in chronic wound pathogenesis, offering potential targets for diagnostic biomarkers and therapeutic intervention.

2.6.4 Epigenetics and Chromatin Dynamics

Epigenetic modulation including DNA methylation, histone modification, and non-coding RNA regulation is increasingly recognised as a key contributor to chronic wound persistence. Unlike genetic mutations, these reversible chromatin modifications can stably alter gene accessibility and transcriptional activity across wound healing phases (Barrientos et al., 2008; Yu et al., 2022).

- *DNA Methylation*: Genome-wide profiling of chronic wound-edge tissue has revealed hypermethylated regions in promoters of genes essential for EMT leading to transcriptional silencing of critical repair pathways (Singh et al., 2022).
- *Histone Modification*: In acute wound healing, dynamic changes in histone marks such as demethylation of Histone H3 Lysine 27 (H3K27) mediated by Jumonji domain-containing protein D3 (Jmjd3) and ubiquitously transcribed tetratricopeptide repeat on chromosome X (UTX) enable expression of reparative genes. Conversely, chronic wounds exhibit aberrations in histone acetylation and methylation, disrupting gene programmes essential for inflammation resolution and regeneration (Yu et al., 2022).
- *Non-Coding RNAs*: Dysregulation of microRNAs (miRNAs) such as miR-21, miR-155, and others disrupts the fine balance of growth factor signaling, redox control, and cellular migration. For instance, miR-155 downregulates fibroblast growth factor 7 (FGF7) and HIF-1α, impairing keratinocyte proliferation and angiogenesis, while miR-21 is

implicated in suppression of inflammatory resolution (Pastar et al., 2021; Yu et al., 2022). These modifications collectively entrap chronic wound tissue in an epigenetically repressive state, resistant to standard therapies.

2.6.5 The Tumour–Wound Paradox: Convergent Pathologies

One of the most provocative insights to emerge from systems biology approaches to wound research is the molecular convergence between chronic wounds and malignant tumours. Despite their contrasting clinical behaviours, chronic wounds failing to close and cancers growing uncontrollably, both conditions share a remarkably similar microenvironmental and molecular landscape (Dvorak, 1986; Schafer and Werner, 2008). Transcriptomic and proteomic analyses have shown that chronic wounds and tumours are characterised by:

- Sustained inflammation, driven by persistent cytokine expression (e.g., IL-6, TNF-α) and activation of JAK/STAT and NF-κB pathways.
- Excessive ECM remodelling, mediated by overexpression of MMPs and fibroblast dysregulation.
- Aberrant angiogenesis, with deregulated VEGF and HIF-1α signaling.
- Immune escape mechanisms, including impaired antigen presentation and tolerogenic macrophage phenotypes (Pastar et al., 2014; Singh et al., 2022).

Moreover, both pathologies demonstrate metabolic reprogramming. For instance, keratinocytes at the edge of chronic wounds exhibit a Warburg-like glycolytic shift, mirroring the metabolic phenotype observed in cancer cells (Basu and Martins-Green, 2022; Schafer and Werner, 2008). These shared features support the conceptual framework in which chronic wounds are regarded as "tumors that do not grow," while tumours behave as "wounds that do not heal," co-opting the regenerative machinery for unchecked expansion (Dvorak, 1986). This convergence has significant therapeutic implications. Anti-cancer agents such as kinase inhibitors, redox modulators, and immunotherapeutics are now being explored for their efficacy in breaking the pathophysiological deadlock of chronic wounds, offering a novel cross-disciplinary approach to treatment (Wilkinson and Hardman, 2020).

2.6.6 Biomarker Discovery and Clinical Translation

Omics-driven research has advanced the identification of predictive and prognostic biomarkers for chronic wound management. Notably, several molecular candidates have been validated in clinical studies and are now under development for point-of-care applications:

- SDC1 and ErbB2 (HER2) have emerged as biomarkers indicative of chronicity and inflammatory activity in non-healing wounds (Basu and Martins-Green, 2022; Langfelder and Horvath, 2008).
- Elevated MMP-9 levels frequently assessed relative to TIMP-1, correlate with excessive ECM degradation and poor healing outcomes. Recent innovations include wireless smart dressings capable of real-time MMP-9 sensing in wound exudate (Deng et al., 2025).
- CXCL10 and IL-6 serve as robust immune profiling markers, with elevated wound-fluid concentrations associated with persistent inflammation and delayed resolution (Rembe et al., 2025).
- Markers of oxidative-stress regulation, including Nrf2 activity and HO-1 expression, reflect the redox balance within the wound microenvironment. Activation of Nrf2 via exosomes or small-molecule agonists has shown promise in accelerating diabetic wound healing in preclinical models (Wang et al., 2022).

These molecular indicators have informed the development of novel diagnostics, such as biosensor-embedded dressings and rapid wound-fluid assays. These tools enable personalised therapeutic strategies by providing clinicians with real-time insights into wound microenvironment dynamics and treatment response.

2.7 THERAPEUTIC INTERVENTIONS TARGETING WOUND SIGNALING PATHWAYS

The recognition that chronic wounds are driven by dysregulated signaling networks has prompted a paradigm shift in treatment from symptomatic care to molecularly targeted therapies. While conventional modalities such as dressings, debridement, and antibiotics remain foundational, they often fail to correct the underlying pathobiology, leading to stalled healing or recurrence. Contemporary strategies thus aim to directly modulate aberrant signaling pathways, restoring the wound's regenerative capacity (Matoori, 2023). Notable pathway-informed interventions include the following:

- PBM
- Stem cell-based therapies
- Nanoparticle-based and oligonucleotide delivery systems
- Natural compounds
- Exosomes and peptide-based platforms
- Combinatorial and smart therapies.

2.7.1 Photobiomodulation (PBM): Light as a Molecular Switch

PBM, formerly termed low-level laser therapy (LLLT), employs low-intensity light—typically in the red to near-infrared range (600–1000 nm)—to promote tissue repair. This non-invasive modality has been shown to reactivate key molecular signaling pathways in chronic wounds, particularly those governing oxidative stress, angiogenesis, and inflammation (Leyane et al., 2021; Rajendran et al., 2021). Low-level laser or LED light therapy has been shown to enhance PI3K/AKT and ERK/MAPK signaling in keratinocytes and fibroblasts, promoting proliferation and migration in chronic wound models (Khan and Arany, 2016; Rupel et al., 2018). In diabetic and pressure ulcer animal models, PBM accelerated wound closure by increasing fibroblast activity, endothelial proliferation, and keratinocyte migration (Leyane et al., 2021; Rajendran et al., 2021). Clinical trials are ongoing to define optimal PBM dosimetry, wavelengths, and treatment regimens tailored to wound type and pathology. PBM exerts its effects through multiple interconnected mechanisms:

Mitochondrial Targeting: Photons are absorbed by cytochrome c oxidase (CCO) in the electron transport chain, increasing ATP production, modulating ROS, and releasing NO, all of which initiate downstream signaling cascades (Kasowanjete et al., 2024).

PI3K/AKT/mTOR Activation: Red light at 660 nm (5 J/cm^2) activates PI3K/AKT, including downstream mTOR and GSK3β, enhancing fibroblast survival and proliferation in diabetic wound models (Jere et al., 2022).

Wnt/β-Catenin Pathway: PBM stabilises β-catenin by inhibiting GSK3β, thereby promoting transcription of β-catenin-dependent genes involved in repair.

Angiogenesis: PBM elevates VEGF expression and improves endothelial proliferation, migration, and tube formation via PI3K/AKT signaling (Ye et al., 2024).

Anti-Inflammatory Impact: PBM downregulates pro-inflammatory cytokines (e.g. IL-6, TNF-α) and suppresses FOXO1 signaling to reduce oxidative stress in diabetic fibroblasts (Leyane et al., 2021).

Nrf2 Antioxidant Response: Indirect stimulation of Nrf2 through mitochondrial modulation and ROS management enhances cellular defence mechanisms, particularly in diabetic wound settings (Polaka et al., 2022).

2.7.2 Stem Cell-Based Therapy: Cellular Reprogramming for Healing

Stem cell therapies offer a promising avenue for chronic wound treatment by replenishing senescent or dysfunctional cells and modulating the wound microenvironment. Multiple stem cell types have been explored, including bone marrow-derived and adipose-derived mesenchymal stem cells (MSCs), induced pluripotent stem cell-derived MSCs (iMSCs), and umbilical cord-derived MSCs. These deliver a spectrum of paracrine factors that modulate inflammation, enhance angiogenesis, and activate PI3K/AKT and Nrf2 pathways (Hamdan et al., 2017; Li et al., 2015). MSCs display robust wound-homing capabilities and adaptive responsiveness to the wound microenvironment, which has translated into measurable healing benefits in human studies. In DFUs, MSC therapy has been associated with significant reductions in wound size and accelerated re-epithelialisation (Zhang et al., 2023).

Stem cells facilitate wound repair through several mechanisms:

Paracrine Signaling: They secrete exosomes and a variety of growth factors such as TGF-β, VEGF, IGF-1 and immunomodulatory molecules (e.g., IL-6, IL-10) that reduce inflammation, promote angiogenesis, and rebalance redox signaling (Liu et al., 2020).

Immune Modulation: MSCs skew macrophage polarisation towards the reparative M2 phenotype and diminish M1 activity; in diabetic models, MSC administration reduces neutrophil infiltration and pro-inflammatory cytokines (Hocking, 2012).

Cell Differentiation: MSCs can directly differentiate into fibroblasts, endothelial cells, and keratinocytes, contributing structurally to granulation tissue (Liang et al., 2021) .

Promotion of Angiogenesis and Collagen Deposition: Both MSCs and iMSC-conditioned media enhance vascularisation via ERK/AKT activation and improve collagen synthesis in chronic wounds (Hocking, 2012).

Oxidative Stress Regulation: MSC-derived exosomal cargo, such as melatonin-stimulated exosomes, modulates the PTEN/AKT pathway and activates Nrf2 signaling, enhancing antioxidant defences (Liu et al., 2020)

2.7.3 Nanoparticle-Based and Oligonucleotide Delivery Systems

Targeted delivery of therapeutic nucleic acids such as siRNAs, miRNA mimics, or mRNA constructs offers precise modulation of gene expression within chronic wounds. However, RNA molecules are vulnerable to enzymatic degradation and immune clearance, particularly in moist wound environments. Nanotechnology overcomes these barriers through: lipid-based nanoparticles (LNPs), polymeric carriers (e.g., PLGA, chitosan), and metallic nanoparticles (e.g., gold, silver) with functionalised surfaces (Hamdan et al., 2017). A few instances of delivery strategies include the following:

- *Small Interfering RNA (siRNA) Targeting MMP-9 (siMMP-9) Delivery*: A thermosensitive hydrogel incorporating siRNA against MMP-9-enabled sustained release over seven days, effectively reducing MMP-9 levels in diabetic rat wounds and promoting collagen deposition and wound closure (Lan et al., 2021).
- *miR-21 Mimics*: miR-21-loaded nanocarriers (e.g., Baicalin-polyethylenimine (BA-PEI) complexes, microvesicle mimics) enhanced fibroblast migration, myofibroblast differentiation, and MAPK/ERK signaling in diabetic wound models (Li et al., 2019b; Peng et al., 2024).
- *Gold Nanoparticle Spherical Nucleic Acids (SNAs)*: SNAs carrying siRNA against monosialodihexosylganglioside (GM3) synthase successfully penetrated skin and enhanced keratinocyte migration, IGF-1 and EGF receptor activation, and wound closure in diabetic mice (Hamdan et al., 2017).
- *Nrf2 Activators*: Nano-formulations delivering Keapl siRNA restore redox balance by activating Nrf2, thereby improving antioxidant responses and promoting regeneration (Liu et al., 2023a).

These delivery platforms enable controlled release, site-specific delivery, and co-delivery of multiple agents to synergistically modulate wound biology (e.g., combining growth factor delivery with antioxidant triggers). As such, they provide an advanced toolkit for targeted, molecular-level intervention in chronic wounds.

2.7.4 Natural Compounds and Phytochemicals: Modulating Endogenous Healing

A diverse array of natural products including polyphenols, terpenoids, flavonoids, and alkaloids has been investigated for their ability to modulate key wound healing signaling pathways (Trinh et al., 2022; Zhou et al., 2021). Few key examples include the following:

- Curcumin activates Nrf2 ("antioxidant response element") and inhibits NF-κB, reducing inflammation and promoting angiogenesis in rat wound models (Rajendran et al., 2021). In full-thickness rat wounds, curcumin-based topical formulations enhanced AKT/mTOR signaling and re-epithelialisation while attenuating NF-κB activity (Danilevicz et al., 2021).
- Resveratrol enhances PI3K/AKT activation, upregulates Nrf2 and manganese superoxide dismutase (Mn-SOD), and accelerates wound closure in oxidative stress models, both *in vitro* (HUVECs) and *in vivo* (burn models) (Zhou et al., 2021).
- Aloe vera stimulates fibroblast migration, proliferation, and collagen deposition in dermal wound models (Trinh et al., 2022).
- *Centella asiatica* (*Gotu Kola*): Triterpenes such as asiaticoside enhance TGF-β/SMAD signaling, increase VEGF expression, and stimulate collagen synthesis and angiogenesis in animal and human studies. A recent systematic review of clinical trials also found evidence for accelerated re-epithelialisation and reduced IL-1β, IL-6, TNF-α, prostaglandin E2 (PGE_2), and COX-2 in wounds treated with *C. asiatica* (Arribas-Lopez et al., 2022).

Many phytochemicals exhibit pleiotropic effects, simultaneously modulating Wnt, MAPKs, Notch, Nrf2, and TGF-β pathways; their low toxicity and broad availability render them attractive as adjunct therapies (Trinh et al., 2022; Zhou et al., 2021).

2.7.5 Exosomes and Peptide-Based Platforms: Cell-Free Regeneration

Exosomes (nano-sized vesicles released by cells) are rich in proteins, RNAs, and lipids that reflect the physiological state of the donor cell. Stem cell-derived exosomes can replicate many of the regenerative effects of whole-cell therapy without the risks of immune rejection or tumorigenicity (Phinney and Pittenger, 2017). Similarly, bioactive peptides, such as LL-37 (cathelicidin), thymosin β4, and RGD-integrin ligands, have been incorporated into wound dressings and hydrogels to stimulate angiogenesis, inhibit bacterial colonisation, and promote ECM synthesis and cell adhesion (Goldstein et al., 2012; Ruoslahti, 1996). These agents can be engineered into biomaterial scaffolds to enhance targeted delivery and prolong activity.

2.7.6 Combinatorial and Smart Therapies: Integrating Multi-Pathway Modulation

Chronic wounds present a complex interplay of pathophysiological factors ranging from persistent inflammation and microbial burden to impaired angiogenesis and dysregulated ECM remodelling. Consequently, single-pathway therapeutic approaches often fall short of achieving complete and sustained healing (Guo and DiPietro, 2010). In response, the field has shifted towards combinatorial and smart therapeutic platforms that integrate multiple modalities to address the multifactorial nature of wound chronicity. Innovative combinations currently under investigation include the following:

- PBM with stem cell therapy, wherein light-based stimulation enhances stem cell survival, homing, and paracrine activity at wound sites (Pourhashemi et al., 2024);
- Nanoparticles co-loaded with phytochemicals, allowing for the targeted, controlled delivery of natural anti-inflammatory and antioxidant agents such as curcumin or resveratrol (Zhang et al., 2024a);
- Peptide-exosome hybrids embedded in responsive hydrogels, offering synchronised release of bioactive molecules based on local wound cues (Zhang et al., 2024b).

Emerging technologies also envision biosensor-integrated dressings capable of monitoring local wound parameters such as pH, oxygen tension, or cytokine levels. Upon detection of pathological changes, these intelligent systems can release appropriate therapeutics in a stimuli-responsive manner, enabling real-time, personalised wound management (Bonnici et al., 2023; Wang et al., 2019). This paradigm shift towards multi-pathway modulation and smart materials holds immense promise for the future of wound care.

2.8 CHALLENGES AND FUTURE PERSPECTIVES

Despite considerable advances in elucidating the molecular mechanisms underlying chronic wound pathology and the development of signaling pathway-targeted therapies, clinical translation remains a significant challenge. While many interventions such as exosome-enriched biomaterials, peptide-loaded scaffolds, and nanocarrier-mediated phytochemical systems have demonstrated impressive efficacy *in vitro* and in animal models, the leap to widespread clinical adoption is often hampered by inconsistencies in outcome, scalability issues, regulatory hurdles, and patient variability (Nunan et al., 2014; Pastar et al., 2014).

One of the major bottlenecks lies in the heterogeneity of chronic wounds, which are influenced by diverse factors such as comorbidities (e.g., diabetes, vascular disease), age, infection status, and genetic predisposition. These variations can affect the responsiveness to a given therapy, thereby limiting the generalisability of trial results (Guo and DiPietro, 2010). Moreover, lack of standardised animal models and variability in clinical trial design including inconsistent endpoints and small sample sizes further complicate the validation process for emerging therapies (Brem and Tomic-Canic, 2007). Additionally, issues of manufacturing complexity, cost-effectiveness, and long-term safety remain unresolved for many advanced therapies, especially those involving live cells, engineered exosomes, or gene-modifying tools. Regulatory frameworks are still evolving and often struggle to keep pace with technological innovation, creating uncertainty for researchers and biotech developers alike (Jacques and Suuronen, 2020).

Looking forward, future perspectives should prioritise the following:

- The development of standardised, predictive preclinical models that more accurately reflect human wound healing physiology.
- Integration of multi-omics and machine learning (ML) approaches to stratify patients and personalise therapy.
- Continued innovation in biosensing and responsive materials, enabling real-time monitoring and on-demand therapeutic delivery.
- Interdisciplinary collaborations between clinicians, bioengineers, and data scientists to accelerate translational pipelines (Bonnici et al., 2023; Zhang et al., 2024b).

Ultimately, the path to next-generation wound care will require not only technological breakthroughs but also a shift towards systems-based, personalised, and adaptive therapeutic strategies that address the multifactorial nature of chronic wounds. Bridging the gap between bench and bedside will depend on thoughtful integration of emerging tools with patient-centric, evidence-driven clinical frameworks.

2.8.1 Translational Gaps: From Bench to Bedside

One of the most persistent challenges in the field of chronic wound research is the translational gap, the disconnection between promising experimental results and limited clinical success. While numerous therapies have demonstrated efficacy in *in vitro* studies or preclinical animal models, their translation into clinically reliable interventions often falls short. This is largely because existing models fail to replicate the complex, multifactorial pathophysiology of human chronic wounds, which are often influenced by systemic factors such as diabetes mellitus, peripheral vascular disease, immunosenescence, and persistent microbial infection (Bonnici et al., 2023; Pastar et al., 2014).

2.8.2 Inter-Patient Variability and Personalised Medicine

Chronic wound healing is a highly individualised process, shaped by an intricate interplay of factors including host genetics, age, immune competence, local and systemic inflammation, wound microbiota, nutritional status, and the presence of underlying comorbidities such as diabetes, obesity, or vascular insufficiency (Basu and Martins-Green, 2022; Guo and DiPietro, 2010). These factors create significant inter-patient variability, often resulting in unpredictable therapeutic outcomes. As a result, interventions that demonstrate efficacy in one patient or wound type may prove ineffective or even detrimental in another. For example, PI3K/AKT-targeted therapies that promote angiogenesis may be advantageous in DFUs where vascularisation is impaired, yet such strategies may exacerbate inflammation or tissue degradation in venous leg ulcers, which often exhibit excessive MMP activity and inflammatory cytokines (Loots et al., 1998; Rousselle et al., 2019). This therapeutic complexity calls for a transition from one-size-fits-all approaches to personalised wound care, guided by molecular profiling and predictive diagnostics.

2.8.3 Complexity of Combinatorial Therapies

While single-target therapies provide a straightforward route to mechanistic specificity, the multifactorial nature of chronic wounds which involve inflammation, impaired angiogenesis, infection, oxidative stress, and ECM degradation often renders mono-therapies insufficient. Consequently, there is growing emphasis on combinatorial therapeutic approaches, which target multiple dysregulated pathways simultaneously. Examples include combining PBM with MSCs and nanoparticle-based drug delivery or integrating bioactive peptides with exosome-enriched hydrogels to enhance regenerative outcomes (Zhang et al., 2024b). However, the shift towards multi-modal therapies introduces a new set of scientific, clinical, and logistical challenges:

- Drug–drug or agent–agent interactions, where the biological or pharmacokinetic effects of one component may alter or interfere with the others (Jacques and Suuronen, 2020).
- Difficulty in attributing efficacy to specific components, making it hard to understand which element drives therapeutic success or failure (Jacques and Suuronen, 2020).
- Complexity in clinical trial design, as multi-agent regimens require factorial or adaptive study models that are more costly, time-consuming, and harder to regulate (Jacques and Suuronen, 2020).
- High cost and limited accessibility, which may hinder widespread implementation, particularly in low-resource settings.

To address these limitations, research has focused on the development of smart biomaterials and multifunctional delivery systems capable of co-delivering multiple therapeutic agents in a controlled, spatially and temporally regulated manner. Advanced platforms such as stimuli-responsive hydrogels, nanocarriers, and exosome-loaded scaffolds are being designed to release drugs based on local wound cues like pH, oxidative stress, or enzymatic activity, thus enhancing both efficacy and safety (Bonnici et al., 2023; Wang et al., 2019). While these platforms offer great promise, their clinical application requires rigorous interdisciplinary collaboration, preclinical validation, and cost-benefit optimisation. The future of combinatorial wound therapy lies in refining these technologies into scalable, patient-specific solutions that balance complexity with practicality.

2.8.4 Scar Formation and Functional Restoration

While the primary goal of most chronic wound therapies has traditionally been rapid wound closure, cosmetic and functional restoration are increasingly recognised as equally critical outcomes. Many wounds that do eventually close particularly those treated under inflammatory or non-regenerative conditions heal with excessive fibrosis, epidermal thinning, contracture, or loss of key skin functions such as barrier integrity, elasticity, and pigmentation (Rousselle et al., 2019). This is particularly problematic in high-mobility areas such as joints, hands, and the face, where contracture can lead to functional impairment and psychosocial distress. It is also of high concern in paediatric and burn patients, where visible or restrictive scarring can impact long-term quality of life and growth (Hariharan et al., 2018). To overcome these challenges, regenerative wound healing strategies—those that restore skin architecture rather than merely replace it with fibrotic tissue are gaining momentum. Key approaches include the following:

- Modulating TGF-β isoform expression, especially by promoting TGF-β3 (associated with scarless healing) while suppressing TGF-β1, which is linked to fibrosis and myofibroblast activation (Ferguson and O'Hare, 2004; Shah et al., 1995).
- Enhancing keratinocyte stem cell activation and epidermal progenitor mobilisation to promote proper stratification and re-epithelialisation (Pontiggia et al., 2022).
- Employing bioprinted or engineered skin substitutes that replicate the complex three-dimensional architecture of both the dermis and epidermis, supporting physiological healing and functional recovery (Varkey et al., 2015).

Such regenerative strategies are not only focused on scar minimisation but also on restoring appendages such as sweat glands, sebaceous glands, and hair follicles features often lost in traditional wound closure. In the future, the integration of tissue-engineering, growth factor modulation, and cell-based therapies may allow for complete functional skin regeneration, particularly when tailored to patient-specific anatomical and physiological needs.

2.8.5 Artificial Intelligence (AI) and Digital Health in Wound Care

The integration of artificial intelligence (AI) and digital health technologies is rapidly transforming the landscape of medical care, and wound management is beginning to reap the benefits of this digital revolution. Given the high volume of chronic wounds and the complexity of their progression, AI offers the potential to augment clinical decision-making, enhance diagnostic accuracy, and improve personalised care pathways (Memon et al., 2025; Reddy et al., 2019). AI algorithms, particularly those based on ML and deep learning (DL), can process large datasets of wound images and clinical metadata to:

- Quantify wound dimensions (area, depth, perimeter) with greater consistency than manual methods (Goyal et al., 2018).

- Predict infection risk, healing outcomes, or complications by analysing patterns in physiological, biochemical, and demographic data (Goyal et al., 2018).
- Recommend evidence-based treatment strategies, drawing from population-level databases and clinical guidelines (Goyal et al., 2018).

Moreover, smartphone-based telemedicine platforms enable remote capture and analysis of wound images, allowing patients in rural or underserved regions to receive expert consultations without needing to travel to tertiary care centres. This is especially valuable for patients with limited mobility or in areas with shortages of trained wound care specialists (Chen et al., 2020). The digital ecosystem is further enhanced by the integration of wearable biosensors, which can continuously monitor critical wound parameters such as oxygen saturation, temperature, pH, moisture levels, and even cytokine or enzyme activity. These devices provide real-time feedback, which can be relayed to clinicians via mobile apps or cloud-based dashboards, enabling proactive intervention before wound deterioration occurs (Bonnici et al., 2023; Wang et al., 2019).

By combining AI, remote monitoring, and biosensing into a unified platform, digital wound care has the potential to create a real-time, data-rich, and personalised care ecosystem that bridges hospital and home, anticipates complications, and improves both clinical outcomes and healthcare efficiency.

2.9 CONCLUSION

Chronic wounds pose an ongoing and intricate medical problem characterised by the body's inability to heal through its well-regulated stages. While acute wounds heal through a sequence of blood clotting, swelling, cell growth, and tissue rebuilding, chronic wounds get stuck in a harmful cycle during the swelling phase because of broken signal pathways, ageing cells, and constant oxidative stress. This chapter has highlighted how cell signaling pathways such as PI3K/AKT, Wnt/β-catenin, TGF-β, Notch, JAK/STAT, MAPKs, and Nrf2 play a key role in normal and abnormal wound healing. When these pathways don't work right in chronic conditions, it leads to poor fibroblast function weak blood vessel growth, keratinocytes that can't move, breakdown of the ECM, and ongoing inflammation. From a system's biological viewpoint, chronic wounds show gene activity that often looks like cancer, with active cancer-causing genes and long-lasting inflammation signals, and changes in how cells use energy. Modern therapeutic strategies aim to restore or replace the dysregulated components of these molecular circuits. PBM, stem cell-based therapies, nanoparticle-assisted gene delivery, exosomes, peptide-based platforms, and naturally derived compounds are all being developed and optimised to modulate specific nodes within these pathways. These innovations offer the potential to shift the chronic wound environment from one of molecular chaos to regenerative order.

However, the road to clinical translation is still fraught with challenges. Inter-patient variability, complex comorbidities, and lack of personalised diagnostic tools hinder treatment efficacy and consistency. The future of wound care is being reshaped by a powerful blend of AI-powered wearable sensors, biomarker-based diagnostics, and intelligent biomaterials. Together, these innovations are paving the way for treatments that are not just advanced but also personalised, responsive, and truly patient-centred. The future of managing long-lasting wounds sits at the crossroads of molecular accuracy, systems integration, and tech breakthroughs. As we keep unravelling the complex biology behind chronic wounds and line it up with cutting-edge multi-pronged treatment plans, we might soon crack one of the toughest nuts in medicine: the stubborn wound that won't heal.

GLOSSARY

Cytokines: These are small signaling proteins released by immune cells; they facilitate inflammation, recruit neutrophils/macrophages, and activate downstream pathways like NF-κB and JAK/STAT.

Haemostasis: The immediate reaction to injury, which includes vasoconstriction and the formation of clots to minimise blood loss.

Inflammation: The second stage of healing, defined by the recruitment of immune cells such as neutrophils and macrophages to eliminate debris and avert infection.

JAK/STAT pathway: This pathway transmits signals from cytokines and growth factors; it is essential for immune modulation and the migration of keratinocytes.

MAPK: This is a significant signaling cascade (which includes ERK, JNK, and p38 kinases) that transmits signals from growth factors and stress, affecting cell proliferation, migration, and inflammation.

MMPs: Proteolytic enzymes that break down ECM components; their activity is controlled by TIMPs and signaling pathways involving TGF-β, AP-1.

NF-κB: A transcription factor that is activated during the inflammatory phase; it regulates the production of cytokines, cell survival, and immune responses.

Nrf2 (nuclear factor erythroid 2-related factor 2): This transcription factor triggers the expression of antioxidant response genes; it protects cells from oxidative harm during the healing process.

PI3K/AKT pathway: This pathway enhances cell survival, angiogenesis, and metabolism during the wound healing process; it is frequently activated by growth factors such as VEGF and IGF.

Proliferation: The third stage characterised by the activation of fibroblasts, angiogenesis, epithelialisation, and the development of granulation tissue.

Remodelling (maturation): The final stage, which entails the reorganisation of the ECM, maturation of collagen from Type III to Type I, and contraction of the wound.

ROS: Serve as signaling molecules at low levels to facilitate angiogenesis and defence mechanisms, but an overabundance of ROS can result in oxidative damage.

TGF-β pathway: This pathway is crucial in all phases of wound healing; it regulates ECM production, fibroblast activation, and scar formation through both SMAD-dependent and independent pathways.

TIMP: These are natural inhibitors of MMPs; they play a crucial role in regulating ECM turnover and ensuring proper tissue remodelling.

TLRs: These pattern-recognition receptors (PRRs) recognise microbial products and activate NF-κB and MAPK signaling pathways to initiate innate immune responses.

Wnt/β-catenin pathway: It governs cell fate, proliferation, and migration, playing a vital role in epithelial regeneration and fibroblast activity during tissue repair.

Wound healing: A dynamic biological process that restores the integrity of skin and tissue following an injury, consisting of overlapping phases: haemostasis, inflammation, proliferation, and remodelling.

REFERENCES

Ahmad, Rehan, Deepak Raina, Colin Meyer, Surender Kharbanda, and Donald Kufe. "Triterpenoid CDDO-Me blocks the NF-κB pathway by direct inhibition of IKKβ on Cys-179." *Journal of Biological Chemistry* 281, no. 47 (2006): 35764–35769.

Ahmed, Syed Minhaj Uddin, Lin Luo, Akhileshwar Namani, Xiu Jun Wang, and Xiuwen Tang. "Nrf2 signaling pathway: Pivotal roles in inflammation." *Biochimica et Biophysica Acta (BBA)-Molecular Basis of Disease* 1863, no. 2 (2017): 585–597.

An, Ning, Rixiang Wang, Lin Li, Bingyu Wang, Huiting Wang, Ganyu Peng, Hua Zhou, and Gen Chen. "Celastrol alleviates diabetic vascular injury via Keap1/Nrf2-mediated anti-inflammation." *Frontiers in Pharmacology* 15 (2024): 1360177.

Armstrong, David G., and Edward B. Jude. "The role of matrix metalloproteinases in wound healing." *Journal of the American Podiatric Medical Association* 92, no. 1 (2002): 12–18.

Arribas-Lopez, Elena, Nazanin Zand, Omorogieva Ojo, Martin John Snowden, and Tony Kochhar. "A systematic review of the effect of *Centella asiatica* on wound healing." *International Journal of Environmental Research and Public Health* 19, no. 6 (2022): 3266. https://www.mdpi.com/1660-4601/19/6/3266?utm

Ashcroft, Gillian S., Xiao Yang, Adam B. Glick, Michael Weinstein, John J. Letterio, Diane E. Mizel, Mario Anzano et al. "Mice lacking Smad3 show accelerated wound healing and an impaired local inflammatory response." *Nature Cell Biology* 1, no. 5 (1999): 260–266.

Baird, Liam, and Albena T. Dinkova-Kostova. "The cytoprotective role of the Keap1–Nrf2 pathway." *Archives of Toxicology* 85 (2011): 241–272.

Barrientos, Stephan, Olivera Stojadinovic, Michael S. Golinko, Harold Brem, and Marjana Tomic-Canic. "Growth factors and cytokines in wound healing." *Wound Repair and Regeneration* 16, no. 5 (2008): 585–601. https://doi.org/10.1111/j.1524-475X.2008.00410.x

Bartolo, Inês, Rui L. Reis, Alexandra P. Marques, and Mariana T. Cerqueira. "Keratinocyte growth factor-based strategies for wound re-epithelialization." *Tissue Engineering Part B: Reviews* 28, no. 3 (2022): 665–676.

Basu, Proma, and Manuela Martins-Green. "Signaling pathways associated with chronic wound progression: A systems biology approach." *Antioxidants* 11, no. 8 (2022): 1506. https://doi.org/10.3390/antiox11081506

Blanpain, Cédric, William E. Lowry, H. Amalia Pasolli, and Elaine Fuchs. "Canonical notch signaling functions as a commitment switch in the epidermal lineage." *Genes & Development* 20, no. 21 (2006): 3022–3035. https://doi.org/10.1101/gad.1477606

Bonnici, Lian, Sherif Suleiman, Pierre Schembri-Wismayer, and Analisse Cassar. "Targeting signalling pathways in chronic wound healing." *International Journal of Molecular Sciences* 25, no. 1 (2023): 50. https://doi.org/10.3390/ijms25010050

Bozulic, Lana, and Brian A. Hemmings. "PIKKing on PKB: Regulation of PKB activity by phosphorylation." *Current Opinion in Cell Biology* 21, no. 2 (2009): 256–261.

Brem, Harold, and Marjana Tomic-Canic. "Cellular and molecular basis of wound healing in diabetes." *The Journal of Clinical Investigation* 117, no. 5 (2007): 1219–1222. https://doi.org/10.1172/JCI32169

Cai, Feiyu, Wenjiao Chen, Ruomei Zhao, and Yi Liu. "Mechanisms of Nrf2 and NF-κB pathways in diabetic wound and potential treatment strategies." *Molecular Biology Reports* 50, no. 6 (2023): 5355–5367.

David R. Calnan and Anne Brunet, "The FoxO Code," *Oncogene* 27, no. 16 (2008): 2276–2288.

Chang, Lili, Jinning Liang, Xiujuan Xia, and Xianjin Chen. "miRNA-126 enhances viability, colony formation, and migration of keratinocytes HaCaT cells by regulating PI3 K/AKT signaling pathway." *Cell Biology International* 43, no. 2 (2019): 182–191.

Chen, Lihong, Lihui Cheng, Wei Gao, Dawei Chen, Chun Wang, and Xingwu Ran. "Telemedicine in chronic wound management: Systematic review and meta-analysis." *JMIR mHealth and uHealth* 8, no. 6 (2020): e15574.

Chen, Mei, Judith Lechner, Jiawu Zhao, Levente Toth, Ruth Hogg, Giuliana Silvestri, Adrien Kissenpfennig, Usha Chakravarthy, and Heping Xu. "STAT3 activation in circulating monocytes contributes to neovascular age-related macular degeneration." *Current Molecular Medicine* 16, no. 4 (2016): 412–423.

Chen, Tong, Yanan You, Hua Jiang, and Zack Z. Wang. "Epithelial–mesenchymal transition (EMT): A biological process in the development, stem cell differentiation, and tumorigenesis." *Journal of Cellular Physiology* 232, no. 12 (2017): 3261–3272.

Chigurupati, Srinivasulu, Thiruma V. Arumugam, Tae Gen Son, Justin D. Lathia, Shafaq Jameel, Mohamed R. Mughal, Sung-Chun Tang et al. "Involvement of notch signaling in wound healing." *PLoS One* 2, no. 11 (2007): e1167.

Clevers, Hans, and Roel Nusse. "Wnt/β-catenin signaling and disease." *Cell* 149, no. 6 (2012): 1192–1205. https://doi.org/10.1016/j.cell.2012.05.012

Cullen, Breda, and Amit Gefen. "The biological and physiological impact of the performance of wound dressings." *International Wound Journal* 20, no. 4 (2023): 1292–1303. https://doi.org/10.1111/iwj.13960

Dalle-Donne, Isabella, Giancarlo Aldini, Marina Carini, Roberto Colombo, Ranieri Rossi, and Aldo Milzani. "Protein carbonylation, cellular dysfunction, and disease progression." *Journal of Cellular and Molecular Medicine* 10, no. 2 (2006): 389–406.

Danilevicz, Chris K., Vivian P. Wagner, Nilson Ferreira, Hugo Bock, Emily F. Salles Pilar, Liana P. Webber, Tuany R. Schmidt et al. "*Curcuma longa* L. effects on Akt/mTOR pathway and NF-κB expression during skin wound healing: An immunohistochemical study." *Applied Immunohistochemistry & Molecular Morphology* 29, no. 10 (2021): e92–e100.

Demidova-Rice, Tatiana N., Michael R. Hamblin, and Ira M. Herman. "Acute and impaired wound healing: Pathophysiology and current methods for drug delivery, part 1: Normal and chronic wounds: Biology, causes, and approaches to care." *Advances in Skin & Wound Care* 25, no. 7 (2012): 304–314. https://doi.org/10.1097/01.ASW.0000416006.55218.d0

Deng, Peixue, Zhenghan Shi, Feiyue Fang, Yi Xu, Li-ang Zhou, Ye Liu, Meng Jin et al. "Wireless matrix metalloproteinase-9 sensing by smart wound dressing with controlled antibacterial nanoparticles release toward chronic wound management." *Biosensors and Bioelectronics* 268 (2025): 116860. https://doi.org/10.1016/j.bios.2024.116860

Dengjel, Jorn, Leena Bruckner-Tuderman, and Alexander Nyström. "Skin proteomics–analysis of the extracellular matrix in health and disease." *Expert Review of Proteomics* 17, no. 5 (2020): 377–391. https://www.unifr.ch/bio/en/assets/public/Research/Joern-Dengjel/Dengjel_2020_ExperProt_SkinReview.pdf?utm

Derynck, Rik, and Erine H. Budi. "Specificity, versatility, and control of TGF-β family signaling." *Science Signaling* 12, no. 570 (2019): eaav5183.

Dvorak, Harold F. "Tumors: Wounds that do not heal." *New England Journal of Medicine* 315, no. 26 (1986): 1650–1659. https://www.nejm.org/doi/abs/10.1056/NEJM198612253152606

Edsberg, Laura E., Jennifer T. Wyffels, Michael S. Brogan, and Kristin M. Fries. "Analysis of the proteomic profile of chronic pressure ulcers." *Wound Repair and Regeneration* 20, no. 3 (2012): 378–401.

Engelman, Jeffrey A., Ji Luo, and Lewis C. Cantley. "The evolution of phosphatidylinositol 3-kinases as regulators of growth and metabolism." *Nature Reviews Genetics* 7, no. 8 (2006): 606–619.

Ferguson, Mark W. J., and Sharon O'Kane. "Scar–free healing: From embryonic mechanisms to adult therapeutic intervention." *Philosophical Transactions of the Royal Society of London. Series B: Biological Sciences* 359, no. 1445 (2004): 839–850. https://doi.org/10.1098/rstb.2004.1475

Gao, Yan-ling, Chang-sheng Liu, Rui Zhao, Lin-lin Wang, Shan-shan Li, Min Liu, Miao Zhang et al. "Effects of PI3K/Akt pathway in wound healing process of mice skin." *Fa Yi Xue Za Zhi* 32, no. 1 (2016): 7–12.

Goldstein, Allan L., Ewald Hannappel, Gabriel Sosne, and Hynda K. Kleinman. "Thymosin β4: A multi-functional regenerative peptide. Basic properties and clinical applications." *Expert Opinion on Biological Therapy* 12, no. 1 (2012): 37–51. https://www.tandfonline.com/doi/abs/10.1517/14712598.2012.634793

Goyal, Manu, Neil D. Reeves, Satyan Rajbhandari, and Moi Hoon Yap. "Robust methods for real-time diabetic foot ulcer detection and localization on mobile devices." *IEEE Journal of Biomedical and Health Informatics* 23, no. 4 (2018): 1730–1741.

Gumede, Dimakatso B., Heidi Abrahamse, and Nicolette N. Houreld. "Targeting Wnt/β-catenin signaling and its interplay with TGF-β and Notch signaling pathways for the treatment of chronic wounds." *Cell Communication and Signaling* 22, no. 1 (2024): 244. https://doi.org/10.1186/s12964-024-01623-9

Guo, Shenyang, and Luisa A. DiPietro. "Factors Affecting Wound Healing." *Journal of Dental Research* 89, no. 3 (2010): 219–229.

Hamdan, Suzana, Irena Pastar, Stefan Drakulich, Emre Dikici, Marjana Tomic-Canic, Sapna Deo, and Sylvia Daunert. "Nanotechnology-driven therapeutic interventions in wound healing: Potential uses and applications." *ACS central Science* 3, no. 3 (2017): 163–175. https://doi.org/10.1021/acscentsci.6b00371

Harding, Keith G., Keith Moore, and Tania J. Phillips. "Wound chronicity and fibroblast senescence–implications for treatment." *International Wound Journal* 2, no. 4 (2005): 364–368.

Hariharan, Narayanan Chandramouli, Rajagopal Sridhar, B. Sankari, V. S. Valarmathy, Ebanesar Asirvatham, and Krishnamurthy Geetha. "Reconstruction of postburn crippled hands: A study of functional outcome." *Indian Journal of Burns* 26, no. 1 (2018): 9–14.

Hayashi, Masaaki, and Jiing-Dwan Lee. "Role of the BMK1/ERK5 signaling pathway: Lessons from knockout mice." *Journal of Molecular Medicine* 82 (2004): 800–808.

Hocking, Anne M. "Mesenchymal stem cell therapy for cutaneous wounds." *Advances in Wound Care* 1, no. 4 (2012): 166–171.

Hu, Qian, Qihui Bian, Dingchao Rong, Leiyun Wang, Jianan Song, Hsuan-Shun Huang, Jun Zeng, Jie Mei, and Peng-Yuan Wang. "JAK/STAT pathway: Extracellular signals, diseases, immunity, and therapeutic regimens." *Frontiers in Bioengineering and Biotechnology* 11 (2023): 1110765.

Huelsboemer, Lioba, Leonard Knoedler, Alejandro Kochen, Catherine T. Yu, Helia Hosseini, Katharina S. Hollmann, Ashley E. Choi et al. "Cellular therapeutics and immunotherapies in wound healing–on the pulse of time?." *Military Medical Research* 11, no. 1 (2024): 23. https://doi.org/10.1186/s40779-024-00528-5

Jacques, Erik, and Erik J. Suuronen. "The progression of regenerative medicine and its impact on therapy translation." *Clinical and Translational Science* 13, no. 3 (2020): 440–450.

Jere, Sandy W., Nicolette N. Houreld, and Heidi Abrahamse. "Role of the PI3K/AKT (mTOR and GSK3β) signalling pathway and photobiomodulation in diabetic wound healing." *Cytokine & Growth Factor Reviews* 50 (2019): 52–59. https://doi.org/10.1016/j.cytogfr.2019.03.001

Jere, Sandy Winfield, Nicolette Nadene Houreld, and Heidi Abrahamse. "Photobiomodulation activates the PI3K/AKT pathway in diabetic fibroblast cells in vitro." *Journal of Photochemistry and Photobiology B: Biology* 237 (2022): 112590. https://pubmed.ncbi.nlm.nih.gov/36334508/

Jiang, Zhongping, Jie Wei, Weize Yang, Wen Li, Feng Liu, Xiaojie Yan, Xiaowei Yan, Niandan Hu, and Jia Li. "MicroRNA-26a inhibits wound healing through decreased keratinocytes migration by regulating ITGA5 through PI3K/AKT signaling pathway." *Bioscience Reports* 40, no. 9 (2020).

Kalogeropoulos, Konstantinos. "Wound exudate degradomics in normal and impaired healing: Exploring the proteomic and proteolytic landscape of human tissue repair in acute and chronic wounds through the lens of wound exudates" (2022).

Kasowanjete, Patricia, Sathish Sundar Dhilip Kumar, and Nicolette N. Houreld. "A review of photobiomodulation on PI3K/AKT/mTOR in wound healing." *Journal of Photochemistry and Photobiology* 19 (2024): 100215.

Khan, Imran, and Praveen R. Arany. "Photobiomodulation therapy promotes expansion of epithelial colony forming units." *Photomedicine and Laser Surgery* 34, no. 11 (2016): 550–555.

Kondoh, Kunio, Kazuya Terasawa, Hiroko Morimoto, and Eisuke Nishida. "Regulation of nuclear translocation of extracellular signal-regulated kinase 5 by active nuclear import and export mechanisms." *Molecular and Cellular Biology* 26, no. 5 (2006): 1679–1690.

Krizanova, Olga, Adela Penesova, Jozef Sokol, Alica Hokynkova, Amir Samadian, and Petr Babula. "Signaling pathways in cutaneous wound healing." *Frontiers in Physiology* 13 (2022): 1030851. https://doi.org/10.3389/fphys.2022.1030851

Lan, Biyun, Liming Zhang, Liqun Yang, Junfeng Wu, Na Li, Chenglin Pan, Xiaoyi Wang et al. "Sustained delivery of MMP-9 siRNA via thermosensitive hydrogel accelerates diabetic wound healing." *Journal of Nanobiotechnology* 19, no. 1 (2021): 130.

Langfelder, Peter, and Steve Horvath. "WGCNA: An R package for weighted correlation network analysis." *BMC Bioinformatics* 9(2008): 1–13. https://doi.org/10.1186/1471-2105-9-559

Lawrence, Toby. "The nuclear factor NF-κB pathway in inflammation." *Cold Spring Harbor Perspectives in Biology* 1, no. 6 (2009): a001651.

Leyane, Thobekile S., Sandy W. Jere, and Nicolette N. Houreld. "Cellular signalling and photobiomodulation in chronic wound repair." *International Journal of Molecular Sciences* 22, no. 20 (2021): 11223. https://doi.org/10.3390/ijms222011223

Li, Jie, Juan Chen, and Robert Kirsner. "Pathophysiology of acute wound healing." *Clinics in Dermatology* 25, no. 1 (2007): 9–18.

Li, Meirong, Yali Zhao, Haojie Hao, Weidong Han, and Xiaobing Fu. "Mesenchymal stem cell–based therapy for nonhealing wounds: Today and tomorrow." *Wound Repair and Regeneration* 23, no. 4 (2015): 465–482. https://doi.org/10.1111/wrr.12304

Li, Min, Haibing Yu, Haiyan Pan, Xueqing Zhou, Qiongfang Ruan, Danli Kong, Zhigang Chu et al. "Nrf2 suppression delays diabetic wound healing through sustained oxidative stress and inflammation." *Frontiers in Pharmacology* 10 (2019a): 1099.

Li, Qian, Hui Zhao, Weimin Chen, Ping Huang, and Jiarui Bi. "Human keratinocyte-derived microvesicle miRNA-21 promotes skin wound healing in diabetic rats through facilitating fibroblast function and angiogenesis." *The International Journal of Biochemistry & Cell Biology* 114 (2019b): 105570.

Liang, Xiaoting, Fang Lin, Yue Ding, Yuelin Zhang, Mimi Li, Xiaohui Zhou, Qingshu Meng et al. "Conditioned medium from induced pluripotent stem cell-derived mesenchymal stem cells accelerates cutaneous wound healing through enhanced angiogenesis." *Stem Cell Research & Therapy* 12, no. 1 (2021): 295.

Liu, Jia, Faping Wang, and Fengming Luo. "The role of JAK/STAT pathway in fibrotic diseases: Molecular and cellular mechanisms." *Biomolecules* 13, no. 1 (2023b): 119.

Liu, Mengqian, Xuerong Wei, Zijun Zheng, Yicheng Li, Mengyao Li, Jiabao Lin, and Lei Yang. "Recent advances in nano-drug delivery systems for the treatment of diabetic wound healing." *International Journal of Nanomedicine* (2023a): 1537–1560. https://www.researchgate.net/publication/369544230_Recent_Advances_in_Nano-Drug_Delivery_Systems_for_the_Treatment_of_Diabetic_Wound_Healing

Liu, Wei, Muyu Yu, Dong Xie, Longqing Wang, Cheng Ye, Qi Zhu, Fang Liu, and Lili Yang. "Melatonin-stimulated MSC-derived exosomes improve diabetic wound healing through regulating macrophage M1 and M2 polarization by targeting the PTEN/AKT pathway." *Stem Cell Research & Therapy* 11(2020): 1–15. https://doi.org/10.1186/s13287-020-01756-x

Liu, Xingjun, Han Hu, and James Q. Yin. "Therapeutic strategies against TGF-β signaling pathway in hepatic fibrosis." *Liver International* 26, no. 1 (2006): 8–22.

Liu, Zhao-Jun, Ying Zhuge, and Omaida C. Velazquez. "Trafficking and differentiation of mesenchymal stem cells." *Journal of Cellular Biochemistry* 106, no. 6 (2009): 984–991. https://doi.org/10.1002/jcb.22091

Long, Min, Montserrat Rojo de la Vega, Qing Wen, Manish Bharara, Tao Jiang, Rui Zhang, Shiwen Zhou et al. "An essential role of NRF2 in diabetic wound healing." *Diabetes* 65, no. 3 (2016): 780–793.

Loots, Miriam A. M., Evert N. Lamme, Jimmy Zeegelaar, Jan R. Mekkes, Jan D. Bos, and Esther Middelkoop. "Differences in cellular infiltrate and extracellular matrix of chronic diabetic and venous ulcers versus acute wounds." *Journal of Investigative Dermatology* 111, no. 5 (1998): 850–857. https://doi.org/10.1046/j.1523-1747.1998.00381.x

Manning, Brendan D., and Lewis C. Cantley. "AKT/PKB signaling: Navigating downstream." *Cell* 129, no. 7 (2007): 1261–1274.

Manning, Brendan D., and Alex Toker. "AKT/PKB signaling: Navigating the network." *Cell* 169, no. 3 (2017): 381–405.

Martin, Paul, and R. Nunan. "Cellular and molecular mechanisms of repair in acute and chronic wound healing." *British Journal of Dermatology* 173, no. 2 (2015): 370–378. https://doi.org/10.1111/bjd.13954

Matoori, Simon. "Breakthrough technologies in diagnosis and therapy of chronic wounds." *ACS Pharmacology & Translational Science* 6, no. 6 (2023): 854–856. https://doi.org/10.1021/acsptsci.3c00060

Memon, Abdul Qadeer, Ashay Shah, Nosheen Akhtar Rana, and Sudhair Abbas Bangash. "AI for wound healing and management." *Prospects* 29, no. 6 (2025): 378–385.

Meng, Xiao-ming, David J. Nikolic-Paterson, and Hui Yao Lan. "TGF-β: The master regulator of fibrosis." *Nature Reviews Nephrology* 12, no. 6 (2016): 325–338. https://doi.org/10.1038/nrneph.2016.48

Miles, Rebecca R., Parth H. Amin, Miguel Barriera Diaz, Jagannath Misra, Erica Aukerman, Amitava Das, Nandini Ghosh et al. "The eIF2 kinase GCN2 directs keratinocyte collective cell migration during wound healing via coordination of reactive oxygen species and amino acids." *Journal of Biological Chemistry* 297, no. 5 (2021).

Monsalve, Eva, Miguel A. Pérez, Antonio Rubio, María José Ruiz-Hidalgo, José J. García-Ramírez, Victoriano Baladron, Juan C. Gómez, Jorge Laborda, and María José M. Díaz-Guerra. "Notch-1 Up-Regulation and Signaling following Macrophage Activation Modulates Gene Expression Patterns Known to Affect Antigen-Presenting Capacity and Cytotoxic Activity." *The Journal of Immunology* 176, no. 9 (2006): 5362–5373.

Nunan, Robert, Keith G. Harding, and Paul Martin. "Clinical challenges of chronic wounds: Searching for an optimal animal model to recapitulate their complexity." *Disease Models & Mechanisms* 7, no. 11 (2014): 1205–1213. https://doi.org/10.1242/dmm.016782

Nusse, Roel, C. Fuerer, W. Ching, K. Harnish, C. Logan, A. Zeng, D. Ten Berge, and Y. Kalani. "Wnt signaling and stem cell control." In *Cold Spring Harbor Symposia on Quantitative Biology*, vol. 73, pp. 59–66. Cold Spring Harbor Laboratory Press, 2008.

Pastar, Irena, Nathan C. Balukoff, Jelena Marjanovic, Vivien Y. Chen, Rivka C. Stone, and Marjana Tomic-Canic. "Molecular pathophysiology of chronic wounds: Current state and future directions." *Cold Spring Harbor Perspectives in Biology* 15, no. 4 (2023): a041243. https://doi.org/10.1089/wound.2013.0473

Pastar, Irena, Jelena Marjanovic, Rivka C. Stone, Vivien Chen, Jamie L. Burgess, Joshua S. Mervis, and Marjana Tomic-Canic. "Epigenetic regulation of cellular functions in wound healing." *Experimental Dermatology* 30, no. 8 (2021): 1073–1089.

Pastar, Irena, Olivera Stojadinovic, Natalie C. Yin, Horacio Ramirez, Aron G. Nusbaum, Andrew Sawaya, Shailee B. Patel, Laiqua Khalid, Rivkah R. Isseroff, and Marjana Tomic-Canic. "Epithelialization in wound healing: A comprehensive review." *Advances in Wound Care* 3, no. 7 (2014): 445–464. https://doi.org/10.1089/wound.2013.0473

Pastore, Saveria, Francesca Mascia, Valentina Mariani, and Giampiero Girolomoni. "The epidermal growth factor receptor system in skin repair and inflammation." *Journal of Investigative Dermatology* 128, no. 6 (2008): 1365–1374.

Pasupuleti, Visweswara Rao, Chandra Sekhar Arigela, Siew Hua Gan, Sirajudeen Kuttulebbai Nainamohamed Salam, Kumara Thevan Krishnan, Nurhanan Abdul Rahman, and Mohammad Saffree Jeffree. "A review on oxidative stress, diabetic complications, and the roles of honey polyphenols." *Oxidative Medicine and Cellular Longevity* 2020, no. 1 (2020): 8878172. https://doi.org/10.1155/2020/8878172

Peng, Huiyu, Fangzhou Du, Jingwen Wang, Yue Wu, Qian Wei, Aoying Chen, Yuhan Duan, Shuaiguang Shi, Jingzhong Zhang, and Shuang Yu. "Adipose-derived stem-cell-membrane-coated PLGA-PEI nanoparticles promote wound healing via efficient delivery of miR-21." *Pharmaceutics* 16, no. 9 (2024): 1113. https://doi.org/10.3390/pharmaceutics16091113

Phinney, Donald G., and Mark F. Pittenger. "Concise review: MSC-derived exosomes for cell-free therapy." *Stem Cells* 35, no. 4 (2017): 851–858. https://doi.org/10.1002/stem.2575

Pi, Xinchun, Chen Yan, and Bradford C. Berk. "Big mitogen-activated protein kinase (BMK1)/ERK5 protects endothelial cells from apoptosis." *Circulation Research* 94, no. 3 (2004): 362–369.

Piechota-Polanczyk, Aleksandra, and Jakub Fichna. "The role of oxidative stress in pathogenesis and treatment of inflammatory bowel diseases." *Naunyn-Schmiedeberg's Archives of Pharmacology* 387, no. 7 (2014): 605–620.

Piek, Ester, Carl-Henrik Heldin, and Peter Ten Dijke. "Specificity, diversity, and regulation in TGF-β superfamily signaling." *The FASEB Journal* 13, no. 15 (1999): 2105–2124.

Polaka, Suryanarayana, Pratik Katare, Bhakti Pawar, Nupur Vasdev, Tanisha Gupta, Kuldeep Rajpoot, Pinaki Sengupta, and Rakesh Kumar Tekade. "Emerging ROS-modulating technologies for augmentation of the wound healing process." *ACS Omega* 7, no. 35 (2022): 30657–30672. https://pubs.acs.org/doi/10.1021/acsomega.2c02675?utm

Pontiggia, Luca, Ingmar A. J. Van Hengel, Agnes Klar, Dominic Rütsche, Monica Nanni, Andreas Scheidegger, Sandro Figi, Ernst Reichmann, Ueli Moehrlen, and Thomas Biedermann. "Bioprinting and plastic compression of large pigmented and vascularized human dermo-epidermal skin substitutes by means of a new robotic platform." *Journal of Tissue Engineering* 13 (2022): 20417314221088513. https://doi.org/10.1177/20417314221088513

Pourhashemi, Erfan, Abdollah Amini, Houssein Ahmadi, Behnaz Ahrabi, Atarodalsadat Mostafavinia, Hamidreza Omidi, Robabeh Asadi et al. "Photobiomodulation and conditioned medium of adipose-derived stem cells for enhancing wound healing in rats with diabetes: An investigation on the proliferation phase." *Lasers in Medical Science* 39, no. 1 (2024): 46.

Rabilloud, Thierry, Mireille Chevallet, Sylvie Luche, and Cécile Lelong. "Two-dimensional gel electrophoresis in proteomics: Past, present and future." *Journal of Proteomics* 73, no. 11 (2010): 2064–2077.

Rajendran, Naresh Kumar, Nicolette Nadene Houreld, and Heidi Abrahamse. "Photobiomodulation reduces oxidative stress in diabetic wounded fibroblast cells by inhibiting the FOXO1 signaling pathway." *Journal of Cell Communication and Signaling* 15, no. 2 (2021): 195–206. https://doi.org/10.1007/s12079-020-00588-x

Rawlings, Jason S., Kristin M. Rosler, and Douglas A. Harrison. "The JAK/STAT signaling pathway." *Journal of Cell Science* 117, no. 8 (2004): 1281–1283.

Reddy, Ché L., Shivani Mitra, John G. Meara, Rifat Atun, and Salim Afshar. "Artificial Intelligence and its role in surgical care in low-income and middle-income countries." *The Lancet Digital Health* 1, no. 8 (2019): e384–e386.

Rembe, Julian-Dario, Waseem Garabet, Matthias Augustin, Joachim Dissemond, Wiebke Ibing, Hubert Schelzig, and Ewa K. Stuermer. "Immunomarker profiling in human chronic wound swabs reveals IL-1 beta/IL-1RA and CXCL8/CXCL10 ratios as potential biomarkers for wound healing, infection status and regenerative stage." *Journal of Translational Medicine* 23, no. 1 (2025): 1–31.

Rousselle, Patricia, Fabienne Braye, and Guila Dayan. "Re-epithelialization of adult skin wounds: Cellular mechanisms and therapeutic strategies." *Advanced Drug Delivery Reviews* 146 (2019): 344–365.

Ruoslahti, Erkki. "RGD and other recognition sequences for integrins." *Annual Review of Cell and Developmental Biology* 12, no. 1 (1996): 697–715. https://doi.org/10.1146/annurev.cellbio.12.1.697

Rupel, Katia, Luisa Zupin, Andrea Colliva, Anselmo Kamada, Augusto Poropat, Giulia Ottaviani, Margherita Gobbo et al. "Photobiomodulation at multiple wavelengths differentially modulates oxidative stress in vitro and in vivo." *Oxidative Medicine and Cellular Longevity* 2018, no. 1 (2018): 6510159. https://doi.org/10.1155/2018/6510159

Rusnati, Marco, Chiara Urbinati, Silvia Bonifacio, Marco Presta, and Giulia Taraboletti. "Thrombospondin-1 as a paradigm for the development of antiangiogenic agents endowed with multiple mechanisms of action." *Pharmaceuticals* 3, no. 4 (2010): 1241–1278. https://www.mdpi.com/1424-8247/3/4/1241?utm

Saha, Sarmistha, Brigitta Buttari, Emiliano Panieri, Elisabetta Profumo, and Luciano Saso. "An overview of Nrf2 signaling pathway and its role in inflammation." *Molecules* 25, no. 22 (2020): 5474.

Samad, Md Abdus, Iftikhar Ahmad, Aakifah Hasan, Mohammad Hassan Alhashmi, Arusha Ayub, Fahad A. Al-Abbasi, Ajoy Kumer, and Shams Tabrez. "STAT3 signaling pathway in health and disease." *MedComm* 6, no. 4 (2025): e70152.

Schafer, Matthias, and Sabine Werner. "Oxidative stress in normal and impaired wound repair." *Pharmacological Research* 58, no. 2 (2008): 165–171.

Shah, Mamta, David M. Foreman, and Mark W. J. Ferguson. "Neutralisation of TGF-β1 and TGF-β2 or exogenous addition of TGF-β3 to cutaneous rat wounds reduces scarring." *Journal of Cell Science* 108, no. 3 (1995): 985–1002. https://doi.org/10.1242/jcs.108.3.985

She, Qing-Bai, David B. Solit, Qing Ye, Kathryn E. O'Reilly, Jose Lobo, and Neal Rosen. "The BAD protein integrates survival signaling by EGFR/MAPK and PI3K/Akt kinase pathways in PTEN-deficient tumor cells." *Cancer Cell* 8, no. 4 (2005): 287–297.

Shi, Zhuojun, Chong Yao, Yujie Shui, Site Li, and Hong Yan. "Research progress on the mechanism of angiogenesis in wound repair and regeneration." *Frontiers in Physiology* 14 (2023): 1284981. https://pmc.ncbi.nlm.nih.gov/articles/PMC10711283/?utm

Singh, Kanhaiya, Yashika Rustagi, Ahmed S. Abouhashem, Saba Tabasum, Priyanka Verma, Edward Hernandez, Durba Pal et al. "Genome-wide DNA hypermethylation opposes healing in patients with chronic wounds by impairing epithelial-mesenchymal transition." *The Journal of Clinical Investigation* 132, no. 17 (2022). https://www.jci.org/articles/view/157279

Skeen, Jennifer E., Prashanth T. Bhaskar, William S. Chen, Xiao-Ding Peng, Chia-Chen Chen, Véronique Nogueira, Annett Hahn-Windgassen, Hiroaki Kiyokawa, and Nissim Hay. "Akt Deficiency Impairs Normal Cell Proliferation and Suppresses Oncogenesis in a p53-Independent and mTORC1-Dependent Manner." *Cancer Cell* 10, no. 4 (2006): 269–280.

Sohn, Sue J., Brieana K. Sarvis, Dragana Cado, and Astar Winoto. "ERK5 MAPK regulates embryonic angiogenesis and acts as a hypoxia-sensitive repressor of vascular endothelial growth factor expression." *Journal of Biological Chemistry* 277, no. 45 (2002): 43344–43351.

Stone, Rivka C., Irena Pastar, Nkemcho Ojeh, Vivien Chen, Sophia Liu, Karen I. Garzon, and Marjana Tomic-Canic. "Epithelial-mesenchymal transition in tissue repair and fibrosis." *Cell and Tissue Research* 365 (2016): 495–506.

Streit, Michael, Paula Velasco, Lucia Riccardi, Lisa Spencer, Lawrence F. Brown, Lauren Janes, Bernhard Lange-Asschenfeldt et al. "Thrombospondin-1 suppresses wound healing and granulation tissue formation in the skin of transgenic mice." *The EMBO Journal* (2000). https://doi.org/10.1093/emboj/19.13.3272

Sun, Weiyong, Kamala Kesavan, Brian C. Schaefer, Timothy P. Garrington, Margaret Ware, Nancy Lassignal Johnson, Erwin W. Gelfand, and Gary L. Johnson. "MEKK2 associates with the adapter protein Lad/RIBP and regulates the MEK5-BMK1/ERK5 pathway." *Journal of Biological Chemistry* 276, no. 7 (2001): 5093–5100.

Suntar, Ipek, Sümeyra Çetinkaya, Emiliano Panieri, Sarmistha Saha, Brigitta Buttari, Elisabetta Profumo, and Luciano Saso. "Regulatory role of Nrf2 signaling pathway in wound healing process." *Molecules* 26, no. 9 (2021): 2424.

Taverna, Domenico, Alonda C. Pollins, Giovanni Sindona, Richard M. Caprioli, and Lillian B. Nanney. "Imaging mass spectrometry for assessing cutaneous wound healing: Analysis of pressure ulcers." *Journal of Proteome Research* 14, no. 2 (2015): 986–996.

Tia, Neelam, Alok Kumar Singh, Poorti Pandey, Chandra Shekhar Azad, Pritee Chaudhary, and Indrajeet Singh Gambhir. "Role of Forkhead Box O (FOXO) transcription factor in aging and diseases." *Gene* 648 (2018): 97–105.

Trinh, Xuan-Tung, Nguyen-Van Long, Le Thi Van Anh, Pham Thi Nga, Nguyen Ngan Giang, Pham Ngoc Chien, Sun-Young Nam, and Chan-Yeong Heo. "A comprehensive review of natural compounds for wound healing: Targeting bioactivity perspective." *International Journal of Molecular Sciences* 23, no. 17 (2022): 9573.

Varkey, Mathew, Jie Ding, and Edward E. Tredget. "Advances in skin substitutes—Potential of tissue engineered skin for facilitating anti-fibrotic healing." *Journal of Functional Biomaterials* 6, no. 3 (2015): 547–563.

Verma, Inder M. "Nuclear Factor (NF)-κB Proteins: Therapeutic Targets." *Annals of the Rheumatic Diseases* 63, no. suppl. 2 (2004): ii57–ii61.

Walczyńska, Katarzyna S., Ling Zhu, and Yujun Liang. "Insights into the role of the Wnt signaling pathway in the regeneration of animal model systems." *The International Journal of Developmental Biology* 67, no. 3 (2023): 65–78.

Wang, Chenggui, Min Wang, Tianzhen Xu, Xingxing Zhang, Cai Lin, Weiyang Gao, Huazi Xu, Bo Lei, and Cong Mao. "Engineering bioactive self-healing antibacterial exosomes hydrogel for promoting chronic diabetic wound healing and complete skin regeneration." *Theranostics* 9, no. 1 (2019): 65.

Wang, Lei, Yuhui Cai, Qingrong Zhang, and Yi Zhang. "Pharmaceutical activation of Nrf2 accelerates diabetic wound healing by exosomes from bone marrow mesenchymal stem cells." *International Journal of Stem Cells* 15, no. 2 (2022): 164–172.

Wang, Ziwen, and Chunmeng Shi. "Cellular senescence is a promising target for chronic wounds: A comprehensive review." *Burns & Trauma* 8 (2020): tkaa021.

Werner, Sabine, and Richard Grose. "Regulation of wound healing by growth factors and cytokines." *Physiological Reviews* 83, no. 3 (2003): 835–870.

Wilkinson, Holly N., and Matthew J. Hardman. "Wound healing: Cellular mechanisms and pathological outcomes." *Open Biology* 10, no. 9 (2020): 200223. https://royalsocietypublishing.org/doi/10.1098/rsob.200223

Wullschleger, Sabine, Robbie Loewith, and Michael N. Hall. "TOR Signaling in Growth and Metabolism." *Cell* 124, no. 3 (2006): 471–484.

Xuan, Yuanhu, Lisha Chi, Haishan Tian, Wanhui Cai, Congcong Sun, Tao Wang, Xuan Zhou et al. "The activation of the NF-κB-JNK pathway is independent of the PI3K-Rac1-JNK pathway involved in the bFGF-regulated human fibroblast cell migration." *Journal of Dermatological Science* 82, no. 1 (2016): 28–37.

Ye, Rongan, Yu He, Wei Ni, Yiqiu Zhang, Ying Zhu, Muqing Cao, Ruida He, and Min Yao. "LLLT accelerates experimental wound healing under microgravity conditions via PI3K/AKT-CCR2 signal axis." *Frontiers in Bioengineering and Biotechnology* 12 (2024): 1387474. https://pmc.ncbi.nlm.nih.gov/articles/PMC11347831/?utm

Yu, Honghao, Yichen Wang, Dawei Wang, Zeming Liu, Yi Yi, Min Wu, Yiping Wu, and Qi Zhang. "Landscape of the Epigenetic Regulation in Wound Healing." *Frontiers in Physiology* 13 (2022): 949498.

Zenz, Rainer, Harald Scheuch, Paul Martin, Carsten Frank, Robert Eferl, Lukas Kenner, Maria Sibilia, and Erwin F. Wagner. "c-Jun regulates eyelid closure and skin tumor development through EGFR signaling." *Developmental Cell* 4, no. 6 (2003): 879–889.

Zhang, Jingyu, Bingkun Zhao, Wuhan Wei, Dan Wang, Haoyu Wang, Aijun Zhang, Changbo Tao, Xueyang Li, Qiang Li, and Peisheng Jin. "Prospective, randomized, and controlled study of a human umbilical cord mesenchymal stem cell injection for treating diabetic foot ulcers." *Journal of Visualized Experiments (JoVE)* 193 (2023): e65045.

Zhang, Ruiying, Hanfeng Li, Weijie Zhang, Tong Zhang, Xin Chen, Weipeng Lu, Bing Zhang, Yihu Wang, and Yanchuan Guo. "Chlorogenic acid/carboxymethyl chitosan nanoparticle-assisted biomultifunctional hyaluronic acid-based hydrogel scaffolds for burn skin repair." *International Journal of Biological Macromolecules* 275 (2024a): 133528.

Zhang, Wei, Lulu Liu, Hui Cheng, Jing Zhu, Xinyi Li, Sheng Ye, and Xiaojing Li. "Hydrogel-based dressings designed to facilitate wound healing." *Materials Advances* 5, no. 4 (2024b): 1364–1394.

Zhang, Wenqian, Lang Chen, Yuan Xiong, Adriana C. Panayi, Abudula Abududilibaier, Yiqiang Hu, Chenyan Yu et al. "Antioxidant therapy and antioxidant-related bionanomaterials in diabetic wound healing." *Frontiers in Bioengineering and Biotechnology* 9 (2021): 707479.

Zhou, Xueqing, Qiongfang Ruan, Ziqing Ye, Zhigang Chu, Maomao Xi, Min Li, Weigang Hu, Xiaoyu Guo, Paul Yao, and Weiguo Xie. "Resveratrol accelerates wound healing by attenuating oxidative stress-induced impairment of cell proliferation and migration." *Burns* 47, no. 1 (2021): 133–139.

3 Biomarkers for Wound Healing

Aarif Ali[1*] *and Javeed Iqbal Ahmad Bhat*[1]
[1] Division of Basic Sciences and Humanities,
Faculty of Horticulture, SKUAST-K, Shalimar,
Srinagar, Jammu & Kashmir, India

3.1 INTRODUCTION TO WOUND HEALING

Wound healing is a multifaceted biological process that encompasses a meticulously regulated sequence of events, which includes hemostasis, inflammation, proliferation, and remodeling. Understanding and tracking this process is crucial for enhancing patient outcomes, particularly in instances of chronic or non-healing wounds. This mechanism is implemented through the communication between various local skin compartments, the supporting extracellular matrix (ECM), and systemic contributors. In a healthy individual under physiological conditions, the cutaneous epidermal repair process is highly effective; however, when this process is hindered, the tissue fails to recover its structural and functional integrity, leading to chronic wounds. With the rising elderly population experiencing a multitude of increasingly prevalent disease states such as diabetes, vascular diseases, and obesity, clinicians especially those in surgical specialties and physicians managing healing disorders often encounter chronic non-healing wounds, including diabetic foot ulcers (DFUs), venous leg ulcers (VLUs), and pressure ulcers (PUs). Physiologic wound healing begins after tissue damage, activating a signaling network among various cell types and skin compartments. Following an injury, hemostasis leads to the creation of a fibrin clot, which acts as both a scaffold and a source of growth factors and chemokines that attract a variety of cells, including inflammatory cells, to move into the wound bed (Martin, 1997; Uccioli et al., 2015). Alterations in the expression of several adhesion molecules on endothelial cells facilitate the recruitment and movement of neutrophils and macrophages from the bloodstream to eliminate potential infections. These inflammatory cells secrete growth factors and cytokines such as interleukins (IL1, IL6), tumor necrosis factor-α (TNFα), platelet-derived growth factor (PDGF), and fibroblast growth factor-2 (FGF2) (Eming et al., 2014). Fibroblasts subsequently multiply and create a provisional matrix, allowing keratinocytes to migrate from the edges of the wound. In reaction to injury, keratinocytes located at the wound perimeter release cytokines to signal that the barrier has been breached. Along with the concurrent rearrangement of integrins and cytoskeletal elements, as well as the expression of proteases like matrix metalloproteinases (MMPs), keratinocytes can clear a pathway for migration at the junction of the fibrin clot and the underlying dermis. This process, known as re-epithelization, is succeeded by keratinocyte proliferation at the wound edge, thereby restoring the stratified epidermis. The integrity of the epithelial barrier is then reinstated, and the remodeling of the underlying matrix continues. Abnormal functioning of standard repair processes can lead to difficulties in healing and the emergence of chronic wounds. Chronic wounds are characterized by insufficient advancement through a timely sequence of anatomical and functional recovery (Lazarus et al., 1994). They can generally be classified into one of three primary etiological categories: vascular insufficiency (such as VLUs), diabetes mellitus (like DFUs), and local pressure effects (for instance, PUs). Moreover, various systemic factors can hinder the healing process (Eming et al., 2014). Nevertheless, the core issues of non-healing wounds share common traits

* Corresponding author: buttaarif31@gmail.com

DOI: 10.1201/9781003644958-3

regardless of the specific wound causes: the presence of prolonged and suboptimal inflammation, concurrent infection, protease deregulation, diminished growth factor activity, stem cell dysfunction, and cellular senescence (Eming et al., 2010; Stojadinovic et al., 2014; Yager et al., 1996).

Extended suboptimal inflammation within the wound microenvironment, increased MMP levels, and decreased levels of their inhibitors all play a role in hindering cell migration and healing (Eming et al., 2010; Loots et al., 1998; Van Asten et al., 2016). This inflammatory condition is exacerbated by polymicrobial infections in the wound, often involving microbes that can form biofilms, which together maintain the influx of pro-inflammatory cells while simultaneously obstructing the host's response to infection (Tarnuzzer and Schultz, 1996; Pastar et al., 2013). Furthermore, impaired angiogenesis and vasculogenesis, due to abnormal regulation and cleavage of growth factors and their receptors, results in inadequate oxygenation and insufficient nutrient delivery to the wound (Ladwig et al., 2002; White and Cutting, 2006). Epidermal stem cell dysfunction, caused by continuous cycling and subsequent depletion, along with deregulated chemokines essential for recruiting bone marrow and endothelial progenitors, results in insufficient healing (Barrientos et al., 2008; Pastar et al., 2014). Additionally, fibroblasts and keratinocytes that are prematurely senescent release increased amounts of MMPs, antiangiogenic factors, and ILs (Jun and Lau, 2010), which disrupt the regulation of their target genes, further contributing to a delay in the healing process. Any of these deregulated molecules and cellular mechanisms could potentially be leveraged for diagnostic applications and tracking the progress of healing.

3.2 BIOMARKERS

The idea of biomarkers has been present since the early days of ayurvedic medicine, approximately in the seventh century, when the sweetness of urine was associated with diabetes, despite the fact that the specific terminology had not yet been established (Karley et al., 2011). Biomarkers are characterized as objective, measurable traits of biological processes. They may consist of molecules, genes, proteins, or physiological alterations that offer insights into normal or pathological processes or reactions to therapeutic interventions. A comprehensive definition from the World Health Organization (WHO) collaborative initiative on chemical safety describes a biomarker as any measurable substance, structure, or process within the body or its products that can affect or forecast the occurrence of an outcome or disease (World Health Organization (WHO), 2001). The use of biomarkers has become essential and well-established in clinical research, often serving as indicators of clinical outcomes across a diverse range of disease conditions and their treatment. In the context of wound healing, biomarkers assist in various aspects as shown in Figure 3.1 respectively.

3.3 CHARACTERISTICS OF AN IDEAL BIOMARKER

Within wound healing, biomarkers are crucial for assessing the healing journey, recognizing complications such as infections or chronic inflammation, and directing treatment strategies. To be effective and clinically relevant in wound healing, a biomarker must fulfill several key criteria:

1. *Specificity*: A biomarker should be associated with a distinct phase or event in the process of wound healing (hemostasis, inflammation, proliferation, or remodeling). For instance, PDGF is pertinent to the early phases of hemostasis and proliferation. Similarly, MMPs are connected to the remodeling of tissue.
2. *Sensitivity*: It must possess the sensitivity to identify minor alterations in the wound environment, including the early signs of infection or slow healing. For instance, c-reactive protein (CRP) and interleukin-6 (IL-6) levels increase rapidly during inflammation, aiding in the detection of infection or heightened inflammation.
3. *Measurable and Quantifiable*: The biomarker must be easily quantifiable through widely available, standardized laboratory methods. This facilitates consistent tracking and comparison over time. For instance, ELISA kits for cytokines (such as IL-1β, TNF-α) or point-of-care tests for wound exudates.

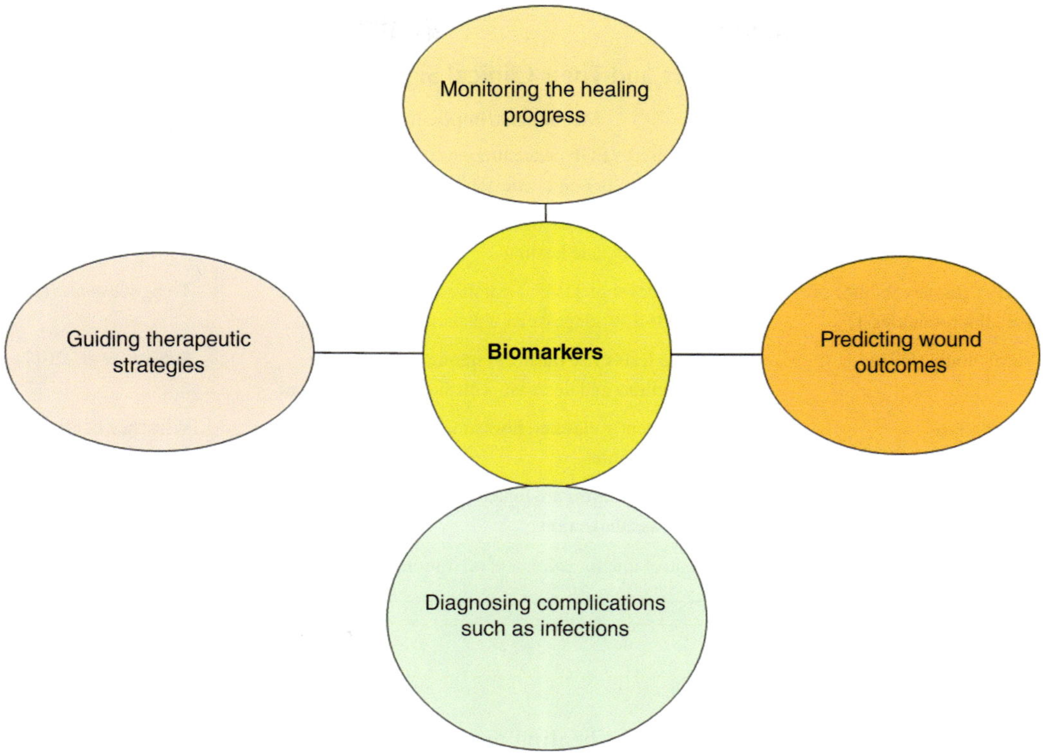

FIGURE 3.1 Role of biomarkers in assisting wound healing.

4. *Predictive Value for Healing Outcomes*: The biomarker is intended to assist in predicting whether a wound will heal properly or face complications (e.g., become chronic). For example, consistently elevated levels of matrix metalloproteinase-9 (MMP-9) alongside low levels of tissue inhibitor of metalloproteinases-1 (TIMP-1) can indicate chronic, non-healing wounds.
5. *Non-Invasive or Minimally Invasive Sampling*: Optimal biomarkers ought to be identifiable in wound exudate, blood, or saliva—minimizing the necessity for biopsies. Biomarkers such as MMPs, vascular endothelial growth factor (VEGF), and cytokines can frequently be measured directly from wound fluid.
6. *Responsive to Therapeutic Interventions*: The biomarker must demonstrate a quantifiable change as a result of treatment, aiding in the assessment of its effectiveness. The effective use of a growth factor-based therapy should elevate the levels of regenerative markers such as transforming growth factor beta (TGF-β) or epidermal growth factor (EGF).
7. *Association with Clinical and Histological Evidence*: Biomarker levels are expected to mirror discernible wound traits and the fundamental tissue activities. This substantiates that the biomarker delivers valuable, real-world clinical perspectives. A reduction in VEGF levels might correlate with lower angiogenesis and suboptimal granulation tissue formation.

3.4 CATEGORIES OF BIOMARKERS

Biomarkers can be classified according to the type of information they offer, which includes predictive, diagnostic, and indicative categories.

TABLE 3.1
Important Predictive Biomarkers and Their Clinical Significance

Biomarker	Mechanism/Implication	References
VEGF (vascular endothelial growth factor)	Increased VEGF concentrations point to a robust potential for angiogenesis	Barrientos et al. (2008)
MMP-9	Increased levels associated with ECM deterioration and inadequate healing	Ladwig et al. (2002)
TIMP-1 (tissue inhibitor of metalloproteinases-1)	A low level of TIMP-1 combined with a high level of MMP-9 suggests an imbalance	Trengove et al. (1999)
Procalcitonin (PCT)	These levels rise during systemic bacterial infections and decrease with successful treatment	Schuetz et al. (2017)
IL-6, TNF-α	Consistently elevated levels may indicate chronic inflammation	Wlaschek et al. (2005)
CD34+ endothelial progenitor cells	Decreased levels are linked to hindered neovascularization	Asahara et al. (1999)
miR-21, miR-146a	Control inflammation and activation of fibroblast	Liu et al. (2021)

3.4.1 Predictive Biomarkers

Predictive biomarkers related to wound healing are quantifiable indicators that assist in evaluating the probability of successful wound recovery and recognizing wounds that might need further treatment. These biomarkers can facilitate the monitoring of healing progress, identify complications such as infections, and possibly inform tailored treatment approaches. These biomarkers are useful for forecasting outcomes or assessing the probability of treatment benefits. They act as a significant resource in customizing treatment approaches for particular groups of patients. Table 3.1 shows some of the important predictive biomarkers along with their clinical significance.

3.4.2 Diagnostic Biomarkers

Diagnostic biomarkers serve as quantifiable biological indicators that help ascertain the type, status, or pathological condition of a wound. These indicators are critical for understanding whether a wound is acute or chronic, infected or non-infected, healing or non-healing, and they support clinicians in distinguishing between inflammation, infection, ischemia, or the presence of biofilm. Table 3.2 shows essential diagnostic biomarkers along with their role in the process of wound healing.

3.4.3 Indicative Biomarkers

Indicative biomarkers are biological molecules that represent the current physiological or pathological condition of a wound, assisting clinicians in comprehending the present situation within the wound environment. In contrast to diagnostic biomarkers, which are utilized to identify a specific disease, or predictive biomarkers, which are employed to anticipate therapy responses, indicative biomarkers reveal the active biological processes such as inflammation, infection, hypoxia, or healing progression. Table 3.3 shows the relevance of indicative biomarkers in wound healing process.

TABLE 3.2
Essential Diagnostic Biomarkers and Their Importance in Wound Healing

Biomarker	Biological Role	Clinical Relevance
IL-6	Pro-inflammatory cytokine	Increased levels suggest ongoing inflammation and a higher likelihood of slow healing. Aids in distinguishing between infected and non-infected wounds
TNF-α (tumor necrosis factor-alpha)	Mediator of inflammation	Elevated levels are indicative of inflammatory wounds that do not heal. Encourages catabolism and leads to tissue damage
MMP-9 (matrix metalloproteinase-9)	ECM degradation enzyme	Elevated MMP-9 levels suggest tissue degradation and prolonged wound healing. This is indicative of a lack of response to conventional treatment.
TIMP-1 (tissue inhibitor of metalloproteinases-1)	Inhibits MMPs	A low TIMP-1 level combined with high MMP-9 indicates a protease imbalance, which results in slower healing
Procalcitonin (PCT)	Systemic inflammatory marker	Assessment for bacterial infections in wounds. Directs antibiotic usage and evaluates sepsis risk
C-reactive protein (CRP)	Acute-phase protein	Increased CRP levels suggest systemic or localized infection. Employed to track inflammatory activity in wounds
Lactate	Marker of hypoxia/ metabolic stress	Increased lactate levels in wound exudate indicate ischemia or inadequate perfusion. This suggests a potential for non-healing
Wound pH	Biochemical environment	A pH greater than 7.5 is linked to chronic, infected wounds, while an acidic pH of less than 6.0 promotes healing
Microbial DNA (e.g., 16S rRNA)	Microbiome detection	Detects harmful colonization or biofilm development. Aids in customizing antimicrobial therapy
Staphylococcus aureus, *Pseudomonas aeruginosa*	Opportunistic pathogens	High levels present ($>10^6$ CFU/g tissue) indicate infection and chronicity associated with biofilms

TABLE 3.3
Indicative Biomarkers and Their Relevance

Biomarker	Biological Role	Indicative Value	Reference
IL-1β, IL-6, TNF-α	Pro-inflammatory cytokines	Elevated levels in wounds that are non-healing or infected, suggesting persistent inflammation	Trengove et al. (1999)
MMP-9	ECM degradation enzyme	Increased levels suggest tissue degradation This indicates that the wound is in a catabolic condition.	Yager et al. (1999)
Lactate	Metabolic stress marker	Elevated levels are observed in hypoxic or ischemic tissues, signifying inadequate perfusion or necrosis	Serena et al. (2016)
Wound pH	Biochemical indicator	A pH level exceeding 7.5 may indicate an infection or a chronic condition A lower acidic pH of approximately 5.5 suggests a healing environment	Gethin (2007)
Reactive oxygen species (ROS)	Oxidative stress markers	Elevated levels of reactive oxygen species (ROS) point to persistent tissue damage or infection	Schafer and Werner (2008)
Neutrophil elastase	Inflammatory enzyme	Increased activity signifies ongoing tissue injury mediated by neutrophils	Wysocki et al. (1999)

3.5 EVALUATION METHODS OF WOUND HEALING BIOMARKERS

Assessing wound healing biomarkers is crucial for comprehending the processes involved in tissue repair and for evaluating the success of therapeutic measures. These techniques vary from conventional histological methods to sophisticated molecular and imaging strategies. Reliable and consistent evaluation techniques enable clinicians and researchers to pinpoint healing phases, identify complications, and implement targeted therapies. Biomarkers can be identified and measured in wound exudates, tissue biopsies, serum, or wound dressings, utilizing a range of qualitative and quantitative techniques. The selection of the method is influenced by the type of biomarker (such as protein, RNA, or enzyme activity), the sample type, and the purpose of the evaluation (whether for diagnosis, prognosis, or treatment monitoring).

3.5.1 Sample Collection Techniques

Comprehending the reasons behind the difficulty in healing certain wounds is crucial for enhancing care and creating more effective treatments. The technique employed for sample collection is a vital part of the research process and can influence the results achieved.

3.5.1.1 Wound Exudate/Fluid Collection

Wound exudate denotes the fluid that seeps from blood vessels into wounds, playing a vital role in the healing process. It consists of a combination of water, electrolytes, nutrients, inflammatory mediators, growth factors, enzymes, and cellular components such as leukocytes and platelets. This exudate is not simply a by-product of inflammation, it also acts as a medium that indicates the status of the wound and the underlying physiological processes involved in healing or infection (Vowden & Vowden, 2004). Wound exudate, the fluid that accumulates in and around a wound, serves as a significant source of biomarkers for evaluating wound healing and detecting possible complications such as infection. Examining the makeup of this fluid can offer valuable information regarding the wound's healing progress, inflammatory response, and the existence of infection. Researchers are actively studying exudate as a potential marker for various wound conditions. Wound exudate can be obtained non-invasively and is rich in cellular components that may reflect a range of physiological processes (Asada et al., 2012; Iizaka et al., 2010; Minematsu et al., 2013). A previous investigation found a connection between pro-inflammatory cytokines in wound exudate and post-operative pain, with a noted reduction in pro-inflammatory concentrations correlating with analgesic administration (Carvalho et al., 2013). Another study indicated that neuropeptide levels in wound exudates from VLUs were associated with the intensity of wound pain, although the correlation coefficients were moderate (rho values $<.05$) (Goto et al., 2016). Moreover, despite several studies presenting simple methods for collecting wound exudate in clinical settings, the precise cellular component or wound material that best indicates wound pain remains uncertain. Wound exudates are generally categorized according to their appearance and texture:

Serous Exudate: A clear, thin, and watery fluid, commonly observed in acute wounds or during the healing process (Collier, 2023).

Sanguineous Exudate: Characterized by fresh bleeding, which suggests vascular injury or trauma.

Serosanguineous: This type consists of a mix of serous fluid and blood, frequently found in wounds that are healing normally.

Purulent Exudate: A thick fluid that can be yellow, green, or brown, usually indicating an infection and containing dead leukocytes and bacteria (Cutting, 2003).

Fibrinous and Malodorous Exudates: These may signal chronic wounds or anaerobic infections (Leaper et al., 2012).

3.5.1.2 Clinical Importance

Increased concentrations of specific cytokines (such as IL-1, IL-6, and TNF-α) found in wound exudate may suggest the presence of non-healing wounds (Yussof et al., 2012). Proteases, including MMPs, are involved in tissue remodeling, and their concentrations can reflect the progress of wound healing or indicate possible complications (Ruf et al., 2017). Additionally, a range of other substances, including growth factors, enzymes, and inflammatory mediators, may act as potential biomarkers (Lindley et al., 2016). Monitoring the quantity, consistency, odor, and color of exudate gives important insights into the healing process of wounds or any complications that may arise. Too much exudate can result in the maceration of the skin around the wound, which can slow down healing, while too little fluid might suggest ischemic conditions. Furthermore, the evaluation of wound fluid biomarkers such as MMPs, cytokines, and growth factors can yield valuable diagnostic and prognostic data (Eming et al., 2007; Trengove et al., 1999).

3.5.1.3 Techniques

The most frequently used techniques include direct aspiration, absorbent pads, vacuum or drainage devices, and occlusive dressings (Harvey et al., 2022).

3.5.1.3.1 Direct Aspiration

Direct aspiration involves extracting wound fluid straight from the wound bed or tissue using a sterile needle and syringe. This approach is particularly advantageous for chronic wounds, abscesses, surgical wounds, or infected locations, where fluid accumulation is notable but may not be draining freely. This method is essential for the non-invasive or minimally invasive evaluation of wound physiology, especially in cases of chronic wounds, infected wounds, and non-healing ulcers. The aspirated wound fluid is rich in important biochemical elements like cytokines, growth factors, proteases, antimicrobial peptides, and immune cells—each of which can act as a biomarker for different phases of wound healing. The capacity to examine these components over time via direct aspiration provides significant diagnostic and therapeutic opportunities in clinical wound management and research. One of the major applications of direct aspiration involves the collection of wound fluid to evaluate biomarkers associated with healing. These include the following: Inflammatory markers: IL-1β, IL-6, TNF-α, Proteases: MMPs (e.g., MMP-2, MMP-9), elastase, Growth factors: VEGF, PDGF, TGF-β, Angiogenic markers: HIF-1α, angiopoietins, Protease inhibitors: TIMPs, Oxidative stress markers: ROS, nitric oxide, ECM fragments: degraded collagen, fibronectin. By employing direct aspiration, these biomarkers can be measured through ELISA, mass spectrometry, Western blotting, or multiplex assays. Increased levels of MMP-9 and decreased levels of TIMP-1 in the aspirated fluid suggest chronic non-healing wounds and have been suggested as possible predictive indicators for delayed healing. Direct aspiration is commonly utilized in the ongoing assessment of chronic wounds such as DFUs, VLUs, PUs (bedsores), and post-surgical infected wounds.

3.5.1.3.2 Absorbent Pads

Among the different sampling techniques available, absorbent pads and dressings have become a patient-friendly, non-invasive, and effective method for gathering wound exudate for biomarker analysis. These dressings are already a key component of wound care management due to their ability to absorb fluids. Their use in biomarker assessment serves a dual function: they maintain wound moisture while simultaneously collecting exudate that is rich in diagnostic information. Absorbent pads are generally made from materials such as polyurethane foams, hydrofibers, hydrocolloids, alginates, or superabsorbent polymers (SAPs) (Jones et al., 2006; Sibbald et al., 2007). These materials are engineered to absorb and hold significant amounts of wound exudate, creating a moist environment that facilitates healing while safeguarding the surrounding skin.

In clinical and research contexts, absorbent pads that have been utilized can be extracted from the wound and processed to obtain the exudate that has built up. This is usually accomplished through an elution technique, where a buffer solution (for instance, phosphate-buffered saline) is employed to rehydrate and flush out the fluid contents. The extract can then be subjected to centrifugation and filtration for subsequent analysis. This approach allows researchers to amass sufficient quantities of exudate for the evaluation of cytokines (such as IL-1β, TNF-α, IL-6), proteases (MMP-2, MMP-9), growth factors (VEGF, PDGF), and other pertinent biomarkers (James et al., 2008; Wiegand et al., 2015). Unlike direct aspiration, which only samples the fluid available at a specific time, absorbent pads enable cumulative collection over time, often providing a more extensive and integrated overview of the wound's molecular environment (Leveen et al., 1973; Schultz et al., 2003).

3.5.1.3.3 Vacuum or Drainage Devices

Vacuum-assisted closure (VAC) systems and surgical drainage devices are extensively utilized in contemporary wound care and post-operative management to eliminate excess exudate, minimize edema, and facilitate healing. These systems also allow for the controlled collection of wound fluid, which contains a wealth of biomarkers including cytokines, proteases, and microbial products. The gathered fluid can serve diagnostic functions or be used to track healing progress and detect infections. Negative pressure wound therapy (NPWT), also referred to as VAC (therapy), entails the application of sub-atmospheric pressure to a wound through a foam or gauze dressing that is sealed with an occlusive film and linked to a vacuum pump. The fluid collected in NPWT canisters has been examined for: cytokines (IL-6, IL-8, TNF-α) (Hasan et al., 2015; Schintler, 2012), proteases (MMP-9, neutrophil, elastase) (Eming et al., 2007), and growth factors (VEGF, PDGF) (Barrientos et al., 2008). These biomarker levels offer valuable information regarding the inflammatory status and healing capacity of the wound.

3.5.1.3.4 Occlusive Dressings

Occlusive dressings are a specific group of wound care products aimed at sealing the wound area to avert external contamination, while also ensuring a moist environment for healing. Typically, these dressings consist of semi-permeable materials, including polyurethane films, hydrocolloids, or composite layers, which permit gas exchange (for instance, oxygen and carbon dioxide) but are resistant to water, bacteria, and other pathogens (Jones et al., 2006; Sibbald et al., 2000).

Occlusive dressings closely adhere to the skin surrounding the wound, trapping moisture at the site, which promotes autolytic debridement. They maintain a warm, moist environment that facilitates epithelial cell migration and tissue regeneration, while also serving as a barrier against bacteria, thereby reducing the risk of external infection. For instance, hydrocolloid dressings are made with gel-forming agents such as pectin, gelatin, and carboxymethylcellulose that interact with wound exudate to create a gel. This gel preserves a moist interface and seals the wound from the external environment (Fonder et al., 2008). Occlusive dressings are utilized for partial-thickness wounds, including abrasions and donor sites, post-surgical wounds and pressure injuries, and VLUs. Circumstances that necessitate non-invasive fluid sampling, like the collection of wound exudate under film dressings (Trengove et al., 1999).

3.5.1.3.5 Wound Tissue Biopsy

Wound tissue biopsy is considered the gold standard method for acquiring direct samples from the wound bed or its edges for diagnostic, histological, molecular, and proteomic evaluations. This technique offers a more thorough depiction of the local wound environment than exudate or swab samples, particularly when analyzing the spatial distribution of biomarkers or pathogens. It includes the following types.

3.5.1.3.5.1 Punch Biopsy A punch biopsy is a technique that is minimally invasive, designed to collect a full-thickness cylindrical piece of skin and subcutaneous tissue for diagnostic or research purposes. It is widely used in dermatology, wound healing research, and pathological analysis due to its straightforward nature, reproducibility, and its effectiveness in preserving the architecture of the skin. Typically, it is performed at the edge of the wound or in the periwound area to capture both healing and non-healing tissue zones. To alleviate discomfort, a local anesthetic like lidocaine is injected.

A punch biopsy reveals the presence of inflammatory cells, fibrosis, granulation tissue, and vascularization. It also identifies delayed healing patterns in chronic wounds. It is also utilized to localize specific proteins include the following: VEGF (angiogenesis), CD31 (endothelial marker), TNF-α, IL-1β (pro-inflammatory cytokines), TGF-β1, collagen I/III (indicators of the remodeling phase) (Eming et al., 2007). Furthermore, it enables the quantification of gene expression associated with growth factors, cytokines, and ECM components through molecular methods like qPCR and microarray (Harding et al., 2002). In addition, punch biopsy samples are utilized to assess protein levels of key wound healing regulators such as MMP-9 and TIMP-1 employing ELISA and Western blotting techniques (Schultz et al., 2003; Trengove et al., 1999).

3.5.1.3.5.2 Incisional Biopsy An incisional biopsy refers to the surgical extraction of a representative portion of a wound or lesion for diagnostic, histopathological, and molecular investigations. Unlike punch biopsies, which remove small circular samples, incisional biopsies enable the collection of larger, deeper, and more diverse tissue sections, often necessary in cases of suspected cancer, atypical wound presentations, or non-healing ulcers. The procedure is carried out under local anesthesia, using a scalpel to cut out an elliptical or rectangular piece of tissue, typically from the edge of the wound to include both viable and necrotic areas (Schultz et al., 2003). The tissue is then preserved according to the intended analysis: formalin fixation for histology and immunohistochemistry (IHC), RNAlater or freezing for gene expression studies, or sterile transport media for microbiological evaluation (Eming et al., 2007).

Incisional biopsies, when examined histologically, demonstrate inflammatory cell infiltration, granulation tissue, fibrosis, angiogenesis, and other aspects pertinent to chronic wound conditions or abnormal healing (Loots et al., 1998). This method is vital for identifying malignant transformations such as Marjolin's ulcers and for differentiating pyoderma gangrenosum from infectious ulcers (Tandara and Mustoe, 2004). Additionally, incisional biopsies allow for molecular analyses like RT-qPCR and Western blotting to measure cytokine expression (e.g., IL-1β, TNF-α) and matrix remodeling indicators (e.g., MMP-9, collagen I/III) (Eming et al., 2007). Although it holds significant clinical value, this technique is more invasive than punch biopsy and carries risks such as delayed healing, pain, and infection, particularly in immunocompromised or diabetic individuals.

3.5.1.3.5.3 Excisional Biopsy An excisional biopsy refers to the complete surgical extraction of a wound, lesion, or ulcerative area, which includes the entire affected tissue along with its surrounding margins (Beard et al., 2018). This technique is frequently utilized when the wound or lesion is small and localized, or when there is a strong clinical suspicion of malignancy, such as in Marjolin's ulcer (a type of squamous cell carcinoma that arises from chronic wounds) or atypical ulcers that do not respond to traditional treatment methods. The excised sample usually contains epidermis, dermis, subcutaneous fat, and sometimes deeper structures, offering a comprehensive tissue sample for both diagnostic and molecular studies. Excisional biopsies are especially crucial in the fields of oncologic dermatology and the evaluation of complex wounds, as they enable a complete histopathological assessment of the tissue architecture. This includes the examination of inflammatory infiltrates, neoplastic cells, fibrosis, vascular proliferation, necrosis, and other pathophysiological aspects. Additionally, the excised tissue can be utilized for IHC to identify protein markers such as VEGF, CD31, TNF-α, and collagen I/III, or for molecular analyses including RT-PCR, Western blotting, and mass spectrometry (Masson-Meyers et al., 2020). While excisional biopsy offers the most detailed information, it is more invasive than incisional or punch biopsies, carries a higher risk of infection and bleeding, and may delay healing, particularly in patients with comorbidities like diabetes or peripheral vascular disease.

3.5.1.4 Factors Affecting Wound Healing and Management

A variety of intrinsic and extrinsic factors influence both the amount and quality of wound fluid: type and size of the wound, degree of inflammation or infection, existing health conditions like diabetes or vascular diseases, and management of external moisture and types of dressings (Thomas, 1990). Effective management requires the selection of suitable dressings that maintain moisture levels, encourage healing, and avert infection. Dressings such as alginates, foams, and hydrofibers are specifically designed to absorb excess exudate while ensuring a moist environment for the wound, which is essential for cellular migration and tissue repair (Dowsett & Ayello, 2004).

3.5.2 Analytical Techniques

Analytical strategies contribute to the characterization of tissue architecture, the identification of crucial biomarkers, and the monitoring of treatment efficacy. Analytical approaches include the following.

3.5.2.1 Enzyme-Linked Immunosorbent Assay

The Enzyme-Linked Immunosorbent Assay (ELISA) is a highly sensitive and quantitative immunoassay that is commonly employed to detect and quantify specific proteins, including wound healing biomarkers such as cytokines, chemokines, and growth factors, found in biological fluids like wound exudate, serum, or tissue lysates. These biomarkers offer valuable information regarding the various stages of wound healing: hemostasis, inflammation, proliferation, and remodeling (Schultz et al., 2003).

Common wound healing biomarkers evaluated by ELISA include the pro-inflammatory cytokines (IL-1β, TNF-α, IL-6), growth factors (VEGF, PDGF, TGF-β1), matrix remodeling enzymes (MMP-2, MMP-9), inhibitors (TIMP-1, TIMP-2), and angiogenesis markers (angiopoietin-2, CD31). These biomarkers are vital for monitoring the healing condition of acute and chronic wounds and for assessing the success of therapeutic strategies (Loots et al., 1998; Trengove et al., 1999). Trengove et al. (1999) assessed IL-1α, IL-1β, IL-6, TNF-α, VEGF, and TGF-β in samples of wound exudates. The results from chronic wounds revealed increased levels of pro-inflammatory cytokines and decreased growth factors, suggesting a delay in the healing process. Loots et al. (1998) conducted an analysis of TGF-β1, VEGF, and IL-6 in tissue homogenate samples from patients with diabetic and venous ulcers. The study's findings showed that chronic wounds were characterized by low VEGF/TGF-β1 levels and high IL-6 concentrations, which reflect a disruption in angiogenesis and an increase in inflammation. Eming et al. (2007) conducted an assessment of IL-1β, TNF-α, IL-10, and MCP-1 in murine wound models. The ELISA data showed changes in cytokine levels over time, indicating a peak in inflammation that gradually resolved. Barrientos et al. (2008) assessed the levels of PDGF, EGF, TGF-β1, VEGF, and IL-1β biomarkers in samples of wound fluids from both humans and animals. The results from the ELISA tests facilitated the monitoring of biomarker variations in relation to treatment responses and the healing process.

3.5.2.2 Western Blot

Western blotting or immunoblotting is a widely utilized analytical technique for detecting specific proteins in a complex mixture. Western blotting serves as a significant technique for the identification and quantification of biomarkers associated with wound healing. It allows to assess the protein expression levels of various molecules that contribute to the wound healing process, such as growth factors, cytokines, and MMPs, thereby providing insights into the mechanisms and progression of wound repair. The method's sensitivity, specificity, and semi-quantitative nature make it an essential tool for studying the molecular mechanisms involved in wound healing. The technique of Western blotting can be utilized to detect proteins that are expressed differently during the process of wound healing, potentially acting as biomarkers. Researchers have the ability to compare protein expression levels in healthy tissue, injured tissue, and tissue from wounds that are healing at different rates to discover possible biomarkers (Patel et al., 2016). Once these biomarkers are identified, Western blotting can be employed to measure their expression levels in various samples. This facilitates the assessment of specific protein abundance and its changes during the wound healing process (Jiang et al., 2025). By monitoring the expression levels of certain biomarkers over time, researchers are able to observe the advancement of wound healing and detect any potential complications or delays in the process. For instance, increased levels of MMPs or pro-inflammatory cytokines in wounds that are not healing may suggest an issue with the healing process (Patel et al., 2016). Western blotting can be utilized to assess the effectiveness of various therapeutic interventions designed to enhance wound healing. Researchers can determine if a specific treatment modifies the expression levels of certain biomarkers in a favorable way, which would indicate its effectiveness (Patel et al., 2016). The technique of Western blotting aids in the detection of increased levels of inflammatory cytokines, such as IL-6 and TNF-α, which are essential mediators during the early stages of wound healing (Zhang et al., 2021). According to Zhang et al. (2021), Western blot was used to assess the expression of IL-6 and TNF-α in wounds of diabetic mice, indicating a slower and more prolonged inflammatory response relative to the control. Proteins such as collagen I, collagen III, MMPs, and TIMPs are crucial targets. Zhang et al. (2020) employed Western blot analysis to measure the levels of MMP-9 and TIMP-1 in samples from chronic wounds, suggesting a dysregulation in ECM remodeling in wounds that do not heal. Yager and Nwomeh (1999) examined the expression levels of MMP-2 and MMP-9 in acute versus chronic wounds. Their research indicated that the activity of MMP-2 and MMP-9 was over 10- and 25-fold greater, respectively, in exudates from PUs lasting at least three months compared to surgical wound fluids.

3.5.2.3 Quantitative PCR (qPCR)

Quantitative PCR (qPCR) serves as an essential method for assessing wound healing biomarkers by measuring the expression levels of particular genes that play a role in the healing process. This technique enables researchers to monitor the fluctuations in gene expression throughout various stages of wound healing, offering valuable insights into the mechanisms that drive tissue repair and highlighting possible therapeutic targets. qPCR has been used to quantify the expression of growth factors such as VEGF and EGF, both of which are vital for cell proliferation and angiogenesis (Hwang et al., 2022). qPCR has been utilized to evaluate the expression levels of cytokines such as IL-6 and TNF-α, both of which are significant in the processes of inflammation and immune response (Patel et al., 2016). Additionally, qPCR has also been used to assess the expression of MMPs and collagen, which are crucial for the remodeling of the ECM during the wound healing process (Lindley et al., 2016). Pastar et al. (2012) explored the gene expression of keratin (KR) (KRT6, KRT16, TGFB1, and IL6 in the context of acute and chronic wounds. Their results demonstrated that the upregulation of keratin genes correlated with active healing, while chronic wounds exhibited a lack of keratinocyte activation, which negatively affected re-epithelialization (Pastar et al., 2012). Salo et al. (1994) investigated the expression of the MMP9 gene in healing compared to non-healing wounds. The results indicated that chronic ulcers showed increased levels of MMP9, which were associated with significant ECM degradation and unfavorable healing results.

3.5.2.4 Mass Spectrometry (LC–MS/MS) and Proteomics

Mass spectrometry, particularly in the form of LC–MS/MS, has established itself as a robust tool in the field of proteomics, enabling comprehensive, high-throughput, and unbiased identification and quantification of proteins involved in the process of wound healing. In contrast to approaches such as ELISA or Western blotting that are designed to detect specific proteins, LC–MS/MS facilitates a global profiling of proteins sourced from wound tissues, fluids, or exudates, which supports the discovery of new biomarkers and mechanisms. Using LC–MS/MS, more than 200 proteins were identified with chronic wounds exhibiting elevated levels of proteases (MMPs), inflammatory markers (S100A8/A9), and stress proteins (Eming et al., 2010). In another study by Kroeze et al. (2012)

employed LC–MS/MS to profile exudates from chronic venous ulcers, revealing a protease imbalance (↑ MMP9, ↓ TIMP1), oxidative stress proteins, and altered cytokine-binding proteins (Kroeze et al., 2012). Identified angiogenic proteins that are dysregulated include VEGFR1, angiopoietins, and thrombospondins (Ridiandries et al., 2018). It has been confirmed that impaired angiogenesis is a characteristic feature of diabetic wounds (Ridiandries et al., 2018). A research study identified crucial proteins associated with oxidative stress, such as catalase, peroxidases, and superoxide dismutase (SOD), as well as those involved in immune modulation, including complement proteins, S100 proteins, and annexins (Bekeschus et al., 2018). Additionally, it emphasized the dynamics driven by neutrophils during the early stages of wound healing (Bekeschus et al., 2018). Wound fluid is ideal for peptidomic analysis as it is closely associated with the wound environment and is easily accessible. A peptidomic strategy, supported by mass spectrometry and bioinformatics, may offer innovative methods for discovering new peptide biomarkers related to wound healing and infection evaluation (Hartman et al., 2021). Antimicrobial peptides derived from hemoglobin and fibrinogen were identified as being enriched in infected wound fluids (Hartman et al., 2021).

LC–MS/MS and proteomics (which encompasses peptidomics and spatial lipidomics) deliver an extensive molecular perspective on wound healing—documenting oxidative, inflammatory, ECM, and antimicrobial alterations. These results present potential biomarker candidates for diagnostics, therapeutic targeting, and phase-specific intervention strategies.

3.5.2.5 Flow Cytometry

Flow cytometry serves as a potent analytical tool that can be employed to evaluate biomarkers related to wound healing, particularly in the investigation of immune cell populations, cytokine expression, and cellular activities in the wound microenvironment. This method enables the concurrent assessment of various parameters at the single-cell level, making it particularly suitable for exploring the dynamic and diverse cellular responses that take place during the different stages of wound healing. By employing specific fluorescently labeled antibodies, scientists can detect and measure immune cells such as macrophages, neutrophils, T cells, and dendritic cells, while also evaluating their activation states and phenotypic alterations over time. Additionally, intracellular cytokine staining and surface marker analysis offer valuable insights into the inflammatory environment and the transition from inflammation to tissue remodeling. Consequently, flow cytometry provides a thorough approach to unraveling the intricate cellular interactions and molecular signals that are essential for effective wound repair.

Flow cytometry facilitates the identification and quantification of various subsets of immune cells, including neutrophils ($CD11b^+$, $Ly6G^+$), macrophages ($F4/80^+$, $CD11b^+$, CD206 for M2 phenotype), and T cells ($CD3^+$, $CD4^+$, $CD8^+$). These immune cells are vital during the inflammatory and proliferative phases of wound healing. Willenborg and Eming (2014) reported that different macrophage phenotypes can modulate wound healing responses by employing flow cytometry to track $F4/80^+$ macrophages in murine models. A research investigation examined the influence of macrophage polarization on the healing of skin wounds (Gao et al., 2023). The study involved isolating bone marrow-derived macrophages (BMDMs) and polarizing them into two distinct phenotypes, M1 and M2, *in vitro* (Gao et al., 2023). These polarized macrophages were then applied to rat skin wounds. Utilizing flow cytometry with $CD45^+CD68^+$ markers and cytokine assays measuring IL-12 and IL-10, the findings revealed that M2 macrophages promote healing, whereas M1 macrophages hinder it (Gao et al., 2023).

Cytokine capture assays, or intracellular cytokine staining, performed through flow cytometry, are capable of assessing crucial mediators of wound healing such as IL-1β, IL-6, TNF-α (pro-inflammatory), IL-10, TGF-β (anti-inflammatory/regulatory), and VEGF, PDGF, and FGF (angiogenesis and tissue regeneration). A study by Lee et al. (2016) used flow cytometry to measure the levels of intracellular IL-10 and TNF-α in macrophages obtained from healing wounds. A study was conducted to thoroughly characterize myeloid cells involved in wound healing in both healthy and healing-impaired diabetic mice. Multi-parametric flow cytometry was utilized to monitor neutrophils, macrophages, dendritic cells, eosinophils, and Langerhans cells at various time points: days 1, 3, 5, 7, 10, and 15 following injury in mice (Joshi et al., 2020). The results indicated a prolonged presence of neutrophils and macrophages in diabetic wounds, underscoring the impaired resolution of inflammation under diabetic conditions (Joshi et al., 2020). Stem cells, such as mesenchymal stem cells (MSCs), are crucial for the process of wound regeneration. Flow cytometry is utilized to characterize these cells based on their surface markers ($CD105^+$, $CD73^+$, $CD90^+$, $CD45^-$, $CD34^-$) and to assess their states of differentiation or activation. Rodrigues et al. (2019) examined MSCs derived from adipose tissue for their potential in wound healing, employing flow cytometry to analyze stem cell surface markers.

To conclude, flow cytometry enriches our insight into the immunological framework of wound healing while also aiding in the formulation of targeted therapies that seek to enhance repair results in both acute and chronic wounds. This technology is vital for developing effective treatment options.

3.5.2.6 Immunohistochemistry (IHC)

IHC serves as a powerful and frequently employed technique for analyzing biomarkers linked to wound healing. This method enables the localization and visualization of distinct proteins within tissue sections, delivering both spatial and temporal context that is vital for grasping the complex cellular dynamics that take place during the wound healing process. IHC provides insights into the specific locations of protein expression (biomarkers) in the wound tissue whether found in the epidermal layers, dermal fibroblasts, blood vessels, or the immune cells that have infiltrated the tissue. Such spatial information is essential for evaluating tissue regeneration and inflammation. IHC assists in identifying and mapping:

- Immune cells (such as macrophages, neutrophils, and T cells).
- Cytokines and growth factors (including IL-6, TNF-α, VEGF, and TGF-β).
- Activity of fibroblasts and myofibroblasts (for example, α-SMA and Collagen I/III).
- Markers of angiogenesis (like CD31 and VEGF).
- Enzymes involved in ECM remodeling (such as MMPs and TIMPs).

A research study was conducted on the immunohistochemical characterization of macrophages involved in skin wound healing in mice (Sindrilaru et al., 2011). The results indicated that M1 macrophages are predominant during the early stages of inflammation (days 1–3), while M2 macrophages show an increase during the tissue regeneration phase (days 7–14) (Sindrilaru et al., 2011). Another study explored the rise in $CD40^+$ fibroblasts associated with hindered wound healing in DFUs (Littig et al., 2022). The results showed that chronic wounds in diabetes are characterized by a sustained presence of M1 macrophages and heightened pro-inflammatory cytokine levels (Littig et al., 2022). A study revealed significantly elevated levels of MMP-1 and MMP-8 in patients suffering from infected ulcers, in contrast to those with uninfected ulcers, which exhibited increased levels of MMP-2 and MMP-9 (Serra et al., 2016). Additionally, it was discovered that patients with infected ulcers had higher concentrations of interleukin (IL)-1, IL-6, IL-8, vascular endothelial growth factor, and tumor necrosis factor-alpha compared to their uninfected counterparts, indicating a potential link between infection, MMP activation, cytokine release, and associated symptoms (Serra et al., 2016). An investigation was conducted on the expression of MMPs in tissue samples taken from chronic dermal wounds induced by radiation through IHC (Gajendrareddy et al., 2013). The findings from the immunohistochemical study indicated that there was an elevated expression of MMP protein in the tissue samples from radiation-induced chronic dermal wounds relative to normal human skin (Gajendrareddy et al., 2013). Determining the age of wounds is a vital component of forensic investigations, providing essential insights into the timing of traumatic incidents. Biomarkers such as VEGF, α-SMA, TNF-α, and TGF-β1 have shown potential in assisting with wound age assessment. A study focused on histopathological analysis uncovered distinct alterations in skin tissue over time, from ulceration to re-epithelialization and tissue regeneration (Soliman et al., 2024).

Immunohistochemical analysis revealed varying levels of VEGF and α-SMA expression, with significant increases observed in the later stages following wounding (Soliman et al., 2024). Molecular analysis demonstrated a notable upregulation of TNF-α and TGF-β1 (Soliman et al., 2024).

Through the in situ mapping of these biomarkers, IHC enriches our knowledge of why wounds heal correctly or fail to do so, providing key insights for the formulation of targeted therapies and diagnostics for both acute and chronic wounds.

3.5.3 Emerging and Specialized Techniques

Wound healing continues to pose a major clinical challenge, particularly in cases of chronic, diabetic, and PUs. While conventional techniques depend on debridement, moisture management, and infection prevention, innovative and developing strategies are increasingly concentrating on bioengineering, molecular modulation, advanced materials, and regenerative medicine to enhance and expedite wound healing results.

3.5.3.1 3D Wound Imaging with Integrated Fluid Sensors

The combination of three-dimensional (3D) wound imaging and fluid sensing technology signifies a groundbreaking development in digital wound monitoring, facilitating accurate, non-invasive, and real-time evaluation of wound dimensions, depth, tissue composition, and biochemical indicators. This integrated method enhances the clinical assessment of wound healing progress and supports tailored, data-informed interventions, particularly for chronic and non-healing wounds. 3D wound imaging systems employ photogrammetry, structured light scanning, or stereo vision to create precise topographical representations of wound surfaces. The advantages of this approach include the following (Kim et al., 2018):

- Facilitates accurate volumetric assessments (area, depth, volume).
- Monitors alterations in wound morphology over time.
- Enhances documentation and minimizes inter-observer variability, and
- Can be incorporated into telemedicine systems.

A prospective pilot study focused on DFUs, utilizing 3D imaging to monitor wound volume revealed a significant correlation ($R > 0.70$, $p = 0.0001$) with the healing process (Malone et al., 2020). This approach exhibited a high intra-class correlation, suggesting dependable and consistent measurements in a clinical environment. Jiang et al. (2023) designed a wireless, flexible bandage that incorporates real-time temperature and impedance monitoring with electrical stimulation. In mouse model studies, the treatment led to roughly 25% faster wound closure, increased vascularization, and improved dermal remodeling. An activation of regenerative genes in immune cells was observed (Jiang et al., 2023). In 2024, a team at Tsinghua University created a dressing that merges unidirectional fluid drainage with a 3D polyaniline-based pH sensor (Wang et al., 2024). Evaluated in rat models and ex vivo skin, it exhibited high sensitivity (61.5 mV/pH) and stability (only 4.8% signal loss over 28 days), facilitating ongoing pH monitoring in conjunction with exudate management (Wang et al., 2024). Khalid et al. (2020) developed electrospun membranes made from silk fibroin and nanodiamond that possess optical thermometry features. In murine wounds, these dressings exhibited remarkable biocompatibility, facilitated normal healing, and were able to non-invasively track temperature variations linked to infection or inflammation (Khalid et al., 2020). Wang et al. (2025) created a wearable device known as iCares, aimed at optimizing the collection of slowly released wound fluid while also sensing reactive oxygen and nitrogen species, pH, and temperature. In experiments with diabetic mice, the sensors successfully monitored changes in these parameters during skin wound infections, which were treated effectively with antibiotics (Wang et al., 2025). In a cohort of 20 human patients with chronic wounds, the data from the iCares multi-parameter sensor were combined with a machine learning algorithm to evaluate wound severity and healing potential (Wang et al., 2025). Advancing this in situ analysis of wound exudate could lead to an objective assessment of wound status, thereby guiding treatment choices for chronic wounds.

Research studies indicate that integrated systems enhance diagnostic precision and improve wound healing results, while also facilitating remote monitoring, which is particularly beneficial for managing chronic wounds in diabetic, elderly, and immobile patients. As wearable electronics, machine learning, and miniaturized biosensors continue to advance, future wound dressings are expected to transform into fully autonomous platforms that can monitor, diagnose, and even treat wounds in real time. Although clinical adoption is still in its early stages, the encouraging results from both preclinical and initial human studies highlight the potential of this technology to revolutionize modern wound management transitioning from passive observation to proactive, data-driven intervention.

3.5.3.2 Hydrogels

Hydrogels are made up of 3D, water-attracting polymer networks that can hold significant amounts of water or biological fluids. In the context of wound healing, hydrogel-based systems not only provide a moist and biocompatible environment but also serve as adaptable platforms for targeted drug delivery and biomarker monitoring. Their adjustable mechanical properties, porosity, and bio-functionality render them ideal for therapeutic and diagnostic applications particularly in monitoring crucial wound healing biomarkers like cytokines, MMPs, growth factors, and microbial indicators. Tamayol et al. (2016) created a microfluidic hydrogel system that can sense pH levels and inflammatory markers to determine infection status. These sensors are capable of detecting biomarkers such as IL-6, TNF-α, MMP9, or VEGF in wound exudate and have the ability to change color, exhibit fluorescence, or generate signals that are associated with healing status or infection (Tamayol et al., 2016). Trengove et al. (1999) utilized hydrogel-absorbed wound fluid to examine cytokine and protease levels, revealing unique profiles between acute and chronic wounds. Mostafalu et al. (2018) created an innovative hydrogel bandage that includes temperature and pH sensors for the real-time monitoring of wounds and automated drug delivery. The bandage combines a stimulus-responsive hydrogel with a thermosensitive drug carrier and a flexible heater managed by a microcontroller (Mostafalu et al., 2018). This setup enables personalized treatment by modifying drug release protocols according to sensor data (Mostafalu et al., 2018).

Hydrogel-based drug delivery systems have emerged as versatile platforms in wound healing, supporting tissue regeneration while also facilitating the real-time monitoring and assessment of essential biomarkers. Their biocompatibility, moisture-retaining properties, and adjustable chemistry render them perfectly suited for interaction with the intricate wound microenvironment. By integrating biosensing features and stimuli-responsive drug release mechanisms, hydrogel dressings can monitor changes in inflammatory markers (such as IL-6, TNF-α), enzymes (like MMP-9), pH levels, and oxidative stress, responding with the precise release of therapeutics. This responsive feedback system transforms conventional wound dressings into smart therapeutic solutions that actively engage in both diagnostics and treatment. While challenges such as sensor stability, cost, and regulatory approval persist, the fusion of hydrogel technology with biomarker science signifies a promising advancement toward personalized, precision wound care in the future.

3.6 CHALLENGES

The application of biomarkers in wound healing presents significant potential; however, various substantial challenges currently hinder their extensive clinical implementation. These obstacles encompass biological, methodological, and translational aspects. Wounds are fundamentally diverse, differing by type (such as DFUs, PUs, VLUs, and burns), stage (acute versus chronic), depth, microbial load, and host factors (including age, comorbidities, nutrition, and vascular supply). This biological intricacy complicates the identification of universal biomarker thresholds or signatures that can be widely utilized. A biomarker that is highly indicative for one type of wound may hold no relevance for another. Therefore, variations specific to the patient and the wound must be considered when developing and interpreting diagnostics based on biomarkers. The techniques for gathering wound fluid, tissue biopsies, or exudate for biomarker analysis have not yet been standardized. Variations in collection methods (such as swab versus aspiration), timing, storage conditions, and sample volumes can result in considerable variability in outcomes. Furthermore, some markers may deteriorate rapidly or adhere to dressings, jeopardizing their integrity prior to analysis. In the absence of standardization, comparisons across studies or clinical environments become unreliable. The wound healing process is dynamic, moving through inflammatory, proliferative, and remodeling phases. Biomarker levels naturally vary during these stages, and their interpretation necessitates a time-sensitive context. Moreover, the wound bed functions as an ecosystem where cytokines, enzymes, immune cells, microbial species, and signaling molecules interact in intricate ways. This interdependence complicates the identification of a single "marker" that accurately represents the wound state. Microbial biomarkers and markers of host response are frequently found at elevated levels in both colonized and infected wounds, complicating the differentiation between harmless colonization and clinically relevant infection. For example, bacterial DNA can be identified through sequencing methods without necessarily indicating an active infection. This diagnostic uncertainty may result in the excessive use of antibiotics or insufficient treatment of infections associated with biofilms. Numerous biomarkers that demonstrate significant predictive or diagnostic potential in research settings have yet to be validated through extensive clinical trials. Regulatory challenges, concerns regarding cost-effectiveness, and the absence of integration into current wound care practices further hinder their adoption in routine clinical use. Moreover, clinicians often do not have access to specialized tools or the necessary training to accurately interpret molecular diagnostic information, which restricts its practical application.

3.7 FUTURE DIRECTIONS

In spite of these obstacles, the domain of wound biomarkers is progressing swiftly. New technologies and collaborative methods present solutions to existing challenges. Miniaturized, real-time biosensors that are incorporated into smart dressings are being created to track pH, temperature, oxygen levels, and specific biomarkers like MMPs and IL-6. These advancements can deliver immediate feedback at the bedside, assisting clinicians in dynamically modifying treatment plans. The future of wound biomarker research is rooted in systems biology. By integrating genomic, transcriptomic, proteomic, and metabolomic data, it is possible to create more detailed molecular profiles of wound healing and infection. This could pave the way for personalized wound care strategies tailored to the molecular characteristics of each wound. The use of machine learning algorithms is on the rise for the analysis of complex biomarker datasets and wound images. Artificial intelligence can aid in pattern recognition and predicting healing outcomes, transforming large biological datasets into actionable clinical insights. It is essential for academic researchers, clinicians, and regulatory bodies to collaborate in order to design and implement strong clinical trials for validating biomarkers. Proving clinical utility, safety, and cost-effectiveness will be essential for obtaining regulatory approval and achieving widespread adoption. As biomarker-driven tools advance, they will establish the basis for personalized wound care—where treatment choices are made not only on the visible characteristics of wounds but also on biological markers that indicate inflammation, infection, perfusion, and healing capacity.

3.8 CONCLUSION

Biomarkers are now recognized as significant tools for grasping, observing, and managing the complex and fluid process of wound healing. These molecular indicators—covering pro-inflammatory cytokines (e.g., IL-6, TNF-α), growth factors (e.g., VEGF, PDGF), proteolytic enzymes (e.g., MMP-2, MMP-9), and oxidative stress markers (e.g., ROS, H_2O_2)—deliver essential insights into the physiological state of a wound during its healing phases: inflammation, proliferation, and remodeling. Recent developments in biosensor technology, hydrogel platforms, nanomaterials, and 3D imaging systems have significantly improved our capacity to detect and analyze these biomarkers in a minimally invasive, real-time, and clinically relevant way. These advancements are transforming wound management from a reactive model to a predictive, personalized, and data-driven strategy. Nevertheless, to fully leverage the clinical potential of wound healing biomarkers, issues such as the standardization of biomarker panels, sampling techniques, and clinical validation need to be resolved. With continuous translational research and the incorporation of smart technologies, biomarker-based wound assessment offers great potential for enhancing outcomes in chronic wounds, diabetic ulcers, burns, and surgical recovery.

GLOSSARY

Biomarker: An observable measure of a biological process, condition, or disease that is employed to determine wound status, progression, or treatment efficacy.

Cytokines: Small proteins that facilitate and control immune responses; often utilized as indicators of inflammation in wounds.

Indicative Biomarker: Represents the current physiological condition of the wound, signaling active processes such as inflammation, hypoxia, or tissue injury.

Matrix Metalloproteinases: A group of enzymes that play a role in the degradation of the extracellular matrix; increased levels (for instance, MMP-9) are associated with non-healing or chronic wounds.

MicroRNAs: Non-coding RNAs that modulate gene expression after transcription; they are emerging as biomarkers for assessing wound healing and responses to regenerative therapies.

Predictive Biomarker: A biomarker that predicts how a wound is likely to respond to a certain therapeutic treatment.

Wound Exudate: The fluid generated by wounds, which includes cells, proteins, and biochemical markers; it is frequently examined for its biomarker content.

Wound Microbiota: The collection of microorganisms found in a wound; its makeup (determined through DNA/RNA analysis) is utilized to detect infections or biofilms.

REFERENCES

Asada, Mayumi, Gojiro Nakagami, Takeo Minematsu, Takashi Nagase, Tomoko Akase, Lijuan Huang, Kotaro Yoshimura, and Hiromi Sanada. "Novel biomarkers for the detection of wound infection by wound fluid RT-PCR in rats." *Experimental Dermatology* 21, no. 2 (2012): 118–122.

Asahara, Takayuki, Tomono Takahashi, Haruchika Masuda, Christoph Kalka, Donghui Chen, Hideki Iwaguro, Yoko Inai, Marcy Silver, and Jeffrey M. Isner. "VEGF contributes to postnatal neovascularization by mobilizing bone marrow-derived endothelial progenitor cells." *The EMBO Journal* 18, no. 14 (1999): 3964–3972.

Barrientos, Stephan, Olivera Stojadinovic, Michael S. Golinko, Harold Brem, and Marjana Tomic-Canic. "Growth factors and cytokines in wound healing." *Wound Repair and Regeneration* 16, no. 5 (2008): 585–601.

Beard , Christopher J., Ponnarasu, Sivaraj, and Schmieder, Gregory J. Excisional biopsy. In: StatPearls [Internet]. Treasure Island (FL): StatPearls Publishing; 2018. Available from: https://www.ncbi.nlm.nih.gov/books/NBK534835/

Bekeschus, Sander, Jan-Wilm Lackmann, Denis Gümbel, Matthias Napp, Anke Schmidt, and Kristian Wende. "A neutrophil proteomic signature in surgical trauma wounds." *International Journal of Molecular Sciences* 19, no. 3 (2018): 761.

Carvalho, Brendan, Harry J. Lemmens, Vicki Ting, and Martin S. Angst. "Postoperative subcutaneous instillation of low-dose ketorolac but not hydromorphone reduces wound exudate concentrations of interleukin-6 and interleukin-10 and improves analgesia following cesarean delivery." *The Journal of Pain* 14, no. 1 (2013): 48–56.

Collier, Mark. "What is, how to manage, and what are the effects of wound exudate? Introducing the 3M exudate pathway." *British Journal of Community Nursing* 28, no. Sup6 (2023): S27–S38.

Cutting, Keith F. "Wound exudate: Composition and functions." *British Journal of Community Nursing* 8, no. Sup3 (2003): S4–S9.

Dowsett, Caroline, and Elizabeth Ayello. "TIME principles of chronic wound bed preparation and treatment." *British Journal of Nursing* 13, no. Sup3 (2004): S16–S23.

Eming, Sabine A., Manuel Koch, Andreas Krieger, Bent Brachvogel, Sandra Kreft, Leena Bruckner-Tuderman, Thomas Krieg, John D. Shannon, and Jay W. Fox. "Differential proteomic analysis distinguishes tissue repair biomarker signatures in wound exudates obtained from normal healing and chronic wounds." *Journal of Proteome Research* 9, no. 9 (2010): 4758–4766.

Eming, Sabine A., Thomas Krieg, and Jeffrey M. Davidson. "Inflammation in wound repair: Molecular and cellular mechanisms." *Journal of Investigative Dermatology* 127, no. 3 (2007): 514–525.

Eming SA, Martin P, Tomic-Canic M. Wound repair and regeneration: Mechanisms, signaling, and translation. *Science Translation Medicine* 6, no. 265 (2014): 265sr6.

Fonder, Margaret A., Gerald S. Lazarus, David A. Cowan, Barbara Aronson-Cook, Angela R. Kohli, and Adam J. Mamelak. "Treating the chronic wound: A practical approach to the care of nonhealing wounds and wound care dressings." *Journal of the American Academy of Dermatology* 58, no. 2 (2008): 185–206.

Gajendrareddy, Praveen K., Christopher G. Engeland, Roger Junges, Michael P. Horan, Isolde G. Rojas, and Phillip T. Marucha. "MMP-8 overexpression and persistence of neutrophils relate to stress-impaired healing and poor collagen architecture in mice." *Brain, Behavior, and Immunity* 28 (2013): 44–48.

Gao, Xuejun, Chao Lu, Yuanxin Miao, Jizhen Ren, and Xia Cai. "Role of macrophage polarisation in skin wound healing." *International Wound Journal* 20, no. 7 (2023): 2551–2562.

Gethin, Georgina. "The significance of surface pH in chronic wounds." *Wounds UK* 3, no. 3 (2007): 52.

Goto, Taichi, Nao Tamai, Gojiro Nakagami, Aya Kitamura, Ayumi Naito, Masayuki Hirokawa, Chisako Shimokawa, Kazuo Takahashi, Junichi Umemoto, and Hiromi Sanada. "Can wound exudate from venous leg ulcers measure wound pain status?: A pilot study." *PloS One* 11, no. 12 (2016): e0167478.

Harding , Keith G., Morris, Hilary L., and Patel, Gopal K. "Healing chronic wounds." *BMJ* 324, no 7330 (2002): 160–163.

Hartman, Erik, Karl Wallblom, Mariena JA Van der Plas, Jitka Petrlova, Jun Cai, Karim Saleh, Sven Kjellström, and Artur Schmidtchen. "Bioinformatic analysis of the wound peptidome reveals potential biomarkers and antimicrobial peptides." *Frontiers in Immunology* 11 (2021): 620707.

Harvey, Joe, Kieran T. Mellody, Nicky Cullum, Rachel EB Watson, and Jo Dumville. "Wound fluid sampling methods for proteomic studies: A scoping review." *Wound Repair and Regeneration* 30, no. 3 (2022): 317–333.

Hasan, Muhammed Y., Rachel Teo, and Aziz Nather. "Negative-pressure wound therapy for management of diabetic foot wounds: A review of the mechanism of action, clinical applications, and recent developments." *Diabetic Foot & Ankle* 6, no. 1 (2015): 27618.

Hwang, Jangsun, Youngmin Seo, Daun Jeong, Xiaoyu Ning, Christian Wiraja, Lixia Yang, Chew Teng Tan et al. "Monitoring wound healing with topically applied optical NanoFlare mRNA nanosensors." *Advanced Science* 9, no. 18 (2022): 2104835.

Iizaka, Shinji, Hiromi Sanada, Takeo Minematsu, Miho Oba, Gojiro Nakagami, Hiroe Koyanagi, Takashi Nagase, Chizuko Konya, and Junko Sugama. "Do nutritional markers in wound fluid reflect pressure ulcer status." *Wound Repair and Regeneration* 18, no. 1 (2010): 31–37.

James, Garth A., Ellen Swogger, Randall Wolcott, Elinor deLancey Pulcini, Patrick Secor, Jennifer Sestrich, John W. Costerton, and Philip S. Stewart. "Biofilms in chronic wounds." *Wound Repair and Regeneration* 16, no. 1 (2008): 37–44.

Jiang, Fuchuan, Sajjad Ahmad, Sadia Kanwal, Yasir Hameed, and Qian Tang. "Key wound healing genes as diagnostic biomarkers and therapeutic targets in uterine corpus endometrial carcinoma: An integrated in silico and in vitro study." *Hereditas* 162, no. 1 (2025): 5.

Jiang, Yuanwen, Artem A. Trotsyuk, Simiao Niu, Dominic Henn, Kellen Chen, Chien-Chung Shih, Madelyn R. Larson et al. "Wireless, closed-loop, smart bandage with integrated sensors and stimulators for advanced wound care and accelerated healing." *Nature Biotechnology* 41, no. 5 (2023): 652–662.

Jones, Vanessa, Joseph E. Grey, and Keith G. Harding. "Wound dressings." *BMJ* 332, no. 7544 (2006): 777–780.

Joshi, Natasha, Lea Pohlmeier, Maya Ben-Yehuda Greenwald, Eric Haertel, Paul Hiebert, Manfred Kopf, and Sabine Werner. "Comprehensive characterization of myeloid cells during wound healing in healthy and healing-impaired diabetic mice." *European Journal of Immunology* 50, no. 9 (2020): 1335–1349.

Jun, Joon-Il, and Lester F. Lau. "The matricellular protein CCN1 induces fibroblast senescence and restricts fibrosis in cutaneous wound healing." *Nature Cell Biology* 12, no. 7 (2010): 676–685.

Karley, Dugeshwar, Deepesh Gupta, and Archana Tiwari. "Biomarkers: The future of medical science to detect cancer." *Journal of Molecular Biomarkers & Diagnosis* 2, no. 5 (2011): 1–7.

Khalid, Asma, Dongbi Bai, Amanda N. Abraham, Amit Jadhav, Denver Linklater, Alex Matusica, Duy Nguyen et al. "Electrospun nanodiamond–silk fibroin membranes: A multifunctional platform for biosensing and wound-healing applications." *ACS Applied Materials & Interfaces* 12, no. 43 (2020): 48408–48419.

Kim, Byoung Soo, Yang Woo Kwon, Jeong-Sik Kong, Gyu Tae Park, Ge Gao, Wonil Han, Moon-Bum Kim, Hyungseok Lee, Jae Ho Kim, and Dong-Woo Cho. "3D cell printing of in vitro stabilized skin model and in vivo pre-vascularized skin patch using tissue-specific extracellular matrix bioink: A step towards advanced skin tissue engineering." *Biomaterials* 168 (2018): 38–53.

Kroeze, Kim L., Liselot Vink, Edith M. de Boer, Rik J. Scheper, Catherine van Montfrans, and Susan Gibbs. "Simple wound exudate collection method identifies bioactive cytokines and chemokines in (arterio) venous ulcers." *Wound Repair and Regeneration* 20, no. 3 (2012): 294–303.

Ladwig, George P., Robson, Martin C., Liu, Rong, Kuhn, Michael A., Muir, Donald F., and Schultz, Gregory S. "Ratios of activated matrix metalloproteinase-9 to tissue inhibitor of matrix metalloproteinase-1 in wound fluids are inversely correlated with healing of pressure ulcers." *Wound Repair Regen* 10 no. 1 (2002): 26–37.

Lazarus, Gerald S., Diane M. Cooper, David R. Knighton, David J. Margolis, Roger E. Percoraro, George Rodeheaver, and Martin C. Robson. "Definitions and guidelines for assessment of wounds and evaluation of healing." *Wound Repair and Regeneration* 2, no. 3 (1994): 165–170.

Leaper, David J., Gregory Schultz, Keryln Carville, Jacqueline Fletcher, Theresa Swanson, and Rebecca Drake. "Extending the TIME concept: What have we learned in the past 10 years." *International Wound Journal* 9 (2012): 1–19.

Lee, Dylan E., Nagi Ayoub, and Devendra K. Agrawal. "Mesenchymal stem cells and cutaneous wound healing: Novel methods to increase cell delivery and therapeutic efficacy." *Stem Cell Research & Therapy* 7(2016): 1–8.

Leveen, Herman H., Gerald Falk, Benjamin Borek, Carlos Diaz, Yetta Lynfield, Benjamin J. Wynkoop, George A. Mabunda, John L. Rubricius, and George C. Christoudias. "Chemical acidification of wounds. An adjuvant to healing and the unfavorable action of alkalinity and ammonia." *Annals of Surgery* 178 no. 6(1973): 745–753.

Lindley, Linsey E., Olivera Stojadinovic, Irena Pastar, and Marjana Tomic-Canic. "Biology and biomarkers for wound healing." *Plastic and Reconstructive Surgery* 138, no. 3S (2016): 18S–28S.

Littig, Joshua Patrick Bungalon, Rebecca Moellmer, Adrienne M. Estes, Devendra K. Agrawal, and Vikrant Rai. "Increased population of CD40+ fibroblasts is associated with impaired wound healing and chronic inflammation in diabetic foot ulcers." *Journal of Clinical Medicine* 11, no. 21 (2022): 6335.

Liu, Shao-Cheng, Oluwaseun Adebayo Bamodu, Kuang-Tai Kuo, Iat-Hang Fong, Chih-Cheng Lin, Chi-Tai Yeh, and Shyi-Gen Chen. "Adipose-derived stem cell induced-tissue repair or wound healing is mediated by the concomitant upregulation of miR-21 and miR-29b expression and activation of the AKT signaling pathway." *Archives of Biochemistry and Biophysics* 705 (2021): 108895.

Loots, Miriam AM, Evert N. Lamme, Jimmy Zeegelaar, Jan R. Mekkes, Jan D. Bos, and Esther Middelkoop. "Differences in cellular infiltrate and extracellular matrix of chronic diabetic and venous ulcers versus acute wounds." *Journal of Investigative Dermatology* 111, no. 5 (1998): 850–857.

Malone, Matthew, Saskia Schwarzer, Annie Walsh, Wei Xuan, Abdulaziz Al Gannass, Hugh G. Dickson, and Frank L. Bowling. "Monitoring wound progression to healing in diabetic foot ulcers using three-dimensional wound imaging." *Journal of Diabetes and Its Complications* 34, no. 2 (2020): 107471.

Martin, Paul. "Wound healing–aiming for perfect skin regeneration." *Science* 276, no. 5309 (1997): 75–81.

Masson-Meyers, Daniela S., Thiago AM Andrade, Guilherme F. Caetano, Francielle R. Guimaraes, Marcel N. Leite, Saulo N. Leite, and Marco Andrey C. Frade. "Experimental models and methods for cutaneous wound healing assessment." *International Journal of Experimental Pathology* 101, no. 1–2 (2020): 21–37.

Minematsu, Takeo, Gojiro Nakagami, Yuko Yamamoto, Toshiki Kanazawa, Lijuan Huang, Hiroe Koyanagi, Sanae Sasaki et al. "Wound blotting: A convenient biochemical assessment tool for protein components in exudate of chronic wounds." *Wound Repair and Regeneration* 21, no. 2 (2013): 329–334.

Mostafalu, Pooria, Ali Tamayol, Rahim Rahimi, Manuel Ochoa, Akbar Khalilpour, Gita Kiaee, Iman K. Yazdi et al. "Smart bandage for monitoring and treatment of chronic wounds." *Small* 14, no. 33 (2018): 1703509.

Pastar, Irena, Olivera Stojadinović, & Tomic-Canic, M. *"Molecular pathology of chronic wounds and the role of keratinocytes." Advances in Wound Care,* 1, no. 3 (2012): 105–112.

Pastar, Irena, Aron G. Nusbaum, Joel Gil, Shailee B. Patel, Juan Chen, Jose Valdes, Olivera Stojadinovic, Lisa R. Plano, Marjana Tomic-Canic, and Stephen C. Davis. "Interactions of methicillin resistant Staphylococcus aureus USA300 and *Pseudomonas aeruginosa* in polymicrobial wound infection." *PloS One* 8, no. 2 (2013): e56846.

Pastar, Irena, Olivera Stojadinovic, Natalie C. Yin, Horacio Ramirez, Aron G. Nusbaum, Andrew Sawaya, Shailee B. Patel, Laiqua Khalid, Rivkah R. Isseroff, and Marjana Tomic-Canic. "Epithelialization in wound healing: A comprehensive review." *Advances in Wound Care* 3, no. 7 (2014): 445–464.

Patel Sanjay, Anjali, Maheshwari, Chandra A. "Biomarkers for wound healing and their evaluation". *Journal of Wound Care* 25, no. 1 (2016): 46–55.

Ridiandries, Anisyah, Joanne TM Tan, and Christina A. Bursill. "The role of chemokines in wound healing." *International Journal of Molecular Sciences* 19, no. 10 (2018): 3217.

Rodrigues, Melanie, Nina Kosaric, Clark A. Bonham, and Geoffrey C. Gurtner. "Wound healing: A cellular perspective." *Physiological Reviews* 99 no. 1 (2019): 665–706.

Ruf, Marie-Thérèse, Arianna Andreoli, Gabriel Vujic, Peter Itin, Gerd Pluschke, and Peter Schmid. "Exudate collection using wound sponges—An easy, non-invasive and reliable method to explore protease activities in ulcers." *Wound Repair and Regeneration* 25, no. 2 (2017): 320–326.

Salo Timo, Marja, Makela, Kylmäniemi M, Autio-Harmainen H, Larjava H. "Expression of matrix metalloproteinase-2 and -9 during early human wound healing." *Laboratory Investigation; a Journal of Technical Methods and Pathology* 70, no. 2 (1994):176–182.

Schafer, Matthias, and Sabine Werner. "Oxidative stress in normal and impaired wound repair." *Pharmacological Research* 58, no. 2 (2008): 165–171.

Schintler, Michael V. Negative pressure therapy: Theory and practice. *Diabetes Metabolism Research and Reviews* 28 (Suppl 1) (2012): 72–77.

Schuetz, Philipp, Robert Birkhahn, Robert Sherwin, Alan E. Jones, Adam Singer, Jeffrey A. Kline, Michael S. Runyon et al. "Serial procalcitonin predicts mortality in severe sepsis patients: Results from the multicenter procalcitonin MOnitoring SEpsis (MOSES) study." *Critical Care Medicine* 45, no. 5 (2017): 781–789.

Schultz, Gregory S., R. Gary Sibbald, Vincent Falanga, Elizabeth A. Ayello, Caroline Dowsett, Keith Harding, Marco Romanelli, Michael C. Stacey, Luc Teot, and Wolfgang Vanscheidt. "Wound bed preparation: A systematic approach to wound management." *Wound Repair and Regeneration* 11(2003): S1–S28.

Serena, Thomas E., Breda M. Cullen, Simon W. Bayliff, Molly C. Gibson, Marissa J. Carter, Lingyun Chen, Raphael A. Yaakov et al. "Defining a new diagnostic assessment parameter for wound care: Elevated protease activity, an indicator of nonhealing, for targeted protease-modulating treatment." *Wound Repair and Regeneration* 24, no. 3 (2016): 589–595.

Serra, Raffaele, Raffaele Grande, Gianluca Buffone, Vincenzo Molinari, Paolo Perri, Aldina Perri, Bruno Amato, Manuela Colosimo, and Stefano de Franciscis. "Extracellular matrix assessment of infected chronic venous leg ulcers: Role of metalloproteinases and inflammatory cytokines." *International Wound Journal* 13, no. 1 (2016): 53–58.

Sibbald R. Gary, Diane, Williamson, Orsted HL, Campbell K, Keast D, Krasner D, Sibbald DS. "Preparing the wound bed—debridement, bacterial balance, and moisture balance." *Ostomy Wound Manage* 46 no. 11 (2000): 14–22.

Sibbald, R. Gary, Heather L. Orsted, Patricia M. Coutts, and David H. Keast. "Best practice recommendations for preparing the wound bed: Update 2006." *Advances in Skin & Wound Care* 20, no. 7 (2007): 390–405.

Sindrilaru, Anca, Thorsten Peters, Stefan Wieschalka, Corina Baican, Adrian Baican, Henriette Peter, Adelheid Hainzl et al. "An unrestrained proinflammatory M1 macrophage population induced by iron impairs wound healing in humans and mice." *The Journal of Clinical Investigation* 121, no. 3 (2011): 985–997.

Soliman, Asmaa Abd Al Aziz, Ola Gaber Haggag, Abeer Abd Elwahab Sharaf Eldin, Omnia Al-saied Abdullah, and Nagah Elsayed Mohammed Ali. "Role of biomarkers in determination of wound age: Histopathological, immunohistochemical and molecular study." *Zagazig Journal of Forensic Medicine and Toxicology* 22, no. 1 (2024): 177–204.

Stojadinovic, Olivera, Irena Pastar, Aron G. Nusbaum, Sasa Vukelic, Agata Krzyzanowska, and Marjana Tomic-Canic. "Deregulation of epidermal stem cell niche contributes to pathogenesis of nonhealing venous ulcers." *Wound Repair and Regeneration* 22, no. 2 (2014): 220–227.

Tamayol, Ali, Mohsen Akbari, Yael Zilberman, Mattia Comotto, Emal Lesha, Ludovic Serex, Sara Bagherifard et al. "Flexible pH-sensing hydrogel fibers for epidermal applications." *Advanced Healthcare Materials* 5, no. 6 (2016): 711–719.

Tandara, Andrea A., and Thomas A. Mustoe. "Oxygen in wound healing—More than a nutrient." *World Journal of Surgery* 28 (2004): 294–300.

Tarnuzzer, Roy W., and Gregory S. Schultz. "Biochemical analysis of acute and chronic wound environments." *Wound Repair and Regeneration* 4, no. 3 (1996): 321–325.

Thomas, Stephen. *Wound Management and Dressings*. Pharmaceutical Press, 1990.

Trengove, Naomi J., Michael C. Stacey, Shawn Macauley, Neil Bennett, Jane Gibson, Frank Burslem, Gillian Murphy, and Gregory Schultz. "Analysis of the acute and chronic wound environments: The role of proteases and their inhibitors." *Wound Repair and Regeneration* 7, no. 6 (1999): 442–452.

Uccioli, Luigi, Valentina Izzo, Marco Meloni, Erika Vainieri, Valeria Ruotolo, and Laura Giurato. "Non-healing foot ulcers in diabetic patients: General and local interfering conditions and management options with advanced wound dressings." *Journal of Wound Care* 24, no. Sup4b (2015): 35–42.

Van Asten, S. A. V., J. La Fontaine, E. J. G. Peters, K. Bhavan, P. J. Kim, and L. A. Lavery. "The microbiome of diabetic foot osteomyelitis." *European Journal of Clinical Microbiology & Infectious Diseases* 35 (2016): 293–298.

Vowden, K., and P. Vowden. "The role of exudate in the healing process: Understanding exudate management." *Trends in Wound Care* 3 (2004): 3–2.

Wang, Canran, Kexin Fan, Ehsan Shirzaei Sani, José A. Lasalde-Ramírez, Wenzheng Heng, Jihong Min, Samuel A. Solomon et al. "A microfluidic wearable device for wound exudate management and analysis in human chronic wounds." *Science Translational Medicine* 17, no. 795 (2025): eadt0882.

Wang, Xueqi, Jing Cheng, and Han Wang. "Chronic wound management: A liquid diode-based smart bandage with ultrasensitive pH sensing ability." *Microsystems & Nanoengineering* 10, no. 1 (2024): 1–15.

White, Richard J., and Keith F. Cutting. "Critical colonization-the concept under scrutiny." *Ostomy Wound Management* 52, no. 11 (2006): 50.

Wiegand, Cornelia, Jörg Tittelbach, Uta-Christina Hipler, and Peter Elsner. "Clinical efficacy of dressings for treatment of heavily exuding chronic wounds." *Chronic Wound Care Management and Research* 2, (2015): 101–111.

Willenborg, Sebastian, and Sabine A. Eming. "Macrophages–sensors and effectors coordinating skin damage and repair." *JDDG: Journal der Deutschen dermatologischen Gesellschaft* 12, no. 3 (2014): 214–221.

Wlaschek, Meinhard, and Karin Scharffetter-Kochanek. "Oxidative stress in chronic venous leg ulcers." *Wound Repair and Regeneration* 13, no. 5 (2005): 452–461.

World Health Organization (WHO). International Programme on Chemical Safety. *Biomarkers in Risk Assessment: Validity and Validation*. Geneva: World Health Organization; 2001.

Wysocki, Annette B., Alan O. Kusakabe, Stella Chang, and Tai-Lan Tuan. "Temporal expression of urokinase plasminogen activator, plasminogen activator inhibitor and gelatinase-B in chronic wound fluid switches from a chronic to acute wound profile with progression to healing." *Wound Repair and Regeneration* 7, no. 3 (1999): 154–165.

Yager, Dorne R., and Benedict C. Nwomeh. "The proteolytic environment of chronic wounds." *Wound Repair and Regeneration* 7, no. 6 (1999): 433–441.

Yager, Dorne R., Liang-Y. Zhang, Hui-Xiu Liang, Robert F. Diegelmann, and I. Kelman Cohen. "Wound fluids from human pressure ulcers contain elevated matrix metalloproteinase levels and activity compared to surgical wound fluids." *Journal of Investigative Dermatology* 107, no. 5 (1996): 743–748.

Yussof, Shah Jumaat Mohd, Effat Omar, Dinker R. Pai, and Suneet Sood. "Cellular events and biomarkers of wound healing." *Indian Journal of Plastic Surgery* 45, no. 02 (2012): 220–228.

Zhang, Nanwen, Qingmei Zheng, Yaduan Wang, Juan Lin, He Wang, Rui Liu, Mengru Yan, Xiaofeng Chen, Juhua Yang, and Xiaole Chen. "Renoprotective effect of the recombinant anti-IL-6R fusion proteins by inhibiting JAK2/STAT3 signaling pathway in diabetic nephropathy." *Frontiers in Pharmacology* 12 (2021): 681424.

Zhang, Yangyu, Lili Qin, Xiaobo Ma, Yueqi Wang, Yanhua Wu, and Jing Jiang. "Coexpression of matrix metalloproteinase-7 and tissue inhibitor of metalloproteinase-1 as a prognostic biomarker in gastric cancer." *Disease Markers* 2020, no. 1 (2020): 8831466.

4 Skin Microbiome

A Guide to Its Interplay with Wound Healing

Farheen Shafique[1*] *and Mahreen ul Hassan*[2]

[1] Department of Zoology, King Abdullah Campus, University of Azad Jammu and Kashmir Muzaffarabad, Pakistan

[2] Shaheed Benazir Bhutto Women University Peshawar, Pakistan

4.1 INTRODUCTION

The human body is home to a dynamic and diverse microbiome, which plays a vital role in maintaining the balance of the skin microbiota and defending against harmful microbes (Johnson et al., 2018; Zhu et al., 2023). The microbiome encompasses the total genetic material of microorganisms, which include bacteria, archaea, viruses, fungi, and protozoa, residing in a specific environment, particularly within and on the human body. This microbiome comprises not just the microbial communities but also their functional environment, which includes structural components, metabolites, and various molecules that are produced and altered by these microbes. The skin microbiota resides on the skin's surface and its appendages, creating a complex ecosystem that plays a role in maintaining skin homeostasis, educating the immune system, and defending against pathogens. As the body's largest organ and primary barrier to the external environment, the skin provides a unique habitat for this microbial population (Gallo, 2017; Zhu et al., 2023). The skin microbiota is integral to supporting skin health, particularly during wound healing, where its presence is crucial for recovery (Zhu et al., 2023; Zielinska et al., 2023). Wound healing is a complex, multi-stage process involving hemostasis, inflammation, proliferation, and remodeling (Zielinska et al., 2023). A variety of cell types, signaling molecules, and extracellular matrix (ECM) components work together to restore skin integrity (Mihai et al., 2024). While the microbiota can aid healing, its role can also hinder recovery under certain circumstances. For example, microbial populations that form biofilms can significantly delay healing by contributing to persistent inflammation and infection (Drago et al., 2019; Mihai et al., 2024; Tomic-Canic et al., 2020; Tuttle, 2015). Recent advances in molecular techniques, such as 16S rRNA gene sequencing and various OMIC technologies, have enhanced our ability to study the microbiome in wounds (Johnson et al., 2018; Mihai et al., 2024; Price et al., 2009). These methods have revealed the complexity of microbial ecosystems and their impact on clinical outcomes (Tomic-Canic et al., 2020). Understanding how the host and microbiome interact during wound healing is essential for developing more effective diagnostic and therapeutic strategies (Mihai et al., 2024; Tomic-Canic et al., 2020).

The exploration of skin microbiota has significantly progressed over the last century, influenced by improvements in microscopy, culture methods, and molecular biology (Santiago-Rodriguez et al., 2023). The documentation of microorganisms on human skin began in the late 19th and early 20th centuries, paralleling the emergence of fundamental microbiological techniques like light microscopy and agar-based cultures. Early microbiologists, including trailblazers such as Robert Koch and

* Corresponding author: farheen.shafiq@gmail.com

DOI: 10.1201/9781003644958-4

Louis Pasteur, established the foundation for recognizing microbes on human surfaces. By the early 1900s, skin swabs uncovered common cultivable bacterial genera like *Staphylococcus epidermidis* (*S. epidermidis*), *Micrococcus*, and *Corynebacterium*. These investigations revealed that the skin is not a sterile surface but a populated and site-specific microbial environment. Nevertheless, the extent of these discoveries was confined to microbes capable of growth in laboratory settings, mainly aerobic, fast-growing bacteria, resulting in a significant portion of microbial diversity remaining unobserved.

During the mid-to-late 20th century, scientists started to recognize the ecological characteristics of the skin microbiota. It became clear that the colonization by microbes was shaped by skin physiology factors such as oil (sebaceous), moisture, pH, and temperature, leading to unique microbial communities in different areas of the body. For instance, sebaceous regions like the forehead were found to contain significant populations of lipophilic bacteria such as *Cutibacterium acnes (C. acnes)*, whereas moist areas like the axilla exhibited a higher diversity, including *Corynebacterium* and *Staphylococcus hominis*. While these findings broadened the ecological understanding of skin microbiology, they were still limited by the constraints of culture-based techniques.

A significant change took place in the early 2000s with the introduction of culture-independent methods, especially 16S ribosomal RNA gene sequencing (Bottger, 1989; Kolbert and Persing, 1999). These innovations enabled researchers to identify and categorize bacteria based on their genetic information, eliminating the necessity for cultivation. The initiation of the Human Microbiome Project (HMP) in 2007 further propelled this evolution (Curtis et al., 2012). By employing high-throughput sequencing on skin samples from healthy subjects, scientists revealed a previously unrecognized variety of microbial communities. It became evident that the skin microbiome was not only more intricate but also highly personalized, with microbial composition shaped by intrinsic factors (age, sex, genetics) and extrinsic factors (climate, hygiene, cosmetics). Additionally, this molecular methodology facilitated the discovery of non-bacterial components within the skin ecosystem. Fungi, particularly Malassezia species, were found to prevail in oily areas. Deoxyribonucleic acid (DNA) viruses, including bacteriophages, as well as eukaryotic viruses, were also identified. Furthermore, the identification of microscopic mites such as Demodex species introduced new aspects to the exploration of skin-associated life.

Currently, the exploration of skin microbiota has progressed from mere characterization to an emphasis on its functions, interactions, and clinical significance. Research indicates that the microbiota plays a role in educating the skin's immune system, aiding in tissue repair, and competing against harmful organisms like *Staphylococcus aureus* (*S. aureus*) (Zhu et al., 2023). The disruption of this equilibrium, known as dysbiosis, is now linked to various dermatological conditions, such as acne, atopic dermatitis, psoriasis, and chronic wounds (CWs). Contemporary methods like metagenomics, metatranscriptomics, and metabolomics are employed to comprehend microbial functions, gene expression, and metabolite generation, providing fresh perspectives on host–microbe relationships. These developments have also ignited interest in therapies focused on the microbiome, such as topical probiotics, bacteriophage treatments, and microbiome transplants designed to restore microbial equilibrium and enhance skin health.

This chapter explores the composition of the skin and wound microbiomes, their impact on healing, the role of biofilms, and the potential for microbiome-based therapies.

4.2 SKIN MICROBIOME

The skin microbiome is vital and dynamic in the wound healing process, serving as both a protector and, in some cases, a possible barrier to recovery. This intricate microbial ecosystem, made up of commensal, symbiotic, and opportunistic microorganisms, closely interacts with the skin's immune responses, barrier functions, and repair mechanisms. Grasping the relationship between skin microbiota and wound healing is crucial for creating innovative therapeutic approaches to enhance tissue regeneration and avert CW development. Under typical physiological circumstances, the skin microbiota plays a role in maintaining epidermal homeostasis, instructs the innate immune system, and provides competition against pathogenic organisms via colonization resistance. The microbes that reside on the skin aid in barrier function, regulate inflammation, and facilitate tissue repair by stimulating keratinocyte proliferation and fostering immune tolerance.

4.2.1 Composition and Characteristics of the Skin Microbiome

The skin microbiome consists of a diverse and site-specific array of microorganisms that reside on the surface of human skin and its appendages. The skin is home to a variety of microorganisms, including bacteria, viruses (collectively known as the virome), fungi (the mycobiome), and other microbes (Gallo, 2017; Johnson et al., 2018; Zhu et al., 2023; Zielinska et al., 2023). The skin microbiome's composition is affected by various host factors, such as age, sex, genetics, and immunity, along with environmental influences like climate, hygiene, and occupation (Sanford and Gallo, 2013). Bacteria represent the most thoroughly researched components of the skin microbiome. Studies using techniques like 16S rRNA gene sequencing have identified at least 19 bacterial phyla and more than 1000 species on healthy skin. The most dominant bacterial phyla are Actinobacteria (51.8%), Firmicutes (24.4%), and Proteobacteria (16.5%), with key genera including *Corynebacteria*, *Cutibacteria* (formerly *Propionibacteria*), and *Staphylococci*. The skin microbiome composition differs across areas due to factors like moisture and lipid content. Sebaceous areas are dominated by *Propionibacterium* and *Staphylococci*, while moist areas often feature *Corynebacteria* and *Staphylococci*. Dry regions display more microbial diversity, with an increased presence of β-proteobacteria and other Gram-negative bacteria. Recent studies also show that deeper layers of the skin microbiome have higher levels of Proteobacteria and Actinobacteria, with lower levels of *Firmicutes*. Fungi play a vital, though less diverse, role in the skin microbiome. The most prevalent fungal genus on human skin is Malassezia, a lipid-dependent yeast that is mainly found in sebaceous-rich areas like the scalp, face, and trunk (Kalan and Grice, 2018; Zhu et al., 2023). Other genera, such as Candida and Aspergillus, can occur transiently or in conditions where the immune system is compromised. The skin's virome comprises eukaryotic viruses, including human papillomaviruses (HPVs) and polyomaviruses, as well as bacteriophages that are viruses that target skin-associated bacteria, notably *Staphylococcus* and *Propionibacterium*. It is believed that these phages play a role in regulating bacterial population dynamics and may affect skin immunity and disease development. The virome of the skin is primarily made up of bacteriophages, which impact microbial competition and interactions between hosts and pathogens (Hannigan et al., 2015).

Microbial communities differ based on their anatomical locations. They are affected by factors such as age, sex, and lifestyle (Byrd et al., 2018). Certain core taxa tend to remain consistent over time in healthy individuals, although changes can happen due to hygiene practices, antibiotic use, or illness. Commensal organisms like *S. epidermidis* offer antimicrobial defense and help regulate immunity, while some may turn opportunistic in immunocompromised situations (Byrd et al., 2018). The skin microbiome shows significant variability among individuals and across different body sites, yet it maintains temporal stability when disease is not present.

4.2.2 Microbiome and the Skin Health

The skin microbiome plays a crucial role in maintaining skin health and supporting physiological functions, including immune defense and tissue repair (Johnson et al., 2018; Tomic-Canic et al., 2020; Yang et al., 2024; Zhu et al., 2023). It collaborates with the skin's immune system to maintain the skin's barrier function, prevent pathogen invasion, and promote tissue repair (Sanford and Gallo, 2013).

Commensal microorganisms support skin health through several key mechanisms:

Barrier Function: The microbiome strengthens the skin's barrier by forming physical and chemical defenses, supporting the epithelial barrier, and modulating immune responses. Commensals cooperate with keratinocytes to maintain skin integrity and produce enzymes like sphingomyelinase, which break down sebum into free fatty acids (FFAs). These FFAs help maintain the skin's acidic pH, inhibiting pathogen growth (Linehan et al., 2018; Meisel et al., 2018; Nakatsuji et al., 2010).

Immune System Support: The microbiome regulates immune functions by influencing both innate and adaptive immune responses. Commensals engage with immune cells from birth, training T cells to recognize potential pathogens and prevent inflammation. *S. epidermidis* (*S. epidermidis*), for example, reduces inflammation via toll-like receptor 2 (TLR2) signaling and stimulates dendritic cells to produce anti-inflammatory cytokines (Laborel-Preneron et al., 2015; Nakatsuji et al., 2017).

Keratinocyte Interaction: Keratinocytes interact with microorganisms through pattern recognition receptors (PRRs), triggering innate immune responses and regulating inflammation. *S. epidermidis* stimulates keratinocytes to produce antimicrobial peptides (AMPs) such as hBD3 and RNase7, which protect the skin from infections and enhance the immune response (Antonsson et al., 2003; Linehan et al., 2018Radek, 2010).

Protection Against Pathogens: The skin microbiome protects against pathogens both indirectly (by outcompeting harmful microbes for space and resources) and directly (by producing antimicrobial compounds like bacteriocins and lipases) (Baldwin et al., 2017; Meisel et al., 2018). Keratinocytes also express AMPs, further strengthening the skin's defense (Midorikawa et al., 2003; Wanke et al., 2011). These AMPs, along with FFAs, form a protective skin barrier (Scharschmidt and Fischbach, 2013).

4.3 MICROBIOME ALTERATIONS IN ACUTE VS. CHRONIC WOUNDS

The microbiome in wounds differs significantly between acute and chronic stages, which impacts the healing process. Understanding these differences is essential for developing effective treatments and managing CW infections.

4.3.1 Microbiome in Acute Wounds

Acute wounds, such as cuts, burns, or surgical incisions, initially harbor a microbiome similar to that of healthy skin. These wounds typically experience a temporary reduction in microbial diversity, but the skin microbiota can recover as the wound heals. Acute wounds generally heal within a few weeks, and microbial populations gradually return to baseline, promoting tissue regeneration and immune function. However, certain acute wounds, like burns, may exhibit unique microbial profiles, often showing an increase in thermophilic microbes and a decrease in species such as *Corynebacterium*, which could raise infection risks. In the absence of complications, acute wounds tend to show microbial compositions akin to healthy skin and heal without significant microbial interference (Plichta et al., 2017).

4.3.2 Microbiome in Chronic Wounds

CWs, such as diabetic foot ulcers (DFUs) and pressure sores, are characterized by disrupted microbial communities that impede healing. These wounds typically harbor more complex microbial populations, with pathogenic species dominating and forming biofilms. The biofilm environment in CWs shields bacteria from immune clearance and antibiotic treatment, leading to prolonged infection and delayed healing.

Reduced Microbial Diversity: CWs often exhibit decreased microbial diversity, with specific pathogens like *S. aureus*, *Pseudomonas aeruginosa* (*P. aeruginosa*), and *Enterococcus faecalis* (*E. faecalis*) becoming predominant. These pathogens form biofilms, which further complicate treatment by protecting them from external threats, contributing to prolonged inflammation and infection (Gardiner et al., 2017).

Biofilm Formation: Biofilms are a hallmark of CWs, providing a protective environment for bacteria. These biofilms secrete exopolysaccharides that aid bacterial adhesion and shield them from antibiotics and immune defenses. Pathogens like *P. aeruginosa* and *S. aureus* within biofilms secrete factors that make them more resistant to treatment, thus complicating wound healing (Liu et al., 2020; Mihai et al., 2024).

Chronic Inflammation: In contrast to acute wounds, where inflammation initiates the healing process, CWs often experience prolonged low-grade inflammation. This sustained inflammatory state, driven by biofilm formation and pathogenic bacteria, impedes the wound's healing by preventing the resolution of inflammation and hindering tissue regeneration (Scales and Huffnagle, 2013).

4.3.3 Factors Influencing the Chronic Wound Microbiome

Several factors, including aging, excessive or inappropriate use of antibiotics, metabolic disorders like diabetes, autoimmune conditions, and immunosuppression, can significantly affect the wound microbiome (Tang et al., 2023; Uberoi et al., 2024). Recognizing the differences between the microbiomes of acute and CWs highlights the importance of tailored treatment strategies. Approaches such as microbiome modulation, including the use of probiotics, may help restore balance and improve healing outcomes in CWs (Han et al., 2019; Ring et al., 2022; Valdez et al., 2005; Zielinska et al., 2023).

4.4 SPECIFIC MICROBIAL SPECIES AND THEIR INFLUENCE ON WOUND HEALING

Microbial species play a critical role in both maintaining skin health and influencing wound healing. The skin is home to a variety of microbes, including bacteria, fungi, and viruses, which help defend against pathogens, regulate immune responses, and support tissue repair. However, the balance of these microorganisms can shift, especially in CWs, leading to impaired healing and increased infection risk. Below, we explore the roles of key microbial species in healthy skin and CWs.

4.4.1 Microbial Species in Healthy Skin and Host Defense

In healthy skin, commensal microorganisms help maintain homeostasis, prevent infection, and promote tissue regeneration. These beneficial microbes support the skin's barrier function and immune response:

S. epidermidis: *S. epidermidis*, a predominant skin commensal, is essential for skin defense. It produces AMPs and phenol-soluble modulins that inhibit pathogenic bacteria like *S. aureus*, thus preventing infections (Cogen et al., 2010a; Iwase et al., 2010). Additionally, *S. epidermidis* modulates immune responses by stimulating keratinocytes to produce more AMPs, which enhance the skin's immune defense.

C. acnes: Previously known as *Propionibacterium acnes*, *C. acnes* is a major component of the skin microbiome, particularly in sebaceous-rich areas. While overgrowth of *C. acnes* can contribute to acne, it generally supports skin health by metabolizing sebum into FFAs. These FFAs lower the skin's pH, creating an environment that inhibits the growth of harmful pathogens (Brown and Shalita, 1998; Youn et al., 2013).

Other Beneficial Bacteria: Species such as *Corynebacterium* and *Micrococcus* are common in healthy skin, helping to maintain the skin's acidic environment and prevent the colonization of harmful microbes. These bacteria also interact with the immune system to promote a balanced immune response, supporting the skin's overall barrier function (Gallo, 2017). In healthy skin, these microbes function symbiotically with host cells, especially keratinocytes, to protect the skin from infections, enhance immune responses, and maintain its structural integrity.

4.4.2 Microbial Species in Chronic Wounds

In CWs, such as DFUs or venous leg ulcers, the microbiome is often disrupted, which leads to delayed healing and an increased risk of infection. The microbial environment in these wounds differs significantly from healthy skin, often favoring pathogenic species:

S. aureus: This common pathogen in CWs, is notorious for its ability to form biofilms, which protect it from antibiotics and the immune system. In CWs, *S. aureus* produces toxins and proteases that damage host tissue and impede wound healing. The biofilm formed by *S. aureus* resists immune clearance, prolonging inflammation and delaying recovery (Kong et al., 2016; Tomic-Canic et al., 2020).

P. aeruginosa: Another major pathogen in CWs, *P. aeruginosa* is a Gram-negative bacterium that forms biofilms and secretes a variety of virulence factors, including exopolysaccharides and toxins. These factors promote biofilm formation, leading to persistent inflammation and tissue destruction. Although *P. aeruginosa* can stimulate angiogenesis (new blood vessel formation) in non-infected wounds, it generally delays healing in CWs, especially when biofilms are present (Arnold et al., 2016; Pastar et al., 2013).

E. faecalis: Commonly found in CWs, *E. faecalis* plays a significant role in polymicrobial infections. This bacterium is resistant to several antibiotics, including vancomycin (VRE), making it difficult to treat. In CWs, *E. faecalis* contributes to biofilm formation, persistent inflammation, and poor healing outcomes (Smith et al., 2016; Zhu et al., 2023). It also modulates the immune response, further hindering wound healing.

Other Pathogens in CWs: CWs are often polymicrobial, harboring a mix of pathogenic bacteria like Enterobacteriaceae, Corynebacterium, and Peptoniphilus. These species form biofilms that protect them from immune responses and antibiotic treatments, complicating the healing process. The presence of these pathogens is associated with prolonged inflammation, reduced oxygenation, and impaired tissue regeneration (Zielinska et al., 2023).

While many of the same microbial species are found in both healthy skin and CWs, their roles differ dramatically. In healthy skin, commensals like *S. epidermidis* and *C. acnes* support immune function and skin health by preventing pathogen colonization. In contrast, in CWs, these species may contribute to infection or biofilm formation, particularly when microbial balance is disrupted. Pathogens like *S. aureus* and *P. aeruginosa* thrive in the inflamed environment of CWs, hindering healing and prolonging infection. Biofilms in CWs further complicate treatment by shielding bacteria from immune clearance and antibiotics.

4.5 MECHANISMS OF MICROBIOME–HOST INTERACTION

The interaction between the skin microbiome and the host is complex and dynamic, influencing skin health and wound healing. The microbiome plays a critical role in immune modulation, barrier function, and inflammation regulation. Disruptions in these interactions, such as dysbiosis, can contribute to various skin conditions, including CWs, and impact healing processes.

4.5.1 Complexity of the Skin Microbiome

The relationship between the skin microbiome and its host is multifaceted, with profound effects on skin health and wound healing. Skin microbes interact with host immune cells and influence key physiological processes. A balanced microbiome supports immune function, prevents pathogen invasion, and promotes tissue repair, whereas dysbiosis leads to chronic inflammation and delayed healing in wounds (Mihai et al., 2024; Sachdeva et al., 2022).

4.5.2 The Skin as a Habitat and Homeostasis

The skin is a dynamic ecosystem that supports a wide variety of microorganisms, despite the challenging conditions it presents, such as low pH, high salinity, and dryness. These microbes help maintain the skin's barrier function, supporting the defense against pathogens. Commensal microorganisms, such as *S. epidermidis* and *C. acnes*, play vital roles in maintaining the skin's acidic pH, producing FFAs that discourage the growth of harmful bacteria and fungi (Flowers and Grice, 2020). The balance between the skin microbiome and the host is essential for healthy skin. Disruptions to this balance, known as dysbiosis, have been linked to skin diseases and CWs. Proper microbiome function ensures the skin's defense mechanisms remain intact, whereas microbial imbalances can increase susceptibility to infections and impair healing (Harris-Tryon and Grice, 2022).

4.5.3 Nature of Microbiome–Host Relationships

The relationship between skin microbes and their host can vary, ranging from commensalism to mutualism. Traditionally, microbes that were not directly associated with disease were categorized as commensals, coexisting peacefully with the host. However, emerging research suggests that many of these microbes actively benefit the host, making them mutualists (Cogen et al., 2010b). These microbes help maintain skin homeostasis and prevent pathogenic microbes from overwhelming the skin's defenses (Zheng et al., 2022).

4.5.4 Contributions to Host Defense and Barrier Function

The skin microbiome is integral to the skin's protective barrier, defending against external threats like pathogens and harmful chemicals. The acid mantle (a thin film of sweat and lipids on the skin) maintains the skin's pH which is slightly acidic in nature. The acidic pH is crucial for inhibiting pathogenic microbial growth. Commensal microbes like *S. epidermidis* and *C. acnes* contribute to this process by breaking down sebum into FFAs, further enhancing the skin's acidity and acting as a deterrent to harmful microorganisms (Byrd et al., 2018; Grice and Segre, 2011).

Additionally, the skin microbiome plays a role in competitive exclusion, a process where resident microbes prevent pathogenic species from establishing themselves on the skin. *S. epidermidis,* for instance, competes for space and resources, and produces AMPs that inhibit the growth of harmful microbes like *S. aureus*, a common skin pathogen. By maintaining this balance, the skin microbiome helps protect the host from infection and supports wound healing.

4.5.5 Modulation of the Immune System

Skin commensals play a significant role in modulating the immune system. They help educate immune cells, regulate skin inflammation, and maintain immune homeostasis. This interaction is crucial for both preventing excessive inflammation and promoting effective wound healing.

Immune Training: Commensals like *S. epidermidis* train immune cells, particularly T cells and macrophages, to recognize harmful pathogens and tolerate benign microbes. This helps prevent chronic inflammation, which can impair healing. For example, *S. epidermidis* induces anti-inflammatory cytokine production, reducing excessive inflammation in wounds (Johnson et al., 2018; Wanke et al., 2011).

Biofilm Influence on Immune Cells: In CWs, biofilms formed by *S. aureus* can alter immune cell behavior, shifting macrophages to an anti-inflammatory M2 phenotype. This impairs immune response and hinders wound healing. Conversely, beneficial commensals can support a more effective immune response, helping the body fight infections and repair tissue (Linehan et al., 2018; Tomic-Canic et al., 2020).

4.5.6 Microbe–Microbe and Metabolite Interactions

The skin microbiome functions as a network of microorganisms that interact with each other in complex ways. These interactions can be cooperative, competitive, or synergistic, influencing both the skin environment and the immune response.

Cooperative Interactions: Some microbial species, such as *S. epidermidis*, help maintain skin health by cooperating with other commensals. For instance, *S. epidermidis* can disrupt *S. aureus* biofilms and enhance immune responses by upregulating perforin-2, a protein that contributes to pathogen clearance (Iwase et al., 2010).

Pathogenic Synergies: In CWs, microbial co-infections can exacerbate inflammation and impair healing. For example, co-infection with *P. aeruginosa* and *S. aureus* leads to more severe tissue damage than when these bacteria are present individually (Alves et al., 2018; DeLeon et al., 2014). These interactions highlight the importance of managing the microbial community to improve healing outcomes.

Microbial Metabolites: Microbes produce a range of metabolites, including fatty acids, short-chain fatty acids (SCFAs), and other byproducts, which influence both microbial growth and host immunity. For example, *P. acnes* helps maintain the skin's acidic pH by metabolizing sebum into fatty acids, which support skin health and microbial balance (Brown and Shalita, 1998; Youn et al., 2013).

4.5.7 Influence of Host Factors

Host-related factors, such as age, genetics, and immune status, significantly influence the skin microbiome and its interactions with the host.

Genetic Factors: Genetic variants, such as those identified through microbiome-genome-wide association studies (mGWAS), affect microbiome composition and immune function. These genetic factors can influence an individual's susceptibility to skin conditions, such as eczema or CWs, by altering microbial interactions and immune responses (Harris-Tryon and Grice, 2022; Meneghin and Hogaboam, 2007).

Impact of Aging and Health Conditions: Aging, diabetes, and other health conditions can disrupt the balance of the microbiome, making the skin more prone to infections and impeding wound healing. The physiological state of the host, including immune competence, plays a crucial role in determining how the microbiome affects healing and inflammation (Harris-Tryon and Grice, 2022; Zielinska et al., 2023).

4.5.8 The Gut–Skin Axis

The gut–skin axis refers to the bidirectional communication between the gut microbiome and skin health. This relationship influences immune balance and contributes to skin diseases like acne, eczema, and psoriasis. Gut-derived metabolites, such as SCFAs, play a significant role in regulating skin inflammation and immune responses. For example, SCFAs stimulate T-regulatory cells (Tregs), which help modulate immune responses in the skin and promote healing. Dysbiosis in the gut can lead to systemic inflammation, exacerbating skin conditions and impairing wound healing. Restoring balance in the gut microbiome through probiotics and dietary changes may reduce inflammation, improve immune responses, and support skin health.

4.6 THERAPEUTIC STRATEGIES TARGETING THE MICROBIOME FOR ENHANCED WOUND HEALING

Various therapeutic strategies are being explored to target the microbiome, especially in relation to skin health and CWs. These strategies aim to manipulate the balance between the host and microorganisms, promoting healing and addressing skin-related diseases (Sachdeva et al., 2022). Here is an overview of these emerging strategies:

4.6.1 Modulating Microbiome Composition and Function

A key approach involves altering the microbial community to restore balance and introduce beneficial functions (Sachdeva et al., 2022). Given the challenges of biofilm formation and antibiotic resistance, alternatives like probiotics and prebiotics are being explored (Sachdeva et al., 2022).

Probiotics: Introducing beneficial microorganisms, such as *Lactobacillus plantarum*, which can reduce *P. aeruginosa* virulence, or *Lactobacillus reuteri*, which aids in wound healing by enhancing mesenchymal stem cells, holds promise for wound management. Clinical applications like *Lactobacillus johnsonii* lotions have shown positive outcomes in reducing *S. aureus* colonization in patients with atopic dermatitis.

Prebiotics: These substances selectively stimulate beneficial microorganisms and are included in various skin care products. Though the potential for prebiotics to aid in wound healing is promising, further research is needed (Sachdeva et al., 2022).

Synbiotics: Combining both probiotics and prebiotics could provide enhanced therapeutic effects (Sachdeva et al., 2022).

Postbiotics: These are metabolites produced by skin-friendly bacteria, like SCFAs (acetic, butyric, and propionic acid). They have antimicrobial effects and can help reduce inflammation in the skin (Mihai et al., 2024).

Skin Microbiome Transplantation: Skin microbiome transplants, similar to fecal transplants for gut dysbiosis, are being explored for skin disease treatment. This could involve transplanting a whole healthy microbiome or using specific probiotics (Sachdeva et al., 2022).

Engineered Microbiota: Synthetic biology is being applied to develop engineered microbiota with desired characteristics for therapeutic purposes (Sachdeva et al., 2022). This involves creating probiotics that can trigger therapeutic responses and restore microbial balance (Sachdeva et al., 2022).

4.6.2 Targeting Pathogenic Communities and Biofilms

Biofilm formation is a major factor delaying healing in CWs. These microbial communities, trapped in a protective matrix, are highly resistant to both antibiotics and the immune system, making treatment difficult. Strategies aimed at targeting biofilms are essential for improving wound healing outcomes.

Wound Debridement: Surgical or sharp debridement is an effective method for removing biofilms and maintaining a healthy wound bed. Reducing microbial diversity and disrupting anaerobic networks, it helps create an environment conducive to healing. Research shows that biofilm removal enhances the wound bed's ability to regenerate, clearing both pathogenic and non-pathogenic microbes from the area (Liu et al., 2020).

Antibiofilm Agents: These agents are designed to break down biofilms and improve the efficacy of antibiotics. Several types of antibiofilm agents are being explored:

Enzymes: Deoxyribonuclease-I (DNase-I) is effective in breaking down extracellular DNA (eDNA), a major component of the biofilm matrix. Dispersin B targets poly-β

1,6-N-acetylglucosamine, another critical biofilm component. These enzymes help reduce biofilm stability, making bacteria more susceptible to treatments.

Chelating Agents: Ethylenediaminetetraacetic acid (EDTA) binds to metal ions like Mg^{2+} and Ca^{2+}, destabilizing the outer cell wall of Gram-negative bacteria. This breaks down biofilms and facilitates the penetration of antibiotics. When combined with topical antibiotics, EDTA has been shown to significantly reduce bacterial counts in infected wounds.

Iron Chelators: Gallium, an FDA-approved compound, disrupts *P. aeruginosa* biofilm formation by inhibiting iron uptake. Similarly, lactoferrin, an iron-binding protein, also reduces the formation of biofilms in *P. aeruginosa*, showing promise for wound healing therapies (Bassler and Losick, 2006).

Topical Antimicrobials: Topical antiseptics and antibiotics, although challenging to use effectively, are still a cornerstone of CW care. Local antibiotic delivery is preferred to minimize systemic effects and target infections directly (Omar et al., 2017). However, biofilm formation often reduces the effectiveness of these treatments, necessitating the development of optimized strategies that enhance antibiotic penetration and effectiveness.

Bacteriophage Therapy: Bacteriophages, viruses that infect bacteria, are emerging as a promising alternative to antibiotics in the treatment of biofilm-related chronic infections. Phages are highly specific, targeting only the harmful bacteria without affecting eukaryotic cells, making them particularly useful in managing CWs caused by *S. aureus* and other biofilm-forming bacteria (Liu et al., 2020). Phage therapy has shown effectiveness in reducing biofilm-related infections, especially in cases resistant to traditional antibiotics (Mihai et al., 2024).

Targeting Quorum Sensing: Quorum sensing (QS) is a bacterial communication process that regulates biofilm formation and virulence factor expression. Inhibiting QS can prevent biofilm formation, decrease bacterial virulence, and improve healing. For example, *L. plantarum* has been shown to inhibit QS in *P. aeruginosa*, making it a potential therapeutic option for CW management. Disrupting QS can also reduce the production of virulence factors that lead to prolonged inflammation and impaired healing in CWs (Johnson et al., 2018).

Combination Therapies: Due to the complex, polymicrobial nature of biofilms, combining antimicrobial treatments with biofilm-disrupting agents and anti-inflammatory therapies is a promising strategy. Studies suggest that using multiple approaches concurrently can target different stages of the infection and inflammation cycle, leading to more effective wound healing (Zhao et al., 2012).

4.6.3 Modulating the Host Response in Chronic Wound Healing

CW healing is a complex process involving the interplay of various biological factors, including the immune system, the microbiome, and biochemical processes. Research focuses on leveraging these factors to improve healing outcomes (Mihai et al., 2024).

- *Host Immune Modulation*: The skin microbiome is crucial in influencing and regulating the immune system of the host, creating a balance between immune tolerance and activation. Through complex interactions between the host and microbes, beneficial microorganisms instruct immune cells, improve the skin's barrier function, and manage inflammatory responses, thereby supporting both innate and adaptive immunity.
  - *Inflammation Control*: Chronic inflammation in wounds delays healing. Targeting inflammatory factors could reduce prolonged inflammation and help regulate bacterial populations. Elevated pro-inflammatory cytokines and high protease levels are characteristic of CWs (Sachdeva et al., 2022).

- *Macrophage Activation*: Macrophages are key targets in wound healing, and strategies aim to boost their recruitment and promote a repair-friendly state. This can be achieved through methods like directly injecting L-1β-activated macrophages or applying granulocyte-macrophage colony-stimulating factor (GM-CSF) locally to the wound.
- *Toll-Like Receptors (TLRs)*: TLRs help activate the immune system in response to microbial patterns (Ramirez et al., 2018; Shanmugam et al., 2017; Stone et al., 2017). *S. epidermidis* and *P. acnes* both interact with TLRs, promoting AMP production and reducing inflammation (Sachdeva et al., 2022).

4.6.4 Diagnostic and 'OMIC' Approaches Informing Therapy

Advanced diagnostic tools, including genomic, proteomic, and metabolomic assays, are enhancing the understanding of wound healing and guiding treatment strategies.

Biomarkers: Identifying metabolites and biomarkers from microbes can help predict infection risks and improve non-invasive diagnostics. Molecular biomarkers are measurable biological substances that reflect normal body processes, disease activity, or how the body responds to treatment. They can serve various purposes, such as guiding debridement, predicting patient outcomes, assessing disease recurrence, tracking healing progress, or evaluating how well a drug is working (Mihai et al., 2024). These biomarkers can be found in swabs, wound fluid (WF) (e.g., MMPs and interleukins) or tissue (such as β-catenin and c-myc), and blood serum (like procalcitonin and MMPs) (Fernandez et al., 2008).

Microbiome Characterization: High-throughput sequencing provides detailed profiles of the wound microbiome, offering insights into virulence, resistance, and potential therapeutic targets. This technology allows clinicians to map microbial populations within wounds, helping to determine which species or biofilm formations are contributing to the chronicity of the wound.

Transcriptomics and Wound Blotting: These approaches allow for a deeper understanding of bacterial behavior (e.g., planktonic or biofilm states) and antimicrobial susceptibility, helping guide treatment decisions. Wound Blotting is an innovative, non-invasive technique that involves placing a nitrocellulose membrane on wound beds to capture and stabilize biomolecules like proteins and mucopolysaccharides for analysis. It helps detect biofilm formation and predict wound healing outcomes. For example, detecting mucopolysaccharides can help foresee slough formation and guide sharp debridement, while analyzing TNF-α patterns can indicate how well a wound is healing (Minematsu et al., 2013).

By evaluating gene expression patterns, it is possible to identify whether bacteria are in a planktonic (free-floating) or biofilm state, as well as their metabolic activity, which is crucial for choosing the right treatment. By analyzing gene expression profiles, it can also identify different bacterial subtypes, such as methicillin-resistant *S. aureus*. This information can greatly influence treatment decisions (Bauer et al., 2013).

4.6.5 Other Influences and Strategies

In addition to microbial manipulation, other factors influencing the microbiome and healing are being considered.

Diet and Lifestyle: Diet and lifestyle play a role in skin diseases. Modifying diet and gut microbiota could improve skin conditions. Research suggests that interventions like prebiotics, probiotics, and changes in dietary patterns can positively influence the skin microbiome and accelerate wound healing (Johnson et al., 2018; Musthaq et al., 2018; Nole et al., 2014; Yu et al., 2020).

Host Factors: Factors like age, sex, genetics, and stress hormones influence the skin microbiome and could be targeted for better wound healing interventions. These host-related factors play a significant role in the wound healing process, either promoting or hindering recovery, depending on their influence on microbial populations.

Genetics: Advances in sequencing technologies and microbiome studies are showing how genetic differences between patients affect their microbiome, influencing both health and disease. For example, specific genetic variants in *ZNF521* and *TLN2* are linked to variations in the levels of *P. aeruginosa* and *S. epidermidis*, two major pathogens in CWs. Wounds with lower microbial diversity, which may be related to genetics, often take longer to heal (Tipton et al., 2017).

Aging: Age affects the skin microbiome, with older individuals generally having more diverse microbial species. Several studies have found that the diversity of skin bacteria increases with age (Juge et al., 2018; Shibagaki et al., 2017). In fact, data from the skin microbiota can more accurately predict a person's age than data from the gut or mouth microbiome (Huang et al., 2020). Skin aging leads to structural and physiological changes, such as reduced moisture, the development of wrinkles, hyperpigmented patches, and altered sebaceous gland activity. This can reduce the nutrients available to beneficial bacteria, making room for opportunistic species to take over, especially if sebum production decreases (Abadias-Granado et al., 2021). In older individuals, there is often a decline in the dominant bacteria on the skin, with an increase in *Corynebacterium* and some Proteobacteria (Juge et al., 2018; Shibagaki et al., 2017). Additionally, the use of cosmetics has been shown to boost skin bacterial diversity, particularly in genera like *Ralstonia*, which may be involved in metabolizing foreign substances (Lee et al., 2018; Staudinger et al., 2011).

Stress Hormones: Glucocorticoids and adrenaline can hinder wound healing and raise the risk of infection. While adrenaline helps in the early stages of healing, prolonged stress can delay later healing stages (Holmes et al., 2015; Luqman and Gotz, 2021; Sivamani et al., 2009; Stojadinovic et al., 2012). Research is focusing on how microbes influence immune responses and how stress hormones affect bacterial growth in CWs (Mihai et al., 2024). Some commensals, like *S. epidermidis*, may counteract the negative effects of stress hormones and promote healing (Luqman and Gotz, 2021). Future treatments summarize could involve agents that block stress hormone pathways to improve wound healing (Stojadinovic et al., 2012).

Biomaterial Scaffolds: New biomaterial scaffolds are being designed to regulate the wound environment and enhance healing. Research has explored how both natural and synthetic scaffolds influence the skin microbiome and wound repair. For instance, synthetic polycaprolactone (PCL) scaffolds have been shown to increase microbiome diversity, helping diabetic wounds heal. On the other hand, natural scaffolds, by promoting the overgrowth of *Escherichia coli* through collagen release, have hindered diabetic wound healing. This suggests that scaffolds can shape the wound microbiome, which plays a crucial role in controlling inflammation and affecting healing outcomes.

Despite the progress in understanding the wound microbiome and its role in healing, many important questions remain. Future research needs to:

- Identify the ideal microbiome composition for optimal wound healing.
- Understand how the microbiome influences the healing process.
- Develop safe and effective strategies for microbiome manipulation, including the use of probiotics, prebiotics, and antibiotics.

- Investigate the complex relationship between the microbiome, host factors (like genetics and immune response), and environmental influences.
- Move from descriptive studies to uncovering the precise mechanisms of host–microbe interactions, focusing on the roles of specific microbial species and strains.
- Examine the roles of non-bacterial microorganisms, such as fungi and viruses, which are common in CWs and can impact healing.
- Standardize research methods, including sampling and analysis techniques, to ensure consistency across studies.
- Ongoing research into balancing beneficial and harmful microbes is essential for advancing CW care and developing personalized therapies.

4.7 CHALLENGES IN UNDERSTANDING AND MANAGING THE WOUND MICROBIOME

The study of microbial interactions in skin health and CWs presents challenges but also opens promising avenues for better diagnostics and therapies. As we gain a deeper understanding of the skin microbiome and its influence on wound healing, various key areas are emerging that have the potential to revolutionize CW management. The skin microbiome is diverse and varies across different skin areas (Sachdeva et al., 2022). While well-studied genera like *Staphylococcus* and *Cutibacterium* are prominent, others, such as *Corynebacterium* and *Kocuria*, are underexplored, as are fungal and viral interactions in the healing process. Determining the ideal microbiota for optimal healing remains a challenge.

A major challenge is distinguishing between pathogens and harmless colonizers in CWs (Johnson et al., 2018). Microbes can shift from being harmless to harmful based on factors like bacterial load, making it difficult to identify true pathogens. Detecting microbes doesn't always indicate infection, as many play beneficial or neutral roles (Sachdeva et al., 2022). Despite advances, understanding how skin microbiota influence wound infections and healing is still limited (Zielinska et al., 2023). Exploring interactions between microbiota and opportunistic pathogens is crucial (Johnson et al., 2018). The relationship between commensals and the immune system, especially in wound repair, is often underestimated. Understanding these dynamics is essential for new therapeutic approaches (Sachdeva et al., 2022).

Current diagnostic methods, like culture-based techniques, miss many microbes, especially anaerobes (Sachdeva et al., 2022). Advanced methods like 16S rRNA sequencing and mNGS offer more insight but struggle with differentiating live, dead, or dormant microbes, and lack strain-level resolution. Interpreting virulence factors and antibiotic resistance genes remains challenging, and reliable point-of-care biofilm detection is still needed.

4.8 FUTURE DIRECTIONS IN CHRONIC WOUND MICROBIOME RESEARCH

Future research should explore how local microbiota influence host responses and interact with pathogens (Zielinska et al., 2023). Understanding the role of skin commensals in pathogen defense, particularly through antimicrobial molecules like Perforin-2, could lead to new treatments for CWs (Sachdeva et al., 2022). High-throughput 'OMIC' techniques are revealing biomarkers to guide wound healing therapies and improve diagnostic methods (Sachdeva et al., 2022). Techniques like transcriptomics and wound blotting are being explored to assess pathogens and their susceptibility (Zielinska et al., 2023), but reliable biofilm detection remains a challenge. As antibiotics struggle with biofilm-related infections, alternative strategies, including probiotics, prebiotics, and biofilm modulation, show promise for promoting healing (Sachdeva et al., 2022). Further research is required to invent the most effective treatments and delivery methods (Sachdeva et al., 2022). Longitudinal studies are necessary to understand microbial dynamics and optimize treatment.

Predictive analytics using machine learning could identify microbial patterns linked to patient outcomes, improving CW management (Sachdeva et al., 2022). Host factors like stress hormones can impair healing and increase infection risk (Sachdeva et al., 2022). Research on how these factors affect the microbiome and healing is essential. Although progress has been made, more research is necessary to fully grasp the interaction between the microbiome and wound healing (Sachdeva et al., 2022).

4.9 CONCLUSION

The skin microbiome is essential for maintaining skin health and promoting wound healing. Disruptions in microbial balance, such as biofilm formation, can impede recovery, especially in CWs. With advancements in molecular technologies, understanding the microbiome's role in healing is becoming clearer. Continued research into microbial interactions and therapies targeting the microbiome offers promise for more personalized and effective treatments for CWs.

GLOSSARY

16S rRNA gene sequencing: A molecular method used to identify and quantify bacteria in a sample based on sequencing the 16S ribosomal RNA gene.

Commensal bacteria: Harmless microorganisms that live with the host organism. They can exist both externally and internally, without leading to disease or offering noticeable benefits.

Cytokines/Chemokines: Signaling molecules involved in the inflammatory response.

Dysbiosis (microbial imbalances): An imbalance in the composition or function of a microbial community.

ECM (extracellular matrix): A network of molecules that provides structural support to cells and tissues.

Host–microbiome crosstalk: The intricate communication and interaction between the host organism and its resident microbial community.

Microbiome: The community of microorganisms (such as bacteria, viruses, and fungi) that live in a particular environment, along with their combined genetic material.

Microbiota: The community of microorganisms residing in a specific environment.

'OMIC' modalities: High-throughput techniques used for broad molecular analysis, such as genomic, proteomic, lipidomic, and metabolomic assays.

Pathogens: Microorganisms capable of causing disease.

Prebiotics: Non-digestible substances that selectively promote the growth and/or activity of beneficial microbes.

Probiotics: Live microorganisms that provide a health benefit to the host when consumed in sufficient quantities.

Symbionts: Organisms living in symbiosis with another organism.

Wound fluid (WF): Fluid present in wounds, which can have pro-inflammatory properties in chronic wounds.

Wound microbiota: A diverse community of microorganisms inhabiting a wound and influencing its physiological and inflammatory function.

REFERENCES

Abadias-Granado, Isabel, Javier Sanchez-Bernal, and Yolanda Gilaberte. "The microbiome and aging." *Plastic and Aesthetic Research* 8 (2021): 27–37.

Alves, Patrícia M., Eida Al-Badi, Cathryn Withycombe, Paul M. Jones, Kevin J. Purdy, and Sarah E. Maddocks. "Interaction between *Staphylococcus aureus* and *Pseudomonas aeruginosa* is beneficial for colonisation and pathogenicity in a mixed biofilm." *Pathogens and Disease* 76, no. 1 (2018): fty003.

Antonsson, Annika, Cornelia Erfurt, Kristina Hazard, Veroniqa Holmgren, Melinda Simon, Akio Kataoka, Shahadat Hossain, Camilla Håkangård, and Bengt Göran Hansson. "Prevalence and type spectrum of human papillomaviruses in healthy skin samples collected in three continents." *Journal of General Virology* 84, no. 7 (2003): 1881–1886.

Arnold, Christopher, Shane Merryman, Aleishia Harris-Arnold, Sean A. McKinney, Chris W. Seidel, Sydney Loethen, Kylie N. Proctor, Longhua Guo, and Alejandro Sánchez Alvarado. "Pathogenic shifts in endogenous microbiota impede tissue regeneration via distinct activation of TAK1/MKK/p38." *elife* 5 (2016): e16793.

Baldwin, Hilary, Neall Bhatia, Adam Friedman, T. Prunty, R. Martin, and S. Seite. "The role of cutaneous microbiota harmony in maintaining a functional skin barrier." *SKIN The Journal of Cutaneous Medicine* 1 (2017): s139.

Nole, Katherine L. Baquerizo, Elizabeth Yim, and Jonette E. Keri. "Probiotics and prebiotics in dermatology." *Journal of the American Academy of Dermatology* 71, no. 4 (201): 814–821.

Bassler, Bonnie L., and Richard Losick. "Bacterially speaking." *Cell* 125, no. 2 (2006): 237–246.

Bauer, Julia, Wafi Siala, Paul M. Tulkens, and Françoise Van Bambeke. "A combined pharmacodynamic quantitative and qualitative model reveals the potent activity of daptomycin and delafloxacin against Staphylococcus aureus biofilms." *Antimicrobial Agents and Chemotherapy* 57, no. 6 (2013): 2726–2737.

Bottger, Erik C. "Rapid determination of bacterial ribosomal RNA sequences by direct sequencing of enzymatically amplified DNA." *FEMS Microbiology Letters* 65, no. 1–2 (1989): 171–176.

Brown, Sonya K., and Alan R. Shalita. "Acne vulgaris." *The Lancet* 351, no. 9119 (1998): 1871–1876.

Byrd, Allyson L., Yasmine Belkaid, and Julia A. Segre. "The human skin microbiome." *Nature Reviews Microbiology* 16, no. 3 (2018): 143–155.

Cogen, Anna L., Kenshi Yamasaki, Jun Muto, Katheryn M. Sanchez, Laura Crotty Alexander, Jackelyn Tanios, Yuping Lai, Judy E. Kim, Victor Nizet, and Richard L. Gallo. "*Staphylococcus epidermidis* antimicrobial δ-toxin (phenol-soluble modulin-γ) cooperates with host antimicrobial peptides to kill group A Streptococcus." *PLoS One* 5, no. 1 (2010b): e8557.

Cogen, Anna L., Kenshi Yamasaki, Katheryn M. Sanchez, Robert A. Dorschner, Yuping Lai, Daniel T. MacLeod, Justin W. Torpey, et al. "Selective antimicrobial action is provided by phenol-soluble modulins derived from *Staphylococcus epidermidis*, a normal resident of the skin." *Journal of Investigative Dermatology* 130, no. 1 (2010a): 192–200.

Curtis, Huttenhower, Gevers Dirk, Knight Rob, Abubucker Sahar, Hallsworth-Pepin Kymberlie, Madupu Ramana, Magrini Vincent et al. "Structure, function and diversity of the healthy human microbiome." *Nature* 486, no. 7402 (2012).

DeLeon, Stephanie, Allie Clinton, Haley Fowler, Jake Everett, Alexander R. Horswill, and Kendra P. Rumbaugh. "Synergistic interactions of *Pseudomonas aeruginosa* and *Staphylococcus aureus* in an in vitro wound model." *Infection and Immunity* 82, no. 11 (2014): 4718–4728.

Drago, Francesco, Lodovica Gariazzo, Margherita Cioni, Ilaria Trave, and Aurora Parodi. "The microbiome and its relevance in complex wounds." *European Journal of Dermatology* 29 (2019): 6–13.

Fernandez, Melissa, James Broadbent, Gary Shooter, Johannes Malda, and Zee Upton. "Development of an enhanced proteomic method to detect prognostic and diagnostic markers of healing in chronic wound fluid." *British Journal of Dermatology* 158, no. 2 (2008): 281–290.

Flowers, Laurice, and Elizabeth A. Grice. "The skin microbiota: Balancing risk and reward." *Cell Host & Microbe* 28, no. 2 (2020): 190–200.

Gaitanis, Georgios, Georgia Tsiouri, Panagiota Spyridonos, Theodoros Stefos, Georgios N. Stamatas, Aristea Velegraki, and Ioannis D. Bassukas. "Variation of cultured skin microbiota in mothers and their infants during the first year postpartum." *Pediatric Dermatology* 36, no. 4 (2019): 460–465.

Gallo, Richard L. "Human skin is the largest epithelial surface for interaction with microbes." *Journal of Investigative Dermatology* 137, no. 6 (2017): 1213–1214.

Gardiner, Melissa, Mauro Vicaretti, Jill Sparks, Sunaina Bansal, Stephen Bush, Michael Liu, Aaron Darling, Elizabeth Harry, and Catherine M. Burke. "A longitudinal study of the diabetic skin and wound microbiome." *PeerJ* 5 (2017): e3543.

Grice, Elizabeth A., and Julia A. Segre. "Interaction of Microbiome and the Innate Immune Response in Chronic Wounds." *Adv Exp Med Biol* 946 (2012): 55–68.

Han, Nannan, Lu Jia, Yingying Su, Juan Du, Lijia Guo, Zhenhua Luo, and Yi Liu. "*Lactobacillus reuteri* extracts promoted wound healing via PI3K/AKT/β-catenin/TGFβ1 pathway." *Stem Cell Research & Therapy* 10 (2019): 1–11.

Hannigan, Geoffrey D., Jacquelyn S. Meisel, Amanda S. Tyldsley, Qi Zheng, Brendan P. Hodkinson, Adam J. SanMiguel, Samuel Minot, Frederic D. Bushman, and Elizabeth A. Grice. "The human skin double-stranded DNA virome: Topographical and temporal diversity, genetic enrichment, and dynamic associations with the host microbiome." *MBio* 6, no. 5 (2015): 10–1128.

Harris-Tryon, Tamia A., and Elizabeth A. Grice. "Microbiota and maintenance of skin barrier function." *Science* 376, no. 6596 (2022): 940–945.

Holmes, Casey J., Jennifer K. Plichta, Richard L. Gamelli, and Katherine A. Radek. "Dynamic role of host stress responses in modulating the cutaneous microbiome: Implications for wound healing and infection." *Advances in Wound Care* 4, no. 1 (2015): 24–37.

Huang, Shi, Niina Haiminen, Anna-Paola Carrieri, Rebecca Hu, Lingjing Jiang, Laxmi Parida, Baylee Russell, et al. "Human skin, oral, and gut microbiomes predict chronological age." *mSystems* 5, no. 1 (2020): 10–1128.

Iwase, Tadayuki, Yoshio Uehara, Hitomi Shinji, Akiko Tajima, Hiromi Seo, Koji Takada, Toshihiko Agata, and Yoshimitsu Mizunoe. "*Staphylococcus epidermidis* Esp inhibits *Staphylococcus aureus* biofilm formation and nasal colonization." *Nature* 465, no. 7296 (2010): 346–349.

Johnson, Taylor R., Belinda I. Gómez, Matthew K. McIntyre, Michael A. Dubick, Robert J. Christy, Susannah E. Nicholson, and David M. Burmeister. "The cutaneous microbiome and wounds: New molecular targets to promote wound healing." *International Journal of Molecular Sciences* 19, no. 9 (2018): 2699.

Jugé, Romain, Pauline Rouaud-Tinguely, Josselin Breugnot, Katia Servaes, Christine Grimaldi, Marie-Paule Roth, Hélène Coppin, and Brigitte Closs. "Shift in skin microbiota of Western European women across aging." *Journal of Applied Microbiology* 125, no. 3 (2018): 907–916.

Kalan, Lindsay, and Elizabeth A. Grice. "Fungi in the wound microbiome." *Advances in Wound Care* 7, no. 7 (2018): 247–255.

Kolbert, Christopher P., and David H. Persing. "Ribosomal DNA sequencing as a tool for identification of bacterial pathogens." *Current Opinion in Microbiology* 2, no. 3 (1999): 299–305.

Kong, Eric F., Jennifer K. Johnson, and Mary Ann Jabra-Rizk. "Community-associated methicillin-resistant *Staphylococcus aureus*: An enemy amidst us." *PLoS Pathogens* 12, no. 10 (2016): e1005837.

Laborel-Préneron, Emeline, Pascale Bianchi, Franck Boralevi, Philippe Lehours, Frédérique Fraysse, Fanny Morice-Picard, Motoyuki Sugai, et al. "Effects of the *Staphylococcus aureus* and *Staphylococcus epidermidis* secretomes isolated from the skin microbiota of atopic children on CD_4^+ T cell activation." *PLoS One* 10, no. 10 (2015): e0141067.

Lee, Hyo Jung, Sang Eun Jeong, Soyoun Lee, Sungwoo Kim, Hyuntak Han, and Che Ok Jeon. "Effects of cosmetics on the skin microbiome of facial cheeks with different hydration levels." *Microbiologyopen* 7, no. 2 (2018): e00557.

Lindley, Linsey E., Olivera Stojadinovic, Irena Pastar, and Marjana Tomic-Canic. "Biology and biomarkers for wound healing." *Plastic and Reconstructive Surgery* 138, no. 3S (2016): 18S–28S.

Linehan, Jonathan L., Oliver J. Harrison, Seong-Ji Han, Allyson L. Byrd, Ivan Vujkovic-Cvijin, Alejandro V. Villarino, Shurjo K. Sen, et al. "Non-classical immunity controls microbiota impact on skin immunity and tissue repair." *Cell* 172, no. 4 (2018): 784–796.

Liu, Chunan, Alise J. Ponsero, David G. Armstrong, Benjamin A. Lipsky, and Bonnie L. Hurwitz. "The dynamic wound microbiome." *BMC Medicine* 18 (2020): 1–12.

Luqman, Arif, and Friedrich Götz. "The ambivalent role of skin microbiota and adrenaline in wound healing and the interplay between them." *International Journal of Molecular Sciences* 22, no. 9 (2021): 4996.

Meisel, Jacquelyn S., Georgia Sfyroera, Casey Bartow-McKenney, Ciara Gimblet, Julia Bugayev, Joseph Horwinski, Brian Kim, et al. "Commensal microbiota modulate gene expression in the skin." *Microbiome* 6 (2018): 1–15.

Meneghin, Alessia, and Cory M. Hogaboam. "Infectious disease, the innate immune response, and fibrosis." *The Journal of Clinical Investigation* 117, no. 3 (2007): 530–538.

Midorikawa, Kazushige, Kazuhisa Ouhara, Hitoshi Komatsuzawa, Toshihisa Kawai, Sakuo Yamada, Tamaki Fujiwara, Kenshi Yamazaki, et al. "*Staphylococcus aureus* susceptibility to innate antimicrobial peptides, β-defensins and CAP18, expressed by human keratinocytes." *Infection and Immunity* 71, no. 7 (2003): 3730–3739.

Mihai, Mara Mădălina, Beatrice Bălăceanu-Gurău, Ana Ion, Alina Maria Holban, Cristian-Dorin Gurău, Marius Nicolae Popescu, Cristina Beiu, et al. "Host–microbiome crosstalk in chronic wound healing." *International Journal of Molecular Sciences* 25, no. 9 (2024): 4629.

Minematsu, Takeo, Gojiro Nakagami, Yuko Yamamoto, Toshiki Kanazawa, Lijuan Huang, Hiroe Koyanagi, Sanae Sasaki, et al. "Wound blotting: A convenient biochemical assessment tool for protein components in exudate of chronic wounds." *Wound Repair and Regeneration* 21, no. 2 (2013): 329–334.

Musthaq, Shenara, Anna Mazuy, and Jeannette Jakus. "The microbiome in dermatology." *Clinics in Dermatology* 36, no. 3 (2018): 390–398.

Nakatsuji, H., Maeda, N., Hibuse, T., Hiuge, A., Hirata, A., Kuroda, Y., ... & Shimomura, I. (2010). Reciprocal regulation of natriuretic peptide receptors by insulin in adipose cells. *Biochemical and biophysical research communications*, *392*(1), 100–105.

Nakatsuji, Teruaki, Tiffany H. Chen, Saisindhu Narala, Kimberly A. Chun, Aimee M. Two, Tong Yun, Faiza Shafiq, et al. "Antimicrobials from human skin commensal bacteria protect against *Staphylococcus aureus* and are deficient in atopic dermatitis." *Science Translational Medicine* 9, no. 378 (2017): eaah4680.

Pastar, Irena, Aron G. Nusbaum, Joel Gil, Shailee B. Patel, Juan Chen, Jose Valdes, Olivera Stojadinovic, Lisa R. Plano, Marjana Tomic-Canic, and Stephen C. Davis. "Interactions of methicillin resistant *Staphylococcus aureus* USA300 and *Pseudomonas aeruginosa* in polymicrobial wound infection." *PLoS One* 8, no. 2 (2013): e56846.

Patel, S., A. Maheshwari, and A. Chandra. "Biomarkers for wound healing and their evaluation." *Journal of Wound Care* 25, no. 1 (2016): 46–55.

Plichta, Jennifer K., Xiang Gao, Huaiying Lin, Qunfeng Dong, Evelyn Toh, David E. Nelson, Richard L. Gamelli, Elizabeth A. Grice, and Katherine A. Radek. "Cutaneous burn injury promotes shifts in the bacterial microbiome in autologous donor skin: Implications for skin grafting outcomes." *Shock* 48, no. 4 (2017): 441–448.

Price, Lance B., Cindy M. Liu, Johan H. Melendez, Yelena M. Frankel, David Engelthaler, Maliha Aziz, Jolene Bowers, et al. "Community analysis of chronic wound bacteria using 16S rRNA gene-based pyrosequencing: Impact of diabetes and antibiotics on chronic wound microbiota." *PLoS One* 4, no. 7 (2009): e6462.

Radek, Katherine A. "Antimicrobial anxiety: The impact of stress on antimicrobial immunity." *Journal of Leukocyte Biology* 88, no. 2 (2010): 263–277.

Ramirez, Horacio A., Irena Pastar, Ivan Jozic, Olivera Stojadinovic, Rivka C. Stone, Nkemcho Ojeh, Joel Gil, Stephen C. Davis, Robert S. Kirsner, and Marjana Tomic-Canic. "*Staphylococcus aureus* triggers induction of miR-15B-5P to diminish DNA repair and deregulate inflammatory response in diabetic foot ulcers." *Journal of Investigative Dermatology* 138, no. 5 (2018): 1187–1196.

Ring, Hans Christian, Jonathan. Thorsen, Kurt. Fuursted, Thomas Bjarnsholt, Line Kolind Bay, Ditte. M. Saunte, Simon. Francis. Thomsen, and Gregor. B. Jemec. "Probiotics in hidradenitis suppurativa: A potential treatment option?" *Clinical and Experimental Dermatology* 47, no. 1 (2022): 139–141.

Sachdeva, Chandni, Kapaettu Satyamoorthy, and Thokur Sreepathy Murali. "Microbial interplay in skin and chronic wounds." *Current Clinical Microbiology Reports* 9, no. 3 (2022): 21–31.

Sanford, James A., and Richard L. Gallo. "Functions of the skin microbiota in health and disease." In *Seminars in immunology* 25, no. 5 (2013), 370–377. Academic Press, 2013.

Santiago-Rodriguez, Tasha M., Brice Le François, Jean M. Macklaim, Evgueni Doukhanine, and Emily B. Hollister. "The skin microbiome: Current techniques, challenges, and future directions." *Microorganisms* 11, no. 5 (2023): 1222.

Scales, Brittan S., and Gary B. Huffnagle. "The microbiome in wound repair and tissue fibrosis." *The Journal of Pathology* 229, no. 2 (2013): 323–331.

Scharschmidt, Tiffany C., and Michael A. Fischbach. "What lives on our skin: Ecology, genomics and therapeutic opportunities of the skin microbiome." *Drug Discovery Today: Disease Mechanisms* 10, no. 3–4 (2013): e83–e89.

Shanmugam, Victoria K., Divya Angra, Hamza Rahimi, and Sean McNish. "Vasculitic and autoimmune wounds." *Journal of Vascular Surgery: Venous and Lymphatic Disorders* 5, no. 2 (2017): 280–292.

Shibagaki, Nakako, Wataru Suda, Cecile Clavaud, Philippe Bastien, Lena Takayasu, Erica Iioka, Rina Kurokawa, et al. "Aging-related changes in the diversity of women's skin microbiomes associated with oral bacteria." *Scientific Reports* 7, no. 1 (2017): 10567.

Sivamani, Raja K., Christine E. Pullar, Catherine G. Manabat-Hidalgo, David M. Rocke, Richard C. Carlsen, David G. Greenhalgh, and R. Rivkah Isseroff. "Stress-mediated increases in systemic and local epinephrine impair skin wound healing: Potential new indication for beta blockers." *PLoS Medicine* 6, no. 1 (2009): e1000012.

Smith, Karen, Andrew Collier, Eleanor M. Townsend, Lindsay E. O'Donnell, Abhijit M. Bal, John Butcher, William G. Mackay, Gordon Ramage, and Craig Williams. "One step closer to understanding the role of bacteria in diabetic foot ulcers: Characterising the microbiome of ulcers." *BMC Microbiology* 16, no. 1 (2016): 54.

Staudinger, Thomas, Allison J. Pipal, and Bernhard Redl. "Molecular analysis of the prevalent microbiota of human male and female forehead skin compared to forearm skin and the influence of make-up." *Journal of Applied Microbiology* 110, no. 6 (2011): 1381–1389.

Stojadinovic, Olivera, Katherine A. Gordon, Elizabeth Lebrun, and Marjana Tomic-Canic. "Stress-induced hormones cortisol and epinephrine impair wound epithelization." *Advances in Wound Care* 1, no. 1 (2012): 29–35.

Stone, Rivka C., Olivera Stojadinovic, Ashley M. Rosa, Horacio A. Ramirez, Evangelos Badiavas, Miroslav Blumenberg, and Marjana Tomic-Canic. "A bioengineered living cell construct activates an acute wound healing response in venous leg ulcers." *Medicine* 9, no. 371 (2017): eaaf8611.

Tang, Qinghan, Nannan Xue, Xiaofeng Ding, Kevin H.-Y. Tsai, Jonathan J. Hew, Ruihan Jiang, Rizhong Huang, et al. "Role of wound microbiome, strategies of microbiota delivery system and clinical management." *Advanced Drug Delivery Reviews* 192 (2023): 114671.

Tipton, Craig D., Marilyn E. Mathew, Richard A. Wolcott, Randall D. Wolcott, Tigga Kingston, and Caleb D. Phillips. "Temporal dynamics of relative abundances and bacterial succession in chronic wound communities." *Wound Repair and Regeneration* 25, no. 4 (2017): 673–679.

Tomic-Canic, Marjana, Jamie L. Burgess, Katelyn E. O'Neill, Natasa Strbo, and Irena Pastar. "Skin microbiota and its interplay with wound healing." *American Journal of Clinical Dermatology* 21, no. Suppl 1 (2020): 36–43.

Tuttle, Marie S. "Association between microbial bioburden and healing outcomes in venous leg ulcers: A review of the evidence." *Advances in Wound Care* 4, no. 1 (2015): 1–11.

Uberoi, Aayushi, Amelia McCready-Vangi, and Elizabeth A. Grice. "The wound microbiota: Microbial mechanisms of impaired wound healing and infection." *Nature Reviews Microbiology* 22, no. 8 (2024): 507–521.

Valdéz, Juan Carlos, Maria C Peral, Mirtha Rachid, M. Santana, and Gabriela Del Valle Perdigon. "Interference of *Lactobacillus plantarum* with *Pseudomonas aeruginosa* in vitro and in infected burns: The potential use of probiotics in wound treatment." *Clinical Microbiology and Infection* 11, no. 6 (2005): 472–479.

Wanke, Ines, Heiko Steffen, Christina Christ, Bernhard Krismer, Friedrich Götz, Andreas Peschel, Martin Schaller, and Birgit Schittek. "Skin commensals amplify the innate immune response to pathogens by activation of distinct signaling pathways." *Journal of Investigative Dermatology* 131, no. 2 (2011): 382–390.

Yang, Yuyan, Jiuzuo Huang, Ang Zeng, Xiao Long, Nanze Yu, and Xiaojun Wang. "The role of the skin microbiome in wound healing." *Burns & Trauma* 12 (2024): tkad059.

Youn, Sung Hwan, Chong Won Choi, Jae Woo Choi, and Sang Woong Youn. "The skin surface pH and its different influence on the development of acne lesion according to gender and age." *Skin research and technology* 19, no. 2 (2013): 131–136.

Yu, Yu., Spencer. Dunaway, Jackson. Champer, Juewon Kim, and Ali Alikhan. "Changing our microbiome: Probiotics in dermatology." *British Journal of Dermatology* 182, no. 1 (2020): 39–46.

Zhao, Ge, Marcia L. Usui, Robert A. Underwood, Pradeep K. Singh, Garth A. James, Philip S. Stewart, Philip Fleckman, and John E. Olerud. "Time course study of delayed wound healing in a biofilm-challenged diabetic mouse model." *Wound Repair and Regeneration* 20, no. 3 (2012): 342–352.

Zheng, Yue, Rachelle L. Hunt, Amer E. Villaruz, Emilie L. Fisher, Ryan Liu, Qian Liu, Gordon Y. C. Cheung, Min Li, and Michael Otto. "Commensal *Staphylococcus epidermidis* contributes to skin barrier homeostasis by generating protective ceramides." *Cell Host & Microbe* 30, no. 3 (2022): 301–313.

Zhu, Yibin, Xi Yu, and Gong Cheng. "Human skin bacterial microbiota homeostasis: A delicate balance between health and disease." *mLife* 2, no. 2 (2023): 107–120.

Zielinska, Małgorzata, Agnieszka Pawłowska, Anna Orzeł, Luiza Sulej, Katarzyna Muzyka-Placzyńska, Arkadiusz Baran, Dagmara Filipecka-Tyczka, et al. "Wound microbiota and its impact on wound healing." *International Journal of Molecular Sciences* 24, no. 24 (2023): 17318.

5 Animal Models for Wound Healing

Nelofer Ereej[1]*, Huma Hameed*[*]*, Sania Niaz*[1]*, Mahtab Ahmad Khan*[1]*, Ana Claudia Paiva-Santos*[2,3]*, Ali Irfan*[4]*, and Hafiz Shoaib Sarwar*[1]

[1] Faculty of Pharmaceutical Sciences, University of Central Punjab (UCP), Lahore Pakistan

[2] Department of Pharmaceutical Technology, Faculty of Pharmacy of the University of Coimbra, University of Coimbra, Coimbra, Portugal

[3] 3REQUIMTE/LAQV, Group of Pharmaceutical Technology, Faculty of Pharmacy of the University of Coimbra, University of Coimbra, Coimbra, Portugal

[4] The Green Institute of Chemical, Biomedical and Environmental Sciences (GICBES), Lahore, Pakistan

5.1 OVERVIEW OF WOUND HEALING

A wound is characterized by a disruption of the underlying tissue or skin characterized by typical anatomical integrity and physiological function commonly resulting from mechanical, chemical, thermal, or biological trauma (Sorg et al., 2017; Sullivan and Myers, 2022). Wound healing is a sophisticated and continuously changing biological process focused on re-establishing tissue structure and function. This process progresses through four connected yet separate phases: hemostasis, inflammation, proliferation, and remodeling. Hemostasis initiates immediately post-injury, involving blood vessel narrowing and the creation of a blood clot via platelet clumping and fibrin accumulation. The inflammatory phase generally spans one to four days, during which specific immune cells (neutrophils and macrophages) move to the wound area to remove harmful microorganism and clear damaged tissue (Ellis et al., 2018). These immune cells also secrete signaling molecules that initiate tissue repair (Masoomikarimi and Salehi, 2022). Between days 4 and 12, the proliferative phase takes place, marked by the activation of fibroblasts, production of collagen, new blood vessel formation, development of granulation tissue, and resurfacing of the wound by epithelial cells (Gonzalez et al., 2016). The final phase, remodeling, focuses on strengthening of the tissue through collagen reorganization and cross-linking in wound contraction and scar development (Potekaev et al., 2021). Even though progress has been made in this area, it is still very hard to correctly simulate the complexity of human chronic wounds in animal models (Flynn et al., 2023). The various phases of wound healing are shown in Table 5.1.

Chronic wounds, often referred to as non-healing wounds, are injuries that do not advance through the typical stages of healing in a timely and organized fashion. Generally, a wound is classified as chronic if it does not heal within four to six weeks despite receiving standard treatment. These wounds present significant clinical challenges, incur high costs, and greatly impact

[*] Corresponding author: huma.hameed@ucp.edu.pk

 DOI: 10.1201/9781003644958-5

TABLE 5.1
Various Phases Involved in Wound Healing

Phase	Disruption in Chronic Wounds
Inflammation	Prolonged inflammatory response persistent neutrophil activity
Proliferation	Impaired fibroblast and keratinocyte function poor angiogenesis
Remodeling	Imbalanced matrix metalloproteins (MMP)/tissue inhibitors of metalloproteins (TIMP) activity Delayed collagen maturation
Immune response	Reduced macrophage polarization to M2 phenotype Impaired bacterial clearance

the quality of life for patients due to pain, infection, immobility, and the potential risk of amputation. Chronic or non-healing wounds in people are usually caused by a number of things, such as poor nutrition, restricted mobility, old age, poor blood flow and long-term condition like diabetes (Avishai et al., 2017). To make sure the study is useful, the animal models closely resemble these pathological traits (Domínguez-Oliva et al., 2023). Research on wound healing relies heavily on these models, which offer important insight into the basic processes behind tissue regeneration (Choudhary et al., 2024). Before moving on to clinical trials, they are also crucial platforms for evaluating the safety and therapeutic potential of novel medicines.

5.2 OVERVIEW OF ANIMAL MODELS

Animal models are vital for wound healing research, enabling scientists to investigate the underlying biological mechanism and assess the effectiveness of novel treatment. These models offer a living system that can examine the distinct stage of the healing process, hemostasis, inflammation, proliferation, and remodeling. These models facilitate detailed investigation of each healing phase hemostasis, inflammation, proliferation, and remodeling helping bridge the gap between preclinical findings and clinical outcomes (Mahmoud et al., 2024). Rodents particularly mice and rats are the most commonly used species due to their ease of care, availability, low cost, and genetic manipulability (Campbell et al., 2015). The rabbit ear dermal ulcer model is another valuable system particularly for studying re-epithelialization and granulation tissue development (Gu et al., 2004; Saeed and Martins-Green, 2023). Unlike body skin, rabbit ear skin adheres to underlying cartilage, which acts as a natural support, reducing wound contraction and promoting a healing process more like to that observed in humans (Hammad, 2021). Among alternative models, pigs are considered the most anatomically and physiologically similar to human, especially in terms of skin structure, including epidermal and dermal thickness, hair follicle density, and vascularization (Khiao In et al., 2019). Their healing of wound patterns, such as minimal contraction and granulation tissue formation, closely resemble those seen in humans (Hakkinen et al., 2011). However, limitation such as thick skin different mechanical properties, high maintenance cost, and limited genetic tools must be considered. The choice of an appropriate animal model depends on specific research objective and the phase of healing under investigation. The primary reason for employing animal models in wound healing studies includes their physiological similarity to human skin, the ability to control the wound environment, ethical consideration before proceeding to human trials and the opportunity to assess histological, biochemical, and molecular responses; furthermore, animal models enable

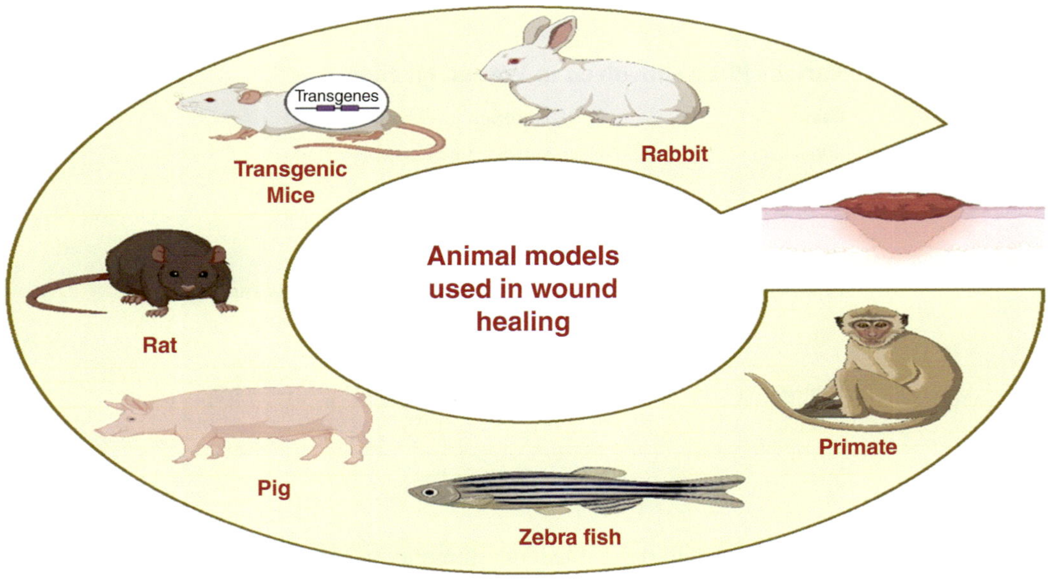

FIGURE 5.1 Animals used in wound healing process. This illustration shows the common animal models (rabbit, transgenic mice, rat, pig, zebrafish, and primate) used to study the biological models of wound healing, providing insight into tissue repair mechanism and potential therapies.

researchers to evaluate the toxicity, biocompatibility, and healing efficacy of new formulation in a controlled and reproducible manner. *In-vivo* animal studies must follow ethical standard included by relevant committees and adherence to 3Rs principle Replacement, Reduction, and Refinement to minimize animal use and enhance welfare (Díaz et al., 2021). All animal producers follow ethical standards such as International Animal Care and Use Committee (IACUC) guidelines to safeguard human treatment (Couto and Cates, 2019).

Animal models are used for the evaluation of therapies i.e., test drugs, dressings, biomaterials, and cell-based therapies as shown in Figure 5.1. They are used to understand the mechanism of wound healing by examining molecular and cellular responses during healing. Here's an overview of the most commonly used animal models and their relevance.

5.3 ANIMALS USED IN WOUND HEALING

Different types of animals used in wound healing are shown in Table 5.2.

5.3.1 Murine

Mice and rats are commonly referred to as murine models. They are commonly used in wound healing research because of their compact size, cost-effective maintenance, and the extensive understanding of their anatomical and physiological characteristics. Rats and mice differ greatly, even if they are comparable in ways that affect how well they work as models for wound healing. For such investigation, rats are frequently thought to be a better option due to their somewhat bigger body size (Dorsett-Martin, 2004). The development of numerous genetically altered mice models made possible by contemporary gene editing technologies, especially Clustered Regularly Interspaced Short Palindromic Repeats (CRISPR) has greatly expanded the research of gene pathways in wound healing. In the intricate process of tissue repair, this has enabled researchers to analyze the distinct functions of individual gene and their interrelated network. Consequently, mouse models have

TABLE 5.2
Use of Various Types of Animals in the Process of Wound Healing

Animals	Why it's Used
Mouse	Mice are the most common animal models in biomedical research due to their small size, ease of handling, fast reproductive cycle, and low cost
Transgenic mice	Genetically engineered mice (e.g., knockout or reporter mice) that allow researchers to study the role of specific genes or protein during wound healing
Rat	Larger than mice, allowing easier wound creation and sample collection: used in diabetic and burn wound models
Rabbit	Used for incision and burn wound models, ear model are popular for angiogenesis studies
Pig	Skin structure (thickness, collagen content, hair distribution) is most similar to humans
Zebrafish	Excellent model for studying epidermal wound healing, angiogenesis, and immune cell behavior
Primates	Closest to human genetically, physiologically, and immunologically
Others	Guinea pigs, dog, sheep, and others

become more popular than rat models in more recent studies on wound healing. However, mouse models are still not flawless because of differences in their skin structure compared to human skin. However, if combinatorial approach is used and one is aware of model's limit and differences when evaluating experimental data, murine models can still make significant contributions to the study of wound healing.

5.3.1.1 Rats

The rats have been a commonly utilized animal model in the investigation of skin wound repair due to their anatomical and physiological similarities to humans, along with practical benefits such as low maintenance cost, availability, and manageable size. Both incisional and excisional wound models are commonly performed on the dorsal surface, which helps ensure consistency while minimizing animal interference. The dorsal location offers a key advantage by limiting the animal's ability to reach the wound site, thus preserving the integrity of the experimental data (Ghanbari et al., 2024). Similar to mice, rats have loose skin, and their wounds primarily heal through contraction rather than re-epithelialization (Udegbunam et al., 2021). Contraction is a faster process as it does not involve the formation of new tissue. However, unlike humans, rats and mice do not form hypertrophic scars or keloids (Berman et al., 2017). Instead, the collagen involved in healing originates from the subcutaneous panniculus carnosus muscle (Lu et al., 2022). While mice may be more economical to maintain, rats provide a larger skin surface, which is particularly useful in wound healing experiments. The preferred model typically involves male Sprague Dawley rats weighing between 250 and 300 g, as they are cost-effective and widely available (Wilson and Islam, 2012).

One commonly used rat model is the ischemic H-shaped double flap developed for evaluating factors that influence flap survival. Despite its usefulness, this model has certain limitations, including a rapid return to normal tissue perfusion, which restricts the window for testing therapeutic interventions. Additionally, incisional wound models pose challenges since tensile strength measurements capture only one dimension of the healing process (Mulisa et al., 2015).

To overcome the limitations of previous models, a refined ischemic flap technique was introduced, utilizing full-thickness excisional wounds within a bipedicle dorsal skin flap. This method incorporates two significant adjustments: the flap is narrowed to promote a random-pattern blood supply, and a silicone sheet is positioned underneath to block reattachment and minimize reperfusion from the underlying tissues. The panniculus carnous muscle is carefully separated just above the fascia to expose the target tissue. The skin flap is then fixed onto the silicon barrier, which restricts but does not entirely prevent wound contraction. This approach maintains a controlled ischemic condition for up to two weeks without inducing tissue necrosis, providing a reliable model for studying delayed or impaired wound healing (Ghanbari et al., 2024). Although rats possess key structural components of human skin such as the epidermis, dermis, basement membrane, and hair follicles, they lack the capacity to form keloids or hypertrophic scars, which are frequently seen in specific human populations, especially those of African or Asian heritage. Despite this shortcoming, rats continue to be essential models in preclinical research. Their close physiological and genetic resemblance to humans makes them valuable for investigating chronic wound conditions and assessing new therapeutic strategies before moving to clinical applications (Gurtner et al., 2008). For example, a study using streptozotocin (STZ)-induced diabetic rats with full-thickness wounds on their backs demonstrated that treatment with a decellularized porcine acellular dermal matrix (ADM) combined with autologous adipose-derived stem cells (ASCs) notably improved tissue repair and stimulated cellular growth. This treatment strategy also proved effective in reducing inflammation, which is a major barrier to healing in chronic diabetic wounds. In another study, researchers tested Rejuveinix (RJX), an intravenous solution packed with essential vitamins and antioxidants on diabetic rats. These rats developed diabetes through a combination of a high-fat diet and STZ, which mimics the disease's progression in humans. RJX was delivered intraperitoneally over a three-week period following the inclusion of full-thickness excisional or burn wounds.

Even at doses more than ten times lower than the maximum clinically tolerated level, RJX significantly improved healing outcomes in both wound models, demonstrating its promise as a novel therapeutic option for diabetic wound care (Uckun et al., 2022).

5.3.1.2 Mouse

Mice *(Mus musculus)* are among the most commonly used animal in wound healing research. Their small size, low cost, ease of handling, and fast breeding make them ideal for laboratory studies. One of their biggest advantages is their genetic similarity to humans and the availability of transgenic and knockout strains, which help researchers study specific genes and molecular pathways involved in tissue repair (Doyle et al., 2012). In wound healing studies, full-thickness excisional or incisional wounds (usually 4–8 mm) are created on the back (dorsal skin) of anesthetized mice using biopsy punches (Rhea and Dunnwald, 2020; Yampolsky et al., 2024). This area is preferred because it reduces the chance of the mice interfering with the wound, often wounds are made on both sides of the back, one side for treatment and other as a control allowing researchers to compare outcomes within the same animal and reduced variability in results. A key difference between mice and humans is how wound heals. In mice, wound closure mainly happens through contraction driven by a muscle layer under the skin called the panniculus carnosus that is not present in human (Naldaiz-Gastesi et al., 2018). This leads to faster healing in mice but does not fully mimic human wound repair. To overcome this, researchers use silicone splint around the wound to limit contraction and encourage healing through granulation and re-epithelialization similar to the way human skin heals (Poelchow, 2022). Different mouse strains are selected based on the research goal. For example, C57BL/6 mice are often chosen for studies focusing on inflammation and immune response due to their well-defined immune system (Gurley et al., 2018; Hackett et al., 2022). BALB/c mice and other strains are also commonly used depending on experimental needs (Andleeb et al., 2022). To study chronic or impaired wound healing, researches use genetically diabetic mice like db/db strain or induce diabetes chemically using streptozotocin (STZ) (Ili and Sari, 2023). Another useful model is the tail wound model, where a rectangular wound (0.3 × 1.0 cm) is created on the upper part of the mouse's tail (Zhou et al., 2019). This area heals more slowly taking up to 21 days providing a longer time window for testing new treatments (Histing et al., 2011). Since tail wound doesn't require euthanizing the mouse for evaluation, researcher can observe the healing process over time in the same animal. Male typically have

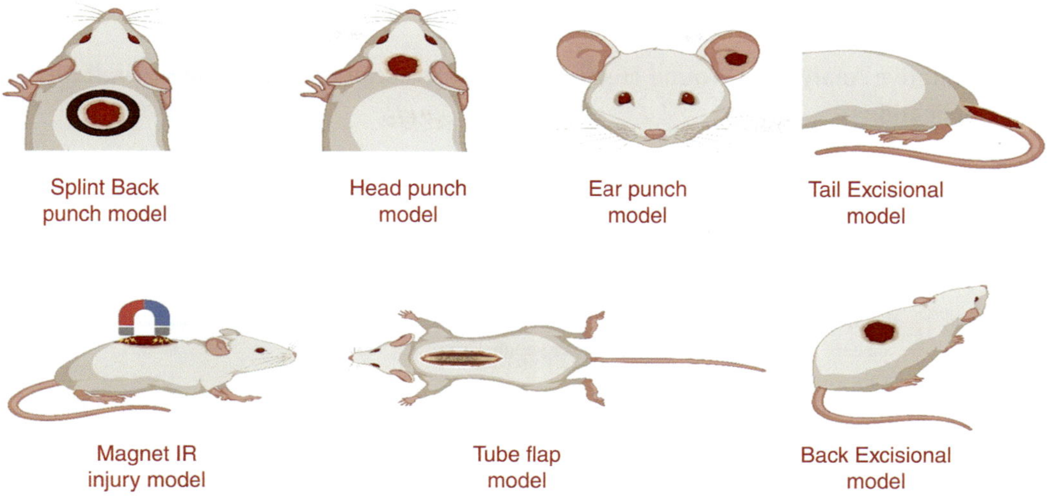

FIGURE 5.2 Murine models. This illustration shows various models (splint back punch, head punch, ear punch, tail excisional, magnet IR injury, tube flap, back excisional) used of study different wounds and healing processes.

thicker and stronger skin due to a thicker dermis while female have thicker epidermis and fat layer (Rahrovan et al., 2018). This difference should be considered when designing experiments. Despite some limitations, mice remain a valuable model for studying wound healing because of their genetic flexibility, cost-effectiveness, and the wealth of tools available for molecular and histological analysis. This continues to play a crucial role in the development of the new therapies and understanding the biology of skin repair (Barakat et al., 2021). Murine models serve as crucial preclinical tools for investigating wound healing mechanism across divers wound characteristic and influencing factors. As shown in Figure 5.2, these models include the splint back punch, head punch, ear punch, tail excisional, magnet infrared (IR) injury, tube flap, and back excisional models, each designed to stimulate specific wound types and locations. Table 5.3 further elaborates on these and additional models such as the back incisional and transverse rectus abdominis myocutaneous (TRAM) flap models, detailing their secretions, advantages, limitations, and applications in wound healing research. These models offer distinct advantages, including the capacity to mimic human-like healing (head punch, splinted back punch), facilitate specific analyses (tail excisional), and model particular clinical conditions like chronic wounds (tail excisional impaired wound healing magnet IR injury) and systemic disease such as diabetes (systemic impairment). However, limitations such as restricted wound areas (head punch), technical complexities (splint back punch, TRAM flap), and physiology disparities with human condition (systemic impairment) must be considered.

5.3.2 Transgenic Mice

Transgenic and knockout mice have significantly advanced our understanding of molecular pathways involved in wound healing (Palmieri et al., 2017). These models allow researchers to explore the roles of various growth factors such as epidermal growth factor (EGF), keratinocyte growth factor (KGF), and transforming growth factor (TGF) which are key to tissue repair, especially in chronic wounds (Bártolo et al., 2022). Beyond growth factors studies also focus on extracellular matrix protein, angiogenic signals, cell death pathways, cytoskeletal elements, signaling cascades, and cell adhesion molecules (Siddhartha and Garg, 2023). As research progress, there is growing interest in examining how systemic condition like diabetes and aging influence chronic wound development. To this end genetically modified mice targeting genes related to oxidative stress, apoptosis and angiogenesis

TABLE 5.3
A Comparative Analysis of Wound Healing Models: Description, Advantages, Limitations, and Applications

Model	Description	Advantages	Limitations	Applications
Head punch model	Full thickness wound on scalp using trephine Heals in ~8 days (Gökkaya and Gorgu, 2020)	Minimal contraction due to skull stability Mimics human healing	Limited wound area	Granulation and re-epithelialization studies
Ear punch model	Full thickness wound on ear heals within a week (Rivera and Spencer, 2007)	Minimal subcutaneous tissue Heals via re-epithelialization	Thin tissue may complicate manipulation	Epithelial healing studies
Tail excisional model	Full thickness wound on proximal tail heals in 2–3 weeks (Stenn and Depalma, 1988).	Minimal contraction Allows histology and molecular analysis	Small wound size Limited space	Chronic wound modeling; non-terminal follow-up
Back incisional model	Linear full-thickness incision on dorsal skin (Boyce et al., 2013)	Models all wound healing phases Good for measuring tensile strength	Rapid re-epithelialization Hard to distinguish granulation tissue	Surgical wound healing; collagen studies
Back punch model	Circular full thickness wound on back with dressing (Yannas et al., 2017)	Simple, widely used Dual wound sites enable internal controls	Up to 90% contraction Less comparable to human skin healing	Preliminary screening; contraction studies
Back splinted punch model	Circular wounds with silicone splint to reduce contraction (Bártolo et al., 2022)	Mimics human-like healing Multiple wounds per animal possible	Technically demanding	Re-epithelialization, granulation, angiogenesis studies
Impaired wound healing	Local ischemia or ischemia-reperfusion injury (Honnegowda et al., 2015)	Models chronic wounds Evaluates roles of growth factors, oxygen, etc.	Requires surgical precision Complex physiology	Chronic wound and IR therapy research
TRAM flap model	Pedicled flap with vessel ligation and silicone barrier (Weng et al., 2010).	Mimics ischemia and necrosis Suitable for angiogenesis research	Technically complex High variability	Ischemic wound healing; vascularization studies
Tube flap model	Bipedicled back skin flap with cylindrical tubing underneath (Fang and Mustoe, 2008).	Prevents vascularization Allows for standard wound creation	Ischemia variability Technical handling	Angiogenesis, epithelialization, and strength assessments
Magnet IR injury model	Compression via external magnet over subdermal magnet to simulate IR injury (Li et al., 2024).	Adjustable IR cycles mimics pressure ulcers and chronic wounds	Requires magnet placement and monitoring	Chronic wound and reperfusion injury modeling
Systemic impairment models	Includes STZ-induced diabetic and db/db genetically diabetic mice (Li et al., 2024).	Mimics systemic factors db/db mouse exhibits delayed healing, obesity-related wound issues	Genetic differences from human diabetes Costly	Diabetic wound therapy, delayed healing mechanisms

are valuable tools. While standard excisional wound models in mice have limitations, they remain useful for demonstrating phenotypic differences, especially when interpreted in the right useful for demonstrating phenotypic differences, especially when interpreted in the right context. Some studies involving gene knockout, like those targeting fibrinogen, tenascin-C, P-selectin, E-selectin, KGF and EGF have shown mixed or inconclusive result (Drew et al., 2001; Matsuda et al., 1999; Mayadas, 1995). These inconsistencies highlight the need for more specific models and approaches to better understand complex wound healing mechanisms. Choosing the right wound model for experiments using transgenic or knockout mice is critical, as each model highlights different aspects of the healing process, for example, the back incisional model is suitable for assessing tensile strength but not re-epithelialization which happens too rapidly in mice (Falanga et al., 2004). In contrast, the ear punch model is ideal for studying re-epithelialization since the wound remains open due to the underlying cartilage (Safaee, 2022). The splinted back punch model is widely regarded as one of the most effective approaches for replicating human wound healing, as it minimizes wound contraction and more accurately mirrors key processes such as re-epithelialization, tissue development, and granulation tissue development (Grambow et al., 2021).

Because wound healing is such a complex process driven by overlapping and compensatory signaling pathways, researchers often rely on specialized models of impaired healing, including double knockout animals, to uncover subtle or hidden biological effects. To gain clearer and more comprehensive insights, scientists typically use a multifaceted approach, combining different wound models with genetically modified animal strains like transgenic or knockout lines. This strategy helps explore the many intricate biological events involved in tissue repair more thoroughly. For example, studies examining the role of growth factors in wound healing might use a mix of head punch, ear punch, and splinted back punch models to observe how these factors work in different wound settings. Similarly, to study new blood vessel formation (angiogenesis), researchers often use impaired healing models such as the TRAM flap or tube flap, along with the splinted punch model, to increase both the clinical relevance and the depth of their findings (Edmunds et al., 2013; Sosnowski et al., 2022; Wang et al., 2013).

A good example of this approach is the study on TGF-α knockout mice (Kim et al., 2022). Although earlier studies showed no significant effect of TGF-α deletion on wound healing, Kim's study used both ear punch and head punch wounds. The ear punch model revealed impaired healing in partial-thickness wounds, while the head punch model showed no effect in full-thickness wounds. This contrast demonstrated that TGF-α plays a crucial role in re-epithelializing partial-thickness wound, but full-thickness wound may activate other compensatory mechanism to overcome its absence. Such finding underscore the importance of model selection and combination in uncovering the true role of various healing factors (Werner and Grose, 2003).

5.3.3 Rabbits

Rabbits are frequently employed in toxicological and pharmacological investigations on wound healing because rabbits have a higher skin permeability than humans and other animals like mice, rats, and pigs. The response of rabbit skin to topical applications of therapeutics and aging is similar to the human skin (Grada et al., 2018). They are advantageous to use when small animal models are required due to larger size as compared to murine, low cost of husbandry, and a short gelation time. The relatively large blood vessels in rabbit ears provide a valuable anatomical feature for the experimental investigation focused on ischemia. The prominent vessels facilitate controlled manipulation and observation of blood flow restriction and its consequences (Grada et al., 2018). Despite their utility in permeability and ischemia studies, rabbit models for wound healing have limitations including the absence of elastic fibers in the dermal connective tissue, a high density of hair follicles similar to rats and the presence of the panniculus carnosus muscle, which contributes to wound contraction, potentially skewing result compared to human healing. Moreover, paucity of specific reagents for rabbit's species, and their limited genetic tractability make them uncommon in the recent years in the study of the processes of wound healing (Grada et al., 2018). Rabbits, especially the New Zealand White (NZW) breed, are frequently utilized in biomedical research focused on skin wound healing. Since rabbits are docile, they have a suitable body size and have skin structure that is quite similar to that of humans; they are widely acknowledged as useful models in wound healing studies. Like human skin, its skin is stratified and made of distinct layers, such as the epidermis, dermis, and subcutaneous tissue. Angiogenesis, re-epithelialization, and collagen remodeling are among the important process that rabbits are perfect for research because of their vast vascular network and human-like healing mechanism (Ostvar et al., 2015).

A major advantage of employing rabbits is their comparatively high surface area, which allows for the development of several uniform wounds on the same animal. This improves the consistency of experimental data and allows for comparison investigations under standardized setting. Additionally, their composure reduces stress and encourages more dependable healing patterns by making them easier to handle throughout crucial procedure like sample collection, dressing application, and wound development. Rabbits are also great models for studying inflammatory responses, infection control, and the immunological-related consequences of wound therapies because of their well-understood immune system (Guoqi et al., 2018). Altogether, these characteristics make rabbits an important preclinical model for evaluating a range of wound care strategies, including skin grafts, tissue-engineered scaffolds, growth factor therapies, and antimicrobial interventions. Rabbits are frequently used in research on diabetic wound healing, burn wound healing, skin graft evaluation, biodegradable scaffold and growth factors delivery system testing and evaluation of antioxidant and antibacterial treatment. Developing successful treatment plans requires a deep understanding of the phase of wound healing as well as the biological element that may obstruct it. This entails adopting systemic therapeutic techniques such as the TIME frameworks which includes tissue management, inflammation/infection control, moisture balance, and wound edge assessment and comprehending both the microscopic and macroscopic components of healing. Integrating this fundamental understanding with proficiency in topical medications, sophisticated therapies, and the appropriate selection of wound dressing is also necessary for successful wound care (Schultz et al., 2003). The rabbit ear model, originally developed by Ahn and Mustoe in 1990, has become a widely accepted method for studying ischemic wounds and investigating the impact of hypoxia on the healing process. The ear is supplied by three primary arteries. Ischemia is induced by ligating the rostral and central arteries through a circumferential incision, disrupting dermal arterial blood flow while preserving venous drainage. A 6-mm punch biopsy down to the cartilage creates a full-thickness wound with minimal vascular support. Because the dermis is firmly attached to the cartilage, wound closure occurs primarily via re-epithelialization and granulation tissue formation, not contraction. This model's main advantage is the ability to generate multiple identical wounds on a single ear, with the opposite ear serving as a control. Additionally, the structural rigidity of the ear allows independent quantification of epithelialization, as contraction is limited by the cartilage's splinting effect. While theoretically replicable in rodents, the model demands technical precision and surgical magnification, which limit its practicality in smaller species. The rabbit ear model has also been instrumental in studying the effect of topical growth factors on chronic wound healing (Ahn and Mustoe, 1990). Although many growth factors have shown promise in preclinical rabbit studies, results in human clinical trials have been less consistent. Nevertheless, rabbit wounds have displayed healing patterns similar to those in humans such as increased scarring with delayed epithelialization and reduced scarring associated with aging, steroid use, or collagen synthesis inhibitors highlighting their relevance for translational research.

5.3.4 Porcine

When selecting an appropriate animal model for wound healing research, several factors must be considered, including availability, cost, and ease of handling, anatomical and physiological relevance, and researcher familiarity. While small mammals like rabbits, guinea pigs, rats, and mice are commonly used due to their affordability and manageability, they exhibit key differences from humans. These differences include denser body hair, thinner skin layers, and a primary wound healing mechanism driven by contraction rather than re-epithelialization. In the study of wound healing, particularly preclinical studies, pigs are commonly used in the recent years due to the similarities in the anatomical and physiological processes between the skin of pigs and human skin. Like the human skin, the skin of pigs is also composed of sparse hair, elastic fibers in the connective tissues, dermal papillae and ridges at epithelial junctions as well as they share similar dermal and epidermal thickness to human skin. The pig skin is strongly attached to the subcutaneous connective tissues, unlike other animals. Their large size makes it possible to create control and experimental wounds on the same animal (Grada et al., 2018). However, the limitations of pig models must be taken into account like other animals when performing experiments for wound healing. While sharing many dermal components with human skin, pig skin presents challenges for wound healing studies due to its significantly denser stratum corneum, poorer dermal vascularization, and elevated epidermal alkaline phosphate, an enzyme associated with collagen turnover. These difference can influence the outcomes of pharmacological and physiological investigation (Krotzsch et al., 2005). A key difference between human and pig skin lies in their sweat and sebaceous glands. Pig skin lacks eccrine sweat gland, a significant source of keratinocyte stem cells crucial for re-epithelialization in human skin and possesses only apocrine sweat glands. This contrasts with human skin, which has both eccrine glands distributed widely and distinct sebaceous glands. Additionally, the biochemical makeup of porcine dermal collagen closely mirrors that of human collagen, making pigs highly relevant for testing wound healing therapies. Neither humans nor pigs have a panniculus carnosus, and both species exhibit similar epidermal turnover rates, keratin patterns, and stratum corneum lipid profiles (Sullivan et al., 2001). Although there are distinctions such as pigs lacking widespread eccrine sweat glands and differences in apocrine gland distribution, the overall structural and physiological similarities make pigs a highly suitable model for human wound healing studies. For example, partial-thickness wounds in both species heal predominantly through re-epithelialization, whereas full-thickness wounds in pigs heal largely by contraction. A refined wound healing model has been developed using Lanyu miniature pigs, which are particularly similar to human skin in both structure and healing response. In this model, researchers created various wounds of different sizes and depths on the pigs' backs to establish optimal testing conditions. It was determined that wounds should be spaced at least 4 cm apart and measure at least 4×4 cm to prevent overlap in healing and to facilitate accurate assessment of treatment efficacy. Deeper wounds required longer healing periods, with complete restoration of skin strength and architecture observed after six months. This model supports ethical research practices by minimizing animal use while providing robust, reproducible data (Kuo et al., 2022).

Despite the advantages, gaps remain in our understanding of certain wound healing mechanisms in pigs, such as the activity of myofibroblasts and the formation of granulation tissue. While pigs and humans share many same features of skin biology, differences in skin stiffness and other mechanical properties could influence cellular behaviors like myofibroblast differentiation. Further standardization of pig breeds and experimental protocols would improve the consistency and relevance of findings. Research focusing on earlier healing stages, especially within the first week post-injury, may provide crucial insights into myofibroblast origin, spatial behavior, and differentiation timing. Pig models have been used to investigate a wide range of cutaneous wound types, including partial- and full-thickness excisions, incisions, burns, ischemic injuries, and laser-induced wounds. In some cases, models of pressure ulcers have been created by inducing limb denervation followed by casting. The pig skin also exhibits transdermal absorption and percutaneous permeability properties that closely resemble those of human skin, making it an ideal model for evaluating topical treatments and transdermal drug delivery systems (Hamilton et al., 2022). However, pigs are more expensive to acquire and maintain than smaller laboratory animals. While their skin is highly comparable to that of humans in many ways, it also differs in some aspects such as higher pH levels, a fattier subcutaneous layer, and thicker, stiffer skin in larger pigs. It is important to consider these distinct when assessing both physiological and therapeutic response. In wound healing research, the pig is still a useful and trustworthy model that provides insights that may be applied directly to clinical settings in humans.

Furthermore, many laboratories are unable to accommodate the larger pigs due to their greater maintenance costs and larger housing requirements. Anesthesia administration and other surgical procedures are particularly difficult (Grada et al., 2018). Notwithstanding these variations, pig models are increasingly being utilized to assess the healing effectiveness of possible therapeutic agents and wound dressings, particularly in light of the growing usage of mini pigs to research the healing of full-thickness wounds because they are considerably easier to handle due to their smaller size; mini pigs have already replaced larger pigs in many types of pig research.

5.3.5 Zebrafish

Aquatic organism faces numerous stressors that can lead to tissue damage. These include environmental shift like ocean acidification, physical trauma from interactions or aquaculture practices (stocking density, abrasion, and handling), chemical pollution (metals, pesticides, industrial waste), and various disease. These factors can induce tissue damage, cell death (apoptosis, necrosis), and mortality. Given that skin and muscle integrity are vital for preventing pathogen invasion and secondary infection, a swift and effective wound healing response is crucial for survival and health of aquatic life. All over the world, researchers are becoming more interested in zebrafish (*Danio rerio*), which have become a well-known model organism for studying a variety of disorder in recent years (He et al., 2020). Zebrafish are topical freshwater teleost species that are indigenous to South Asia's Himalayan region. They belong to the class Actinopterygii and the family Cyprinidae (Arunachalam et al., 2013). Zebrafish are a very intriguing model for scientific study because of their extraordinary capacity for organ regeneration. Their embryos are transparent, rapidly developing externally and easy to observe (Chen et al., 2020). This facilitates scientist observation of the life cycle of zebrafish; the larvae's main organs are already working by five days after fertilization. Furthermore, female zebrafish can produce hundreds of embryos per week and they breed all year long (Veldman and Lin, 2008).

Moreover, their small size, rapid reproduction rate, short developmental and reproductive cycles, transparent embryos, and significant genetic and physiological similarities to humans contribute to their growing popularity among researchers (Beffagna, 2019). Due to their notable similarities with humans, zebrafish are often regarded as a standard vertebrate model. As a result, numerous laboratories utilize their biological systems to explore unexplained diseases and support drug discovery efforts (Antinucci and Hindges, 2016; Arjmand et al., 2020; Hason and Bartunek, 2019; He et al., 2021; Kanungo et al., 2014; Zang et al., 2017, 2018). This involves creating disease-specific assays and identifying gene counterparts associated with human's illness (Rubinstein, 2003).

Zebrafish are very useful in this type of research because they are small, transparent, and their immune system shares similarities with humans. Researchers can tag their immune cells with fluorescent markers, making it possible to watch those cells move inside the fish in real time under a microscope (Subaramaniyam et al., 2023). Research has revealed that hydrogen peroxide (H_2O_2), widely known for its use in wound disinfection, also plays a critical role as an early signaling molecule that directs immune cells to the sites of injury. In zebrafish models, tissue damage triggers a rapid release of H_2O_2 within five minutes, creating a chemical gradient that extends from the wound to the nearby blood vessels. This gradient serves as a guiding signal, enabling immune cells to navigate to the injured area. To visualize this process, scientists used a fluorescent probe called HyPer, which lights up in the presence of H_2O_2. The fluorescent signal appeared before any immune cell movement, suggesting that H_2O_2 acts as an early "danger signal" to initiate the immune response. When researchers blocked H_2O_2 production using genetic tools or chemical inhibitors, neutrophils failed to migrate to the wound, highlighting H_2O_2's essential role in immune cell recruitment. The enzyme responsible for H_2O_2 generation in this context is Duox, found in the outermost skin and fin layers of the zebrafish. Duox is part of the nicotinamide adenine dinucleotide phosphate (NADPH) oxidase enzyme family, which typically generates reactive oxygen species during normal cell functions. Disrupting Duox activity through targeted genetic modifications stops H_2O_2 production and prevents immune cells from accumulating at the injury site. This clearly highlights the critical role of Duox in initiating the wound healing process (Niethammer et al., 2009). In mammals, especially adults, wound healing is a complex process involving blood clotting, inflammation, skin cell migration (re-epithelialization), new tissue formation (granulation), formation of new blood vessels (neovascularization), and finally, scar formation. But in embryos, wounds heal more simply, without inflammation or scars. The researchers wanted to see whether adult zebrafish could serve as a good model to study wound healing, since these fish are easier and cheaper to work with compared to mammals (Richardson et al., 2013).

To do this, they developed a method where they used a laser to make full-thickness wounds (damaging both the outer skin and inner layers) on the side (flank) of adult zebrafish. They found that the zebrafish healed these wounds very quickly, and interestingly, without making a blood clot on the surface, which usually happens in mammals. Within 24 hours, the wound was covered again with a new layer of skin (called the neoepidermis), even though the initial wound had removed all layers of skin, including scales and fat cells below the skin. This shows that the first step of skin healing, re-epithelialization, happens very fast in zebrafish. The researchers also observed a strong inflammatory response. They used special zebrafish with glowing immune cells (green and red under a microscope) to track inflammation. Within hours, neutrophils and macrophages (types of white blood cells) moved to the wound site, though the very first skin cells that migrated to cover the wound seemed to do so without waiting for the immune cells to arrive. Over the next few days, the number of immune cells increased and then slowly decreased again, with mostly macrophages remaining at day 4 after wounding (Richardson et al., 2013).

Alongside this inflammation, the zebrafish formed granulation tissue under the new skin just like mammals. This tissue included fibroblasts (cells that make collagen), macrophages, and new blood vessels. The granulation tissue reached its peak around day 4 and began to shrink after that (Alhajj and Goyal, 2022). By day 10, most of it was gone.

5.3.6 Non-Human Primates

Non-human primates (NHPs) are considered highly relevant models for wound healing studies due to their close evolutionary relationship with humans. Their anatomical, physiological, and immunological features make them particularly advantageous over commonly used small animals such as rodents and rabbits (Barrenas et al., 2019). Wound healing follows four overlapping stages: hemostasis, inflammation, proliferation, and remodeling, which are conserved across mammalian species. Nevertheless, differences exist in signaling mechanisms, cellular behavior, and tissue responses across species. NHPs exhibit skin characteristics that closely resemble those of humans, such as comparable epidermal thickness, collagen organization within the dermis, and the presence of both hair follicles and sweat glands (Arakawa et al., 2019). In contrast to rodents, primates do not have a well-developed panniculus carnosus muscle, a structure that enables rapid wound contraction in rodents. This anatomical distinction makes the wound healing process in primates more reflective of human skin regeneration (Bahri et al., 2019). In addition, the immune system of NHPs closely mirrors human innate and adaptive responses, which is particularly significant during the inflammatory phase of wound healing. Primate models also develop scars in a manner similar to humans, making them valuable for studying fibrotic conditions like hypertrophic scars and keloids, pathologies that are difficult to replicate in highly regenerative species such as rabbits. Moreover, primates are well-suited for investigating complex wound healing issues linked to systemic illnesses like diabetes and vascular disorders. Their relatively long lifespan further supports prolonged observation, which is essential for assessing the durability and safety of therapeutic interventions over time (Lorenz et al., 1993; Mistry et al., 2022; Wang et al., 2020).

To investigate various aspects of wound healing, NHPs have been widely used in wound healing making possible the detailed studies of immune cell recruitment, growth factor expression vascular endothelial growth factor (VEGF), platelet-derived growth factor (PDGF), and TGF beta (TGF-β), angiogenesis, and extracellular remodeling. For exploring wound healing impairment caused by metabolic disease such as diabetes that compromised cellular growth, delayed healing, and dysregulated growth factors, these models are very beneficial. When it comes to developing effective antimicrobial treatments for challenging conditions like wound infections especially those complicated by biofilm formation and immune system responses, primate models play a vital role.

These models closely mimic human biology, making them especially useful for studying complex processes such as fibrosis. They provide an important platform for exploring molecular pathways involved in scarring and are essential for testing new therapies aimed at reducing excessive scar formation and improving wound healing outcomes (Babensee, 2020). Primate models have been key in evaluating cutting-edge treatments like stem cell therapy, innovative biomaterials, gene-based interventions, and targeted delivery of growth factors. Due to their close similarity to humans, these models are also ideal for studying wound closure techniques, surgical care, and post-operative management in settings that closely reflect real-life clinical situations.

Despite their benefits, using monkey models in wound healing research has a number of drawbacks. Due to their genetic proximity to humans, necessitating stringent oversight, and adherence to welfare standards, ethical considerations are significant. These studies have high cost associated with animal procurement, specialized care, housing, and longer study durations. Skilled personnel are required for the handling of primates. The use of sedation more frequently can influence immune response and healing, potentially confounding results (Hocum Stone et al., 2021; Tardif et al., 2013). As such, while invaluable in certain contexts, the use of primates is generally reserved for studies where alternative models are insufficient.

Genetic diversity within primate populations may also introduce variability, affecting reproducibility unlike the consistency offered by inbred rodent strains. Furthermore, while their slower healing rates make them more reflective of human healing, they also extend the duration of studies. Some species-specific differences in collagen deposition and scar formation have been observed, which may affect the direct applicability of findings to human patients (Strobel et al., 2015).

5.4 ADVANTAGES AND DISADVANTAGES OF USING ANIMAL MODELS

Animal models are crucial for improving our knowledge of the process underlying wound healing and evaluating the effectives of cutting-edge therapeutics approaches. Rats, pig, rabbits, and other animal are used as models to stimulate many facets to human wound recovery. Each model has unique benefits, such as cost-effectiveness, convince of handling, or physiological and anatomical resemblance to human skin. In contrast, porcine models are considered more relevant clinically because of their physiological and anatomical similarities to human skin, including dermal thickness and dermal structure. Limitations do occur through variations in immunological response, healing processes (such as contraction versus re-epithelialization) and moral considerations. Depending on the particular research goals, the model will be selected. Mice and rats are popular small animal models as they are readily available, inexpensive to maintain, and genetically manipulable, nevertheless their wound healing process is dominated by contraction, unlike human wound healing which is dominated by re-epithelialization. However, pigs are expensive to maintain and require specialized facilities and handling. Rabbits are useful for certain wound infection and vascular studies, though their skin structure differs significantly from humans. Zebrafish are used in the research of wound healing due to their transparent embryos, which allow real-time visualization of molecular and cellular processes. However, their regenerative mechanisms and skin structure differ significantly from mammals, limiting their direct translational relevance. Primates may offer closer physiological relevance but raise significant ethical and regulatory concerns. Therefore, selecting an appropriate animal model requires balancing scientific objectives with ethical, logistic, and translational considerations. The advantages and disadvantages of different animal models in wound healing are summarized in Table 5.4.

TABLE 5.4
Advantages and Disadvantages of Animal Models Used in Wound Healing Research

Animal Model	Advantages	Disadvantages	References
Rodents (rats and mice)	• Easy handling • Inexpensive • Possibility of rapid study cycles due to short duration of healing • Modified strains available for mechanistic studies	• Have panniculus carnosus muscle layer (skin structural differences as compared to humans) • Wound healing by contraction instead of re-epithelization • Extent of experimental manipulation is limited due to the smaller size	Masson-Meyers et al. (2020)
Rabbits	• Multiple standardized wounds are possible by ear model • Healing mechanism is similar to humans (re-epithelization) • Suitable for the study of hypertrophic scar formation	• Pattern of hair follicles and thickness of skin differ from humans • Specific reagents are limited for the species	Martin and Nunan (2015), Flynn et al. (2023)
Pigs	• Close resemblance of anatomy and physiology with human skin • Best suited for the study of deeper and larger wounds • Content of elastin and collagen is similar to humans	• Cost of housing and maintenance is high • Limited availability of molecular reagents • High risk of infection •	Oryan et al. (2017)
Guinea pigs	• Cost-effective • Small size • Vitamin C required for the synthesis of collagen similar to humans	• Wound healing by contraction • Not used commonly causing smaller knowledge base	Silva-Gomes et al. (2015)
Zebrafish	• Inexpensive • Small • Distinctive phase studies are possible due to non-overlapping wound healing phases	• Less established protocols limit its use • Dissimilar skin structure as compared to humans	Lin et al. (2023)
Primate	• Physiological similarity to humans • Similar immune response as compared to humans, allowing better modeling of immune-related aspects of wound healing	• Subjected to stringent ethical regulations limit their availability • High cost of maintenance and housing • Slower rate of wound healing as compared to humans	Utsumi et al. (2017)

5.5 BIOMEDICAL SIGNIFICANCE

Using small animal models for wound healing studies has significant biomedical relevance.

5.5.1 Understanding Wound Healing: Cellular and Molecular Mechanisms

Wound healing is a dynamic and orchestrated biological process involving a series of overlapping phases: hemostasis, inflammation, proliferation, and remodeling. Small animal models such as mice and rats have been instrumental in elucidating the underlying cellular and molecular mechanisms due to their genetic tractability and physiological similarities in skin repair. Neutrophils are the first immune cells to infiltrate the wound, attracted by chemokines such as C-X-C motif chemokine ligand 1 (CXCL1) and C-X-C motif chemokine ligand 8 (CXCL8). They play a key role in microbial clearance and release of pro-inflammatory cytokines (e.g., **IL-1β**, **TNF-α**) and proteases to digest damaged tissue (Brem and Tomic-Canic, 2007). Their early removal by apoptosis is critical to prevent prolonged inflammation.

Monocyte-derived macrophages dominate later stages of the inflammatory phase. Initially polarized to the M1 phenotype, they release nitric oxide and reactive oxygen species. During the transition to the proliferative phase, they adopt an M2 phenotype, releasing IL-10 and TGF-β, which aid in fibroblast recruitment, angiogenesis, and matrix remodeling (Koh and DiPietro, 2011). Activated fibroblasts migrate into the wound bed under the influence of TGF-β and PDGF, secreting collagen I, fibronectin, and proteoglycans to form the granulation tissue scaffold (Burd et al., 2007). Keratinocytes proliferate and migrate over the wound bed in response to **EGF** and **FGF**, re-establishing the epidermal barrier. They are also involved in crosstalk with immune cells through cytokine secretion (Pastar et al., 2014). Endothelial cells stimulated by **VEGF** and **angiopoietins** initiate **angiogenesis**, restoring perfusion to the regenerating tissue and facilitating metabolic support during repair (Tonnesen et al., 2000).

5.5.2 Molecular Signaling Pathways in Wound Repair

TGF-β plays a dual role in wound healing by modulating inflammation and promoting ECM synthesis. It signals through **Smad2/3** to activate fibroblasts and stimulate collagen deposition. Aberrant signaling can lead to fibrosis (Border and Noble, 1994). NF-κB is activated early in response to injury and controls the transcription of pro-inflammatory mediators like **IL-6**, **IL-8**, and **COX-2**. It is critical for leukocyte recruitment and sustaining inflammatory responses (Leavy, 2009). **ERK1/2** promotes keratinocyte proliferation and migration; **JNK and p38** mediate responses to stress and regulate apoptosis, inflammation, and cytokine production; and Wnt/β-Catenin pathway regulates **stem cell activation**, keratinocyte migration, and dermal remodeling. Wnt ligands are upregulated during wound healing, promoting re-epithelialization and granulation tissue formation (Tríbulo et al., 2012).

5.5.3 Inflammatory Response and Immune Modulation

The inflammatory phase is initiated by danger-associated molecular patterns (DAMPs) and pathogen-associated molecular patterns (PAMPs), which activate Toll-like receptors (TLRs) on immune cells. The proper resolution of inflammation requires a shift from pro-inflammatory to anti-inflammatory signals, characterized by a rise in M2 macrophages, increased IL-10, and reduced TNF-α levels (Patel, 2018; Willenborg and Eming, 2014). Failure to resolve inflammation leads to chronic wounds, often seen in diabetic and ischemic conditions, where small animal models like db/db mice are used to study impaired healing (Mieczkowski et al., 2022; Tesch and Lim, 2011; Galiano et al., 2004).

5.6 CHALLENGES ASSOCIATED WITH THE USE OF SMALL ANIMAL MODELS IN WOUND HEALING RESEARCH

Although small animal models, particularly mice and rats, have been instrumental in advancing our understanding of wound healing, several limitations and challenges must be acknowledged when translating findings to clinical practice.

5.6.1 Anatomical and Physiological Differences

A limitation of using small animals, particularly rodents, in wound healing research is the noticeable difference in their skin structure and repair mechanisms compared to humans. In rodents, the skin is loosely connected to underlying muscle layer that promotes rapid wound contraction, a process that plays only a minor role in human healing (Sullivan et al., 2001). While human wound repair primarily depends on re-epithelialization and granulation tissue formation, rodents rely heavily on wound contraction, making it challenging to directly translate their healing rates and processes to human clinical scenarios.

5.6.2 Immunological Differences

Rodents exhibit notable differences from humans in their innate and adaptive immune systems, including variations in cytokine expression patterns, leukocyte populations, and their functions, as well as macrophage plasticity and polarization. These variations can lead to **differences in inflammatory responses**, which are critical in wound healing outcomes. For example, the balance between **M1** and **M2 macrophages** can vary, influencing the progression from inflammation to tissue repair (Willenborg and Eming, 2014).

5.6.3 Genetic Homogeneity Versus Human Diversity

Most laboratory animals used are inbred strains, such as C57BL/6 mice, which exhibit genetic uniformity and reduce variability in experimental outcomes. However, this does not mimic the **genetic heterogeneity** seen in human populations. As a result, findings may not account for variations in wound healing due to age, sex, comorbidities, or genetic predisposition, which are common in clinical settings (Seok et al., 2013).

5.6.4 Differences in Wound Etiology and Comorbidities

Wounds in small animal models are typically **surgically induced**, clean, and sterile, which differs from clinical wounds that are often chronic, infected, complicated by systemic conditions (e.g., diabetes, vascular disease). While diabetic models like **db/db mice** or streptozotocin-induced rats are available, they only partially mimic the complexity of chronic wounds in humans (Galiano et al., 2004).

5.6.5 Ethical Considerations and Welfare

Although small animals are considered more ethically manageable than larger mammals, their use still raises concerns related to pain management, stress from restraint or wound dressing changes, mortality in long-term studies. These factors can confound experimental results and need to be strictly regulated by institutional animal care guidelines (Campos-Junior et al., 2011).

5.6.6 Limitations in Imaging and Monitoring

Due to their small size, **real-time imaging, wound depth measurement**, and **histological sampling** can be technically challenging. Specialized equipment (e.g., multiphoton microscopy, Doppler imaging) is often required, which may not be universally accessible (Holt et al., 2018).

5.6.7 Translational Limitations

The ultimate challenge lies in translating findings from small animals to humans. Therapeutics that are successful in rodents frequently **fail in human trials** due to the above-mentioned discrepancies. For improved translational accuracy, researchers often complement small animal work with **large animal models (e.g., pigs, dogs)** or **ex vivo human skin models**, especially in late-stage preclinical testing (Seaton et al., 2015).

5.7 CONCLUSION

Animal models play a vital role in *in-vivo* studies, serving as an essential step in translating biomedical research findings to human application. Animal models contribute significantly not only to human healthcare but also to the welfare of veterinary patients. They are crucial in various aspects of biomedical research, including drug development, toxicity evaluation, pharmacokinetic analysis, and preclinical testing of medical and tissue engineering devices designed for human use. While laboratory animal models are more economical and suitable for high-throughput testing compared to larger animals, the development of large animal models is essential for generating preclinical data and evaluating the clinical potential of vascular grafts, orthopedic implants, and bone plates due to their closer resemblance to human anatomy and physiology. Regardless of the approach taken in utilizing animal models for biomedical research, adherence to the 3Rs principles (Replacement, Reduction, and Refinement) is imperative.

GLOSSARY

Adeno-Associated Virus (AAV): A small, non-pathogenic virus used as a vector for gene delivery in transgenic animal models.

Adipose-Derived Stem Cells (ASCs): Stem cells obtained from fat tissue, which can differentiate into various cell types and aid in tissue repair.

Angiogenesis: The physiological process involving the growth of new blood vessels from pre-existing vessels.

Apoptosis: A process of programmed cell death that occurs in multicellular organisms.

Biofilm: A complex community of microorganisms that adhere to a surface, often contributing to chronic wound infections.

Biomaterials: Materials used in medical devices that interact with biological systems.

Bipedicle Flap: A surgical flap that is attached to the body at two points, ensuring blood supply from both ends.

Burn Wound: Tissue damage resulting from heat, electricity, radiation, or chemical exposure.

Cell-Based Therapies: Medical treatments that involve the use of living cells to repair or replace damaged tissue.

Chronic Wounds: Wounds that fail to proceed through the normal healing process and do not heal in a predictable timeframe.

Collagen: The main structural protein found in skin and other connective tissues.

Dermis: The layer of skin between the epidermis and subcutaneous tissues, consisting of connective tissue and containing blood vessels, hair follicles, and glands.

Differentiation: The process by which cells become specialized in structure and function.

Fibrin: When the thrombin is involved in the blood clotting process, fibrinogen is converted into an insoluble protein.

Fibroblasts: Collagen and other extracellular matrix components are produced by cell in connective tissue.

Gene Knockout: Genetic technique in which a gene is inactivated or deleted.

Growth Factors: Naturally occurring substances, usually proteins, capable of stimulating cellular growth, proliferation, and differentiation.

Inflammation: A localized physical condition in which part of the body becomes reddened, swollen, hot, and often painful, especially as a reaction to injury or infection.

***In-Vivo*:** Experiments conducted within a living organism.

Ischemia: An inadequate blood supply to an organ or part of the body, especially to the heart.

Lentiviral Vectors: Viruses used as a tool to deliver genetic material into cells.

Neutrophils: A type of white blood cell that is one of the first responders to inflammation at the site of injury.

Non-Healing Wounds: Wounds that do not proceed through the normal healing process.

Streptozotocin (STZ): A chemical compound that is toxic to insulin-producing cells in the pancreas, used to induce diabetes in animal models.

TRAM Flap: A surgical procedure where tissue from the abdomen, along with its blood supply, is used to reconstruct another part of the body.

Transgenic Animal Models: Animals that have had their genetic makeup altered by the introduction of a gene from another species.

REFERENCES

Ahn, Sang Tae, and Mustoe, Thomas A. 1990. Effects of ischemia on ulcer wound healing: a new model in the rabbit ear. *Annals of Plastic Surgery* 24 (1): 17–23. doi:10.1097/00000637-199001000-00004.

Alhajj, Mandy, and Amandeep Goyal. 2022. "Physiology, granulation tissue." In *StatPearls [Internet].* StatPearls Publishing.

Andleeb, Saiqa, Sadia Nazer, Suliman Yousef Alomar, Naushad Ahmad, Imran Khan, Abida Raza, Uzma Azeem Awan, and Sadaf Azad Raja. 2022. "Wound healing and anti-inflammatory potential of Ajuga bracteosa-conjugated silver nanoparticles in Balb/c mice." *bioRxiv*:2022.09. 21.508872.

Antinucci, Paride, and Robert Hindges. 2016. "A crystal-clear zebrafish for in vivo imaging." *Scientific Reports* 6 (1): 29490.

Arakawa, Nami, Daisuke Utsumi, Kenzo Takahashi, Akiko Matsumoto-Oda, Atunga Nyachieo, Daniel Chai, Ngalla Jillani, Hiroo Imai, Yoko Satta, and Yohey Terai. 2019. "Expression changes of structural protein genes may be related to adaptive skin characteristics specific to humans." *Genome Biology and Evolution* 11 (3): 613–628.

Arjmand, Babak, Akram Tayanloo-Beik, Najmeh Foroughi Heravani, Setareh Alaei, Moloud Payab, Sepideh Alavi-Moghadam, Parisa Goodarzi, Mahdi Gholami, and Bagher Larijani. 2020. "Zebrafish for personalized regenerative medicine; A more predictive humanized model of endocrine disease." *Frontiers in Endocrinology* 11: 396.

Arunachalam, Muthukumarasamy, Manickam Raja, Chinnian Vijayakumar, Punniyam Malaiammal, and Richard L Mayden. 2013. "Natural history of zebrafish (*Danio rerio*) in India." *Zebrafish* 10 (1): 1–14.

Avishai, Eden, Kristina Yeghiazaryan, and Olga Golubnitschaja. 2017. "Impaired wound healing: Facts and hypotheses for multi-professional considerations in predictive, preventive and personalised medicine." *EPMA Journal* 8: 23–33.

Babensee, Julia E. 2020. "Inflammation, wound healing, the foreign-body response, and alternative tissue responses." In *Biomaterials Science*, 737–746. Elsevier.

Bahri, Ola A, Neia Naldaiz-Gastesi, Donna C Kennedy, Antony M Wheatley, Ander Izeta, and Karl JA McCullagh. 2019. "The panniculus carnosus muscle: A novel model of striated muscle regeneration that exhibits sex differences in the mdx mouse." *Scientific Reports* 9 (1): 15964.

Barakat, May, Luisa A DiPietro, and Lin Chen. 2021. "Limited treatment options for diabetic wounds: Barriers to clinical translation despite therapeutic success in murine models." *Advances in Wound Care* 10 (8): 436–460.

Barrenas, Fredrik, Kevin Raehtz, Cuiling Xu, Lynn Law, Richard R Green, Guido Silvestri, Steven E Bosinger, Andrew Nishida, Qingsheng Li, and Wuxun Lu. 2019. "Macrophage-associated wound healing contributes to African green monkey SIV pathogenesis control." *Nature Communications* 10 (1): 5101.

Bártolo, Inês, Rui L Reis, Alexandra P Marques, and Mariana T Cerqueira. 2022. "Keratinocyte growth factor-based strategies for wound re-epithelialization." *Tissue Engineering Part B: Reviews* 28 (3): 665–676.

Beffagna, Giorgia. 2019. "Zebrafish as a smart model to understand regeneration after heart injury: How fish could help humans." *Frontiers in Cardiovascular Medicine* 6: 107.

Berman, Brian, Andrea Maderal, and Brian Raphael. 2017. "Keloids and hypertrophic scars: Pathophysiology, classification, and treatment." *Dermatologic Surgery* 43: S3–S18.

Border, Wayne A, and Nancy A Noble. 1994. "Transforming growth factor β in tissue fibrosis." *New England Journal of Medicine* 331 (19): 1286–1292.

Boyce, Mary K, Troy N Trumble, Cathy S Carlson, Donna M Groschen, Kelly A Merritt, and Murray P Brown. 2013. "Non-terminal animal model of post-traumatic osteoarthritis induced by acute joint injury." *Osteoarthritis and Cartilage* 21 (5): 746–755.

Brem, Harold, and Marjana Tomic-Canic. 2007. "Cellular and molecular basis of wound healing in diabetes." *The Journal of Clinical Investigation* 117 (5): 1219–1222.

Burd, Andrew, Chi H Kwok, Siu C Hung, Hui S Chan, Hua Gu, Wai K Lam, and Lin Huang. 2007. "A comparative study of the cytotoxicity of silver-based dressings in monolayer cell, tissue explant, and animal models." *Wound Repair and Regeneration* 15 (1): 94–104.

Campbell, Karl J, Joe Beek, Charles T Eason, Alistair S Glen, John Godwin, Fred Gould, Nick D Holmes, Gregg R Howald, Francine M Madden, and Julia B Ponder. 2015. "The next generation of rodent eradications: Innovative technologies and tools to improve species specificity and increase their feasibility on islands." *Biological Conservation* 185: 47–58.

Campos-Junior, Paulo HA, Cristina A Silva, Joao GV Grazia, Milena BP Soares, Ricardo R Santos, and Joao HM Viana, and. 2011. "Use of ultrasound biomicroscopy to evaluate induced ovarian follicular growth and ovulation in mice." *Laboratory Animals* 45 (4): 254–258.

Chen, Hong, Feng Ding, Zhan Zhou, Xiaojun He, and Jianliang Shen. 2020. "FRET-based sensor for visualizing pH variation with colorimetric/ratiometric strategy and application for bioimaging in living cells, bacteria and zebrafish." *Analyst* 145 (12): 4283–4294.

Choudhary, Vivek, Mrunal Choudhary, and Wendy B Bollag. 2024. "Exploring skin wound healing models and the impact of natural lipids on the healing process." *International Journal of Molecular Sciences* 25 (7): 3790.

Couto M, Cates C. *Laboratory guidelines for animal care. In: Vertebrate Embryogenesis: Embryological, Cellular, and Genetic Methods*. New York: Humana Press; 2019. p. 407–430.

Díaz, Lorenza, Elena Zambrano, María E Flores, Mariela Contreras, Jose C Crispín, Gabriela Alemán, César Bravo, Alejandra Armenta, Victor J Valdés, and Armando Tovar. 2021. "Ethical considerations in animal research: The principle of 3R's." *Revista de investigacion clinica* 73 (4): 199–209.

Domínguez-Oliva, Adriana, Ismael Hernández-Ávalos, Julio Martínez-Burnes, Adriana Olmos-Hernández, Antonio Verduzco-Mendoza, and Daniel Mota-Rojas. 2023. "The importance of animal models in biomedical research: Current insights and applications." *Animals* 13 (7): 1223.

Dorsett-Martin , Wendy A. 2004. Rat models of skin wound healing: a review. *Wound Repair and Regeneration* 12 (6): 591–599. doi: 10.1111/j.1067-1927.2004.12604.x

Doyle, Alfred, Michael P McGarry, Nancy A Lee, and James J Lee. 2012. "The construction of transgenic and gene knockout/knockin mouse models of human disease." *Transgenic Research* 21: 327–349.

Drew AF, Liu H, Davidson JM, Daugherty CC, Degen JL. 2001. Wound-healing defects in mice lacking fibrinogen. *Blood* 97 (12): 3691–3698. doi:10.1182/blood.v97.12.3691.

Edmunds, Marie-Claire, Stephen Wigmore, and David Kluth. 2013. "In situ transverse rectus abdominis myocutaneous flap: A rat model of myocutaneous ischemia reperfusion injury." *Journal of Visualized Experiments:* JoVE (76): 50473.

Ellis, Samantha, Elaine J Lin, and Danielle Tartar. 2018. "Immunology of wound healing." *Current Dermatology Reports* 7: 350–358.

Falanga Vincent, Deborah Schrayer, Jin Cha, Joseph Butmarc, Peter Carson, Abraham B. Roberts, and Kim, Sung-Joon. 2004. Full-thickness wounding of the mouse tail as a model for delayed wound healing. *Wound Repair and Regeneration* 12 (2): A20. doi: 10.1111/j.1067-1927.2004.012316.x

Fang, Robert C, and Thomas A Mustoe. 2008. "Animal models of wound healing: Uility in transgenic mice." *Journal of Biomaterials Science, Polymer Edition* 19 (8): 989–1005.

Flynn, Kiley, Nouf N Mahmoud, Shahriar Sharifi, Lisa J Gould, and Morteza Mahmoudi. 2023. "Chronic wound healing models." *ACS Pharmacology & Translational Science* 6 (5): 783–801.

Galiano, Robert D, Joseph Michaels, V, Michael Dobryansky, Jamie P Levine, and Geoffrey C Gurtner. 2004. "Quantitative and reproducible murine model of excisional wound healing." *Wound Repair and Regeneration* 12 (4): 485–492.

Ghanbari, Mahboubeh, Salkovskiy Yury, Carlson Mark A. 2024. "The rat as an animal model in chronic wound research: an update." *Life Sciences* 351, 122783. doi: 10.1016/j.lfs.2024.122783

Gökkaya, Ali, and Metin Gorgu. 2020. "Treatment of scalp defects with a combination of trephination and platelet-rich plasma." *Journal of Tissue Viability* 29 (3): 211–215.

Gonzalez, Ana Cristina de Oliveira, Tila Fortuna Costa, Zilton de Araújo Andrade, and Alena Ribeiro Alves Peixoto Medrado. 2016. "Wound healing—A literature review." *Anais brasileiros de dermatologia* 91 (5): 614–620.

Grada A, Mervis J, Falanga V. 2018. "Research techniques made simple: animal models of wound healing." *Journal of Investigative Dermatology* 138 (10): 2095–2105.

Grambow, Eberhard, Heiko Sorg, Christian GG Sorg, and Daniel Strüder. 2021. "Experimental models to study skin wound healing with a focus on angiogenesis." *Medical Sciences* 9 (3): 55.

Gu, Da-Ling, Dong-Wook Kang, David Looper, Daniel Maneval, and Zepeda, Miguel. A rabbit ear excessive scarring model using growth factor stimulators. *Wound Repair and Regeneration*. 2004;12(2):A9.

Guoqi, Guoqi, Zhirui Li, Song Wang, Tongtong Li, Lihai Zhang, Licheng Zhang and Tang, Peifu T. 2018. "Negative pressure wound therapy reduces the motility of *Pseudomonas aeruginosa* and enhances wound healing in a rabbit ear biofilm infection model." *Antonie Van Leeuwenhoek* 111 (9): 1557–1570.

Gurley, Susan B, Sujoy Ghosh, Stacy A Johnson, Kengo Azushima, Rashidah Binte Sakban, Simi E George, Momoe Maeda, Timothy W Meyer, and Thomas M Coffman. 2018. "Inflammation and immunity pathways regulate genetic susceptibility to diabetic nephropathy." *Diabetes* 67 (10): 2096–2106.

Gurtner, Geoffrey C, Sabine Werner, Yann Barrandon, and Michael T Longaker. 2008. "Wound repair and regeneration." *Nature* 453 (7193): 314–321.

Hackett, Justin, Heather Gibson, Jeffrey Frelinger, and Adam Buntzman. 2022. "Using the collaborative cross and diversity outbred mice in immunology." *Current Protocols* 2 (9): e547.

Hakkinen, Lari, Hannu Larjava, and Leeni Koivisto. 2011. "Granulation tissue formation and remodeling." *Endodontic Topics* 24 (1): 94–129.

Hamilton, David W., John T. Walker, Danielle Tinney, Madeline Grynyshyn, Ahmed El-Warrak, Elizabeth Truscott and Flynn, Lauren E. 2022. "The pig as a model system for investigating the recruitment and contribution of myofibroblasts in skin healing." *Wound Repair and Regeneration* 30 (1):45–63.

Hammad, Mira. 2021. *Reconstruction of Auricular Cartilage Using Natural-Derived Scaffolds with an in Vivo Application in Rabbit Model*. Normandie Université.

Hason, Martina, and Petr Bartůněk. 2019. "Zebrafish models of cancer—New insights on modeling human cancer in a non-mammalian vertebrate." *Genes* 10 (11): 935.

He, Xiaojun, Zaian Deng, Wei Xu, Yahui Li, Chuchu Xu, Hong Chen, and Jianliang Shen. 2020. "A novel dual-response chemosensor for bioimaging of exogenous/endogenous hypochlorite and hydrazine in living cells, *Pseudomonas aeruginosa* and zebrafish." *Sensors and Actuators B: Chemical* 321: 128450.

He, Xiaojun, Feng Ding, Xiaoshuai Sun, Yujing Zheng, Wei Xu, Lisong Ye, Hong Chen, and Jianliang Shen. 2021. "Renovated multifunctional colorimetric/fluorometric sensor for simultaneous detection, imaging of pH variance and antimicrobial therapies." *Sensors and Actuators B: Chemical* 332: 129496.

Histing, Tina, Patric Garcia, Jörg H Holstein, Moritz Klein, Romano Matthys, Reto Nuetzi, Roland Steck, Michael W Laschke, Tim Wehner, and Ronny Bindl. 2011. "Small animal bone healing models: standards, tips, and pitfalls results of a consensus meeting." *Bone* 49(4):591–599.

Hocum Stone, Laura, Scott Hunter Oppler, Julia L Nugent, Sarah Gresch, Bernhard J Hering, Michael P Murtaugh, Rebecca L Hegstad-Davies, Sabarinathan Ramachandran, and Melanie L Graham. 2021. "Serum cytokine profiles in healthy nonhuman primates are blunted by sedation and demonstrate sexual dimorphism as detected by a validated multiplex immunoassay." *Scientific Reports* 11 (1): 2340.

Holt, Andrew W, Jennifer S McDaniel, Gregory T Bramblett, Elof Eriksson, Anthony J Johnson, and Gina L Griffith. 2018. "Use of an ocular wound chamber for the prevention of exposure keratopathy in a guinea pig model." *Wound Repair and Regeneration* 26 (5): 351–358.

Honnegowda, Thittamaranahalli Muguregowda, Pramod Kumar, Echalasara Govindarama Padmanabha Udupa, Sudesh Kumar, Udaya Kumar, and Pragna Rao. 2015. "Role of angiogenesis and angiogenic factors in acute and chronic wound healing." *Plastic and Aesthetic Research* 2: 243–249.

Ili, Petrit and Fatma Sari. 2023. "Egg yolk oil accelerates wound healing in streptozotocin induced diabetic rats." *Biotech Histochem* 98(2):94–111. doi: 10.1080/10520295.2022.2115554.

Kanungo, Jyotshna, Elvis Cuevas, Syed F Ali, and Merle G Paule. 2014. "Zebrafish model in drug safety assessment." *Current Pharmaceutical Design* 20 (34): 5416–5429.

Khiao In, Maneenooch, Kenneth C Richardson, Anna Loewa, Sarah Hedtrich, Sabine Kaessmeyer, and Johanna Plendl. 2019. "Histological and functional comparisons of four anatomical regions of porcine skin with human abdominal skin." *Anatomia, Histologia, Embryologia* 48 (3): 207–217.

Kim, Eunbin, Seoyoon Ham, Bok Ki Jung, Jin-Woo Park, Jihee Kim, and Ju Hee Lee. 2022. "Effect of baicalin on wound healing in a mouse model of pressure ulcers." *International Journal of Molecular Sciences* 24 (1): 329.

Koh, Timothy J, and Luisa Ann DiPietro. 2011. "Inflammation and wound healing: The role of the macrophage." *Expert Reviews in Molecular Medicine* 13: e23.

Krötzsch, Elke, Ricardo M. Salgado, Daniel Caba, Abraham Lichtinger, Luis Padilla, and Macro Di Silvio. 2005. "162 Alkaline phosphatase activity is related to acute inflammation and collagen turnover during acute and chronic wound healing." *Wound Repair and Regeneration* 13(2):A28–A48.

Kuo, Tzu-Yen, Chih-Cheng Huang, Shu-Jen Shieh, Yi-Bing Wang, Min-Jung Lin, Ming-Chih Wu, and Li-Li Huang. 2022. "Skin wound healing assessment via an optimized wound array model in miniature pigs." *Sci Rep* 12(1):445. doi: 10.1038/s41598-021-03855-y.

Leavy, Olive. 2009. "The identification of THEMIS." *Nature Reviews Immunology* 9 (9): 604–605.

Li, Xiaodong, Xiaodong Xue, and Peilin Xie. 2024. "Smart dressings and their applications in chronic wound management." *Cell Biochemistry and Biophysics* 82 (3): 1965–1977.

Lin, Bangchang, Jiahui Ma, Yimeng Fang, Pengyu Lei, Lei Wang, Linkai Qu, Wei Wu, Libo Jin, Da Sun. 2023. "Advances in Zebrafish for Diabetes Mellitus with Wound Model." *Bioengineering (Basel)* 10(3). doi: 10.3390/bioengineering10030330.

Lorenz, H Peter, David J Whitby, Michael T Longaker, and N Scott Adzick. 1993. "Fetal wound healing the ontogeny of scar formation in the non-human primate." *Annals of Surgery* 217 (4): 391–396.

Lu, Lu, Dandan Liu, Jianghui Ying, Zuochao Yao, Qiang Hou, Hui Wang, Fazhi Qi, Wenjie Luan, and Hua Jiang. 2022. "Denervation affected skin wound healing in a modified rat model." *The International Journal of Lower Extremity Wounds*:15347346221090758.

Mahmoud, Nouf N, Khawla Hamad, Aya Al Shibitini, Sarah Juma, Shahriar Sharifi, Lisa Gould, and Morteza Mahmoudi. 2024. "Investigating inflammatory markers in wound healing: Understanding implications and identifying artifacts." *ACS Pharmacology & Translational Science* 7 (1): 18–27.

Martin, Paul and Rachael Nunan. 2015. "Cellular and molecular mechanisms of repair in acute and chronic wound healing." *Br J Dermatol* 173(2):370–8. doi: 10.1111/bjd.13954.

Masoomikarimi, Masoomeh, and Majid Salehi. 2022. "Modulation of the immune system promotes tissue regeneration." *Molecular Biotechnology* 64 (6): 599–610.

Masson-Meyers, Danielle S., Tania A. M. Andrade, Guilherme F. Caetano, Fernanda R. Guimaraes, Marcelo N. Leite, Silvia N. Leite, and Marco A. C. Frade. 2020. "Experimental models and methods for cutaneous wound healing assessment." *Int J Exp Pathol* 101(1–2):21–37. doi: 10.1111/iep.12346.

Matsuda, Akira, Akihiko Yoshiki, Yoshiaki Tagawa, Hiroshi Matsuda, and Masaru Kusakabe. 1999. "Corneal wound healing in tenascin knockout mouse." *Invest Ophthalmol Vis Sci* 40(6):1071–80.

Mayadas, Tanya N. 1995. "Gene knockout on P-selectin: Its biology and function." *Trends Cardiovasc Med* 5 (4):149–57. doi: 10.1016/1050-1738(95)00057-g.

Mieczkowski, Mateusz, Beata Mrozikiewicz-Rakowska, Michał Kowara, Marcin Kleibert, and Leszek Czupryniak. 2022. "The problem of wound healing in diabetes—From molecular pathways to the design of an animal model." *International Journal of Molecular Sciences* 23 (14): 7930.

Mistry, Riyam, Mark Veres, and Fadi Issa. 2022. "A systematic review comparing animal and human scarring models." *Frontiers in Surgery* 9: 711094.

Mulisa, Eshetu, Kaleab Asres, and Ephrem Engidawork. 2015. "Evaluation of wound healing and anti-inflammatory activity of the rhizomes of *Rumex abyssinicus* J.(Polygonaceae) in mice." *BMC Complementary and Alternative Medicine* 15 (1): 10.

Naldaiz-Gastesi, Neia, Ola A Bahri, Adolfo Lopez de Munain, Karl JA McCullagh, and Ander Izeta. 2018. "The panniculus carnosus muscle: An evolutionary enigma at the intersection of distinct research fields." *Journal of Anatomy* 233 (3): 275–288.

Niethammer, Philipp, Christoph Grabher, A. Thomas Look, and Timothy J. Mitchison. 2009. "A tissue-scale gradient of hydrogen peroxide mediates rapid wound detection in zebrafish." *Nature* 459(7249):996–9. doi: 10.1038/nature08119.

Oryan, Ahmad, Ehsan Alemzadeh, and Ali Moshiri. 2017. "Burn wound healing: present concepts, treatment strategies and future directions." *J Wound Care* 26(1):5–19. doi: 10.12968/jowc.2017.26.1.5.

Ostvar, Omid, Saeed Shadvar, Ehsan Yahaghi, Kamran Azma, Amir F. Fayyaz, Khosro Ahmadi, and Iraj Nowrouzian. 2015. "Effect of platelet-rich plasma on the healing of cutaneous defects exposed to acute to chronic wounds: a clinico-histopathologic study in rabbits." *Diagn Pathol* 10:85. doi: 10.1186/s13000-015-0327-8.

Palmieri, Bruno, Monica Vadalà, and Concetta Laurino. 2017. "Review of the molecular mechanisms in wound healing: new therapeutic targets?" *Journal of Wound Care* 26(12):765–775. https://doi.org/10.12968/jowc.2017.26.12.765

Pastar, Irena, Olivera Stojadinovic, Natalie C Yin, Horacio Ramirez, Aron G Nusbaum, Andrew Sawaya, Shailee B Patel, Laiqua Khalid, Rivkah R Isseroff, and Marjana Tomic-Canic. 2014. "Epithelialization in wound healing: A comprehensive review." *Advances in Wound Care* 3 (7): 445–464.

Patel, Seema. 2018. "Danger-associated molecular patterns (DAMPs): The derivatives and triggers of inflammation." *Current Allergy and Asthma Reports* 18: 1–12.

Poelchow, Fiona. 2022. "A Study on Post-burn Healing: Optimising Scar Outcome Through the Use of a Silicone-based Film-forming Wound Dressing."

Potekaev, Nikolai N, Olga B Borzykh, German V Medvedev, Denis V Pushkin, Marina M Petrova, Artem V Petrov, Diana V Dmitrenko, Elena I Karpova, Olga M Demina, and Natalia A Shnayder. 2021. "The role of extracellular matrix in skin wound healing." *Journal of Clinical Medicine* 10 (24): 5947.

Rahrovan, Sanaz, Fanny Fanian, Parvin P. P. H. Mehryan, Philippe Humbert, and Alireza Firooz. 2018. "Male versus female skin: what dermatologists and cosmeticians should know." *International journal of women's dermatology* 4(3):122–130. https://doi.org/10.1016/j.ijwd.2018.03.002

Rhea, Lindsey, and Martine Dunnwald. 2020. "Murine excisional wound healing model and histological morphometric wound analysis." *Journal of Visualized Experiments: JoVE* (162): doi: 10.3791/61616.

Richardson, Rebecca, Krasimir Slanchev, Christopher Kraus, Philipp Knyphausen, Sabine Eming, and Matthias Hammerschmidt. 2013. "Adult zebrafish as a model system for cutaneous wound-healing research." *Journal of Investigative Dermatology* 133 (6): 1655–1665.

Rivera, Albert E, and James M Spencer. 2007. "Clinical aspects of full-thickness wound healing." *Clinics in Dermatology* 25 (1): 39–48.

Rubinstein, Amy L. 2003. "Zebrafish: From disease modeling to drug discovery." *Current Opinion in Drug Discovery and Development* 6 (2): 218–223.

Saeed, Shayan, and Manuela Martins-Green. 2023. "Animal models for the study of acute cutaneous wound healing." *Wound Repair and Regeneration* 31 (1): 6–16.

Safaee, Fatemeh. 2022. "Assessing Ear Pinna Repair in the MRL/MpJ Mouse Strain."

Schultz, Gregory S., Robert G. Sibbald, Vincenzo Falanga, Elizabeth A. Ayello, Christopher Dowsett, Keith Harding, Marco Romanelli, Marcia C. Stacey, Luc Teot, and Wolfgang Vanscheidt. 2003. "Wound bed preparation: a systematic approach to wound management." *Wound Repair Regen* 11 Suppl 1:S1-28. doi: 10.1046/j.1524-475x.11.s2.1.x.

Seaton, Max, Anne Hocking, and Nicole S Gibran. 2015. "Porcine models of cutaneous wound healing." *ILAR Journal* 56 (1): 127–138.

Seok, Junhee, H Shaw Warren, Alex G Cuenca, Michael N Mindrinos, Henry V Baker, Weihong Xu, Daniel R Richards, Grace P McDonald-Smith, Hong Gao, and Laura Hennessy. 2013. "Genomic responses in mouse models poorly mimic human inflammatory diseases." *Proceedings of the National Academy of Sciences* 110 (9): 3507–3512.

Siddhartha, Rohit, and Minal Garg. 2023. "Interplay between extracellular matrix remodeling and angiogenesis in tumor ecosystem." *Molecular Cancer Therapeutics* 22 (3): 291–305.

Silva-Gomes, R., E. Marcq, G. Trigo, C. M. Gonçalves, A. Longatto-Filho, A. G. Castro, J. Pedrosa, and A. G. Fraga. 2015. "Spontaneous Healing of Mycobacterium ulcerans Lesions in the Guinea Pig Model." *PLoS Neglected Tropical Diseases* 9(12):e0004265. doi: 10.1371/journal.pntd.0004265.

Sorg, Heiko, Daniel J Tilkorn, Stephan Hager, Jörg Hauser, and Ursula Mirastschijski. 2017. "Skin wound healing: An update on the current knowledge and concepts." *European Surgical Research* 58 (1–2): 81–94.

Sosnowski, Paweł, Piotr Sass, Paulina Słonimska, Rafał Płatek, Jolanta Kamińska, Jakub Baczyński Keller, Piotr Mucha, Grażyna Peszyńska-Sularz, Artur Czupryn, and Michał Pikuła. 2022. "Regenerative drug discovery using ear pinna punch wound model in mice." *Pharmaceuticals* 15 (5): 610.

Stenn, Kurt S, and Louis Depalma. 1988. "Re-epithelialization." In *The Molecular and Cellular Biology of Wound Repair*, 321–335. Springer.

Strobel, Marjorie C, Laura G Reinholdt, Rachel D Malcolm, and Kathleen Pritchett-Corning. 2015. "Genetic monitoring of laboratory mice and rats." In *Laboratory Animal Medicine*, 1403–1416. Elsevier.

Subaramaniyam, Udayadharshini, Rethi Saliya Allimuthu, Shanu Vappu, Divya Ramalingam, Ranjini Balan, Biswaranjan Paital, Niranjan Panda, Prasana Kumar Rath, Nirmaladevi Ramalingam, and Dipak Kumar Sahoo. 2023. "Effects of microplastics, pesticides and nano-materials on fish health, oxidative stress and antioxidant defense mechanism." *Frontiers in Physiology* 14: 1217666.

Sullivan, Justine Victoria, and Simon Myers. 2022. "Skin structure and function, wound healing and scarring." In *Plastic Surgery-Principles and Practice*, 1–14. Elsevier.

Sullivan, Tory P, William H Eaglstein, Stephen C Davis, and Patricia Mertz. 2001. "The pig as a model for human wound healing." *Wound Repair and Regeneration* 9 (2): 66–76.

Tardif, Suzette D, Kristine Coleman, Theodore R Hobbs, and Corrine Lutz. 2013. "IACUC review of nonhuman primate research." *ILAR Journal* 54 (2): 234–245.

Tonnesen, M. G., X. Feng, and R. A. Clark. 2000. "Angiogenesis in wound healing." *Journal of Investigative Dermatology Symposium Proceedings* 5(1):40–46. doi: 10.1046/j.1087-0024.2000.00014.x.

Tríbulo, Celeste, María Guadalupe Barrionuevo, Tristán H Agüero, Sara S Sanchez, Nora B Calcaterra, and Manuel J Aybar. 2012. "ΔNp63 is regulated by BMP4 signaling and is required for early epidermal development in Xenopus." *Developmental Dynamics* 241 (2): 257–269.

Uckun, Fatih M, Cemal Orhan, Mehmet Tuzcu, Ali Said Durmus, Ibrahim H Ozercan, Michael Volk, and Kazim Sahin. 2022. "RJX improves wound healing in diabetic rats." *Frontiers in Endocrinology* 13: 874291.

Udegbunam, Sunday O, Sunday Ogbobe, Nnamdi H Okereke, Adaobi S Enejere, Ijeoma R Udegbunam, and Toochukwu H Ezeobialu. 2021. "Assessment of wound contraction, re-epithelialization and histological changes in full thickness excision wounds of rats treated with different concentrations of hydrogen peroxide." *Tropical Journal of Pharmaceutical Research* 20 (8): 1623–1629.

Utsumi, Daisuke, Akiko Matsumoto, Ngalla Jillani, Daniel Chai Chivastasi, Atunga Nyachieo, and Kenzo Takahashi. 2017. "Cutaneous wound healing is exclusively slow and poor in human compared to those of other mammals including evolutionarily related primates." *Journal of Dermatological Science* 86 (2): e54–e55.

Veldman, Matthew B, and Shuo Lin. 2008. "Zebrafish as a developmental model organism for pediatric research." *Pediatric Research* 64 (5): 470–476.

Wang, Xusheng, Jianfeng Ge, Edward E Tredget, and Yaojiong Wu. 2013. "The mouse excisional wound splinting model, including applications for stem cell transplantation." *Nature Protocols* 8 (2): 302–309.

Wang, Zheng-Cai, Wan-Yi Zhao, Yangyang Cao, Yan-Qi Liu, Qihang Sun, Peng Shi, Jia-Qin Cai, Xiao Z Shen, and Wei-Qiang Tan. 2020. "The roles of inflammation in keloid and hypertrophic scars." *Frontiers in Immunology* 11: 603187.

Weng, Rui, Qingfeng Li, Hua Li, Mei Yang, and Lingling Sheng. 2010. "Mimic hypoxia improves angiogenesis in ischaemic random flaps." *Journal of Plastic, Reconstructive & Aesthetic Surgery* 63 (12): 2152–2159.

Werner, Sabine, and Richard Grose. 2003. "Regulation of wound healing by growth factors and cytokines." *Physiological Reviews* 83 (3): 835–870.

Willenborg, Sebastian, and Sabine A Eming. 2014. "Macrophages – Sensors and effectors coordinating skin damage and repair." *JDDG: Journal der Deutschen Dermatologischen Gesellschaft* 12 (3): 214–221.

Wilson, Rachel D, and Md Shahidul Islam. 2012. "Fructose-fed streptozotocin-injected rat: an alternative model for type 2 diabetes." *Pharmacological Reports* 64 (1): 129–139.

Yampolsky, Matan, Ido Bachelet, and Yaron Fuchs. 2024. "Reproducible strategy for excisional skin-wound-healing studies in mice." *Nature Protocols* 19 (1): 184–206.

Yannas, Ioannis V, Dimitrios S Tzeranis, and Peter TC So. 2017. "Regeneration of injured skin and peripheral nerves requires control of wound contraction, not scar formation." *Wound Repair and Regeneration* 25 (2): 177–191.

Zang, Liqing, Lisette A Maddison, and Wenbiao Chen. 2018. "Zebrafish as a model for obesity and diabetes." *Frontiers in Cell and Developmental Biology* 6: 91.

Zang, Liqing, Yasuhito Shimada, and Norihiro Nishimura. 2017. "Development of a novel zebrafish model for type 2 diabetes mellitus." *Scientific Reports* 7 (1): 1461.

Zhou, S., W. Wang, S. Zhou, G. Zhang, J. He, and Q. Li. 2019. "A novel model for cutaneous wound healing and scarring in the rat." *Plastic and Reconstructive Surgery* 143(2):468–477. doi: 10.1097/prs.0000000000005274.

6 Psychological Aspects of Wound Healing

Abiza Wali[1], *Faizan Mohi Din Bhat*[2], *Aarif Ali*[3*], *Shameem Ahmad Shameem*[4], *Javeed Iqbal Ahmad Bhat*[3], *and Mashooq Ahmad Dar*[2]

[1] Department of Biochemistry, Government Degree College Boys, Khwaja Bagh Baramulla, Jammu & Kashmir, India

[2] Department of Clinical Biochemistry, Government Degree College Boys, Khwaja Bagh Baramulla, Jammu & Kashmir, India

[3] Division of Basic Sciences and Humanities, Faculty of Horticulture, Sher-e-Kashmir University of Agricultural Sciences and Technology of Kashmir, Shalimar, Srinagar, Jammu & Kashmir, India

[4] Division of Environmental Sciences, Sher-e-Kashmir University of Agricultural Sciences and Technology of Kashmir, Shalimar, Srinagar, Jammu & Kashmir, India

6.1 PSYCHOLOGICAL IMPACT OF CHRONIC WOUNDS

Wound healing represents a complex biological process that includes various stages, such as inflammation, tissue formation, and remodeling. This process is meticulously regulated by the immune system, hormonal balance, and a range of cellular mechanisms. Wounds that do not heal in an orderly and timely fashion through the normal phases of wound healing are referred to as chronic wounds. Chronic wounds fail to progress beyond the inflammation phase of healing. Although, at the molecular level, causes of chronic wounds vary, they all have some common characteristics, such as high levels of proteases, pro-inflammatory cytokines, reactive oxygen species (ROS), and senescent cells, persistent infection, and a lack of stem cells, which are usually dysfunctional. Microorganisms and platelet-derived factors, like transforming growth factor-b (TGF-b) or extracellular matrix (ECM) fragment molecules, stimulate the continuous influx of immune cells as a result of repeated tissue injury; this amplifies and prolongs the pro-inflammatory cytokine cascade, which raises the levels of proteases. In chronic wounds, the levels of protease are far more than those of their specific inhibitors, which results in the destruction of the ECM and the degradation of growth factors and their receptors. In addition to keeping the wound from progressing into the proliferative phase, the proteolytic degradation of ECM draws in additional inflammatory cells, intensifying the inflammatory cycle (McCarty and Percival, 2013). Immune cells form ROS, which in low concentrations protects against microbes. However, the predominant hypoxic and inflammatory environment in chronic wounds increases ROS production, which degrades ECM proteins and causes cell damage. Due to this sequence of events,

* Corresponding author: buttaarif31@gmail.com

DOI: 10.1201/9781003644958-6

there is an enhanced stimulation of proteases and inflammatory cytokines (Schreml et al., 2010). Furthermore, chronic wounds are not affected by typical wound healing signals as they are characterized by senescent cell populations with impaired proliferative and secretory capacities (Schultz et al., 2003).

Chronic wounds such as diabetic foot ulcers, venous leg ulcers, and pressure injuries are characterized by their inability to heal within the expected duration, generally longer than four to six weeks (Martinengo et al., 2019). Besides the physical pain and medical challenges they present, chronic wounds also have a deep psychological effect on patients, impacting their mental health, emotional wellness, and quality of life. Nevertheless, psychological conditions like depression, anxiety, and social isolation can severely disrupt these processes, causing delays or impairments in healing.

6.1.1 Depression

Depression is one of the most frequent psychological responses associated with chronic wounds (Zhou and Jia, 2016). Depression is not simply a feeling of sadness; it is a medical condition that affects both mental and physical well-being. The ongoing nature of the wound, along with pain, restricted movement, and extended treatment plans, can foster feelings of hopelessness, sadness, and a loss of autonomy. Research has demonstrated that up to 30–50% of patients with chronic wounds may experience clinical depression symptoms (Finlayson et al., 2017). For instance, about 37% of the participants in a study of 216 patients with chronic wounds had depressive symptoms (Yan et al., 2021). In particular, a meta-analysis of individuals with diabetic foot ulcer found that 47% of them had depression (Jiang et al., 2022).

Moreover, depression can negatively affect the outcomes of wound healing through both biological factors and reduced adherence to treatment (Kiecolt-Glaser et al., 2002a). Therefore, persistent, non-healing wounds are more strongly associated with people who suffer from depression. For example, according to a study using a Hospital Anxiety and Depression Scale (HADS), higher depression scores (>14) were significantly associated with chronic wounds at a one-month follow-up (Kumar et al., 2020). One significant biological pathway by which depression impacts wound healing is through the dysregulation of the immune response. By dysregulating the sympathetic nervous system and hypothalamic–pituitary–adrenal (HPA) axis, depression affects vital physiological processes involved in wound healing, such as inflammation and immunological function (Herbert and Cohen, 1993). At wound sites, cell-mediated immunity is hampered by elevated stress hormones such as glucocorticoids, which cause healing to be delayed. Depression correlates with increased cortisol levels, a stress hormone that diminishes immune function and hinders the release of pro-inflammatory cytokines essential during the initial phase of wound healing (Glaser and Kiecolt-Glaser, 2005). This can lead to a compromised inflammatory response, hindering the wound's proper progression through the healing stages.

Furthermore, depression fosters a condition of chronic low-grade inflammation, which disturbs the necessary balance for effective healing. Rather than resolving inflammation and advancing to tissue repair, wounds in individuals suffering from depression may remain inflamed for extended periods, resulting in slower tissue regeneration (Gouin and Kiecolt-Glaser, 2011). A pivotal study conducted by Kiecolt-Glaser et al. (1995) offered compelling empirical support for this relationship. The researchers performed a controlled experiment where they induced small biopsy wounds in older adults. Participants displaying symptoms of depression demonstrated significantly slower healing of their wounds compared to those who did not show such symptoms. This research underscored the clinical importance of psychological health in physical recovery. Depression also influences healing through behavioral factors. Individuals suffering from depression may face issues such as poor sleep, decreased appetite, and a general lack of motivation, which can result in malnutrition, inadequate hygiene, and neglect in wound care (Christian et al., 2007). Wound healing is delayed due to nutritional deficiencies, which decrease fibroblast proliferation, collagen production, tensile strength, and angiogenesis. Weakened function of the immune system due to reduced T-cell function in nutritional deficiency also elevates the risk of infection (Stechmiller, 2010). Furthermore, sleep disturbances frequently observed in individuals with depression adversely affect cellular repair processes and the functioning of the immune system (Kumar et al., 2020).

Moreover, people suffering from depression frequently experience weakened social support networks, which can intensify both emotional suffering and physical health issues (Platsidaki et al., 2017; Renner and Erfurt-Berge, 2017). Social isolation and feelings of loneliness have also been linked to delayed wound healing (Gouin and Kiecolt-Glaser, 2011). In summary, the data clearly indicates that depression can significantly impact the body's healing capabilities. It is essential to address mental health in patients with wounds, particularly those dealing with chronic or post-surgical wounds, as part of a holistic approach to medical care. By simultaneously treating mental and physical health, healthcare professionals can improve both psychological health and physical recovery.

6.1.2 Anxiety

Anxiety, much like depression, has been shown to negatively influence physical health, including the body's healing capabilities for wounds. Alongside depression, anxiety is notably common in individuals with chronic wounds. Numerous patients express concerns regarding the potential for infection, the possibility of limb amputation, the recurrence of their wounds, or the financial strain associated with continuous care. Although wound healing is primarily a physiological process that encompasses inflammation, tissue formation, and remodeling, it is increasingly recognized that psychological states such as anxiety can greatly delay or disrupt this process. Anxiety triggers the activation of the HPA axis and the sympathetic nervous system, which leads to an increase in cortisol and catecholamine production (Kiecolt-Glaser et al., 2002a). High cortisol levels inhibit immune functions that are critical for wound healing, such as leukocyte activity and the production of pro-inflammatory cytokines (Gouin and Kiecolt-Glaser, 2011). These immune responses are vital during the early stages of wound healing when the body clears out pathogens and starts the tissue repair process. In both clinical and experimental research, individuals exhibiting high levels of anxiety have shown a slower healing process for both acute and chronic wounds. For instance, patients who underwent surgical procedures and reported elevated preoperative anxiety levels often faced delayed healing of their post-surgical wounds, along with a heightened risk of infections and complications (Broadbent et al., 2003). Additionally, anxiety correlates with increased systemic inflammation, indicated by higher levels of C-reactive protein (CRP) and pro-inflammatory cytokines. This ongoing low-grade inflammation can disrupt the carefully regulated stages of wound healing, especially the shift from inflammation to proliferation (Segerstrom and Miller, 2004).

In addition to biological pathways, anxiety-related behavioral changes also play a role in poor wound healing outcomes. Those suffering from anxiety are more prone to experience sleep issues, inadequate nutrition, and lower adherence to medical care or wound hygiene, all of which can hinder healing (Christian et al., 2007). Furthermore, individuals with anxiety may resort to coping strategies such as smoking or alcohol consumption, which can further impede the healing process. In conclusion, anxiety affects wound healing through a mix of neuroendocrine, immune, and behavioral factors. Acknowledging the impact of anxiety on wound care underscores the importance of a comprehensive medical approach that incorporates psychological support as a fundamental aspect of wound management strategies.

6.1.3 Social Isolation

Wound healing represents a complex biological process influenced by both physiological conditions and psychosocial factors. Among these, social isolation defined as the absence of social contacts or interactions has been recognized as a significant factor that can negatively affect the body's ability to heal wounds properly. Those who are socially isolated often endure chronic psychological stress, which activates the HPA axis and elevates cortisol levels. High cortisol levels can suppress the immune response, hindering the production of pro-inflammatory cytokines and reducing the functionality of immune cells such as macrophages and neutrophils, which are crucial during the early stages of wound healing (Gouin et al., 2008).

Social support aids in decreasing stress and cortisol levels while increasing levels of oxytocin, which suppresses the activity of HPA, leading to improved wound healing (Christian et al., 2007; Detillion et al., 2004). The patients suffering from chronic venous ulcers, some of them demonstrated wound improvement when engaged in group nursing home activities (Brown, 2008). The effects of social support on outcomes are inconsistent in chronic wound healing (Brown, 2008; Robinson et al., 2017). Family involvement in wound care education can also have a significant effect on wound healing by enabling caregivers to be more involved in the patient's wound care and identifying early indicators of infection and poor wound healing (Kirkland-Kyhn et al., 2022).

In a groundbreaking study, Kiecolt-Glaser et al. (2005) explored how loneliness and social support affect wound healing in older adults. Those who reported higher loneliness and lower perceived social support showed a marked delay in the healing of small skin wounds that were experimentally induced. This research offered some of the initial compelling evidence that social isolation can have a direct effect on physiological healing processes. The biological mechanism behind this effect involves the disruption of inflammatory responses. An appropriate inflammatory response is essential for healing; however, social isolation may result in either diminished or extended inflammation, which hampers the transition to the tissue regeneration phase (Cole, 2013). Consequently, this can postpone wound closure and elevate the risk of infection or other complications. Patients suffering from chronic wounds often curtail their social interactions because of mobility limitations, stigma, or worries about odor. This reduction in social contact can lessen emotional support and considerably heighten psychological distress, thereby worsening the negative feedback loop between mental and physical health (Vowden and Vowden, 2016). Patients with chronic wounds have far lower levels of social participation and interactions (Klein et al., 2021). Studies have repeatedly shown that social isolation hinders wound healing (Detillion et al., 2004). According to a systematic review by Robinson et al. (2017), social support improves wound healing, and the magnitude of the effects varies depending on the sort of assistance the patient receives.

Moreover, social isolation is often associated with unhealthy behaviors like poor nutrition, lack of physical activity, smoking, and insufficient adherence to medical treatments. These behaviors can further weaken immune response and hinder tissue repair (Uchino et al., 1996). Individuals experiencing social isolation may also miss out on essential support, such as assistance with changing wound dressings, transportation to healthcare appointments, or reminders for medication, elements that are particularly crucial in the context of post-surgical or chronic wound management. In conclusion, social isolation negatively affects wound healing through both psychoneuroimmune pathways and associated behavioral issues. Therefore, fostering social connections should be regarded as an essential aspect of thorough wound care, particularly for elderly or chronically ill individuals. Initiatives aimed at encouraging social interaction and alleviating loneliness could significantly improve healing results.

6.1.4 Chronic Wounds and Sleep Disturbances

Chronic wounds present not only physical and emotional difficulties but also greatly interfere with sleep quality and patterns. Sleep is crucial for tissue repair, immune function, and mental health. Nevertheless, those suffering from chronic wounds frequently experience challenges in falling asleep, maintaining sleep, or obtaining restorative sleep, which can further hinder wound healing and overall well-being. A major factor contributing to sleep disturbances in patients with chronic wounds is pain, particularly nociceptive and neuropathic pain, which often worsens at night when distractions are minimal and the body is at rest (Vanderwee et al., 2008). The persistent discomfort, along with episodes of intense or burning pain, can result in frequent awakenings, disrupted sleep, and decreased sleep duration.

In a cross-sectional study, it was discovered that more than 70% of patients with chronic wounds faced sleep disturbances, frequently citing pain and discomfort from dressings as the main reasons (Upton and Upton, 2015). Patients also reported that regular wound care during the night (such as addressing leaking dressings or providing pressure relief) further interrupted their ability to achieve uninterrupted sleep. In addition to physical issues, psychological distress is a significant factor. Those with chronic wounds often suffer from anxiety, depression, or fear of their condition worsening, all of which have been associated with insomnia and diminished sleep quality (Finlayson et al., 2017). Sleep disturbances, in turn, create a detrimental cycle by weakening immune function and hindering wound healing, resulting in extended suffering and an increased psychological burden.

Furthermore, chronic wounds often necessitate the use of bulky or uncomfortable dressings, which can hinder mobility in bed and lead to discomfort in sleep positions. For individuals with pressure injuries or limited mobility, the requirement for regular repositioning during the night—either initiated by themselves or assisted by caregivers—contributes to sleep fragmentation (Vowden and Vowden, 2016). To summarize, chronic wounds considerably affect sleep quality through a combination of pain, psychological stress, physical limitations, and frequent interruptions. Addressing sleep disturbances should be a fundamental component of chronic wound management, as improving sleep may positively impact both emotional well-being and wound healing outcomes.

6.2 MECHANISMS AND IMPACT OF PSYCHOSOCIAL FACTORS ON WOUND HEALING

The psychological states affect wound recovery due to mechanisms involving inflammation, immune function, neuroendocrine pathways, and the perception and modulation of pain. The neuroendocrine system plays a significant role in associating psychological stress with the process of wound healing. Psychological stress leads to the release of hormones, such as cortisol and catecholamines, that can seriously affect the healing process, for example, cortisol delays wound healing by suppressing the inflammatory response important for the initial phase of wound repair (Gouin and Kiecolt-Glaser, 2011). In the same way, catecholamines hinder healing by decreasing the flow of blood to the wound site. Particularly, a study assessing the healing of punch biopsies indicated a direct association between wound healing and stress, regardless of other health behaviors. The study revealed that individuals with slower healing rates had elevated stress and cortisol levels upon waking as compared to their faster-healing counterparts (Ebrecht et al., 2004).

The impact of psychosocial factors on wound healing is also analysed by the bidirectional communication between the immune and nervous systems, referred to as neuroimmune interactions. By stimulating the sympathetic–adrenal–medullary axis and the HPA axis, stress leads to the release of neuroendocrine hormones that impair the function of the immune system (Glaser and Kiecolt-Glaser, 2005). This impairment exhibits modified cytokine production, including tumor necrosis factor (TNF) and interferons, dysregulating both the immune response and wound repair. Cytokine profile modification can shift the balance in favour of inflammation, which ironically may slow the recovery by extending inflammation (Segerstrom and Miller, 2004). Therefore, pro-inflammatory cytokines play a vital part in neuroimmune interactions as they stimulate the secretion of stress hormones, by releasing corticotropin-releasing hormone (CRH), which in turn activates the HPA axis. Also, in response to increased levels of stress hormones, there is an attenuated response to cortisol due to the downregulation of receptors that bind neuroendocrine hormones, which further promotes inflammation and risk of tissue damage from pro-inflammatory cytokines (Vileikyte, 2007). The ability of the immune system to react to wounds is also impacted by psychological stress. Important immune cells, such as lymphocytes and macrophages, that are necessary for wound cleaning, prevention of infection, and tissue repair, can have their activity suppressed by stress (Dhabhar, 2009; Segerstrom and Miller, 2004). Inflammatory mediators have a complex function in wound healing; stress-induced changes in pro- and anti-inflammatory mediators make tissue regeneration more difficult and may even encourage chronic inflammation, which is harmful to the healing process (Sivamani et al., 2009).

It has also been confirmed that pain is a major cause of delayed healing due to the associated inflammatory responses, as well as psychosocial impacts that may include behavioral changes secondary to pain-related stress (Snyder, 2006). Pain associated with a chronic wound includes a psychological stressor that stimulates the HPA axis, promoting the production of glucocorticoid (cortisol) and vasopressin (McGuire et al., 2006). This has been linked with increased perceived stress and subsequent delayed healing (Ebrecht et al., 2004). A lower pain threshold has been associated with higher levels of stress and anxiety because these conditions can make a person more aware of somatic signals, which can lead to a vicious cycle of pain, stress, or anxiety, which in turn can make the pain worse (Woo, 2010). Similarly, it has been shown that during a dressing change, the expected discomfort can exacerbate the actual pain (Woo, 2010). Snyder (2006) also found that individuals with wounds experience significant stress due to problems with body image and odour. It was also found that secondary social isolation and pessimistic emotions resulted from the inability to perform everyday tasks.

Thus, when studying predictors of delayed wound healing, it is necessary to consider the physiology behind psychosocial factors involved in healing. Decreased wound healing rates have been exhibited by patients with depression (Jian et al., 2019). Stress and depression also affect the immune system, which leads to elevated wound severity and delayed recovery (Walburn et al., 2017; Wang et al., 2024).

Psychosocial factors also play an essential role in the outcomes of chronic wounds. Poor health outcomes are often linked to lower educational attainment, which might be linked to socioeconomic status and low health literacy. On the other hand, quick recovery is linked to higher educational attainment. Slower wound healing is associated with higher levels of stress, sadness, and negative views of illness (Walburn et al., 2017). Individuals with greater stress levels are more likely to consume tobacco and alcohol, make poorer nutritional choices, exercise less, and have sleep disruptions than individuals with decreased levels of stress. Excessive alcohol use hinders the healing process by slowing collagen deposition and cell migration at wound sites. Additionally, smoking is linked to slower wound healing in surgical as well as non-surgical wounds. Lack of sleep reduces the production of growth hormones and delays the regeneration of the skin barrier. Inadequate exercise might also cause wounds to heal more slowly. Additionally, the healing process may be slowed down by inadequate consumption of glucose, polyunsaturated proteins, and certain vitamins (Gouin and Kiecolt-Glaser, 2012). Without proper education or health literacy, patients might not be aware of the detrimental effects of actions on health outcomes.

Living with a wound may cause one to lose their independence, mobility, and job, which can have a detrimental effect on their socioeconomic standing and mental health. When there is a perceived loss of control, adaptive abilities are diminished, which might be a sign of problems down the road. Understanding the physiology of psychosocial healing variables is crucial when researching factors that predict delayed wound healing. Individuals suffering from depression exhibit reduced rates of wound healing (Jian et al., 2019). Recovery may be directly impacted by psychological stress because it triggers stress response systems, including the sympathetic–adrenal–medullary axis and the HPA, which raise the synthesis of catecholamines and glucocorticoids (Gouin and Kiecolt-Glaser, 2012; Lin et al., 2014). Chronic wounds also aggravate the detrimental effects of psychosocial issues on the quality of life of dermatological patients (Balikji et al., 2022; Popescu et al., 2023). These psychological challenges not only hinder the healing process but also create a cycle of distress that can lead to further complications in managing dermatological conditions. Consequently, chronic wounds have a substantial negative impact on people's physical, mental, and financial well-being as well as on societies (Zhu et al., 2022). Inadequate health literacy and socioeconomic position are two characteristics that may be connected to lower educational attainment, which is often linked to inferior health outcomes. Higher education levels, on the other hand, are linked to quicker recovery periods, and discomfort can result from patients feeling unable to cope with their wounds, which has an adverse effect on the body's natural healing process (Ousey and Edward, 2014). By reducing the psychological effects of wounds, social support may contribute to better outcomes in the healing of chronic wounds. Through indiscriminate methods, financial circumstances may potentially influence wound healing. While effective bandaging adherence accelerates wound healing, access to appropriate care is hampered by a lack of funds and appropriate transportation (Walburn et al., 2009). Multidisciplinary approaches to stress reduction, quitting smoking, diet, medication administration, and organized exercise regimens have shown promise for improving patient happiness and quality of life (Reinboldt-Jockenhöfer et al., 2021). While addressing cofactors such as genetic vulnerability, obesity, lack of exercise, smoking, diabetes, hypertension, hyperlipidaemia, and stress, psychosocial treatment can help reduce fear and anxiety (Reinboldt-Jockenhöfer et al., 2021).

6.3 ASSESSMENT OF COMPREHENSIVE BIOPSYCHOSOCIAL THERAPIES

Wound healing is significantly hampered by psychological stress. Unrecognized psychological stress can exacerbate a patient's skin condition and significantly reduce their quality of life (Connor, 2017). Taking a psychosocial approach in treating patients with chronic wounds can enhance results by treating wounds holistically. The psychological and social elements that might affect a patient's healing process must also be considered in addition to the physical components of wound care. Social resources and mental health assistance can be included in treatment plans to help patients' overall wellness and promote more successful and long-lasting rehabilitation.

The contribution of integrated biopsychosocial therapies to the enhancement of chronic wound healing has been examined in a number of case reports, literature reviews, and case–control studies. Numerous approaches, including mind–body therapies, social support programs, and psychological interventions, show promise in the treatment of chronic wounds and the psychiatric comorbidities associated with them. There is growing evidence that psychological emotions have a significant impact on physiological functions, such as wound healing (Gouin and Kiecolt-Glaser, 2011). Numerous psychological therapies, including mindfulness-based stress reduction, cognitive–behavioral therapy, and relaxation techniques, have shown promise in a range of medical conditions and have accelerated post-surgery recovery.

6.3.1 Psychosocial Interventions

A study by Bae and colleagues (2001) discussed individuals with systemic lupus erythematosus (SLE) who have access to social assistance showed better results and recuperate from a wide range of clinical disorders like cancer (Spiegel and Sephton, 2001), chronic pain (Penttinen et al., 2002), and cardiovascular disease (Grace et al., 2002). According to Cacioppo et al. (2002), lack of social support is linked to reduced mental or physical health. Investigations on the neuroendocrinological and immunological processes behind the psychosocial influence on health have mostly concentrated on how immune function is hampered by unfavourable social interactions. There is still uncertainty regarding the exact processes by which advantageous social influences enhance immune function and overall health (Kiecolt-Glaser et al., 2002b). The potential of constructive social contact to reduce the adverse impacts of stress on the healing of wounds is largely unknown. Determining and describing the ways in which social variables support wound healing may result in the creation of treatments that speed up the healing process and, eventually, enhance recuperation from disease or surgery (Padgett et al., 1998). Social support dramatically changes how the human cardiovascular and corticosteroid systems react to challenges in the workplace (Thorsteinsson and James, 1999). It has been found that psychological treatments can alter the expectations of cancer patients, surgical patients, and healthy persons regarding pain and recovery. They can reduce physiological arousal, stress, anxiety, HPA activity, and catecholamine levels (Antoni et al., 2000). Psychological therapies have the potential to consistently boost immunity in a range of diseases. Wahbeh et al. (2009) recently found that the relaxation techniques had the ideal effects on immunological function. These treatments may improve wound healing and have a major impact on patient groups, including surgical patients; even though sources have not included wound healing as an endpoint, they have demonstrated that psychological therapies can reduce postoperative pain and anxiety (Good, 1996; Kekecs et al., 2014; Nelson et al., 2013). There have only been a few studies done on each kind of psychological intervention, and since the outcomes include multiple categories of wounds, it is challenging to identify the most appropriate intervention for each specific wound. At least a few studies supported the therapeutic benefits of psychological interventions on the healing of all types of wounds, with the exception of laser ablation. The evidence supporting surgical wounds was consistently robust. Numerous surgical procedures have been addressed, and recovery was positively impacted by hypnosis, supportive treatments, and relaxation. The therapies had a greater chance of lowering elevated stress levels since surgical patients could have experienced the highest levels of anxiety and stress. Patients who have various kinds of clinical wounds may also experience higher levels of stress and anxiety. There were at least some encouraging findings from both trials that examined psychosocial treatments for persistent leg ulcers. Likewise, the two studies that looked into psychosocial treatments for stomach ulcers had encouraging findings. It is difficult to determine which aspects of psychological therapies are beneficial because many of them have similar characteristics. For example, structured social support and short-term empathetic care exhibited several common features, while relaxation-based therapies often integrated multiple methods such as controlled breathing, mental imagery, biofeedback, and gradual muscle relaxation. The processes by which psychosocial therapies promote better recovery are currently not well demonstrated by the available data. Despite the fact that many studies did not assess these variables, treatments are believed to lower stress, bad mood, and stress hormones. Future research ought to track alterations in the HPA and sympathetic–adrenal–medullary axis systems as well as shifts in mood and stress. Interventions may influence recovery in ways other than stress alleviation. Research examining the link between stress and the development or progression of peptic ulcers suggests that stress can lead to alterations in behavior, such as bad eating, drinking, smoking, and sleep deprivation, which may make healing more difficult by altering the secretion of stomach acid (Levenstein, 1998). Psychosocial therapies have demonstrated positive impact rates. particularly those therapies that include relaxation techniques when used on surgical patients and ulcers. However, therapies that have not yet undergone clinical testing should be handled with care. Research is going on to see if there are consistent results across broader groups with various wound forms and compliance with therapies. There are several psychological repercussions that people with wounds have been seen to experience. These conditions may be specific (associated with a certain kind of wound). This is valid for the wider psychosocial effects of wounds as well. So, the individual's medical experience is expected to vary depending on whether the wound results from trauma, such as a burn or surgical lesion, or chronic and persistent, like a diabetic foot ulcer or venous leg ulcer. These psychological effects can be brought on by the wound itself, the discomfort it causes, and other medical repercussions. For instance, it has been discovered that the exudate and malodour that are typical signs of chronic wounds might exacerbate depressive and anxious feelings (Hareendran et al., 2005; Herber et al., 2007). The social burden associated with wounds can contribute to emotional or mental distress, making these issues worse. People who have exudate leakage and bad breath frequently feel ashamed and struggle to keep their dignity and look (Hyde et al., 1999; Walshe, 1995). Patients may develop maladaptive coping mechanisms as a result of such events, which occasionally exacerbate existing injuries (Lo et al., 2008). These measures might include covering wounds to prevent leaks, removing bandages to spread exudate, and restricting fluid intake in the hopes of lowering exudate production. These exudates and malodour experiences may also cause a patient's look, as well as their footwear and wardrobe choices, to alter (Persoon et al., 2004). It was also reported that patients may also experience shame and a lack of self-worth as a result of this (Herber et al., 2007). It makes them avoid social interactions and activities, leading to social isolation (Jones et al., 2008). Furthermore, an individual's overall well-being may decline as a result of this. Therefore, there is evidence that a loop of psychological issues and physical issues might eventually delay the healing process. People dealing with chronic wounds frequently express that they feel unable to continue with their regular activities and social interactions (Woo et al., 2009). In order to reduce unpleasant experiences, patients could also need to restrict their activities (Woo, 2010), which would lower their social engagement (Solowiej et al., 2010). Researchers observed that patients frequently avoid activities such as walking, shopping, and exercising due to the pain associated with these activities (Herber et al., 2007). In addition, patients often refrain from participating in specific activities out of concern that these may aggravate their wounds or lead to further complications (Persoon et al., 2004). As a result, individuals with chronic wounds may have substantial limitations in their everyday activities that we typically take for granted due to pain and mobility constraints. Additionally, restricted movement may make it impossible to work (Faria et al., 2011). Patients frequently consider these limitations to be among the worst parts of having a wound, and these issues may significantly affect an individual's psychological well-being (Hamer et al., 1994). In socially isolated patients, it has been observed that patients would restrict their participation in recreational pursuits, including walking, swimming, gardening, and travel (Chase et al., 2000; Hareendran et al., 2005; Krasner, 1998). It was noticed that over 50% of the patients they talked to had stopped engaging in certain activities. Another reason for this decrease in social connections and activities might be a lack of energy brought on by sleep loss, which can happen as a result of the excruciating pain of a chronic wound (Harlin et al., 2009; Upton and Andrews, 2013a, 2013b; Upton and Andrews, 2015). Furthermore, social isolation and a feeling of disconnection from society can have serious effects on an individual's overall well-being and can be exacerbated by the helplessness and worry that come with being unable to carry out everyday tasks (Brown, 2005a, 2005b). Further evidence of the significance of behavioral and psychological elements in wound healing comes from intervention trials that enhance healing results by lowering psychological stress. Improved post-surgical results are often reflected in reduced complications and quicker hospital discharge have been linked to behavioral stress management therapies prior to surgery, according to meta-analyses of clinical research (Johnston and Vogele, 1993; Montgomery et al., 2002). Social assistance is linked to improved health outcomes (House et al., 1988). Compared to rats kept alone, monogamous rodents kept in pairs recovered a typical punch biopsy wound more quickly in animal experiments (Glasper and DeVries, 2005). Oxytocin and vasopressin, two hypothalamic peptides, have been shown to influence social bonding and physiological stress responses in both human and animal studies. During a social support task, those who interacted with the partner more positively had greater levels of plasma oxytocin. The normal blister wounds healed more quickly when there was a higher quantity of oxytocin in the blood. Moreover, quicker healing was associated with higher plasma vasopressin levels in women but not in males (Gouin et al., 2010). Psychosocial disorders can significantly affect a patient's mental well-being and overall quality of life, making them significant. But this is not the only thing that makes them relevant. These psychological issues have been linked to the healing process in a substantial and detrimental way. Research has shown that people who experience depression or anxiety related to their physical health typically recuperate significantly more slowly than those who have an optimistic mental state (Moffatt et al., 2011). Positive psychological traits may help shield individuals from the harmful impacts of illness. For instance, people who look for information or guidance on their wound and its care might learn that accomplishing it increases their sense of control over their health and relieves their distress (Moffatt et al., 2011).

6.3.2 Cognitive Behavioral Therapy (CBT)

Redirecting unfavourable ideas toward more goal-oriented thinking is the purpose of cognitive behavioral therapy (CBT), a type of psychotherapy. For patients with chronic wounds, CBT improves coping strategies, changes negative thought patterns, and develops problem-solving abilities, all of which contribute to better healing and improved overall well-being (Yu et al., 2024). Because of its remarkable efficacy, CBT has been widely used to treat a variety of psychological discomfort and post-traumatic stress disorder (PTSD) conditions. Patient's emotions and behaviors are improved by CBT, which actively helps people identify and get rid of harmful ideas and routines (Cherkin et al., 2016; Jiskoot et al., 2020; Meyerhoff et al., 2018; Milgrom et al., 2016; Sprich et al., 2016; Wood et al., 2021). However, the utility of CBT for older patients with severe burns has received very little investigation, particularly when it comes to emotional rehabilitation. Studies are being conducted on how a CBT-based psychosocial regimen affects the emotional recovery of elderly people who have sustained severe burns. After suffering severe burns, the focus is on strengthening patients' emotional health and improving their day-to-day quality of life by reducing common negative feelings including worry, sadness, and self-doubt using methodical intervention approaches. According to the Numeric Pain Rating Scale and shoulder range of motion, CBT has been shown to alleviate chronic shoulder pain and mobility that resulted from a traumatic event. CBT-based psychological therapies can be quite effective in enhancing senior patients' emotional health. The fundamental concept of CBT is that patterns of reaction to experiences impact feelings and actions. Elderly individuals who have unfavourable cognitive patterns, such as overgeneralization, self-blame, or helplessness, might benefit from CBT. Compared to traditional treatment, CBT-based therapy significantly improved patients' social abilities. Elderly individuals who suffer severe burns may experience worry and anxiety while they heal. Through the instruction of coping mechanisms, including deep breathing, relaxation training, and systematic desensitization, the CBT regimen assists patients in facing these fears and worries (Van Loenen et al., 2022). Self-efficacy and self-esteem were greater among patients getting CBT-based treatment than among the controls. The CBT care protocol offers problem-solving skills and approaches to help older patients overcome obstacles and hurdles during their recuperation. CBT might help older patients feel less emotionally distressed by enhancing their coping mechanisms and self-efficacy through the teaching of problem-solving techniques such as goal-setting, problem-solving, action planning, and outcome evaluation (Visla et al., 2023). Patients who received CBT-based nursing had a greater quality of life and much stronger pain reduction after treatment compared to those who received conventional care. Through the development of positive self-awareness, self-care practices, such as self-acceptance and self-esteem enhancement, and proactive self-care behaviors, CBT supports both physical and psychological rehabilitation. CBT-based psychological interventions offer effective methods and techniques for enhancing the mental health of senior patients who have sustained severe burns. CBT-based care can aid in the healing process and improve the psychological well-being and quality of life of senior citizens by assisting them in altering harmful thought patterns, managing anxiety and fear, improving problem-solving skills, and increasing self-awareness and self-care. The efficacy and adaptability of CBT should be assured, nonetheless, by emphasizing individual variations and the customization of therapeutic regimens. CBT presents a potential adjunctive strategy for wound healing treatment, especially when it comes to addressing the psychological stress that may impede healing. CBT helps patients eliminate stress-related behaviors, anxiety, and depression that impair tissue regeneration and immunological function by assisting them in creating healthy thinking patterns and coping mechanisms. Incorporating CBT into wound treatment facilitates quicker and more efficient healing, in addition to promoting mental well-being. CBT is a key component of holistic wound care, since the mind–body connection is becoming more and more clear in clinical studies.

6.3.3 Social Support Programs

Patients with chronic wounds have considerably limited social connections and engagement (Christian et al., 2007). Living with chronic wounds can be difficult and limit a patient's everyday activities in a number of ways. A wound becomes chronic if it doesn't heal enough in a timely manner (Lazarus et al., 1994). Patient's social participation may be hampered by the financial burden and activity restrictions experienced by many individuals with chronic wounds (Blome et al., 2014; Douglas et al., 2016); the relationship between social involvement and health has been described. They specify many types of social engagement. Research has consistently shown that social isolation hinders the healing process of wounds (Detillion et al., 2004). Evidence suggests that social ties improve resilience. Social support, or the presence of helpful and loving connections, has been shown to promote favourable outcomes in a range of chronic conditions; therefore, this association between resilience and social support may offer a way to control resilience. Robinson et al. (2017) showed that social support can promote wound healing, with the magnitude of the effects varying depending on the kind of support the patient receives. Social connections: relationships with members of one immediate or general social circle. Engaging in activities with people for personal enjoyment while utilizing social contact is known as informal social involvement. Volunteering is when someone works for the benefit of others in organizations (like their neighbourhood or church). They proposed that the following factors act as a mediator between various types of social engagement and the health benefits they provide. Social support refers to aid and support provided to the individual when required. The sense of reciprocity and trust that each member of the society has is known as social cohesiveness. Social engagement may be advantageous in two ways: either by reducing the impact of certain stresses during stressful periods or by consistently improving health regardless of stressors (Berkman, 1995; Cohen and Wills, 1985); it is asserted that social support could boost health by giving the person a feeling of closeness and belonging. However, social assistance may have detrimental effects on patients as well as support givers. Informal caregiving can be taxing and have a negative impact on health from the provider's point of view (Kuzuya et al., 2011). The challenge of social engagement is especially significant for those with chronic wounds, as older individuals often experience delayed healing and a higher prevalence of such wounds (Heyer et al., 2016; Ousey et al., 2013; Wicke et al., 2009). Reduced social participation, potentially escalating to full social isolation, represents the most significant concern (Bukov et al., 2002). Since social involvement is a complicated concept that significantly affects patients' well-being, it should be considered in order to fully understand how chronic wounds affect patients' lives. The primary source of social support is family members, who offer emotional support as well as direct and indirect wound care. The most common sources of transportation, social connection, and emotional support are friends, neighbours, and co-workers (Lacerda et al., 2014). Results have shown that, despite the broad advantages of family and friends, some people and patients were unable to get the support they desired (Bandeira et al., 2018). Patients consciously avoided asking for assistance so as not to burden others (Brod, 1998). In addition to psychological factors like stress and anxiety having a negative correlation with social relationships, the only clinical condition that showed fewer social ties was wound odour (Septiana and Gayatri, 2019; Upton et al., 2015). Due to psychological effects of the wound, patients were unable to engage in social activities because they felt humiliated, could not engage in things they used to like, or acted more withdrawn (Jackson et al., 2018). These limitations may result in monetary difficulties, a decline in social networks, or guilt as the patient's burden placed limitations on others. However, other patients stated that they continued to work at desk jobs or had a busy social life (Firth et al., 2011). The healing of wounds is a highly coordinated and dynamic process that involves social support programs, which are essential but frequently underappreciated. Structured psychosocial support, community involvement, and emotional support have a significant impact on recovery outcomes in addition to the physiological interventions. These initiatives not only reduce stress, anxiety, and depression, all of which are recognized for hindering the healing process and may also contribute to an increase in treatment compliance and provide patients a feeling of agency. Incorporating family, peer groups, and community-based therapies can lead to more comprehensive and long-lasting recovery from wounds. Social support is crucial, as it is becoming more and more evident that helping the individual in order to heal their wounds is also necessary.

6.3.4 Mind–Body Therapies

Mind–body treatments (MBTs) focus on the interaction between the mind and the body, with the intent to use the mind to influence physical functions and directly affect health (Medicine, 2016). These treatments may help counteract some of the harmful physiological effects associated with stress. They could also curb negative behaviors. In certain situations, like biofeedback, they could cause advantageous physiological alterations. The healing of wounds is positively impacted by mind–body therapies, such as yoga, biofeedback, hypnosis, and relaxation techniques. The most effective relaxation techniques that show the biggest gains (Robinson et al., 2017). Contemplative practice, awareness-based therapeutic approach, guided visualization exercises, hypnotherapy, body–mind monitoring technique, and other techniques are examples of MBTs (Medicine, 2016). In addition to being reasonably priced and low risk, these are beneficial for both physical and mental health (Stahl et al., 2015). Seven days following laparoscopic cholecystectomy, researchers found that combining preoperative music therapy, deep breathing exercises, progressive muscle relaxation, and guided imagery significantly enhanced postoperative outcomes, leading to higher hydroxyproline deposition and lower levels of overall stress (Broadbent et al., 2012). Improvements in wound healing are also demonstrated by yoga and expressive writing therapies (Robinson et al., 2017). Yoga's ability to normalize HPA activity and promote mindfulness makes it a powerful therapeutic tool. Hypnosis and other mind–body therapies have demonstrated encouraging results in the healing of wounds. Compared to usual surgical site care methods, hypnotized postoperative patients demonstrated objectively better healing process assessed in a randomized clinical trial (Ginandes et al., 2003). When considered collectively, these results highlight the significance of treating chronic wounds using a multidisciplinary approach. Because integrated biopsychosocial interventions addressing psychiatric comorbidities and alleviate the negative impact of psychological distress on wound healing, these interventions play a crucial role in enhancing recovery outcomes for patients with chronic wounds. The purpose of mind–body therapies is to affect physical function and health by concentrating on the relationship between the mind and body. These interventions can help mitigate some of the harmful physiological effects associated with stress, and they can also alleviate negative behaviors. Sometimes they can also lead to positive physiologic changes, such as with biofeedback. MBTs address the complex relationship between psychological health and physical recovery, making them a viable adjunct to traditional wound care. Stress reduction, immune system enhancement, and tissue regeneration have all been shown to be possible using techniques including yoga, meditation, guided imagery, biofeedback, and relaxation training. Chronic stress and mental disorders have been shown to affect wound healing by lowering circulation, decreasing immunological responses, and disrupting hormonal balance. MBTs aim to promote patient resilience, restore homeostasis, and hasten the healing process by addressing these psychological variables. Integration of mind–body practices into holistic wound care protocols may result in better results, more patient satisfaction, and a greater understanding of the importance of mental health in physical healing.

6.4 CONCLUSION

Chronic wounds are injuries that may persist for years or do not heal completely. These wounds resist normal healing and demonstrate continuous inflammation. Many studies have examined the negative influence of psychological disorders on chronic wound healing. Psychiatric conditions like stress and anxiety not only affect the neuroendocrine pathway but also the immune system of a chronic wound patient, which further deteriorates the wound repair process. These conditions lead to pain, sleep disturbances, behavioral changes, and social isolation, which aggravate the quality of life of the patient. Thus, psychosocial factors have a major impact on the outcomes of chronic wounds. Therefore, a multidisciplinary approach should be taken to improve the outcomes of chronic wound patients with psychological disorders. This approach should integrate wound care specialists with mental health professionals, including psychologists, psychiatrists, and social workers, to address both the physical and psychological dimensions of healing. Early screening for psychological distress, such as anxiety, depression, and PTSD, is crucial for timely intervention and improved patient management. Tailored psychological interventions, including cognitive-behavioral therapy, stress reduction techniques, and support groups, can mitigate the negative effects of stress on immune function and promote a more conducive environment for wound repair. Furthermore, addressing pain management, sleep hygiene, and social support systems are integral components of comprehensive care, enhancing patient well-being and fostering better adherence to treatment regimens. By recognizing the profound interplay between mind and body, a holistic and integrated care model can significantly accelerate healing, reduce recurrence rates, and ultimately enhance the overall quality of life for individuals suffering from chronic wounds.

GLOSSARY

Acute Stress: A temporary stress reaction to a direct threat. In contrast to chronic stress, acute stress can have limited or even beneficial effects on wound healing if it is not sustained.

Anxiety: A psychological condition marked by feelings of worry or fear.

Behavioral Interventions: Methods like relaxation training, mindfulness practices, and cognitive behavioral therapy (CBT) are designed to alleviate stress and enhance healing results.

Chronic Stress: Prolonged exposure to stressors that can adversely impact immune function, heighten inflammation, and considerably hinder wound healing.

Depression: A mood disorder that can disrupt immune function and extend the duration of wound healing, potentially due to behavioral changes and biological factors.

Motivation: The impetus to behave in specific ways. Patients with greater motivation tend to follow wound care and recovery plans more closely, leading to improved results.

Neuropeptides: Tiny protein-like substances (such as substance P and endorphins) that affect mood, pain, and inflammation that are crucial in the stress-related regulation of wound healing.

Psychoeducation: Delivering insights about psychological aspects of health, which helps patients in grasping the mind–body relationship and promotes enhanced wound management.

Psychosocial Factors: A comprehensive term that includes psychological and social influences, such as relationships, socioeconomic status, and mental health—that affect health and recovery.

Psychological Trauma: Emotional injuries resulting from troubling experiences. Unresolved trauma may cause persistent stress responses, affecting immune function and tissue healing.

Stress: A physical and mental reaction to perceived challenges or threats.

REFERENCES

Antoni, Michael H., Dean G. Cruess, Stacy Cruess, Susan Lutgendorf, Mahendra Kumar, Gail Ironson, Nancy Klimas, Mary Ann Fletcher, and Neil Schneiderman. "Cognitive–behavioral stress management intervention effects on anxiety, 24-hr urinary norepinephrine output, and T-cytotoxic/suppressor cells over time among symptomatic HIV-infected gay men." *Journal of Consulting and Clinical Psychology* 68, no. 1 (2000): 31.

Bae, Sang-Cheol, Hideki Hashimoto, Elizabeth W. Karlson, Matthew H. Liang, and Lawren H. Daltroy. "Variable effects of social support by race, economic status, and disease activity in systemic lupus erythematosus." *The Journal of Rheumatology* 28, no. 6 (2001): 1245–1251.

Balikji, Jessica, Maarten M. Hoogbergen, Johan Garssen, and Joris C. Verster. "Mental resilience, mood, and quality of life in young adults with self-reported impaired wound healing." *International Journal of Environmental Research and Public Health* 19, no. 5 (2022): 2542.

Bandeira, Luciana Alves, Maxuel Cruz dos Santos, Êrica Rosalba Mallmann Duarte, Andrea Gonçalves Bandeira, Deise Lisboa Riquinho, and Letícia Becker Vieira. "Social networks of patients with chronic skin lesions: Nursing care." *Revista Brasileira de Enfermagem* 71 (2018): 652–659.

Berkman, Lisa F. "The role of social relations in health promotion." *Biopsychosocial Science and Medicine* 57, no. 3 (1995): 245–254.
Blome, Christine, Katrin Baade, Eike Sebastian Debus, Patricia Price, and Matthias Augustin. "The "Wound-QoL": A short questionnaire measuring quality of life in patients with chronic wounds based on three established disease-specific instruments." *Wound Repair and Regeneration* 22, no. 4 (2014): 504–514.
Broadbent, Elizabeth, Arman Kahokehr, Roger J. Booth, Janine Thomas, John A. Windsor, Christina M. Buchanan, Benjamin RL Wheeler, Tarik Sammour, and Andrew G. Hill. "A brief relaxation intervention reduces stress and improves surgical wound healing response: A randomised trial." *Brain, Behavior, and Immunity* 26, no. 2 (2012): 212–217.
Broadbent, Elizabeth, Keith J. Petrie, Patrick G. Alley, and Roger J. Booth. "Psychological stress impairs early wound repair following surgery." *Biopsychosocial Science and Medicine* 65, no. 5 (2003): 865–869.
Brod, M. "Pilot study-quality of life issues in patients with diabetes and lower extremity ulcers: Patients and care givers." *Quality of Life Research* 7, no. 4 (1998): 365–372.
Brown, Annemarie. "Chronic leg ulcers, part 2: Do they affect a patient's social life?" *British Journal of Nursing* 14, no. 18 (2005a): 986–989.
Brown, Annmarie. "Chronic leg ulcers, part 1: Do they affect a patient's social life?" *British Journal of Nursing* 14, no. 17 (2005b): 894–898.
Brown, Annemarie. "Does social support impact on venous ulcer healing or recurrence?" *British Journal of Community Nursing* 13, no. Supl (2008): S8–S15.
Bukov, Aleksej, Ineke Maas, and Thomas Lampert. "Social participation in very old age: Cross-sectional and longitudinal findings from BASE." *The Journals of Gerontology Series B: Psychological Sciences and Social Sciences* 57, no. 6 (2002): P510–P517.
Cacioppo, John T., John M. Ernst, Mary H. Burleson, Martha K. McClintock, William B. Malarkey, Louise C. Hawkley, Ray B. Kowalewski et al. "Lonely traits and concomitant physiological processes: The MacArthur social neuroscience studies." *International Organization of Psychophysiology* 35 (2002): 143–154.
Chase, Susan K., Robin Whittemore, Nancy Crosby, Daniel Freney, Paula Howes, and Tania J. Phillips. "Living with chronic venous leg ulcers: A descriptive study of knowledge and functional health status." *Journal of Community Health Nursing* 17, no. 1 (2000): 1–13.
Cherkin, Daniel C., Karen J. Sherman, Benjamin H. Balderson, Andrea J. Cook, Melissa L. Anderson, Rene J. Hawkes, Kelly E. Hansen, and Judith A. Turner. "Effect of mindfulness-based stress reduction vs cognitive behavioral therapy or usual care on back pain and functional limitations in adults with chronic low back pain: A randomized clinical trial." *JAMA* 315, no. 12 (2016): 1240–1249.
Christian, Lisa M., Jennifer E. Graham, David A. Padgett, Ronald Glaser, and Janice K. Kiecolt-Glaser. "Stress and wound healing." *Neuroimmunomodulation* 13, no. 5–6 (2007): 337–346.
Cohen, Sheldon, and Thomas A. Wills. "Stress, social support, and the buffering hypothesis." *Psychological Bulletin* 98, no. 2 (1985): 310.
Cole, Steven W. "Social regulation of human gene expression: Mechanisms and implications for public health." *American Journal of Public Health* 103, no. S1 (2013): S84–S92.
Connor, Cody J. "Management of the psychosocial comorbidities of dermatological conditions: Practitioners' guidelines." *Clinical, Cosmetic and Investigational Dermatology* 10 (2017): 117–132.
Detillion, Courtney E., Tara KS Craft, Erica R. Glasper, Brian J. Prendergast, and A. Courtney DeVries. "Social facilitation of wound healing." *Psychoneuroendocrinology* 29, no. 8 (2004): 1004–1011.
Dhabhar, Firdaus. S. "Effects of psychological stress on skin immune function: Implications for immunoprotection versus immunopathology." *Neuroimmunology of the Skin: Basic Science to Clinical Practice* 16 (2009): 113–123.
Douglas, Heather, Andrew Georgiou, and Johanna Westbrook. "Social participation as an indicator of successful aging: An overview of concepts and their associations with health." *Australian Health Review* 41, no. 4 (2016): 455–462.
Ebrecht, Marcel, Justine Hextall, Lauren-Grace Kirtley, Alice Taylor, Mary Dyson, and John Weinman. "Perceived stress and cortisol levels predict speed of wound healing in healthy male adults." *Psychoneuroendocrinology* 29, no. 6 (2004): 798–809.
Faria, Elaine, Leila Blanes, Bernardo Hochman, Marcos Mesquita Filho, and Lydia Ferreira. "Health-related quality of life, self-esteem, and functional status of patients with leg ulcers." *Wounds* 23, no. 1 (2011): 4–10.
Finlayson, Kathleen, Christine Miaskowski, Kimberly Alexander, Wei-Hong Liu, Bradley Aouizerat, Christina Parker, Diane Maresco-Pennisi, and Helen Edwards. "Distinct wound healing and quality-of-life outcomes in subgroups of patients with venous leg ulcers with different symptom cluster experiences." *Journal of Pain and Symptom Management* 53, no. 5 (2017): 871–879.

Firth, Jill, E. Andrea Nelson, Michelle Briggs, and Claudia Gorecki. "A qualitative study to explore the impact of foot ulceration on health-related quality of life in patients with rheumatoid arthritis." *International Journal of Nursing Studies* 48, no. 11 (2011): 1401–1408.

Ginandes, Carol, Patricia Brooks, William Sando, Christopher Jones, and John Aker. "Can medical hypnosis accelerate post-surgical wound healing? Results of a clinical trial." *American Journal of Clinical Hypnosis* 45, no. 4 (2003): 333–351.

Glaser, Ronald, and Janice K. Kiecolt-Glaser. "Stress-induced immune dysfunction: Implications for health." *Nature Reviews Immunology* 5, no. 3 (2005): 243–251.

Glasper, Erica R., and A. Courtney DeVries. "Social structure influences effects of pair-housing on wound healing." *Brain, Behavior, and Immunity* 19, no. 1 (2005): 61–68.

Good, Marion. "Effects of relaxation and music on postoperative pain: A review." *Journal of Advanced Nursing* 24, no. 5 (1996): 905–914.

Gouin, Jean-Philippe, C. Sue Carter, Hossein Pournajafi-Nazarloo, Ronald Glaser, William B. Malarkey, Timothy J. Loving, Jeffrey Stowell, and Janice K. Kiecolt-Glaser. "Marital behavior, oxytocin, vasopressin, and wound healing." *Psychoneuroendocrinology* 35, no. 7 (2010): 1082–1090.

Gouin, Jean-Philippe, Liisa Hantsoo, and Janice K. Kiecolt-Glaser. "Immune dysregulation and chronic stress among older adults: A review." *Neuroimmunomodulation* 15, no. 4–6 (2008): 251–259.

Gouin, Jean-Philippe, and Janice K. Kiecolt-Glaser. "The impact of psychological stress on wound healing: Methods and mechanisms." *Immunology and Allergy Clinics of North America* 31, no. 1 (2011): 81.

Gouin, Jean-Philippe, and Janice K. Kiecolt-Glaser. "The impact of psychological stress on wound healing: Methods and mechanisms." *Critical Care Nursing Clinics* 24, no. 2 (2012): 201–213.

Grace, Sherry L., Susan E. Abbey, Zachary M. Shnek, Jane Irvine, Renée-Louise Franche, and Donna E. Stewart. "Cardiac rehabilitation I: Review of psychosocial factors." *General Hospital Psychiatry* 24, no. 3 (2002): 121–126.

Hamer, C, Nicky Cullum, and Brenda. H. Roe. "Patients' perceptions of chronic leg ulcers." *Journal of Wound Care* 3, no. 2 (1994): 99–101.

Hareendran, Asha., Andrew Bradbury, John. Budd, George Geroulakos, Richard. Hobbs, Joyce. Kenkre, and Tanya. Symonds. "Measuring the impact of venous leg ulcers on quality of life." *Journal of Wound Care* 14, no. 2 (2005): 53–57.

Harlin, Stephen L., Ryan D. Harlin, Thomas I. Sherman, Courtney M. Rozsas, M. Shuja Shafqat, and William Meyers. "Using a structured, computer-administered questionnaire for evaluating health-related quality of life in patients with chronic lower extremity wounds." *Ostomy Wound Manage* 55, no. 9 (2009): 30.

Herber, Oliver R., Wilfried Schnepp, and Monika A. Rieger. "A systematic review on the impact of leg ulceration on patients' quality of life." *Health and Quality of Life Outcomes* 5, no. 1 (2007): 44.

Herbert, Tracy B., and Sheldon Cohen. "Depression and immunity: A meta-analytic review." *Psychological Bulletin* 113, no. 3 (1993): 472.

Heyer, Kristina, Katharina Herberger, Kerstin Protz, Gerd Glaeske, and Matthias Augustin. "Epidemiology of chronic wounds in Germany: Analysis of statutory health insurance data." *Wound Repair and Regeneration* 24, no. 2 (2016): 434–442.

House, James S., Karl R. Landis, and Debra Umberson. "Social relationships and health." *Science* 241, no. 4865 (1988): 540–545.

Hyde, Cheryl, Betty Ward, Jan Horsfall, and Giselle Winder. "Older women's experience of living with chronic leg ulceration." *International Journal of Nursing Practice (Wiley-Blackwell* 5, no. 4 (1999).

Jackson, Debra E., Lisa A. Durrant, Marie Hutchinson, Charlotte A. Ballard, Stephen Neville, and Kim Usher. "Living with multiple losses: Insights from patients living with pressure injury." *Collegian* 25, no. 4 (2018): 409–414.

Jian, Jin, Huang Yi-Heng, Zhu Bang-Hui, Chen Jian-Hua, Zhang Xu-Dong, Xiao Shi-Chu, Luo Peng-Fei, Hu Xiao-Yan, and Xia Zhao-Fan. "Effects of depression on healing and inflammatory responses of acute wounds in rats." *Wound Repair and Regeneration* 27, no. 5 (2019): 462–469.

Jiang, Fu-Hui, Xiao-Man Liu, Hai-Rong Yu, Yan Qian, and Hong-Lin Chen. "The incidence of depression in patients with diabetic foot ulcers: A systematic review and meta-analysis." *The International Journal of Lower Extremity Wounds* 21, no. 2 (2022): 161–173.

Jiskoot, Geranne, Reinier Timman, Annemerle Beerthuizen, Alexandra Dietz de Loos, Jan Busschbach, and Joop Laven. "Weight reduction through a cognitive behavioral therapy lifestyle intervention in PCOS: The primary outcome of a randomized controlled trial." *Obesity* 28, no. 11 (2020): 2134–2141.

Johnston, Marie, and Claus Vögele. "Benefits of psychological preparation for surgery: A meta-analysis." *Annals of Behavioral Medicine* 15, no. 4 (1993): 245–256.

Jones, June E., Jude Robinson, Wally Barr, and Caroline Carlisle. "Impact of exudate and odour from chronic venous leg ulceration." *Nursing Standard (through 2013)* 22, no. 45 (2008): 53.

Kekecs, Zoltán, Tamás Nagy, and Katalin Varga. "The effectiveness of suggestive techniques in reducing postoperative side effects: A meta-analysis of randomized controlled trials." *Anesthesia & Analgesia* 119, no. 6 (2014): 1407–1419.

Kiecolt-Glaser, Janice K., Timothy J. Loving, Jeffrey R. Stowell, William B. Malarkey, Stanley Lemeshow, Stephanie L. Dickinson, and Ronald Glaser. "Hostile marital interactions, proinflammatory cytokine production, and wound healing." *Archives of General Psychiatry* 62, no. 12 (2005): 1377–1384.

Kiecolt-Glaser, Janice K., Phillip T. Marucha, Ana M. Mercado, William B. Malarkey, and Ronald Glaser. "Slowing of wound healing by psychological stress." *The Lancet* 346, no. 8984 (1995): 1194–1196.

Kiecolt-Glaser, Janice K., Lynanne McGuire, Theodore F. Robles, and Ronald Glaser. "Emotions, morbidity, and mortality: New perspectives from psychoneuroimmunology." *Annual Review of Psychology* 53, no. 1 (2002a): 83–107.

Kiecolt-Glaser, Janice K., Lynanne McGuire, Theodore F. Robles, and Ronald Glaser. "Psychoneuroimmunology: Psychological influences on immune function and health." *Journal of Consulting and Clinical Psychology* 70, no. 3 (2002b): 537.

Kirkland-Kyhn, Holly, Stephanie Anne Generao, Oleg. Teleten, and Heather. M. Young. "Teaching wound care to family caregivers: An overview of methods to use to promote wound healing." *Home Healthcare Now* 40 (2022): 8–13.

Klein, Toni Maria, Valerie Andrees, Natalia Kirsten, Kerstin Protz, Matthias Augustin, and Christine Blome. "Social participation of people with chronic wounds: A systematic review." *International Wound Journal* 18, no. 3 (2021): 287–311.

Krasner, Diane. "Painful venous ulcers: Themes and stories about their impact on quality of life." *Ostomy/ Wound Management* 44, no. 9 (1998): 38–46.

Kumar, Navin, Farhanul Huda, Rishit Mani, Tanuj Singla, Ashikesh Kundal, Jyoti Sharma, and Bhargav Gajula. "Role of hospital anxiety and depression on the healing of chronic leg ulcer: A prospective study." *International Wound Journal* 17, no. 6 (2020): 1941–1947.

Kuzuya, Masafumi, Hiromi Enoki, Jun Hasegawa, Sachiko Izawa, Yoshihisa Hirakawa, Hiroshi Shimokata, and Iguchi Akihisa. "Impact of caregiver burden on adverse health outcomes in community-dwelling dependent older care recipients." *The American Journal of Geriatric Psychiatry* 19, no. 4 (2011): 382–391.

Lacerda, Flavia Karine Leal, Evanilda Souza de Santana Carvalho, Edna. Maria de. Araújo, Nadjane. Barbosa Miranda, Ana. Luisa. Dias, and Terezinha Andrade Almeida. "Women with sickle anemia living with leg ulcers and pain." *Journal of Nursing UFPE on Line* 8, no. 7 (2014): 2054–2060.

Lazarus, Gerald S., Diane M. Cooper, David R. Knighton, David J. Margolis, Roger E. Percoraro, George Rodeheaver, and Martin C. Robson. "Definitions and guidelines for assessment of wounds and evaluation of healing." *Wound Repair and Regeneration* 2, no. 3 (1994): 165–170.

Levenstein, Susan. "Stress and peptic ulcer: Life beyond helicobacter." *BMJ* 316, no. 7130 (1998): 538.

Lin, Tzu-Kai, Mao-Qiang Man, Juan-Luis Santiago, Tiffany C. Scharschmidt, Melanie Hupe, Gemma Martin-Ezquerra, Jong-Kyung Youm et al. "Paradoxical benefits of psychological stress in inflammatory dermatoses models are glucocorticoid mediated." *Journal of Investigative Dermatology* 134, no. 12 (2014).

Lo, Shu-Fen, Wen-Yu Hu, Mark Hayter, Shu-Chuan Chang, Mei-Yu Hsu, and Li-Yue Wu. "Experiences of living with a malignant fungating wound: A qualitative study." *Journal of Clinical Nursing* 17, no. 20 (2008): 2699–2708.

Martinengo, Laura, Maja Olsson, Ram Bajpai, Michael Soljak, Zee Upton, Artur Schmidtchen, Josip Car, and Krister Järbrink. "Prevalence of chronic wounds in the general population: Systematic review and meta-analysis of observational studies." *Annals of Epidemiology* 29 (2019): 8–15.

McCarty, Sara M., and Steven L. Percival. "Proteases and delayed wound healing." *Advances in Wound Care* 2, no. 8 (2013): 438–447.

McGuire, Lynanne, Kathi Heffner, Ronald Glaser, Bradley Needleman, William Malarkey, Stephanie Dickinson, Stanley Lemeshow et al. "Pain and wound healing in surgical patients." *Annals of Behavioral Medicine* 31, no. 2 (2006): 165–172.

Medicine, Section On Integrative. "Mind-body therapies in children and youth." *Pediatrics* 138, no. 3 (2016): e20161896.

Meyerhoff, Jonah, Michael A. Young, and Kelly J. Rohan. "Patterns of depressive symptom remission during the treatment of seasonal affective disorder with cognitive-behavioral therapy or light therapy." *Depression and Anxiety* 35, no. 5 (2018): 457–467.

Milgrom, Jeannette, Brian G. Danaher, Alan W. Gemmill, Charlene Holt, Christopher J. Holt, John R. Seeley, Milagra S. Tyler, Jessica Ross, and Jennifer Ericksen. "Internet cognitive behavioral therapy for women with postnatal depression: A randomized controlled trial of MumMoodBooster." *Journal of Medical Internet Research* 18, no. 3 (2016): e499.

Moffatt, Christine. J., Lynda Mapplebeck, Susie Murray, and Phillip A. Morgan. "The experience of patients with complex wounds and the use of NPWT in a home-care setting." *Journal of Wound Care* 20, no. 11 (2011): 512–527.

Montgomery, Guy H., Daniel David, Gary Winkel, Jeffrey H. Silverstein, and Dana H. Bovbjerg. "The effectiveness of adjunctive hypnosis with surgical patients: A meta-analysis." *Anesthesia & Analgesia* 94, no. 6 (2002): 1639–1645.

Nelson, Elizabeth A., Michelle M. Dowsey, Simon R. Knowles, David J. Castle, Michael R. Salzberg, Kaveh Monshat, Anthony J. Dunin, and Peter FM Choong. "Systematic review of the efficacy of pre-surgical mind-body based therapies on post-operative outcome measures." *Complementary Therapies in Medicine* 21, no. 6 (2013): 697–711.

Ousey, Karen, and Karen-Leigh Edward. "Exploring resilience when living with a wound—An integrative literature review." *Healthcare* 2, no. 3 (2014): 346–355.

Ousey, Karen, John Stephenson, Simon Barrett, Brenda King, Nicky Morton, Kim Fenwick, and Caryn Carr. "Wound care in five English NHS trusts: Results of a survey." *Wounds UK* 9, no. 4 (2013): 20–28.

Padgett, David A., Phillip T. Marucha, and John F. Sheridan. "Restraint stress slows cutaneous wound healing in mice." *Brain, Behavior, and Immunity* 12, no. 1 (1998): 64–73.

Penttinen, Jyrki, Nina Nevala-Puranen, Olavi Airaksinen, Markku Jääskeläinen, Harri Sintonen, and Jorma Takala. "Randomized controlled trial of back school with and without peer support." *Journal of Occupational Rehabilitation* 12, no. 1 (2002): 21–29.

Persoon, Anke, Maud M. Heinen, Carien JM Van Der Vleuten, Michette J. De Rooij, Peter CM Van De Kerkhof, and Theo Van Achterberg. "Leg ulcers: A review of their impact on daily life." *Journal of Clinical Nursing* 13, no. 3 (2004): 341–354.

Platsidaki, Eftychia, Anargyros Kouris, and Christos Christodoulou. "Psychosocial aspects in patients with chronic leg ulcers." *Wounds: A Compendium of Clinical Research and Practice* 29, no. 10 (2017): 306–310.

Popescu, Valentin, Victor Cauni, Marius Septimiu Petrutescu, Maria Madalina Rustin, Raluca Bocai, Cristina Rachila Turculet, Horia Doran et al. "Chronic wound management: From gauze to homologous cellular matrix." *Biomedicines* 11, no. 9 (2023): 2457.

Reinboldt-Jockenhöfer, Finja, Zeynep Babadagi, Heinz-Dieter Hoppe, Alexander Risse, Christos Rammos, Anna Cyrek, Christine Blome, Sven Benson, and Joachim Dissemond. "Association of wound genesis on varying aspects of health-related quality of life in patients with different types of chronic wounds: Results of a cross-sectional multicentre study." *International Wound Journal* 18, no. 4 (2021): 432–439.

Renner, Regina, and Cornelia Erfurt-Berge. "Depression and quality of life in patients with chronic wounds: Ways to measure their influence and their effect on daily life." *Chronic Wound Care Management and Research* 4 (2017): 143–151.

Robinson, Hayley, Sam Norton, Paul Jarrett, and Elizabeth Broadbent. "The effects of psychological interventions on wound healing: A systematic review of randomized trials." *British Journal of Health Psychology* 22, no. 4 (2017): 805–835.

Schreml, Stephan, Rolf Markus Szeimies, Lukas Prantl, Sigrid Karrer, Michael Landthaler, and Philipp Babilas. "Oxygen in acute and chronic wound healing." *British Journal of Dermatology* 163, no. 2 (2010): 257–268.

Schultz, Gregory S., R. Gary Sibbald, Vincent Falanga, Elizabeth A. Ayello, Caroline Dowsett, Keith Harding, Marco Romanelli, Michael C. Stacey, Luc Teot, and Wolfgang Vanscheidt. "Wound bed preparation: A systematic approach to wound management." *Wound Repair and Regeneration* 11 (2003): S1–S28.

Segerstrom, Suzanne C., and Gregory E. Miller. "Psychological stress and the human immune system: A meta-analytic study of 30 years of inquiry." *Psychological Bulletin* 130, no. 4 (2004): 601.

Septiana, Marinda Navy, and Dewi Gayatri. "The impact of discomfort: Physical and psychological to social interaction in diabetic ulcer patients in Jakarta–Indonesia." *Enfermeria Clinica* 29 (2019): 407–412.

Sivamani, Raja K., Christine E. Pullar, Catherine G. Manabat-Hidalgo, David M. Rocke, Richard C. Carlsen, David G. Greenhalgh, and R. Rivkah Isseroff. "Stress-mediated increases in systemic and local epinephrine impair skin wound healing: Potential new indication for beta blockers." *PLoS Medicine* 6, no. 1 (2009): e1000012.

Snyder, Robert J. "Venous leg ulcers in the elderly patient: Associated stress, social support, and coping." *Ostomy/Wound Management* 52, no. 9 (2006): 58–66.

Solowiej, Kazia, Victoria Mason, and Dominic Upton. "Psychological stress and pain in wound care, part 3: Management." *Journal of Wound Care* 19, no. 4 (2010): 153–155.

Spiegel, David, and Sandra E. Sephton. "Psychoneuroimmune and endocrine pathways in cancer: Effects of stress and support." *Seminars in Clinical Neuropsychiatry* 6, no. 4 (2001): 252–265.

Sprich, Susan E., Steven A. Safren, Daniel Finkelstein, Jocelyn E. Remmert, and Paul Hammerness. "A randomized controlled trial of cognitive behavioral therapy for ADHD in medication-treated adolescents." *Journal of Child Psychology and Psychiatry* 57, no. 11 (2016): 1218–1226.

Stahl, James E., Michelle L. Dossett, A. Scott LaJoie, John W. Denninger, Darshan H. Mehta, Roberta Goldman, Gregory L. Fricchione, and Herbert Benson. "Relaxation response and resiliency training and its effect on healthcare resource utilization." *PLoS One* 10, no. 10 (2015): e0140212.

Stechmiller, Joyce K. "Understanding the role of nutrition and wound healing." *Nutrition in Clinical Practice* 25, no. 1 (2010): 61–68.

Thorsteinsson, Einar B., and Jack E. James. "A meta-analysis of the effects of experimental manipulations of social support during laboratory stress." *Psychology and Health* 14, no. 5 (1999): 869–886.

Uchino, Bert N., John T. Cacioppo, and Janice K. Kiecolt-Glaser. "The relationship between social support and physiological processes: A review with emphasis on underlying mechanisms and implications for health." *Psychological Bulletin* 119, no. 3 (1996): 488.

Upton, Dominic, and Abbye Andrews. "Negative pressure wound therapy: Improving the patient experience part 2 of 3." *Journal of Wound Care* 22, no. 11 (2013a): 582–591.

Upton, Dominic, and Abbye Andrews. "Negative pressure wound therapy: Improving the patient experience part 1 of 3." *Journal of Wound Care* 22, no. 10 (2013b): 552–557.

Upton, Dominic, and Abbye Andrews. "Pain and trauma in negative pressure wound therapy: a review." *International wound journal* 12, no. 1((2015): 100–105.

Upton, Dominic, and Penney Upton. *Psychology of Wounds and Wound Care in Clinical Practice*. Springer, 2015.

Upton, Dominic, Penney Upton, and R. Alexander. "Contribution of the leg club model of care to the well-being of people living with chronic wounds." *Journal of Wound Care* 24, no. 9 (2015): 397–405.

Vanderwee, Katrien, Maria Grypdonck, and Tom Defloor. "Alternating pressure air mattresses as prevention for pressure ulcers: A literature review." *International Journal of Nursing Studies* 45, no. 5 (2008): 784–801.

Van Loenen, Inge, Willemijn Scholten, Anna Muntingh, Johannes Smit, and Neeltje Batelaan. "The effectiveness of virtual reality exposure–based cognitive behavioral therapy for severe anxiety disorders, obsessive-compulsive disorder, and posttraumatic stress disorder: Meta-analysis." *Journal of Medical Internet Research* 24, no. 2 (2022): e26736.

Vileikyte, Loretta. "Stress and wound healing." *Clinics in Dermatology* 25, no. 1 (2007): 49–55.

Visla, Andreea, Mathias Allemand, and Christoph Flückiger. "Within-and between-patients associations between self-efficacy, outcome expectation, and symptom change in cognitive behavioral therapy for generalized anxiety disorder." *Journal of Clinical Psychology* 79, no. 1 (2023): 86–104.

Vowden, Peter, and Kathryn Vowden. "The economic impact of hard-to-heal wounds: Promoting practice change to address passivity in wound management." *Wounds International* 7, no. 2 (2016): 10–15.

Wahbeh, Helané, Ashley Haywood, Karen Kaufman, and Heather Zwickey. "Mind-body medicine and immune system outcomes: A systematic review." *The Open Complementary Medicine Journal* 1 (2009): 25.

Walburn, Jessica, Kavita Vedhara, Matthew Hankins, Lorna Rixon, and John Weinman. "Psychological stress and wound healing in humans: A systematic review and meta-analysis." *Journal of Psychosomatic Research* 67, no. 3 (2009): 253–271.

Walburn, Jessica, John Weinman, Sam Norton, Matthew Hankins, Karen Dawe, Bolatito Banjoko, and Kavita Vedhara. "Stress, illness perceptions, behaviors, and healing in venous leg ulcers: Findings from a prospective observational study." *Psychosomatic Medicine* 79, no. 5 (2017): 585–592.

Walshe, Catherine. "Living with a venous leg ulcer: A descriptive study of patients' experiences." *Journal of Advanced Nursing* 22, no. 6 (1995): 1092–1100.

Wang, Na, Lin Pei, Mengqian Zhang, Gang Wang, Shumao Zheng, Xiaoli Kou, and Hu Chen. "RETRACTED: The impact of psychological interventions on surgical site wound healing post-surgery in psoriasis patients: A meta-analysis." *International Wound Journal* 21, no. 4 (2024): e14509.

Wicke, Corinna, Andreas Bachinger, Stephan Coerper, Stefan Beckert, Maria B. Witte, and Alfred Königsrainer. "Aging influences wound healing in patients with chronic lower extremity wounds treated in a specialized wound care center." *Wound Repair and Regeneration* 17, no. 1 (2009): 25–33.

Woo, Kevin. "Wound-related pain: Anxiety, stress and wound healing." *Wounds UK* 6, no. 4 (2010): 92–98.

Woo, Kevin Y., Patricia M. Coutts, Patricia Price, Keith Harding, and R. Gary Sibbald. "A randomized cross-over investigation of pain at dressing change comparing 2 foam dressings." *Advances in Skin & Wound Care* 22, no. 7 (2009): 304–310.

Wood, Jeffrey J., Karen Sze Wood, A Chuen Cho, Kashia A. Rosenau, Maria Cornejo Guevara, Chardée Galán, Alicia Bazzano, Ari S. Zeldin, and Gerhard Hellemann. "Modular cognitive behavioral therapy for autism-related symptoms in children: A randomized controlled trial." *Journal of Consulting and Clinical Psychology* 89, no. 2 (2021): 110.

Yan, Ruihong, Kara Strandlund, Hai Ci, Yinyan Huang, Yixuan Zhang, and Yujuan Zhang. "Analysis of factors influencing anxiety and depression among hospitalized patients with chronic wounds." *Advances in Skin & Wound Care* 34, no. 12 (2021): 638–644.

Yu, Hong-Mei, Ling Wang, and Cui-Zhen Pan. "The impact of cognitive behavioural therapy-based psychological intervention on emotional improvement in elderly patients with extensive burns." *International Wound Journal* 21, no. 2 (2024): e14594.

Zhou, Kehua, and Peng Jia. "Depressive symptoms in patients with wounds: A cross-sectional study." *Wound Repair and Regeneration* 24, no. 6 (2016): 1059–1065.

Zhu, Xiaoli, Maja Magdalena Olsson, Ram Bajpai, Krister Järbrink, Wern Ee Tang, and Josip Car. "Health-related quality of life and chronic wound characteristics among patients with chronic wounds treated in primary care: A cross-sectional study in Singapore." *International Wound Journal* 19, no. 5 (2022): 1121–1132.

7 Herbal Remedies for Wound Healing

Zuha Shahid[1], *Huma Hameed*[1*], *Mahtab Ahmad Khan*[1], *Ana Cláudia Paiva-Santos*[2,3], *Muhammad Zaman*[1], *Anam Hameed*[4], *and Tooba Mehboob*[1]

[1] Faculty of Pharmaceutical Sciences, University of Central Punjab (UCP), Lahore, Pakistan

[2] Department of Pharmaceutical Technology, Faculty of Pharmacy of the University of Coimbra, University of Coimbra, Coimbra, Portugal

[3] REQUIMTE/LAQV, Group of Pharmaceutical Technology, Faculty of Pharmacy of the University of Coimbra, University of Coimbra, Coimbra, Portugal

[4] Department of Human Nutrition and Dietetics, Faculty of Rehabilitation and Allied Health Sciences, Riphah International University, Gulberg III, Lahore, Pakistan

7.1 INTRODUCTION TO HERBAL MEDICINE

Herbal healing or phytomedicine or botanical medicine is a healing method in which plants and plant extracts are used for the prevention and treatment of diseases. It is the most ancient type of medicine, having existed for thousands of years in different traditions and cultures, like ancient Egypt, China, India, and many indigenous traditions in various corners of the world. The basic premise of herbal medicine revolves around the concept that plants contain natural substances that can act therapeutically in the body of humans. Traditional medicine usually processes the whole plants or specific parts like leaves, flowers, roots, bark, or seeds, which ooze an array of bioactive compounds. These active compounds, which include alkaloids, flavonoids, glycosides, essential oils, and saponins, work in synchronization to provide the desired benefits. The synergism is mostly advanced as an advantage since it may produce fewer side effects and a broader range of action, unlike any isolated compound (Heinrich et al., 2017).

Herbal medicine practitioners consider not only the person's physical symptoms but the individual's overall constitution, lifestyle, and emotional state. Treatments tend to be quite individualized and may include teas, tinctures, capsules, poultices, or topical creams in different herbal preparations. Although these remedies are often described as "natural" – and, therefore, one assumes, safe – it is really important to know that herbal medicines are potent and may have interactions with conventional drugs or contraindications with certain health conditions (Ekor, 2014). It's always a good idea to chat with a qualified healthcare professional, especially one who knows their way around herbal medicine, before diving into any herbal treatments. With the rising interest in natural health solutions, there's been a surge in scientific research looking into how effective and safe traditional herbal remedies really are. This is helping to connect the dots between ancient wisdom and what modern science understands today (Zhang, 2013).

* Corresponding author: huma.hameed@ucp.edu.pk

 DOI: 10.1201/9781003644958-7

7.1.1 Historical and Cultural Significance

Uses of plants for therapeutic purposes are among the oldest forms of medical treatment, reaching back thousands of years. Herbal medicines are an integral part of traditional healing systems worldwide, from traditional Chinese medicine through Ayurveda to indigenous healing systems. Modern healthcare's invigorated interest in herbal medicine springs from the options available with the concept of safety and therapy of many ailments, including the healing of wounds. Such an interest also reflects a paradigm shift toward natural and holistic healthcare, which now places increasing value on the complexities of plant-based remedies.

The use of several medicinal plants for wound treatment and healing from infections was recorded by ancient civilizations, such as those of Mesopotamia, Egypt, Greece, India, and China. Here they are: the Egyptian medical papyri, particularly the Ebers Papyrus (c. 1550 BCE), indicate several remedies from plants for an array of ailments, including details for wound dressings and ointments (Klein, 1905). The Hippocratic Corpus (5th–4th century BCE) emphasized the role of herbal medicine in maintaining health and treating diseases, advocating for the use of specific plants based on their observed effects, while Ayurveda classified plants according to their therapeutic properties and energetic qualities, providing a systematic approach to herbal remedies (Heinrich, 2015). Traditional Chinese Medicine incorporated herbs in complex formulations to restore balance and promote healing, often combining multiple plants to address the root causes of illness (Zhang et al., 2012). Native American, African, and Middle Eastern cultures have all created such rich traditions of herbal medicine that, besides local flora for wound healing and other health conditions, they have often relied on empirical knowledge transferred from one generation to another (Fabricant and Farnsworth, 2001). Historical texts emphasize that the whole world recognizes plants as great healing sources.

7.1.2 Bioactive Compounds and Mechanisms of Action

Herbal remedies exert their actions via various bioactive compounds, including alkaloids, flavonoids, tannins, glycosides, and essential oils. These compounds possess antimicrobial, anti-inflammatory, antioxidant, and tissue-regenerating capacities with multiple biological targets to influence the healing process. Flavonoids and polyphenols in many herbs scavenge free radicals and minimize oxidative stress, thereby enhancing wound repair and reducing tissue damage (Patra et al., 2018). Some of the herbs will stimulate immune responses and thus synergize to hasten healing through collagen synthesis and fibroblast proliferation, which play an important role in the process of tissue regeneration. With the support of modern phytochemistry and pharmacology, many traditional herbal remedies have seen scientific validation, telling the story of the mechanisms involved in their treatment. The most studied herb regarding wound healing is *Aloe vera* because of its polysaccharides, enzymes, and growth factors, which together enhance epithelialization and decrease inflammation (Newman and Cragg, 2020). *Curcuma longa* (turmeric) is another herb containing curcumin, known to enhance wound healing due to its strong anti-inflammatory as well as antimicrobial properties, and as an important adjunct in wound management (Posadzki, 2013). The backing of herbal medicine with evidence gives it a real opportunity to be an integrated part of modern healthcare practice.

7.2 MEDICINAL PLANTS IN WOUND HEALING

Medicinal plants have been essential in both traditional and contemporary wound healing because of their bioactive compounds that possess anti-inflammatory, antimicrobial, antioxidant, and tissue-regenerative characteristics. These plants support different phases of wound healing by influencing cytokines, growth factors, and enzymes. Some of the medicinal plants involved in wound healing are mentioned as follows:

7.2.1 *Aloe vera*

Wound healing is perhaps the most common use of *A. vera*, owing to its multifarious therapeutic properties (Surjushe et al., 2008). Primarily, it deals with a plethora of bioactive compounds, which include polysaccharides, glycoproteins, flavonoids, and anthraquinones. These components act together by stimulating fibroblast proliferation, a vital step in the formation of new connective tissue resulting from the process of wound repair (Maenthaisong et al., 2007). They also stimulate angiogenesis, the formation of new blood vessels, which supply oxygen and nutrients to the healing tissue (Hashemi et al., 2015). The cumulative effects of these compounds hasten re-epithelialization, which is the restorative action for the epithelial layer of the skin to close the wound (Singer and Clark, 1999).

Besides tissue regenerative properties, *A. vera* has high antimicrobial and anti-inflammatory activities of therapeutic relevance (Reuter et al., 2008). Therefore, these properties are useful in curbing infection risk, a common complication of wound healing, and inflammation, which may retard healing (Sari and Rahmawati, 2023). Many clinical and preclinical studies support *A. vera* in treating numerous types of wounds, including burns, which usually require extensive skin regeneration (Radha and Laxmipriya, 2015). It has also been useful for surgical wound healing, where the healing process should be well managed to avoid complications (Chithra et al., 1998). Also, *A. vera* has shown promise in the treatment of chronic wounds such as diabetic ulcers, which are very hard to heal, and pressure sores resulting from prolonged pressure on the skin (Reynolds and Dweck, 1999).

Polysaccharides such as acemannan in *A. vera* gel are of prime importance in immune modulation and healing (Surjushe et al., 2008). Acemannan reduces oxidative stress, which damages cells and rescues the normal process of wound repair by scavenging the free radicals (Hamman, 2008). Acemannan stimulates collagen synthesis, which is the crucial component of the extracellular matrix supporting the healing tissue (Jettanacheawchankit et al., 2009). Such augmented collagen formation results in the making of a strong and resilient scar (Hashemi et al., 2015). Acemannan's immunomodulatory effect also prevents the hyper-inflammatory response, ensuring tissue destruction during healing (Davis et al., 1989).

Therefore, *A. vera* has been formulated in many pharmaceutical and cosmetic products to exploit its therapeutic effects (Reuter et al., 2008). These products include topical creams and gels that can be directly applied to the wound site to transmit the active compounds into the tissue affected (Surjushe et al., 2008). *A. vera*-impregnated dressings help in the sustained dispensing of the active ingredients, maintaining a moist environment conducive for healing over time (Radha and Laxmipriya, 2015). There are also oral supplements and juices made from *A. vera* for encouraging systemic healing and better health (Boudreau and Beland, 2006). All these kinds of *A. vera* formulations address different wound types as well as a patient's needs, ensuring that its therapy is optimally harnessed in a particular consciousness of the best wound healing outcomes for its patients (Reynolds and Dweck, 1999).

The molecular mechanisms underlying *A. vera*'s wound healing are complex and multilateral, relying on many signaling pathways. *A. vera* proved capable of stimulating the production of important growth factors such as fibroblast growth factors (FGF) and vascular endothelial growth factors (VEGFs) relevant to tissue regeneration and angiogenesis (Hamman, 2008). It also plays a role in regulating the expression of cytokines, which are the main signaling molecules responsible for the regulation of inflammation and immune responses (Budai et al., 2013). Thus, these molecular pathways are modulated by *A. vera* for optimum induction toward wound healing and tissue repair (Hashemi et al., 2015).

However, some reactivity could be the minor effect included under the good safety profile for *A. vera* (Ernst, 2000). This includes possible allergic reactions, most notably skin irritation and contact dermatitis, especially when the product is used for prolonged periods or excessive applications (Reuter et al., 2008). It is wise to use *A. vera* products according to the recommendations of the manufacturer and consult a health provider in cases of any adverse reactions (Guo and Mei, 2016). There is a significant variance in the quality and purity of *A. vera* products; thus, it is important to buy from well-recognized manufacturers that practice good methods of production (Radha and Laxmipriya, 2015).

Thus, *A. vera* is one treasured natural resource in wound healing that offers diverse therapeutic benefits with an array of bioactive compounds (Sánchez et al., 2020). It has the ability to boost fibroblast proliferation, further making it vascular applicable for angiogenesis, accelerated re-epithelialization, and further reduction in inflammation and infection, as such rendering the treatment versatile and highly effective in any injury type (Reynolds and Dweck, 1999). Further research investigation into the molecular mechanisms and clinical applications of *A. vera* will feed and continue to grow the paradigm of insights for the potential therapeutic applications of *A. vera*, providing the foundation for future development of wound healing approaches.

7.2.2 *Crocus sativus*

Crocus sativus, or saffron, is the name that medicine calls this flower, and due to its different medicinal interpretations, it is regarded highly in ancient and modern medicine. It has families of bioactive compounds like crocin, crocetin, and safranal, which contribute to its medicinal effect. They exert a very strong anti-inflammatory, antioxidant, and wound-healing activity (Abdullaev, 2006). Studies have shown that one of the mechanisms by which saffron promotes healing in wounds is by enhancing collagen biosynthesis, which leads to shortening of repair time (Soheilifar et al., 2024). In addition, being a good antioxidant in the healing process, it alleviates oxidative stress in all of its forms (Mehdizadeh et al., 2013). Oxidative stress damages cellular components and prolongs healing time; therefore, saffron, due to its antioxidant capacities, acts to reduce the damage (Rahiman et al., 2018). According to studies, saffron extracts possess antimicrobial properties that prevent the growth of various microbial pathogens, thus promoting wound healing and preventing infection (Moshiri et al., 2015). This action is necessary to provide an environment for a clean wound that would promote healing (Javadi et al., 2013). Moreover, flavonoids and phenolic compounds found in saffron are suggested to enhance cell regeneration and angiogenesis, which is the development of new blood vessels and an important factor for tissue repair (Cristani et al., 2024). These substances help in reducing scarring and promoting better tissue repair by providing oxygen and nutrients to the wound repair site (Soheilifar et al., 2024).

Experimental research has demonstrated that saffron is an effective promoter and enhancer of wound healing. In such research, saffron was demonstrated to reduce inflammatory markers that impede healing, usually through excessive inflammation (Zhao et al., 2021). Besides this, saffron also enhances growth factors, particularly the VEGF, which is crucial in angiogenesis and tissue regeneration (Soheilifar et al., 2024). Therefore, through the appropriate modulation of these growth factors, an optimal environment is established to facilitate the closure and remodeling of a wound. The capacity of saffron to reduce levels of excess pro-inflammatory cytokines, such as tumor necrosis factor-alpha (TNF-α) and interleukin-6 (IL-6), which are pivotal in inflammation, forms the foundation of its anti-inflammatory activity. Saffron regulates inflammation to reduce pain and swelling, preventing further damage to the tissue and allowing the tissue to heal the wound more easily (Poursamimi et al., 2020). This is why saffron is useful in healing injuries. Its antioxidant effectiveness stems from the ability to liberate cells from free radicals and reactive oxygen species (ROS) that are present at the time of injury to tissues. This is why saffron protects cells from oxidative injury and restores cell functions (Sadeghnia et al., 2017). The range of bacteria and fungi, as well as the most common pathogens of wounds, such as *Staphylococcus aureus* and *Candida albicans* does not limit the antibacterial action of saffron. It renders great service in assisting as a natural antiseptic in the care of wounds (Nanda and Madan, 2021). Some examples of flavonoids and phenolic compounds in saffron, such as kaempferol and quercetin, aid in collagen formation, which is a structural protein that lends strength and support to the skin and all connective tissues. Increased production of collagen allows wounds to close with the least possible scarring (Hosseinzadeh et al., 2009). Saffron increases the secretion of VEGF for the formation of new blood vessels, which guarantees the supply of oxygen and nutrients to the tissues undergoing proliferation. This is crucial for fibroblast proliferation, as the fibroblasts synthesize collagen and other proteins for the extracellular matrix (Moshiri et al., 2015). Studies have shown that saffron aids in accelerating the wound healing process by shortening contraction time and speeding up epithelialization, which is the formation of new skin tissues (Javadi et al., 2013). The acceleration and the provoked actions of wound healing processes happen due to complete wound healing (Javadi et al., 2013). The anti-inflammatory activities of saffron are thought to influence the release of some interfering markers and growth factor boosters through several signaling pathways, predominantly the NF-κB and MAPK-induced pathways, which are the main regulatory pathways involved in cell proliferation and inflammation (Cristani et al., 2024).

Saffron is being researched for possible application in advanced wound-care formulations – embedding hydrogels and nanoparticles – to augment bioavailability and efficacy. This research aims to achieve the maximally therapeutic effects by targeting the wound with the bioactive compounds of saffron (Soheilifar et al., 2024). Therefore, the employment of nanoparticles as carriers for saffron extracts will optimize their stability, increase solubility, and enhance transdermal delivery, improving therapeutic efficacy for wound healing (Javadi et al., 2023). Hydrogels impregnated with saffron extracts promote a moist environment, which enhances healing and reduces infections. Long-term therapeutic effects are also delivered by the gradual release of bioactive compounds from saffron (Pagano et al., 2022).

7.2.3 *Arctium lappa*

Humans utilized the burdock for wounds even prior to its name, *Arctium lappa*. The plant has been referenced as a key remedy in folklore medicine across the world for its outstanding anti-inflammatory, antimicrobial, and tissue-regenerating activities (Chan et al., 2011). The aerial parts of the burdock, particularly the root that grows underground, are richly gifted with a complex array of bioactive compounds like lignans, flavonoids, and polyphenols that all contribute to its vast potential in wound healing (Predes et al., 2011). It has been found that the *A. lappa* root extract has a complex effect on the wound-healing process, strongly activating fibroblast activity, necessary for collagen formation and other elements of the extracellular matrix (Lin et al., 2002). Furthermore, it enhances the extracellular matrix deposition that serves as a scaffold for the tissue repair and regeneration process (Purcaru et al., 2018). The extract, additionally, has the advantage of accelerating wound closure through improved keratinocyte migration, which is the first step in re-epithelialization for the damaged skin (Knipping et al., 2008).

The significant antimicrobial properties of *A. lappa* above make it very useful for wound care. They indeed make it highly effective against bacterial infection usually found with chronic wounds (Jiménez-Monreal et al., 2009). The inhibition of the growth of pathogenic bacteria by burdock helps to prevent infection hindrances in the healing process and worsens the wound (Tanga, 2021). Studies have shown that burdock extracts have the potential to modulate cytokine levels to a significant extent, thus reducing inflammation at the wound site (Li et al., 2024). This is very critical in developing optimal conditions for repairing tissues. Granulation tissue formation, which comprises the proliferation of new blood vessels and connective tissue, is also induced by *A. lappa* and is one of the major processes of wound healing (Pathak and Mazumder, 2024). In fact, this increased vasculature and tissue remodeling help to close and repair the damaged tissues quickly (de Souza et al., 2022).

Traditional applications of *A. lappa* in cases of wound healing have been amply validated through modern pharmacological studies, providing whole new scientific evidence for their therapeutic use (Chan et al., 2011). Some studies focused on the mechanism of action of burdock in wound healing and revealed its ability to influence some cellular and molecular pathways that function in tissue repair (Predes et al., 2011). For instance, *in vitro* studies proved that lignans from burdock can exert powerful antioxidant and anti-inflammatory effects that are critical in reducing oxidative stress and inflammation in the tissues of wounds (Lin et al., 2002). Besides, flavonoids, which are other constituents of the plant, are also relevant to their antimicrobial and tissue-rebuilding abilities through enhancing cell proliferation and collagen synthesis (Purcaru et al., 2018). The polyphenols in *A. lappa* also significantly contribute to wound healing by promoting angiogenesis and the stimulation of growth factor synthesis, which are vital for tissue regeneration (Knipping et al., 2008). The synergistic effect of these compounds enhances burdock's effectiveness in wound healing (Jiménez-Monreal et al., 2009). To substantiate this assertion, most modern wound healing modifications have employed *A. lappa*, and it has been demonstrated to be safe and effective in tissue repair in clinical and vitro research (Tanga, 2021). These simple disagreements reinforce the medicinal potential of burdock and support its therapeutic potential in wound therapy (Li et al., 2024).

The topical use of *A. lappa* in creams, ointments, and extracts has been adapted to suit different wounds and treatment Regimens (Pathak and Mazumder, 2024). Additionally, it has holistic wound healing properties; these include curing inflammation, infection, and tissue damage, and hence it is an all-encompassing therapeutic agent (de Souza et al., 2022). More studies into the pharmacological and clinical applications of *A. lappa* would help to develop a better understanding of its therapeutic abilities and motivate advanced design patterns for wound healing (Chan et al., 2011). The integration of conventional and modern scientific methods into *A. lappa* in wound healing is what makes this herb special among herbal remedies (Predes et al., 2011).

7.2.4 *Curcuma longa*

C. longa or turmeric has a long history in traditional medicine, where it has been widely used for centuries due to its multipurpose therapeutic value, especially its anti-inflammatory, antimicrobial, and tissue-healing properties (Hewlings and Kalman, 2017). Its extensive application across various cultures is a testament to its efficacy in the treatment of a wide range of illness types, particularly wound healing and skin disorders (Akbik et al., 2014). *C. longa* rhizome extract is a treasure trove of bioactive molecules like curcuminoids, particularly curcumin, which are the key factors for its outstanding wound-healing activity (Jagetia and Aggarwal, 2007). These molecules act synergistically to enhance tissue repair and regeneration, rendering turmeric an excellent natural drug for all types of wounds (Mohanty and Sahoo, 2017).

Extensive study has explored the definite mechanisms whereby *C. longa* mediates wound healing, with reports implying that it remarkably enhances fibroblast function, an imperative initial step toward synthesizing collagen and other elements of the extracellular matrix (Jagetia and Rajanikant, 2012). Improved activity of the fibroblast ensures greater levels of collagen synthesis that enhance the physical strength of healing tissue as well as expedite closure of the wound (Goel et al., 2008). Also, *C. longa* is able to promote extracellular matrix deposition, acting as a cellular migration scaffold and tissue remodeling site during the process of healing (Maquart et al., 1988). Such increased matrix deposition facilitates the creation of a strong and stable scar tissue that prevents dehiscence of the wound and supports the best tissue function (Adhami et al., 2008). In addition, the turmeric rhizome extract efficiently enhances keratinocyte migration, an important process in re-epithelialization, which is necessary for the recovery of the protective barrier of the skin (Tønnesen et al., 2002). This enhanced keratinocyte migration speeds up the closure of the wound surface, minimizing the risk of infection and facilitating fast healing (Aggarwal et al., 2003).

The antimicrobial activity of *C. longa* is especially important in wound healing, as it acts as a natural barrier against bacterial infections that are typically seen in chronic wounds (Negi et al., 1999). Through the inhibition of pathogenic bacterial growth, turmeric prevents infection that can hinder the healing process and worsen wound severity (Ramasamy et al., 2015). The ability of turmeric to kill a broad range of microorganisms makes it a multifaceted antimicrobial agent, effective in managing various wound infections (Epstein et al., 2010). Studies have shown that turmeric extracts have the potential to regulate cytokine levels by efficiently modulating the inflammatory response at the wound site and enhancing the healing process (Chainani-Wu, 2003). This inflammatory modulation is crucial for avoiding damage to tissues and maintaining orderly progression in the healing cascade (Imran et al., 2017). *C. longa* has also been shown to cause granulation tissue formation, a critical wound healing process characterized by the development of new blood vessels and connective tissue (Shimada et al., 2018). This enhanced tissue remodeling and angiogenesis make rapid healing and repair of damaged tissue and promote the development of healthy scar tissue possible (Buhrmann et al., 2011).

The traditional applications of *C. longa* in wound treatment have been thoroughly validated by contemporary pharmacological studies, providing a sound scientific rationale for its therapy and revalidating it in contemporary wound therapy (Jagetia and Aggarwal, 2007). The research has also investigated the exact mechanisms through which turmeric increases wound healing, determining its ability to influence various cellular and molecular pathways of tissue repair (Hewlings and Kalman, 2017). For example, curcuminoids present in turmeric have been shown to have potent antioxidant and anti-inflammatory activities, contributing significantly to inhibiting oxidative stress and inflammation of wound tissues (Akbik et al., 2014). The activities provide for the prevention of damage to cells and the creation of an appropriate environment for tissue regeneration (Mohanty and Sahoo, 2017). Curcumin, which is the principal curcuminoid, is credited with its tissue-regenerating and antimicrobial activities through its function in stimulating cell proliferation and collagen synthesis (Jagetia and Rajanikant, 2012). These compounds enhance the structural strength of the healing tissue and improve the process of wound closure (Goel et al., 2008). The diverse compounds in *C. longa* also help in wound healing by supporting angiogenesis and stimulating the production of growth factors, which are important for tissue regeneration (Maquart et al., 1988). Turmeric thus emerges as a useful therapeutic agent due to its holistic support of wound healing, covering several aspects of tissue repair (Adhami et al., 2008). Clinical trials and *in vitro* tests have also confirmed the application of *C. longa* in modern formulations for wound healing, showing its efficacy and safety in initiating the process of tissue regeneration (Tønnesen et al., 2002). The consistency of the results of these tests indicates the therapeutic value of turmeric and validates its reputation as a good natural healing medicine for wounds (Aggarwal et al., 2003).

7.2.5 *Althaea officinalis*

The *Althaea officinalis*, also referred to as marshmallow root, is among the medicinal plants of traditional medicine widely known in most societies for its relatively good healing and soothing properties (Miraj and Alesaeidi, 2016). In addition to its root, this plant has a vast storehouse of complex bioactive compounds, e.g., mucilage, flavonoids, and phenolic acids, among others, that are credited with its considerable therapeutic potential (Zargari, 1991). Excess mucilage in *A. officinalis* forms a barrier, which possesses hydrating properties on the wound site, thereby reducing inflammation and producing a soothing effect (Rezaei et al., 2015). Hydration is the prime requirement for the formation of a moist environment for wound healing. Experiments have proved that the presence of this hydration accelerates the healing process by causing cell migration and proliferation (Wichtl, 2004). Flavonoids and phenolic acids contained in marshmallow root are additional sources of contributing anti-inflammatory and antioxidant activities toward the alleviation of the oxidative-stress-and-inflammation-associated events in wound tissues (Blumenthal et al., 2000).

A. officinalis extracts are reported to promote epithelialization, where the new skin cells develop over the wound surface and accelerate wound closure (Jafari-Sales et al., 2015). *A. officinalis* extracts have been reported in other studies to induce wound contraction, which is essential for healing because it decreases the wound size due to the contraction of the surrounding tissues (Bisset and Wichtl, 1994). The stimulation of collagen formation, a structural protein giving strength and support to the regenerating tissue, facilitates above-mentioned contraction (Kianitalaei et al., 2019). Furthermore, improved collagen production speeds up the closure of wounds but additionally boosts the tensile strength of the healed tissue, thereby lowering the chance of scarring and injury recurrence (Bone and Mills, 2012).

A. officinalis also stands guard as a protective barrier against microbial infections that occur easily when wounds are treated; it is among the common complications found (Xue et al., 2023). Marshmallow root provides a physical barrier and antimicrobial properties that prohibit bacterial colonization and infection. Thus, an optimal condition for tissue repairing is created (Miraj and Alesaeidi, 2016). The use of marshmallow root in folk traditions has been proven with modern pharmacological scientific research for the treatment of burns, ulcers, and surgical wounds (Zargari, 1991). Studies scientifically have explored the exponded mechanisms by which *A. officinalis* acts in the promotion of wound healing that can affect a variety of cellular and molecular tissue-repairing pathways (Rezaei et al., 2015).

The mucilage of the marshmallow root is demulcent in nature as it forms a soothing layer on irritated tissues and so relieves inflammation (Wichtl, 2004). Flavonoids that are found in *A. officinalis*, along with demulcent properties anti-inflammatory and antioxidant activities in that they inhibit the synthesis of inflammatory mediators and scavenge free radicals (Blumenthal et al., 2000). Another group of compounds found in marshmallow root is phenolic acids, which help the antimicrobial and tissue regenerative properties by inducing cell proliferation and synthesizing collagen (Jafari-Sales et al., 2015). These bioactive components exert combined synergistic effect in enhancing the overall wound-healing ability of *A. officinalis* (Bisset and Wichtl, 1994).

Also supporting the modern wound-healing formulations with *A. officinalis*, these clinical trials and *in vitro* have proven safety and efficiencies of the species in tissue repair (Kianitalaei et al., 2019). Great healing and anti-inflammatory properties make it very applicable in the treatment of dry and irritated wounds, including burns, eczema, and psoriasis (Bone and Mills, 2012). The repetitive conclusions of the aforementioned studies point toward the therapeutic potential of marshmallow root and supporting its effectiveness as a valuable herbal medicine in the management of wounds (Xue et al., 2023). *A. officinalis* exists in various dosage forms of creams, ointments, and extracts and thus can be adjusted according to different types of wounds and treatments (Miraj and Alesaeidi, 2016). Ongoing research into the pharmacological and clinical properties of *A. officinalis* will continue to enhance this understanding and assist with the development of new wound-healing methods (Zargari, 1991). Its common history of use, alongside scientific verification, firmly positions *A. officinalis* as a category within the realm of significant herbs used in herbal drug for wound care (Rezaei et al., 2015).

7.2.6 *Calendula officinalis*

C. officinalis is an extremely unique plant that has played a leading role in folk medicine due to its therapeutic potential in the healing of wounds (Della Loggia et al., 1994). This wonder plant has been utilized for centuries in the treatment of an array of skin injuries – minor and severe – and therefore still remains a vital therapy (Jiménez-Medina et al., 2006). Its remarkable activities are due to the large number of bioactive compounds such as flavonoids, triterpenoids, carotenoids, and volatile oils, which in combination bestow diverse pharmacological activities on it (Parente et al., 2012). These compounds synergistically bestow anti-inflammatory, antimicrobial, and antioxidant activities to form a healthy habitat for optimal healing of wounds (Ashwlayan et al., 2018). Scientific studies have consistently demonstrated that *C. officinalis* preparations effectively promote fibroblast proliferation, a process necessary in collagen and other elements of the extracellular matrix synthesis involved in healing tissues. Increased fibroblast activity is the cause of the production of granulation tissue, an essential precursor to the production of new tissue and wound healing (Givol et al., 2019). *C. officinalis* induces collagen synthesis that provides strength and stability to healing tissue and hence averts excessive scarring (Della Loggia et al., 1994). Wound healing is also accelerated by the extract through the promotion of new epithelial cell migration along the wound surface, where re-epithelialization covers the surface and thus allows closure of the wound as well as restoration of the skin barrier (Jiménez-Medina et al., 2006). Working together, these mechanisms make *C. officinalis* highly effective in the management of several types of wounds (Parente et al., 2012).

C. officinalis has some antimicrobial activities that prevent healing by preventing infection, which remains a common complication (Ashwlayan et al., 2018). By preventing the growth of pathogenic fungi and bacteria, marigold minimizes infection chances and leave the path for tissue regeneration open (Leach, 2008). Antioxidant activity helps in reducing oxidative stress at the wound site, removing free radicals that otherwise would hinder healing and increase tissue damage (Givol et al., 2019). Due to the antioxidant activity, the process works by preserving the cells from harm and makes them grow, also enabling closure of the wound (Della Loggia et al., 1994). Its other antimicrobial and antioxidant activities are now attributed to its multi-modality to healing of a wound (Jiménez-Medina et al., 2006).

C. officinalis has been proved through innumerable clinical studies as successful in treating surgical wounds, diabetic ulcers, and minor burns (Parente et al., 2012). Studies show that marigold accelerates healing, reduces inflammatory process, and prevents scarring for the better outcome to patients (Ashwlayan et al., 2018). Thus, formulations on the basis of *C. officinalis* such as ointments, creams, and hydrogel dressings were produced to maximize the therapeutic potential and improve the delivery to a wound site (Leach, 2008). These formulations assure a long-lasting therapeutic effect as well as promote better wound healing due to the controlled release of bioactive compounds (Givol et al., 2019). As such combinatorial properties of anti-inflammation and regeneration, *C. officinalis* represents an important herbal remedy for wound healing and offers a natural but effective alternative to allopathic treatment (Della Loggia et al., 1994). Future research and innovations using therapies based on *C. officinalis* would, therefore, enhance therapeutic usability along with new, improved strategies for wound management (Jiménez-Medina et al., 2006).

7.2.7 Jojoba

Simmondsia chinensis, commonly called jojoba, is a hardy desert shrub containing a unique wax-like oil that interestingly has structural resemblance with human sebum (Ranzato et al., 2011). This peculiar feature permits jojoba oil to blend seamlessly with the lipid barrier naturally present in skin, thereby imparting a high degree of moisturizing and protective effects (Habashy et al., 2005). These features account for the oil's high content of long-chain fatty acids, particularly gadoleic acid, which gives it emollient and skin-softening properties (Lin et al., 2018). In addition, jojoba oil provides a rich source of vitamin E, a potent antioxidant with the phenomenal capacity to scavenge free radicals and shield cells from oxidative damage – a wound-healing requirement (Tietel et al., 2021). In addition, phytosterols in jojoba oil provide anti-inflammatory and skin-healing activities, thus enhancing its therapeutic value (Samarasinghe et al., 2024). The blend of these elements ultimately endows jojoba oil with broad levels of moisturizing, anti-inflammatory, and antimicrobial activities that contribute to wound healing (Gad et al., 2021).

The ability to close wounds has been established in numerous studies to occur by several mechanisms, where fibroblast migration is increased (Ranzato et al., 2011). Fibroblasts are the cells that produce collagen and other extracellular matrix components, thereby aiding in tissue repair and regeneration (Habashy et al., 2005). Jojoba oil stimulates fibroblast activity and thus increases the deposition of new connective tissue, thereby speeding the closure of wounds (Lin et al., 2018). Meanwhile, jojoba oil reduces oxidative stress levels at the site of the wound, which minimizes cellular damage and establishes a favorable healing environment (Tietel et al., 2021). Free radical activity, especially lipid peroxidation, which can interfere with the healing process, is inhibited by the antioxidant activity of vitamin E (Samarasinghe et al., 2024). Put together, these properties of jojoba oil also confer antimicrobial action to its healing potential by effectively preventing the development of microbial infections that would otherwise hinder wound healing (Gad et al., 2021). Wound healing with jojoba oil reduces the chances of infection by inhibiting the growth of pathogenic bacteria and fungi, thus providing a clean, healthy healing environment (Ranzato et al., 2011).

The emollient nature of jojoba oil plays a special advantage in its property to maintain optimal hydration for skin – the most important factor in proper wound healing (Habashy et al., 2005). Adequate hydration prevents the drying of the area, which results in cracking, itching, and delayed healing (Lin et al., 2018). Jojoba oil thus forms a protective layer on the surface skin to prevent transepidermal water loss in order to retain skin moisture (Tietel et al., 2021). This barrier also protects the wound from environmental stressors and external irritants, thus facilitating healing (Samarasinghe et al., 2024). The oil can penetrate deep into the skin due to its capacity to emulate human sebum, thus making its moisturizing and therapeutic properties available in the deeper epidermis (Gad et al., 2021). This penetration of the dip into the skin avoids dryness of the skin and renders it soft, thus making natural healing easier (Ranzato et al., 2011).

Jojoba oil is a very common example used in most of the pharmaceutical and cosmetic formulations, including creams, ointments, and dressings, for wound healing and skin regeneration (Habashy et al., 2005). Its compatibility and versatility also render it a valuable ingredient of topical formulations intended for wound healing (Lin et al., 2018). Clinical trials and *in vitro* studies indicated the effectiveness of jojoba oil in repairing tissues in conjunction with reducing inflammation in various wound types (Tietel et al., 2021). Its ability to modulate inflammation and support tissue regeneration can make it a good candidate for the treatment of wound applications, such as chronic wounds or even burns (Samarasinghe et al., 2024). Non-comedogenic also makes it relevant for use on acne skin since it doesn't block pores or aggravate skin conditions (Gad et al., 2021). Continued exploration of the therapeutic properties of jojoba oil will create new strategies and formulations for wound healing (Ranzato et al., 2011).

On the contrary, the property of jojoba oil, being devoid of triglycerides, makes it oxidation rancidity proof, thus affording a much longer and also uniform therapeutic life (Habashy et al., 2005). Thesaurus stability is for topical formulations; most requirements are going to long periods of storage and application (Lin et al., 2018). This makes the oil comprehensive in non-irritating and hypoallergenic properties, making it suitable to be used on sensitive skin (Tietel et al., 2021). Jojoba oil being biocompatible with human skin permits its use in a wide range of applications in wound care safely and efficiently (Samarasinghe et al., 2024). Besides, jojoba being such a shrub is a desert, really makes it becomes an ethical source of therapeutic oils (Gad et al., 2021). As the scientific support for wound healing by herbal means grows, it is becoming increasingly important historically and in modern medicine (Ranzato et al., 2011).

7.2.8 *Plantago major*

Plantago major, or the broadleaf plantain, holds an extensive place in traditional medicine, especially for its remarkable wound-healing activity (Samuelsen, 2000). This seemingly modest plant has formed an important constituent of folk remedies in several cultures, where its leaves were used externally to treat cuts, burns, and other skin loci of injury (Triastuti et al., 2022). The therapeutic efficacy of *P. major* depends upon the complex phytochemical composition and a very wide variety of bioactive compounds (Stanisavljević et al., 2008). This encompasses the well-known cohort of flavonoids, which have had very much of a reputation when antioxidant-based cytoprotection of free radical-mediated injuries and inflammation is concerned (Bouali et al., 2024). In addition, tannins possess wound-contracting and infection-inhibiting activity owing to their astringent and antimicrobial properties (Stanisavljević et al., 2008). Iridoid glycosides, such as aucubin, possess anti-inflammatory and analgesic activities, thereby alleviating pain and aiding in healing (Kartini et al., 2021). Moreover, *P. major* mucilage creates an impermeable barrier over any wound, hence trapping moisture and allowing tissue regeneration (Samuelsen, 2000). The resultant curative action due to its various bioactive constituents is the antimicrobial, anti-inflammatory, and antioxidant activities of *P. major*, and this makes it a potential natural drug for wound healing (Triastuti et al., 2022).

P. major can induce contraction of the wound, an essential process that reduces the distance between the edges of the wound in order to make the wound smaller (Stanisavljević et al., 2008). Thus, the contraction is made possible by the capacity of the plant to cause the proliferation of fibroblasts, which then become accountable for collagen and other extracellular matrix components synthesis (Bouali et al., 2024). *P. major* also strengthens tissue growth through enhanced collagen synthesis, increasing tensile strength and wound integrity to be healed (Stanisavljević et al., 2008). These synergistic effects combine to promote healing, resulting in rapid wound closure and minimal scarring (Kartini et al., 2021). Antimicrobial activity in *P. major* is critical in preventing infections of wounds. This is a common complication that can retard healing considerably (Samuelsen, 2000). Thus, this plant acts to inhibit pathogenic bacteria and fungi and lower infection risks and ensure a healthy and clean wound environment (Triastuti et al., 2022). The antioxidant activity of *P. major* also counteracts oxidative stress at the point of injury, which keeps cells safe from free radical-induced damage (Stanisavljević et al., 2008). Inhibition of oxidative stress thus provides an optimal environment for tissue repair and regeneration (Bouali et al., 2024).

P. major is due to being a highly researched plant in both *in vitro* and *in vivo* models, with considerable research favoring its wound-healing activity (Stanisavljević et al., 2008). Positive *in vitro* study indicates that the plant possesses the ability to induce the proliferation of fibroblasts, enhance collagen production, and hinder bacterial growth (Kartini et al., 2021). Despite *in vivo* studies conducted on animal models that reported that topical application of *P. major* extracts improved tissue regeneration and healing of wounds (Samuelsen, 2000), experiments validate the anecdotal reports and the empirical findings supporting the use of *P. major* for wound healing (Triastuti et al., 2022). The repetition of such findings lends legitimacy to the medicinal potential of *P. major* and its position in natural wound healing (Stanisavljević et al., 2008). The range of the different forms of *P. major* that may be applied in practice, such as poultices, extracts, and ointments, offers room for differences in application for different wounds and forms of treatment (Bouali et al., 2024). Poultices are simply created by grinding leaves and applying the paste to the wound, thus providing an easy and effective entry into the plant's therapeutics (Stanisavljević et al., 2008). The second process, extracting *P. major* by macerating the plant with a solvent such as water or alcohol, brings together and concentrates bioactive compounds that will enable the practitioner's precise dosage (Kartini et al., 2021). Ointments are the formulations that mix *P. major* extracts with a base like petroleum jelly or beeswax and act by forming a protective layer and facilitating slow release of the plant therapeutic component (Samuelsen, 2000).

Traditionally handed down knowledge existed about *P. major*, indicating an eminent age of its utilization in different cultures (Triastuti et al., 2022). Such long usage in many societies gives more reinforcement, enhanced by the modern scientific findings, putting *P. major* as one of the components of utmost importance in herbal medicine for the treatment of wounds (Stanisavljević et al., 2008). The overall effect of the plant on diverse processes in wound healing, including inflammation, infection, and tissue repair, declares it to be a broad-spectrum therapeutic agent (Bouali et al., 2024). Ongoing assessment and perspective of the pharmacological and potential clinical uses of *P. major* will assist in further valuing its therapeutic potential (Stanisavljević et al., 2008). Thereafter, this research can contribute to the creation of new plant-derived wound care strategies and preparations with *P. major* as the primary component (Kartini et al., 2021). The ease of cultivation and natural abundance of the plant in the wild render it a resource full of promise for therapeutic chemicals (Samuelsen, 2000). Extremely minimal risk of side effects with the topical use of *P. major* renders it an accepted and safe remedy in the management of wounds (Triastuti et al., 2022).

P. major wound healing capacity is good as studies continue to establish and identify its bioactive compounds (Stanisavljević et al., 2008). Technological progress in the extraction and purification methods improves the ability to purify and synthesize the individual bioactive compounds responsible for the activity of the plant of medicinal interest (Bouali et al., 2024). This data will allow for the development of standardized herbal products and will allow for better quality and consistency of *P. major*-derived wound products (Stanisavljević et al., 2008). The future of incorporating *P. major* into conventional medicine is interdisciplinary, combining traditional knowledge and contemporary scientific approaches (Kartini et al., 2021). This includes clinical trials and the establishment of standard quality control protocols during the training of health practitioners in the appropriate application of *P. major* in wound care (Samuelsen, 2000). The increased scientific basis for the wound-healing activity of *P. major* supports its use in research and traditional medicine (Triastuti et al., 2022).

7.2.9 Andrographis paniculata

Andrographis paniculata is also referred to as the "king of bitters", a familiar medicinal plant of integral medicine and broadly utilized in traditional Asian medicine because of its extensive array of pharmacological properties, including the property that promotes wound healing. It belongs to the family Acanthaceae, whose geographical distribution is widespread across South to Southeast Asia. Bioactive constituents of *A. paniculata*, andrographolide and its derivatives, most particularly, are major contributors to its therapeutic potential in tissue repair and wound healing.

Diterpenoids, flavonoids, and polyphenols are the most common bioactive molecules in *A. paniculata*, the best of which is andrographolide. Andrographolide has been extensively studied for its anti-inflammatory, antimicrobial, antioxidant, and immunomodulatory activities. All these activities are essential for wound healing through the modulation of various cellular events like fibroblast proliferation, collagen synthesis, and angiogenesis. As evidenced from research, andrographolide will be able to exhibit potent anti-inflammatory activity by inhibiting the pro-inflammatory cytokines such as TNF-α, IL-6, and interleukin-1 beta (IL-1β) (Raghavan et al., 2018). Inhibition of the above inflammatory mediators will cause a reduction of the inflammation in the injury region and thus trigger tissue regeneration. In addition, its antioxidant activity helps in scavenging ROS, thereby avoiding oxidative stress damage in the cells that are involved in wound healing (Yadav, 2021).

The greatest complication that occurs while healing a wound is the presence of microorganisms that impede tissue healing and most probably convert lesions to chronic wounds. *A. paniculata* has displayed significant antimicrobial activity against different bacterial and fungal pathogens of wound infection, including *S. aureus*, *Pseudomonas aeruginosa*, and *Escherichia coli* (Khan et al., 2022). The inclusion of andrographolide and flavonoids in the plant extract is responsible for its bacteriostatic as well as bactericidal activity, inhibiting biofilm formation and ensuring wound sterility. The preclinical evaluations have established that topically applied *A. paniculata* extract causes epithelialization and granulation tissue formation in experimental wound models. Enhanced fibroblast proliferation and collagen synthesis lead to faster wound closure and greater tensile strength of regenerated tissue (Kumar et al., 2025). Furthermore, the angiogenic activity of *A. paniculata* actively stimulates the formation of new blood vessels to ensure adequate oxygenation and nourishment of the healing tissue (Li et al., 2021).

In the context of traditional medicine, *A. paniculata* has other additional uses for the purpose of wound treatment, skin infections, and inflammatory conditions. This herb has been known to some Ayurvedic and Chinese medical practitioners to make such medicinal formulations that act as a cure to speed up the time of healing for ulcers, cuts, and burns. Modern-day research does endorse the traditional claims regarding wound care efficacy. Clinical testing of *A. paniculata* use has already proven promising for chronic wound patients' management. Furthermore, in patients treated with such formulations containing *A. paniculata*, the wound contraction, pain, and infection occurred in low proportions as compared to conventional treatments (Worakunphanich et al., 2021). More long-term clinical trials are thus necessary to verify these results and set the appropriate clinical dosages for therapeutic purposes.

A. paniculata thus serves the therapeutic purposes for which it is used as a natural option in lieu of synthetic wound care agents, given its strong healing capability. Further therapeutic efficacy may also be obtained through utilizing its bioactive constituents in modern advanced drug delivery systems such as nanocarriers and hydrogels. Such future research is also likely to elucidate the molecular mechanisms of wound healing by itself and its potential stacking with other medicinal plants. In summary, *A. paniculata* is implied to assist in wound healing due to several properties, including anti-inflammatory, antimicrobial, antioxidant, and angiogenic properties. The huge application it has with both traditional and modern medicine sets a precedent for it being valued as an herbal cure in wound treatment. More research and clinical validation will optimize its use in dermatological and wound healing therapeutics.

7.2.10 Pine

History of pinus (*Pinus* spp.) traditional therapy is long, and the plant is considered useful as a remedy because it possesses a variety of pharmacological effects, particularly that of potent wound healing (Batista et al., 2022). Thanks to its high phytochemical composition, rich in various bioactive compounds, pine is therapeutically very promising. Polyphenols, flavonoids, and terpenoids are some of the bioactive compounds whose synergistic action is what actually guarantees the plant's antimicrobial, anti-inflammatory, and antioxidant activities, all vital organs in wound management (Batista et al., 2022). They are also isolated from pine needles and bark through their essential oils and are cited to have high levels of α-pinene and β-pinene, features which have been extensively studied and documented to advance wound healing by modulating inflammatory responses and enhancing tissue repair (Batista et al., 2022). This means that the above monoterpenes will alter the action of inflammatory mediators to favor a balanced immune response and not hinder the healing process.

Studies have consistently proven that the extract of pine bark has a strong wound healing potency, increasing both fibroblast proliferation and collagen synthesis (Maimoona et al., 2011). Fibroblasts, the primary cells involved in producing collagen and other extracellular matrix components, are vital to the structure and strength of newly created tissue (Maimoona et al., 2011). By inducing the activity of fibroblasts, pine bark extract has immediate action on the process of granulation tissue formation and tries to close a wound. Clinical studies have also found that proanthocyanidins from pine trees, which are a group of polyphenols, speed up contraction and re-epithelialization of various conditions in wound models (Maimoona et al., 2011). These proanthocyanidins also enhance the migration of keratinocytes, the outermost skin cells, which is required for the epidermal barrier to be re-established. Pine extracts, however, have also been found active against several pathogens such as *S. aureus* and *P. aeruginosa*, two bacteria most often implicated in wound infections (Jacob et al., 2024). Supply by inhibiting their growth in wounds: pine extracts are meant to minimize the risk of infection, thus providing a healthy environment for wounds to heal. Oxidative stress is a prime factor in the development of chronic wounds, and thus, pine polyphenols' antioxidant activities are critical in reducing such stress (Choudhury et al., 2023). It can damage components of cells and prolong inflammation in the wound that ultimately hinders healing. An imbalance between free radicals and antioxidants leads to oxidative stress. It impedes wound healing by damaging cellular components and prolonging inflammation. Pine polyphenols scavenge free radicals, protecting cells from oxidative damage, and thus promote tissue repair.

Earlier, the traditional application of pine in wound healing was the preparation of pine resin-based ointments and decoctions (Batista et al., 2022). These medicinal formulations are effective in treating different classes of wounds, such as cuts, burns, and ulcerative wounds. The resin is a complex mixture of terpenoids and resin acids, which seals the wound to avoid infection and stimulate healing (Maimoona et al., 2011). However, contemporary pharmaceutical research has started incorporating pine extracts into advanced drug delivery systems such as hydrogels and nanoformulations for effective bioavailability and therapeutic amplification. Hydrogels create a moist environment that promotes healing, and these can be loaded with pine extracts to ensure continuous release of the active components. Nanoformulations such as liposomes and nanoparticles could also help improve penetration of the extracts across the deeper tissue strata, thus maximizing their therapeutic efficacy.

7.2.11 *Punica granatum*

It is a memorialized medical plant that has been used for ages in traditional medicine, especially in the healing of wounds. It is called *Punica granatum* or pomegranate (Fontoura et al., 2022). Its fruit, which contains the peel, seeds, and juice, is endowed with bioactive compounds that include polyphenols, flavonoids, tannins, and anthocyanins, and these help in exhibiting antimicrobial, anti-inflammatory, and antioxidant effects (Fontoura et al., 2022). These compounds act in synergy by multiple mechanisms to facilitate wound healing. Studies have shown that extracts of pomegranate modulate the impact of inflammatory cytokines, enhance collagen deposition, and induce angiogenesis for wound healing (Fontoura et al., 2022). Thus, the pomegranate extract can reduce excessive inflammation by regulating the production of inflammatory mediators and creating an environment conducive to tissue repair. Again, increased collagen deposition serves to strengthen the newly formed tissue, while angiogenesis ensures an adequate supply of blood and nutrients to the wound site.

Pomegranate extracts have been found to display extensive germicidal efficacy against an array of pathogenic killers, including some multidrug-resistant bacterial strains comprising *S. aureus* and *E. coli*. The antimicrobial activity is mainly due to the negatively charged components, polyphenols and tannins, which disrupt the bacterial cell membrane and prevent further growth. Additionally, the fruit's high tannin concentration hastens wound closure through contraction, approximating the edges of the wound, which, in turn, improves the wound healing rate. Due to prompt closure, scarring diminishes. Also, pomegranate-derived polyphenols have been shown to induce some growth factors involved in tissue regeneration, including VEGF and transforming growth factor-beta (TGF-β) (Karole et al., 2019). These growth factors enhance the proliferation of endothelial cells and fibroblasts, which in turn stimulate collagen formation and promote angiogenesis, respectively.

Pomegranate in traditional medicine is used for wound healing in the form of poultice and pastes (Fontoura et al., 2022). These traditional preparations by crushing the fruit or the peel were used for treating many types of wounds – pyogenic infections, abrasions, cuts, burns, and ulcers. In recent years, research has turned toward the delivery of pomegranate extracts via advanced wound dressing and hydrogels for better clinical applications (do Nascimento et al., 2020). Bioactive dressings with pomegranate extract provide sustained release of the therapeutic agents, causing faster wound healing and reducing the chances of infection. On the other hand, pomegranate extracts can be loaded into hydrogels that maintain a moist environment and promote tissue regeneration. The sustained exploration of pomegranate for wound healing is expected to offer novel therapeutic strategies and formulations.

7.2.12 *Echinacea*

Echinacea (*Echinacea* spp.), long used in traditional healing, is also cherished by herbalists for its unique immune-modulating and wound-healing properties (Russo et al., 2025). This flowering genus native to North America has realized such a value in Native American medicine for centuries, concerning the treatment of different ailments such as wounds, infections, and inflammatory conditions. The clinical efficacy of Echinacea lies in a complex composition of phytochemicals that manifest in a strikingly diverse array of bioactive compounds. Alkamides, caffeic acid derivatives, polysaccharides, and flavonoids interact and synergistically contribute toward anti-inflammatory, antimicrobial, and antioxidant activities – all of which are vital in effective wound management (Russo et al., 2025). Alkamides have been shown, for example, to alter immune responses as well as reduce inflammation, while caffeic acid derivatives have shown extremely potent antioxidant and antimicrobial properties. Moreover, polysaccharides induce the activity of immune cells and support tissue regeneration, while flavonoids lend antioxidant and anti-inflammatory activity to the plant (Russo et al., 2025). Echinacea extracts in many studies have shown that wound healing is accelerated by promoting fibroblast proliferation and collagen synthesis, particularly in the case of *Echinacea purpurea*, one of the most common species. The fibroblasts, the cells that create collagen and extracellular matrix components, are key to the construction and tensile strength of the newly formed tissue. By stimulating fibroblasts, therefore, the extracts of Echinacea would facilitate the formation of granulation tissue and closing of the wound (Russo et al., 2025).

Forms of intense acute inflammation usually contain the loss of contracting wounds (Ezzat et al., 2025). According to clinical investigations, it has been confirmed that topical Echinacea extracts speed up the process of contraction as well as epithelialization for wounds. Contraction of the wound draws its edge together, which is important in the size reduction of the injury and the developing scar. The epithelialization process, where new epithelial cells form over the wound surface, restores the skin's barrier function against the outside environment and thus helps prevent infection (Ezzat et al., 2025). It was said that the plant's active components work on the expression of the cytokines, which modulate excessive inflammatory restoration and, on the other hand, enable the formation of new blood vessels. Cytokines are a group of signal molecules that play an important role in the regulation of immune response, thus they represent a great importance in the healing of wounds (Ezzat et al., 2025). However, the balancing of inflammatory response attributes with a lot of damaging tissues, thus promoting repair, is achieved through modulation of the expression of cytokines by Echinacea extracts. Besides that, angiogenesis is needed for providing either oxygen and nutrients to the wound site, which is helpful for tissue regeneration (Ezzat et al., 2025). Moreover, Echinacea's antimicrobial properties are important in preventing wounds from bacterial infections, particularly against *S. aureus* and *E. coli*, which are common pathogens of wounds (Burlou-Nagy et al., 2023). By killing the growth of the bacteria, Echinacea extracts reduce the possibility of infection in the wound and favor a clean, healthy environment for the wound. This antimicrobial effect is especially significant for the management of chronic wounds that are almost always complicated by bacterial colonization (Burlou-Nagy et al., 2023).

Traditional medicine has used Echinacea in wound healing by using crushed parts of plants or by making decoction applications to expose the wound and the skin infections (Fahimirad et al., 2025). Traditionally, these preparations were made by crushing fresh or dried parts of the plant and applying them directly to the injury for different skin traumas, such as cuts, burns, and bites by insects. The above-mentioned decoctions were given to wash wounds and to support healing by boiling the parts of the plant in water. More recent advances in pharmaceuticals have sought the addition of Echinacea extracts in advanced wound dressings and gels to their therapeutic effects (Fahimirad et al., 2025). Echinacea impregnated wound dressings release the therapeutic agents in an extended fashion, thereby hastening wound healing and reducing the risk of infection. Echinacea extract-based formulation gels present moist wound conditions with good tissue regeneration. These advancements increase the bioavailability of bioactive Echinacea compounds and therefore maximize their efficacy as therapies (Fahimirad et al., 2025). It would be anticipated that scientific reports on Echinacea's wound healing activity should culminate in the design of new treatments and products to further their therapeutic uses. The use of Echinacea in traditional medicine needs a multidisciplinary strategy combining traditional with scientific methods for ensuring safety and efficacy in wound healing. Continued research into the pharmacological activity and clinical applications of Echinacea will continue to advance our understanding of its therapeutic potential and assist in the development of new wound-healing strategies (Fahimirad et al., 2025).

7.2.13 *Centella asiatica*

It is accurate to note that Gotu Kola, also known as *Centella asiatica*, has a very long and ancient history of being used in conventional medicine, especially for the healing of skin lesions owing to its wound-healing nature (Diniz et al., 2023). An herbaceous perennial, growing extensively in Asia's marshlands, Gotu Kola has been the mainstay of Ayurvedic medicine and Chinese traditional medicine practices for centuries as a remedy for various skin ailments, with tissue repair as the chief objective. The herb's phytotherapeutic activity is attributed to the rich phytochemical content and intricate mixture of triterpenoids such as asiaticoside, madacassoside, and asiatic acid (Diniz et al., 2023).

The triterpenoids have been shown to possess various useful activities in wound healing processes such as collagen synthesis, fibroblast proliferation, and angiogenesis (Diniz et al., 2023). Collagen is the major connective tissue protein; it sustains and facilitates the strength and integrity of the wound. The role of the fibroblast is to generate collagen and other extracellular matrix and is indispensable for healing tissue and regeneration. Angiogenesis or the growth of blood vessels is required in order to furnish adequate nutrients and oxygen to the wounds to produce regeneration.

C. asiatica has been consistently proven through studies to enhance the process of wound healing by modulating inflammatory responses and increasing tensile strength of the repaired tissue (Arribas-López et al., 2022). The plant modulates the synthesis of inflammatory mediators such as cytokines and chemokines, thus creating a conducive environment for tissue repair and preventing excessive inflammation. Besides, *C. asiatica* also increases collagen fibre deposition and cross-linking and thereby makes the newly formed tissue stronger and increases tensile strength further. The *in vivo* and clinical reports clearly show significant facilitation of wound contraction, especially in chronic wounds and burns (Arribas-López et al., 2022). Wound contraction refers to the mechanism whereby the wound margins are drawn toward each other, closing up the injured surface area and healing. *C. asiatica* also significantly benefits from chronic wound healing, colloquially linked with excessive inflammation and defective tissue repair. The herb is also well known to hasten the healing of burns, showing great tissue destruction and inflammation. Moreover, *C. asiatica* has protective action against wound infection and oxidative stress-induced tissue damage through its antimicrobial and antioxidant activity (Ramli et al., 2020). Its antimicrobial activity, as a result of its triterpenoids and other bioactive molecules, acts against pathogenic bacteria and fungi, hence reducing the possibility of wound infection. *C. asiatica* antioxidants combat free radicals in the body and defend living cells, thus minimizing oxidative stress and supporting the tissues' healing process (Ramli et al., 2020). Buildup of oxidative stress when there is an imbalance between free radicals and antioxidants disrupts proper healing of wounds by harming cellular structures and sustaining an inflammatory condition.

Healing of wounds has historically entailed the preparation of *C. asiatica* crushed leaves, pastes, and infusions for the management of wounds (Diniz et al., 2023). For example, fresh leaves would either be crushed or immersed in water to prepare these compounds for topical use in the healing of wounds, ranging from cuts to burns and ulcers (Arribas-López et al., 2022). The pastes and crushed leaves directly deposit the therapeutic compounds of the plant on the wound area, whereas infusions remove water-soluble bioactive compounds. Yet, contemporary pharmaceutical science has reversed the trend by developing drug delivery systems like hydrogels, creams, and nanoparticles, filled with *C. asiatica* extracts, to enhance the efficiency of wound healing (Ramli et al., 2020). For instance, hydrogels that have conditions of moisture favoring healing of wounds can be supplemented with extracts of *C. asiatica* and release active ingredients in a sustained manner. Creams and ointments with *C. asiatica* extracts form a protective film over the wound and aid in tissue regeneration. Nanoparticles such as liposomes and polymeric nanoparticles can further increase *C. asiatica* extract penetration into deeper tissues, exhibiting the maximum therapeutic effectiveness (Ramli et al., 2020). Further research on *C. asiatica* wound-healing activity would pave the way for new avenues toward novel therapy and formulation.

Synergism between the individual triterpenoids of *C. asiatica* enhances the overall wound healing activity (Diniz et al., 2023). Asiaticoside, for instance, can initiate collagen synthesis as well as support angiogenesis, whereas madecassoside possesses good anti-inflammatory and antioxidant activity (Arribas-López et al., 2022). Yet another significant triterpenoid is asiatic acid, which can stimulate the proliferation of fibroblasts and extracellular matrix deposition (Ramli et al., 2020). Targeting different aspects of tissue repair and regeneration, these molecules have an overall effect in a multidirectional way in wound healing. It is a useful therapeutic in this regard as different wounds can be treated due to its ability to modulate various cellular and molecular signal pathways utilized in the repair of the wound. Additional data regarding the therapeutic application of *C. asiatica* would be obtained by continuous investigation of the pharmacological properties and clinical applications of the plant. It would actually assist greatly toward their standardization and very stringent clinical trials for drug efficacy and safety in contemporary wound care.

7.2.14 *Rosmarinus officinalis*

Rosemary (*Rosmarinus officinalis*) has been therapeutically employed for centuries; current scientific research has established its great potential for wound healing. The fragrant herb cultivated in the Mediterranean region boasts extremely wide phytochemical patterning containing rosmarinic acid, carnosic acid, and other flavonoid classes that make it therapeutically effective. These compounds show synergy against inflammation, infection, and oxidative stress, all pushing tissue regeneration and wound healing (de Macedo et al., 2020). Rosmarinic acid is a strong phenolic compound in the modulation of inflammatory activity and the inhibition of oxidative stress, whereas carnosic acid is a strong antioxidant and neuroprotector. The rosemary flavonoids, luteolin, and apigenin, are the ones with anti-inflammatory and antimicrobial activity. The synergistic activity of these bioactive compounds unveils the potential of rosemary as a folk medicine for the management of wounds.

Laboratory studies have shown that extracts of *R. officinalis* augment many aspects significant in effective wound healing, such as collagen formation, fibroblast proliferation, and keratinocyte migration. Collagen, being a primary structural protein of connective tissue, provides strength and integrity to a wound. Thus, the promotion of new tissue development and closure of wounds is promoted by rosemary extracts in the stimulation of collagen synthesis. Fibroblasts play an essential role in wound healing and regeneration since they produce collagen and other extracellular matrix components (Li Pomi et al., 2023). Rosemary extracts proved to induce fibroblast growth, thereby enhancing the pace of granulation tissue formation and enhancing wound healing. Keratinocytes constitute the topmost skin cell layer and play a crucial role in re-epithelialization, re-establishing the epidermal barrier (Li Pomi et al., 2023). Through the promotion of keratinocyte migration by rosemary extracts, wound closure is achieved quickly and with minimal scarring. The antimicrobial activity of rosemary also prevents infection of wounds by inhibiting the growth of the majority of pathogenic fungi and bacteria (Sahlabgi et al., 2025). Additional antimicrobial activity is provided by other constituent components such as carnosic acid and rosmarinic acid, which induce microbial cell membrane disruption and further inhibit microbial growth. Therefore, by inhibiting infections, rosemary facilitates a clear and healthy environment for improved wound healing.

The traditional use of rosemary in wound healing has involved the use of infused oils and poultices (de Macedo et al., 2020). Infused oils, which are produced by steeping rosemary leaves in a carrier oil such as olive or coconut oil, can extract and concentrate the therapeutic properties of the plant. These oils can be applied directly to the wound for the release of active ingredients in sustained kinetics. Apply simple but highly effective ways for the therapeutic benefits of the plant, crushed freshly and dried rosemary leaves, and applied to the wound to create poultices (Li Pomi et al., 2023). Recent pharmaceutical research has evaluated the introduction of rosemary extracts in advanced wound dressings and hydrogels in clinical settings (Gavan et al., 2022). Bioactive dressings soaked with rosemary extracts allow continuous diffusion of therapeutic compounds, hastening wound healing effects while preventing chances of infection. These types of hydrogels can be enhanced by rosemary extracts to improve their therapeutic activity because these hydrogels maintain a moist microclimate and facilitate tissue regeneration. Continued exploration of the effects of rosemary on wound healing is, therefore, expected to come up with new therapeutic strategies and formulations.

The antioxidant nature of rosemary is highly important in wound healing for combating some of the oxidative stress and free radical damage to cells (de Macedo et al., 2020). Wounds may fail to heal due to oxidative stress, which refers to the condition in which there are more free radicals than antioxidants. Damage to cellular components and a prolonged inflammatory period are impediments to wound healing. Highly potent antioxidant activity, such as rosmarinic acid and carnosic acid, scavenges free radicals and protects cells from oxidative damage for a good environment for tissue repair (Li Pomi et al., 2023). Rosemary, accordingly, mediates the production of some of the inflammatory mediators, for instance, cytokines and prostaglandins, thus having an anti-inflammatory effect, which further aids in wound healing (Sahlabgi et al., 2025). By decreasing excessive inflammation, rosemary balances the immune responses and lessens tissue destruction. In fact, rosemary has addressed many aspects of wound healing, such as inflammation, infection, and even tissue regeneration, which makes it a very active and useful therapeutic agent (Gavan et al., 2022). Further research on the pharmacological domain and clinical application of rosemary is likely to make great inroads toward understanding its therapeutic strength and developing novel wound healing techniques (de Macedo et al., 2020).

7.2.15 *Allium sativum*

Allium sativum, or garlic, has gained much reputation and attention with its medicinal properties throughout history in traditional healing systems, where it greatly contributed to wound healing (Jikah et al., 2024). The medicinal properties of garlic can be attributed to its complex phytochemistry, which is endowed with a wide variety of bioactive constituents. These bioactive constituents include, but are not limited to, allicin, sulfur-containing compounds, and flavonoids, and contribute to the antimicrobial, anti-inflammatory, and antioxidant activities of garlic (Jikah et al., 2024). These multidimensional properties make garlic one of the best-known natural treatments for improving wound healing and countering infections, with promise in both traditional and contemporary methods of wound care (Jikah et al., 2024). Garlic possesses all these modes of healing wounds through synergistic effects: preventing infections, reducing inflammation, stimulating tissue regeneration, and shielding against oxidative damage.

Studies provide consistent evidence for a variety of modes of action whereby garlic extract, among others, stimulates fibroblast proliferation, promotes collagen deposition, and modifies inflammatory response, thus greatly enhancing wound healing (Taha, 2022). Fibroblasts synthesize collagen and other extracellular matrix components, thus ensuring the structural integrity and strength of newly formed tissue. By stimulating fibroblast activity, garlic hastens the formation of granulation tissue and thus wound closure. Garlic formulations have been reported to enhance the rate of wound closure and augment the tensile strength of regenerated tissue, further suggesting that the new tissue is significantly stronger and more resistant (Taha, 2022). The enhanced property of tensile strength, being a vital property, helps in the prevention of dehiscence of the wound and works toward its proper functioning. Allicin, one of the major bioactive components of garlic, has been reported to show major antimicrobial activity against a wide variety of wound pathogens, including *S. aureus, P. aeruginosa*, and *E. coli*, thus reducing the potential for wound infection (Bhatwalkar et al., 2021). The action of garlic harming these bacterial cells establishes a clean, healthy environment for the wound to heal. Such aggressive antimicrobial activity plays an important role in the prevention and treatment of infections that can further aggravate wound healing and develop into chronic wounds.

There have been clinical and preclinical studies reporting the effectiveness of garlic extracts in terms of their antioxidant role in treating oxidative-stressed wounds through their ability to neutralize high levels of ROS, thus preventing any damage to the tissues and aiding cell repair mechanisms (Alam, 2016). It is further mentioned that oxidative stress, which results from an imbalance between free radicals and antioxidants, impedes wound repair due to damage to cell components and prolonged inflammation (Alam, 2016). Scavenging free radicals protects garlic cells against oxidative damage and promotes tissue repair. Garlic has been used conventionally through poulticing and ointments for treating a variety of injuries such as cuts, burns, or even ulcerous wounds, indicating its ancient use in traditional medicines (Jikah et al., 2024). These are typical formulations whose preparation involves crushing garlic cloves and then mixing them with a compatible base. They were applied to the surface of the skin for wound healing and pest control. New inventions have been explored in future research for devising advanced drug delivery systems, such as bioactive wound dressings and hydrogel carrying garlic for enhancing the efficacy of wound healing (Hamid, 2024). Bioactive wound dressings, enriched with garlic extract, assure a sustained release of therapeutic compounds that accelerate wound healing and lower infection risks. In fact, with therapeutic efficacy enhancement, hydrogels can be loaded with garlic extracts to promote optimal wound closure through maintaining a moist environment and facilitating tissue regeneration. These advanced formulations represent a remarkable step forward in the direction of garlic use for wound healing, integrating traditional knowledge with modern pharmaceutical technology.

Garlic's comprehensive wound-healing effect actually goes beyond its antimicrobial and anti-inflammatory effects: garlic also stimulates angiogenesis, the growth of new blood vessels, which would be necessary for the distribution of oxygen and nutrients to the wound area (Taha, 2022). This vascularization would further support the growth of new tissue and hasten the closure of wounds. Such sulfur-containing compounds in garlic, including alliin and allicin, have stimulated the synthesis of growth factors important in angiogenesis and tissue remodeling, such as VEGF and TGF-β (Bhatwalkar et al., 2021). Besides this, garlic has also been demonstrated to enhance migration and proliferation of keratinocytes – the cells that make up the outermost layer of the skin – thus aiding re-epithelialization and restoration of the epidermal barrier (Alam, 2016). This re-epithelialization process will further reduce scarring and promote its smooth, even surface.

While garlic-derived products have been vigorously researched for safety and tolerance, the clinical trials testify that they are efficacious in wound healing, mainly without any side effects (Hamid, 2024). A little irritation or allergic reactions, in general, are tolerated for topical garlic applications; however, skin irritation or even contact dermatitis could still arise in some individuals following prolonged use or excess application of garlic. Hence, an important recommendation is for garlic formulations to be used under the supervision of a healthcare practitioner. The current research on and fabrication of garlic wound-healing bioproducts will likely give rise to the development of new therapeutic strategies or formulations that fully exploit the efficacy of the marvelous botanical (Jikah et al., 2024). Combining traditional knowledge with modern scientific methods will further elucidate the therapeutically relevant actions of garlic and help develop efficacious products for wound care (Taha, 2022). The integration would be consistent with a paradigm that broadly addresses healthcare matters, in this case, for the enhancement of wound healing and improved patient outcomes (Bhatwalkar et al., 2021). Table 7.1 summarizes the general information about the above-explained herbal products.

TABLE 7.1
Medicinal Plants Used in Wound Healing and Their Therapeutic Properties

Scientific Name	Other Names	Mechanism of Action	Dosage Forms Available	Advantages	Disadvantages
Aloe vera (L.) Burm.f.	Aloe, Ghrita-kumari (India)	Enhances collagen synthesis, anti-inflammatory, antimicrobial (Hamman, 2008)	Gels, creams, ointments	Natural, easily available, effective for burns and wounds	Can cause allergic reactions in some individuals
Crocus sativus L.	Saffron, Kesar (India), Zafran (Middle East)	Antioxidant, enhances re-epithelialization, anti-inflammatory (Akbari et al., 2018)	Extracts, creams, herbal pastes	Promotes tissue regeneration, reduces inflammation	Expensive, adulteration common
Arctium lappa L.	Burdock, Gobo (Japan)	Stimulates fibroblast activity, antimicrobial (Chan et al., 2011)	Extracts, ointments	Rich in polyphenols, promotes skin regeneration	Potential for contact dermatitis
Curcuma longa L.	Turmeric, Haldi (India)	Curcumin enhances collagen deposition, anti-inflammatory, antibacterial (Chainani-Wu, 2003)	Powders, gels, creams	Strong anti-inflammatory, speeds up healing	Poor bioavailability, requires enhancers like piperine

(Continued)

TABLE 7.1 *(Continued)*
Medicinal Plants Used in Wound Healing and Their Therapeutic Properties

Scientific Name	Other Names	Mechanism of Action	Dosage Forms Available	Advantages	Disadvantages
Althaea officinalis L.	Marshmallow, Guimauve (France)	Mucilage forms a protective layer, anti-inflammatory, promotes granulation tissue (Xue et al., 2021)	Creams, infusions	Soothing effect, supports tissue regeneration	Limited clinical trials
Calendula officinalis L.	Marigold, Pot marigold	Stimulates angiogenesis, anti-inflammatory, antimicrobial (Ashwlayan et al., 2018)	Ointments, creams, oils	Widely available, well-tolerated	Possible allergies
Simmondsia chinensis (Jojoba)	Jojoba, Goat nut	Anti-inflammatory, enhances skin repair, antimicrobial (Habashy et al., 2005)	Oils, creams	Long shelf-life, excellent skin penetration	May not be suitable for all wounds
Plantago major L.	Broadleaf plantain, Waybread	Antimicrobial, anti-inflammatory, promotes fibroblast proliferation (Samuelsen, 2000)	Extracts, ointments	Effective for chronic wounds, readily available	Potential for mild skin irritation
Andrographis paniculata (Burm.f.) Nees	Kalmegh (India), Green chiretta	Anti-inflammatory, stimulates keratinocytes, antibacterial (Kumar et al., 2021)	Extracts, topical formulations	Potent wound-healing effects	Bitter taste limits oral use
Pinus species	Pine, Scots pine, Chir pine	Resin has antimicrobial and anti-inflammatory effects, promotes tissue regeneration (Tümen et al., 2018)	Resin-based ointments	Natural antiseptic, readily available	Some species cause allergic reactions
Punica granatum L.	Pomegranate, Anar (India)	Enhances collagen synthesis, antimicrobial, antioxidant (Arun and Singh, 2012)	Extracts, gels, ointments	Rich in bioactive compounds	Potential for mild irritation
Echinacea purpurea L.	Coneflower, Purple coneflower	Enhances immune response, stimulates fibroblasts (Xu et al., 2021)	Extracts, creams	Boosts immune response, reduces infection risk	Can cause allergic reactions in some individuals
Centella asiatica (L.) Urb.	Gotu kola, Brahmi (India)	Enhances collagen synthesis, anti-inflammatory, improves circulation (Gohil et al., 2010)	Extracts, creams, ointments	Effective in scar reduction, well-researched	Possible gastrointestinal discomfort when ingested
Rosmarinus officinalis L.	Rosemary	Antioxidant, antimicrobial, improves epithelialization (Rašković et al., 2014)	Oils, creams, herbal extracts	Strong antimicrobial properties, aromatic	Potential skin irritation
Allium sativum L.	Garlic, Lahsun (India)	Allicin has antimicrobial, anti-inflammatory, and angiogenic properties (Bayan et al., 2014)	Extracts, creams, herbal preparations	Potent antimicrobial, natural remedy	Strong odor, can cause skin irritation

7.3 CHALLENGES AND SOLUTIONS IN THE USE OF NATURAL PRODUCTS AS WOUND HEALING AGENTS

The healing of wounds with herbal remedies has received a large amount of attention due to the vastness of potent bioactive compounds, affordability, and perceived low side effects (Yadav, 2021). Nevertheless, notwithstanding the advantages, certain factors stand in the way of their extensive clinical use. These issues vary widely; they range from standardization of products and the intrinsic variability in bioactive constituents to stability of formulations and regulatory issues, and the requirements of well-designed and rigorous clinical evaluations (Kumar et al., 2025). While these matters are complex, ongoing scientific endeavors, with enhanced quality control and formulation work, remain paramount for the optimal application of plants like *A. vera, C. sativus, A. lappa, C. longa, A. officinalis, C. officinalis, Jojoba, P. major, A. paniculata, Pinus* spp., *P. granatum, Echinacea, C. asiatica, R. officinalis*, and *A. sativum* in wound management today as shown in Figure 7.1 (Thakur et al., 2011).

One of the foremost challenges in the utilization of natural products for wound healing is the persistent lack of standardization. The variability in the phytochemical composition of the herbal extracts may be pronounced due to a host of factors, such as geographic area, cultivation conditions, the exact time of collection, and the methods of extraction (Yadav, 2021). This inherent variability extremely limits the consistency and predictable efficacy of herbal formulations and creates a daunting obstacle to their reliable clinical use. In order to counter this challenge effectively, sophisticated analytical techniques must be employed. Advanced procedures like HPLC and mass spectrometry should be pursued with systematic application to assure quantification of bioactive constituents in herbal extracts and maintain batch-to-batch consistency (Kumar et al., 2025). This rigorous analytical approach will facilitate the production of standardized herbal products with predictable therapeutic outcomes. Furthermore, the use of chemometric methods may assist in the identification of key biomarkers and the development of quality control parameters (Atanasov et al., 2015).

The bioavailability and stability of the extracts from the plants should also be among those issues resolved. Most bioactive compounds available from the medicinal plants, curcumin from *C. longa* and andrographolide from *A. paniculata*, are poorly soluble and degrade very quickly.

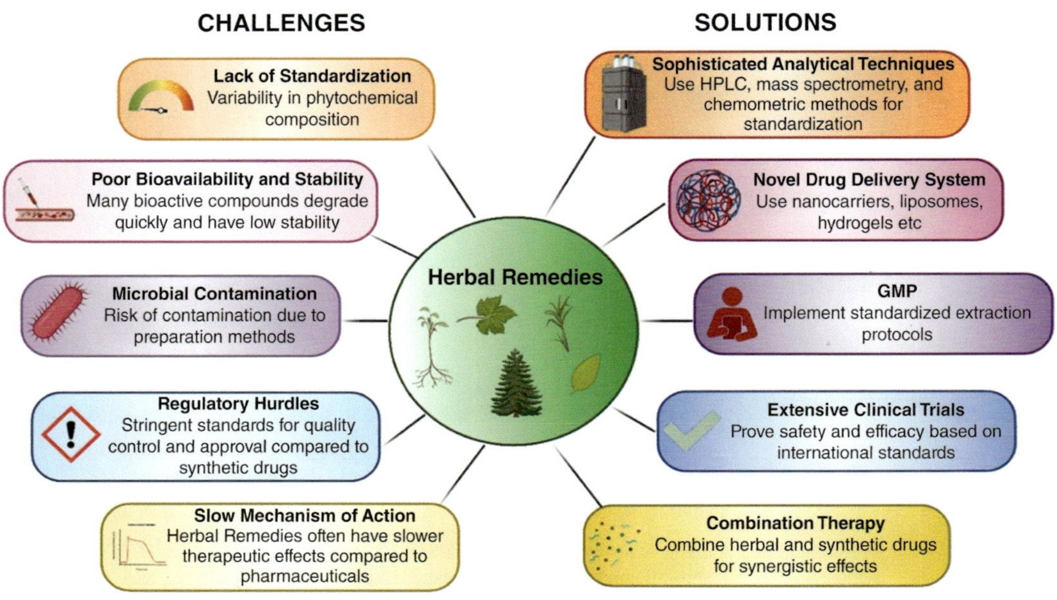

FIGURE 7.1 Challenges and solutions of plants in wound healing.

As a result, it affects its wound healing applications (Thakur et al., 2011). The remedy for these will be making innovative drug delivery systems viable for actual usage in supplementing such forms of herbal therapy. These technologies – nanocarriers, liposomes, and targeted hydrogel formulations – can improve bioavailability and stability of bioactive compounds while improving their biomolecular therapeutics activity (Li et al., 2021). Nanocarrier, for example, encapsulated bioactive nutrients to prevent degradation, liposomes aided in cellular uptake and targeted delivery, while hydrogels would sustain a wet environment in most parts of the predicted state to enable a sustained release directly to the wound area (Langer, 1990). In addition, absorption and stability for some of these compounds can be enhanced with the use of permeation enhancers, besides being optimized through prodrug strategies (Mitra and Mitra, 2003).

Microbial contamination and variation in antimicrobial activity are additional significant challenges with herbal remedies for wound care. Of various plants, some, for example, *A. sativum* and *P. granatum*, give strong defense mechanisms against general wound pathogens; differences in the methods used for preparation can lead to varying results (Al-Defiery et al., 2021). Faced with this identification, proper good manufacturing practices (GMP) and well-standardized extraction protocols would be necessary. These would ensure the antimicrobial efficacy of these natural products would not be hindered in their use and that safety for clinical applications remained ensured. The risk of contamination would also be reduced and quality standards ensured in the final product (Zhang, 2000). The validated sterilization techniques, gamma irradiation or ethylene oxide sterilization, are also effective for ensuring microbial safety for herbal products (Block, 2001).

Like this, the regulatory challenges also become a great barrier to getting these herbal remedies into a clinical setting for wound healing. Unlike synthetic medicines, standards are imposed on the herbal products for better regulation of quality control or approval for them (Ahmed et al., 2023). To surpass such regulatory hurdles, it would, therefore, take extensive clinical trials to be able to build effective pharmacovigilance systems. These efforts will then be able to generate the necessary evidence needed to authenticate the safety and efficacy of such plants as medicinal ones in wound healing, for example, *A. vera* and *C. asiatica* (Anand et al., 2024). Moreover, create clear and sound regulatory frameworks for herbal products according to international standards, for instance, those of the World Health Organization (WHO), toward ensuring proper quality and safety (Calixto, 2000).

The slow mechanisms of herbal remedies as to action vis-à-vis orthodox pharmaceutical therapeutics are other limitations acknowledged. Healing by nature is holistic and often needs a prolonged application before one can see remarkable therapeutic effects (Moses et al., 2023). To overcome this limitation, one must think about combination therapies that encourage the addition of herbal extracts with synthetic agents to derive a more effective and balanced approach to wound healing (Rates, 2001). Hybrid wound dressings, both with natural and conventional healing substances, would provide a synergistic effect, hence overall improvement during healing (Newman and Cragg, 2020). For example, adding herbal extracts to growth factors or antimicrobial peptides would speed up wound closure and enhance tissue regeneration.

A standardized polyherbal formulation is one that combines synergistic medicinal plants. These are alternatives to boosting efficacy in the wound healing effects (Karole et al., 2019): with the inclusion of *C. asiatica, Echinacea*, and *C. officinalis*, fibroblast proliferation with collagen synthesis will be accelerated and thus will result in closing wounds quickly. Apart from this, using modern research collaborations with practitioners of traditional medicine boosts knowledge transfer from studying scientifically verified herbal therapies (Balkrishna et al., 2024). This could act as a link between conventional knowledge in Eastern studies and evidence-based medicine (Heinrich, 2015).

Advances in biotechnology and genetic engineering bring generous opportunities in improving the wound healing potential of medicinal plants. It is worthwhile knowing that this biomanipulation of the important phytochemicals would result in plant drugs, which have improved bioavailability properties and therapeutic effects (Gonçalves and Romano, 2018). Bringing awareness through public education campaigns would fill in the gap between traditional knowledge and evidence-based medicine, thus building more acceptance and integration of herbal wound healing products into modern healthcare (Farnsworth, 1988). In fact, despite the realities like standardization, bioavailability, microbial contamination, regulatory approval, and slowness in efficacy that restrict the prevalence of herbal remedies in modern clinics for wound healing, there are several models of strategic solutions to these barriers. By means of advanced formulation techniques, stringent quality control measures, regulatory advances, and interdisciplinary collaborations, it is possible to integrate these herbal medicines into modern wound care practices. So, research and innovation have a bright future in the development of effective and reliable herbal-based agents for wound healing.

7.4 OUTLOOK AND FUTURE DEVELOPMENT

The increasing demand for herbal treatments in wound healing is accompanied by an increasing amount of evidence of their viability, ease, and fewer side effects than synthetic drugs. Herbal medicine is further developing through distinct biotechnology, nanotechnology, and pharmacological studies, so as to enhance its use for therapeutic purposes as traditional and modern knowledge merge. The section relates the future direction of herbal wound healing agents, emphasizing critical innovations.

7.4.1 Advancements in Herbal Medicine Research and Development

Lately, scientific scrutiny related to herbal wound healing agents has changed into an almost craze, with researchers investigating their molecular action. Optimization of their therapeutic efficacy is dependent upon identifying the bioactive compounds responsible for anti-inflammatory, antimicrobial, and tissue regenerative functions (Newman and Cragg, 2020). For example, studies performed recently on *C. asiatica* suggest that its major triterpenoids, asiaticoside and madecassoside, may be responsible for stimulating collagen synthesis and angiogenesis in the process of wound healing (Vitale et al., 2022). This knowledge allows us to explore the signaling pathways modulated by these compounds, such as the TGF-β and VEGF pathways, to further understand their possible therapeutic effects (Wu et al., 2012). All this knowledge facilitates the design of targeted therapy and optimizing the extraction purification process.

This leads to the application of nanotechnology, which shall be a promising way to improve bioavailability and stability of the herbal compounds. Plant extracts encapsulated in nanoparticles, liposomes, and hydrogels have displayed improved wound penetration along with the sustained release of active components (Hajialyani et al., 2018). Curcumin-loaded nanofibers were found, for instance, to induce increased contraction and epithelialization rates of wounds in preclinical models (Akrami-Hasan-Kohal et al., 2020). Moreover, the incorporation of stimulus-responsiveness-based nanomaterials allows triggered release of plant molecules based on the particular condition of the wound, for instance, pH or temperature (Hajialyani et al., 2018). This type of site-specific delivery lowers systemic exposure and increases therapeutic efficiency at the site of the wound. Additionally, polymer-drug conjugates have been reported to increase bioavailability and circulation time of herbal medicines (Li et al., 2025).

Another development of equal significance is the introduction of herbal compounds with biomaterials for the development of bioactive wound dressings. Herbal-infused natural polymer films, hydrogels, and scaffolds such as chitosan and alginate exhibit good wound healing characteristics based on their biocompatibility and humid environment (Zhou et al., 2022). Growth factors such as epidermal growth factor (EGF) or platelet-derived growth factor (PDGF) combined with herbal extracts can further enhance tissue regeneration (Devescovi et al., 2008). Moreover, the electrospinning of such nanofibers, with herbal compounds, gives a large surface area for drug release and imitates the extracellular matrix, thereby enhancing cell adhesion and proliferation (Zeng et al., 2024). Such developments brought about a paradigm shift from the use of herbs for topical applications toward new biomedical technologies that enable more convenient and effective options for wound care.

7.5 CONCLUSION

Evidence of the healing properties of herbal medicines for healing wounds is present in ancient knowledge and in the history of recent scientific studies. The medicinal herbs *A. vera*, *C. asiatica*, and *C. longa* play different roles, such as the antimicrobial, anti-inflammatory, and tissue-regenerative activity of the plants that would make them excellent alternatives to, or adjuncts of, traditional means of treatment. At the same time, variability in the quality of the products, coupled with insufficient regulatory control and the necessity for large-scale clinical trials, act as significant hurdles to the complete integration of these herbal remedies into contemporary medicine. In order to realize the potential provided by these herbal medicines, it therefore becomes necessary to address these issues through the help of modern research, enhanced methods for standardization, and ecologically sound harvesting. The future of herbal wound healing lies in the standardization of herbal preparations, the novel compositions by nanotechnology-facilitated extracts, and an interdisciplinary convergence of traditional healers, scientists, and medical practitioners. Following the trace between ethnopharmacology and evidence-based medicine will ensure that herbal remedies become accepted and accessible for wound care in the global healthcare scenario.

GLOSSARY

Abrasions: Minor skin injuries resulting from rubbing or scraping.

Alkaloids: Organic compounds of natural origin present in plants that tend to possess strong physiological activity.

Allicin: A sulfur-containing molecule in garlic, which has strong antimicrobial and antioxidant activities.

Allopathic Medicine: Western or traditional medicine, defined by the administration of pharmacologically active drugs or physical therapy to treat or suppress signs or pathophysiological processes of illnesses or disorders.

Angiogenesis: Creation of new blood vessels to provide oxygen and nutrients to tissues, specifically during wound healing.

Anthocyanins: Soluble pigments found in plants that create red, purple, and blue colors and are antioxidant and anti-inflammatory agents.

Anthraquinones: Plant-derived organic compounds, such as *Aloe vera*, which are laxative and antimicrobial in nature.

Antimicrobial: Having the ability to kill or inhibit the growth of microorganisms, including bacteria, fungi, and viruses.

Antioxidant: A compound that prevents oxidation, particularly one applied to neutralize the destructive impact of free radicals in living tissues.

Ayurveda: An ancient system of Indian medicine, with a focus on holistic equilibrium and the administration of herbal medications.

Bioactive Compounds: Plant substances that exert a biological action on living organisms, which are responsible for their medicinal activity.

Carotenoids: Pigments responsible for yellow, orange, or red colors in plants, which also function as antioxidants.

Crocetin: A dicarboxylic acid carotenoid present in saffron, which possesses antioxidant and anti-inflammatory activities.

Crocin: A carotenoid pigment in saffron responsible for its color and antioxidant activity.

Curcumin: A yellow pigment in turmeric (*Curcuma longa*), which is anti-inflammatory and antioxidant.

Diterpenoids: Class of organic compounds derived from terpenes and commonly occurring in plant resins and essential oils.

Flavonoids: Plant antioxidants and anti-inflammatory compounds.

Glycosides: Non-sugar chemical compounds having a sugar attached to the non-sugar part.

Phenolics: A class of chemical compounds that contain a hydroxyl group (-OH) bonded directly to an aromatic hydrocarbon group.

Phenolic Acids: Plant compounds with antioxidant and antimicrobial properties.

Phytochemistry: The study of the chemical compounds found in plants.

Phytosterols: Plant sterols that have structural similarities to cholesterol, with anti-inflammatory properties.

Phytotherapy: The use of plants or plant-derived substances for medicinal purposes; also known as herbal medicine or botanical medicine.

Polyherbal Formulations: Medicinal preparations combining multiple herbs for synergistic therapeutic effects.

Polysaccharides: Complex carbohydrates made up of many sugar molecules, often with biological activity.

Resin: A solid or semi-solid substance discharged from plants and commonly used in traditional medicine because of its healing and protective functions.

Safranal: A monoterpene aldehyde present in saffron and responsible for its odor and antioxidant activity.

Saponins: Plant natural detergents, some of which possess antimicrobial and anti-inflammatory activity.

Tannins: Plant astringent chemicals involved in wound healing due to their tissue-contracting and infection-preventing properties.

Terpenoids: A large and varied group of organic plant products, some of which possess antimicrobial and anti-inflammatory activity.

Therapeutic Properties: The medicinal or curative properties of a substance.

Traditional Chinese Medicine (TCM): A traditional form of medicine that originated in China and is based on balance and the application of herbal medicines.

Traditional Medicine: Health practices, methods, knowledge, and beliefs embracing plant, animal, and mineral-based medicines, spiritual therapies, manual procedures, and exercises, used singly or in mixtures to treat, diagnose, and prevent disease or maintain health.

Triterpenoids: A class of natural compounds found in plants, known for their diverse biological activities, including wound healing.

REFERENCES

Abdullaev, Fikrat. 2006. "Biological properties and medicinal use of saffron (Crocus sativus L.)." *Proceedings of the Second International Symposium on Saffron Biology and Technology*, Acta Horticulturae 739, pp. 339–345. Mashhad, Iran; 28–30 October 2006. https://www.actahort.org/books/739/739_44.htm.

Adhami, Vaqar M, Deeba N Syed, Naghma Khan, and Farrukh Afaq. 2008. "Phytochemicals for prevention of solar ultraviolet radiation-induced damages." *Photochemistry and Photobiology* 84 (2): 489–500.

Aggarwal, Bharat B, Anushree Kumar, and Alok C Bharti. 2003. "Anticancer potential of curcumin: Preclinical and clinical studies." *Anticancer Research* 23 (1/A): 363–398.

Ahmed, Sidra Nisar, Mushtaq Ahmad, Mohammad Zafar, Ghulam Yaseen, Nadia Iqbal, Neelum Rashid, Samina Kousar, and Adeela Haroon. 2023. "Herbal drugs: Safety, cost-effectiveness, regulation, current trends, and future directions." In *Bioprospecting of Tropical Medicinal Plants*, 1479–1493. Springer.

Akbari, Ghaidafeh, Seyyed ali Mard, and Ali Veisi. 2018. "A comprehensive review on regulatory effects of crocin on ischemia/reperfusion injury in multiple organs." *Biomedicine & Pharmacotherapy* 99: 664–670.

Akbik, Dania, Maliheh Ghadiri, Wojciech Chrzanowski, and Ramin Rohanizadeh. 2014. "Curcumin as a wound healing agent." *Life Sciences* 116 (1): 1–7.

Akrami-Hasan-Kohal, Mohammad, Lobat Tayebi, and Marjan Ghorbani. 2020. "Curcumin-loaded naturally-based nanofibers as active wound dressing mats: Morphology, drug release, cell proliferation, and cell adhesion studies." *New Journal of Chemistry* 44 (25): 10343–10351.

Alam, Khorshed. 2016. "Medicinal plant *Allium sativum*. A review." *Journal of Medicinal Plant Studies* 4 (6): 72–79.

Al-Defiery, ME, AK Al-Muttairi, HH Saeed, and RK Hadi. 2021. "Antimicrobial activity of garlic and pomegranate peel extracts against some pathogenic bacteria." *IOP Conference Series: Earth and Environmental Science* 722.

Anand, Shubham, Devendra Dhanorya, Manoj Kumar Bajhaiya, Anantika Rahi, Anuj Kumar, Rekha Rani, and Sapna Yadav. 2024. "A comprehensive overview: Advancements in pharmacovigilance and herbal drug development." *World Journal of Pharmaceutical Research* 13(13):1324–1337. https://www.wjpr.net/abstract_show/26244?utm

Arribas-López, Elena, Nazanin Zand, Omorogieva Ojo, Martin John Snowden, and Tony Kochhar. 2022. "A systematic review of the effect of *Centella asiatica* on wound healing." *International Journal of Environmental Research and Public Health* 19 (6): 3266.

Arun, Neelam, and DP Singh. 2012. "*Punica granatum*: A review on pharmacological and therapeutic properties." *International Journal of Pharmaceutical Sciences and Research* 3 (5): 1240.

Ashwlayan Vrish Dhwaj, Amrish Kumar, Mansi Verma, Vipin Kumar Garg, and SK Gupta. 2018. "Therapeutic potential of *Calendula officinalis*." *Pharmacy & Pharmacology International Journal* 6 (2): 149–155.

Atanasov, Atanas G, Birgit Waltenberger, Eva-Maria Pferschy-Wenzig, Thomas Linder, Christoph Wawrosch, Pavel Uhrin, Veronika Temml, Limei Wang, Stefan Schwaiger, and Elke H Heiss. 2015. "Discovery and resupply of pharmacologically active plant-derived natural products: A review." *Biotechnology Advances* 33 (8): 1582–1614.

Balkrishna, Acharya, Nidhi Sharma, Deepika Srivastava, Ankita Kukreti, Stuti Srivastava, and Vedpriya Arya. 2024. "Exploring the safety, efficacy, and bioactivity of herbal medicines: Bridging traditional wisdom and modern science in healthcare." *Future Integrative Medicine* 3 (1): 35–49.

Batista, João VC, Annekathrin Uecker, Carla Holandino, Fabio Boylan, Jakob Maier, Jörg Huwyler, and Stephan Baumgartner. 2022. "A scoping review on the therapeutic potential of resin from the species *Larix decidua* Mill. [Pinaceae] to treat ulcerating wounds." *Frontiers in Pharmacology* 13: 895838.

Bayan, Leyla, Peir Hossain Koulivand, and Ali Gorji. 2014. "Garlic: A review of potential therapeutic effects." *Avicenna Journal of Phytomedicine* 4 (1): 1.

Bhatwalkar, Sushma Bagde, Rajesh Mondal, Suresh Babu Naidu Krishna, Jamila Khatoon Adam, Patrick Govender, and Rajaneesh Anupam. 2021. "Antibacterial properties of organosulfur compounds of garlic (*Allium sativum*)." *Frontiers in Microbiology* 12: 613077.

Bisset, Norman Grainger, and Max Wichtl. 1994. *Herbal Drugs and Phytopharmaceuticals: A Handbook for Practice on a Scientific Basis*. Medpharm Scientific Publishers.

Block, Seymour Stanton. 2001. *Disinfection, Sterilization, and Preservation*. Lippincott Williams & Wilkins.

Blumenthal, Mark, Alicia Goldberg, and Josef Brinckmann. 2000. *Herbal Medicine: Expanded Commission E Monographs*. American Botanical Council.

Bone, Kerry, and Simon Mills. 2012. *Principles and Practice of Phytotherapy: Modern Herbal Medicine*. Elsevier Health Sciences.

Bouali, Anouar, Ylenia Spissu, Antonio Barberis, Angela Fadda, Emanuela Azara, Germano Orrù, Alessandra Scano, Giuseppe Palmieri, Guy D'hallewin, and Héla El Ferchichi Ouarda. 2024. "Phytochemical evaluation and exploration of some biological activities of aqueous and ethanolic extracts of two species of the genus *Plantago* L." *PLoS One* 19 (2): e0298518.

Boudreau, Mary D, and Frederick A Beland. 2006. "An evaluation of the biological and toxicological properties of *Aloe barbadensis* (Miller), *Aloe vera*." *Journal of Environmental Science and Health Part C* 24 (1): 103–154.

Budai, Marta, Andrea Varga, Sandor Milesz, Jozsef Tőzsér, and Szilvia Benkő. 2013. "*Aloe vera* downregulates LPS-induced inflammatory cytokine production and expression of NLRP3 inflammasome in human macrophages." *Molecular Immunology* 56 (4): 471–479. https://doi.org/10.1016/j.molimm.2013.05.005.

Buhrmann, Constanze, Ali Mobasheri, Franziska Busch, Constance Aldinger, Ralf Stahlmann, Azadeh Montaseri, and Mehdi Shakibaei. 2011. "Curcumin modulates nuclear factor κB (nf-κB)-mediated inflammation in human tenocytes in vitro: Role of the phosphatidylinositol 3-kinase/Akt pathway." *Journal of Biological Chemistry* 286 (32): 28556–28566.

Burlou-Nagy, Cristina, Florin Bănică, Rodica Anamaria Negrean, Tünde Jurca, Laura Grațiela Vicaș, Eleonora Marian, Ildikó Bácskay, Rita Kiss, Pálma Fehér, and Simona Ioana Vicaș. 2023. "Determination of the bioactive compounds from *Echinacea purpurea* (L.) Moench leaves extracts in correlation with the antimicrobial activity and the in vitro wound healing potential." *Molecules* 28 (15): 5711.

Calixto, Joao Batista. 2000. "Efficacy, safety, quality control, marketing and regulatory guidelines for herbal medicines (phytotherapeutic agents)." *Brazilian Journal of Medical and Biological Research* 33: 179–189.

Chainani-Wu, Nita. 2003. "Safety and anti-inflammatory activity of curcumin: A component of tumeric (*Curcuma longa*)." *The Journal of Alternative & Complementary Medicine* 9 (1): 161–168.

Chan, Yuk-Shing, Long-Ni Cheng, Jian-Hong Wu, Enoch Chan, Yiu-Wa Kwan, Simon Ming-Yuen Lee, George Pak-Heng Leung, Peter Hoi-Fu Yu, and Shun-Wan Chan. 2011. "A review of the pharmacological effects of *Arctium lappa* (burdock)." *Inflammopharmacology* 19 (5): 245–254.

Chithra, Parvathy, Gopalakrishnan B. Sajithlal, and Gowri Chandrakasan. 1998. "Influence of aloe vera on the healing of dermal wounds in diabetic rats." *Journal of Ethnopharmacology* 59 (3): 195–201. https://doi.org/10.1016/S0378-8741(97)00124-4. https://www.sciencedirect.com/science/article/pii/S0378874197001244.

Choudhury, Abinash, Preet Amol Singh, Neha Bajwa, Subhransu Dash, and Preeti Bisht. 2023. "Pharmacovigilance of herbal medicines: Concerns and future prospects." *Journal of Ethnopharmacology* 309: 116383.

Cristani, Mariateresa, Andrea Citarella, Federica Carnamucio, and Nicola Micale. 2024. "Nano-formulations of natural antioxidants for the treatment of liver cancer." *Biomolecules* 14 (8): 1031. https://www.mdpi.com/2218-273X/14/8/1031.

Davis, Richard H., Michael G. Leitner, Joseph M. Russo, and Mary E. Byrne. 1989. "Anti-inflammatory activity of *Aloe vera* against a spectrum of irritants." *Journal of the American Podiatric Medical Association* 79 (6): 263–276.

de Macedo, Lucas Malvezzi, Érica Mendes dos Santos, Lucas Militão, Louise Lacalendola Tundisi, Janaína Artem Ataide, Eliana Barbosa Souto, and Priscila Gava Mazzola. 2020. "Rosemary (*Rosmarinus officinalis* L., syn *Salvia rosmarinus* Spenn.) and its topical applications: A review." *Plants* 9 (5): 651.

de Souza, Ariádine Reder Custodio, Thais Latansio de Oliveira, Pâmela Dias Fontana, Milena Carvalho Carneiro, Marcos Lúcio Corazza, Iara Jose de Messias Reason, and Lorena Bavia. 2022. "Phytochemicals and biological activities of burdock (*Arctium lappa* L.) extracts: A review." *Chemistry & Biodiversity* 19 (11): e202200615.

Della Loggia, Roberto, Aurelia Tubaro, Silvio Sosa, Heinz Becker, and Otto Isaac. 1994. "The role of triterpenoids in the topical anti-inflammatory activity of *Calendula officinalis* flowers." *Planta Medica* 60 (06): 516–520.

Devescovi, Valentina, Elisa Leonardi, Gabriela Ciapetti, and Elisabetta Cenni. 2008. "Growth factors in bone repair." *La Chirurgia degli organi di movimento* 92: 161–168.

Diniz, Lúcio Ricardo Leite, Leonardo Luiz Calado, Allana Brunna Sucupira Duarte, and Damião Pergentino de Sousa. 2023. "*Centella asiatica* and its metabolite asiatic acid: Wound healing effects and therapeutic potential." *Metabolites* 13 (2): 276.

do Nascimento, Marismar F, Juliana C Cardoso, Tarsizio S Santos, Lívia A Tavares, Tatiana N Pashirova, Patricia Severino, Eliana B Souto, and Ricardo LC de Albuquerque-Junior. 2020. "Development and characterization of biointeractive gelatin wound dressing based on extract of *Punica granatum* Linn." *Pharmaceutics* 12 (12): 1204.

Ekor, Martins. 2014. "The growing use of herbal medicines: Issues relating to adverse reactions and challenges in monitoring safety." *Frontiers in Pharmacology* 4: 177.

Epstein, Jenny, Guillermo Docena, Thomas T MacDonald, and Ian R Sanderson. 2010. "Curcumin suppresses p38 mitogen-activated protein kinase activation, reduces IL-1β and matrix metalloproteinase-3 and enhances IL-10 in the mucosa of children and adults with inflammatory bowel disease." *British Journal of Nutrition* 103 (6): 824–832.

Ernst, Edzard. 2000. "Adverse effects of herbal drugs in dermatology." *British Journal of Dermatology* 143 (5): 923–929.

Ezzat, Marwa Ibrahim, Mai Mohamed Abdelhafez, Asmaa Khaled Al-Mokaddem, and Shahira Mohamed Ezzat. 2025. "Targeting TGF-β/VEGF/NF-κB inflammatory pathway using the polyphenols of *Echinacea purpurea* (L.) Moench to enhance wound healing in a rat model." *Inflammopharmacology* 33 (4): 2151–2164. https://doi.org/10.1007/s10787-025-01681-6.

Fabricant, Daniel S, and Norman R Farnsworth. 2001. "The value of plants used in traditional medicine for drug discovery." *Environmental Health Perspectives* 109 (suppl 1): 69–75.

Fahimirad, Shohreh, Mohammadreza Memarzadeh, Hasan Jafari, Morteza Sanagooiezadeh Isfahani, Amir Almasi-Hashiani, and Hamid Abtahi. 2025. "Enhanced wound healing and antibacterial efficacy of a novel chitosan quaternary ammonium salt gel incorporating *Echinacea purpurea* extract." *Carbohydrate Research* 552: 109445.

Farnsworth, Norman R. 1988. "Screening plants for new medicines." *Biodiversity* 15 (3): 81–99.

Fontoura, Gomes, Guilherme Martins, Liliane dos Santos Rodrigues, Aramys Silva Reis, Andresa Aparecida Berretta, and Márcia Cristina Gonçalves Maciel. 2022. "Therapeutic potential of *Punica granatum* in wound healing: An overview." *Revista Colombiana de Ciencias Químico-Farmacéuticas* 51 (2): 992–1008.

Gad, Heba A, Autumn Roberts, Samirah H Hamzi, Haidy A Gad, Ilham Touiss, Ahmed E Altyar, Osama A Kensara, and Mohamed L Ashour. 2021. "Jojoba oil: An updated comprehensive review on chemistry, pharmaceutical uses, and toxicity." *Polymers* 13 (11): 1711.

Gavan, Alexandru, Liora Colobatiu, Daniela Hanganu, Catalina Bogdan, Neli Kinga Olah, Marcela Achim, and Simona Mirel. 2022. "Development and evaluation of hydrogel wound dressings loaded with herbal extracts." *Processes* 10 (2): 242.

Givol, Or, Rachel Kornhaber, Denis Visentin, Michelle Cleary, Josef Haik, and Moti Harats. 2019. "A systematic review of *Calendula officinalis* extract for wound healing." *Wound Repair and Regeneration* 27 (5): 548–561.

Goel, Ajay, Ajaikumar B Kunnumakkara, and Bharat B Aggarwal. 2008. "Curcumin as "Curecumin": From kitchen to clinic." *Biochemical Pharmacology* 75 (4): 787–809.

Gohil, Kashmira J, A Patel Jagruti, and K Gajjar Anuradha. 2010. "Pharmacological review on *Centella asiatica*: A potential herbal cure-all." *Indian Journal of Pharmaceutical Sciences* 72 (5): 546.

Gonçalves, Sandra, and Anabela Romano. 2018. "Production of plant secondary metabolites by using biotechnological tools." *Secondary Metabolites-Sources and Applications* 5: 81–99.

Guo, Xiaoqing, and Nan Mei. 2016. "*Aloe vera*: A review of toxicity and adverse clinical effects." *Journal of Environmental Science and Health, Part C* 34 (2): 77–96.

Habashy, R Ramy, Ashraf B Abdel-Naim, Amani E Khalifa, Mohammed M Al-Azizi. 2005. "Anti-inflammatory effects of jojoba liquid wax in experimental models." *Pharmacological Research* 51 (2): 95–105.

Hajialyani, Maryam, Deepak Tewari, Enrique Sobarzo-Sánchez, Seyed Mohammad Nabavi, Mohammad Hossein Farzaei, and Mohammad Abdollahi. 2018. "Natural product-based nanomedicines for wound healing purposes: Therapeutic targets and drug delivery systems." *International Journal of Nanomedicine* 13: 5023–5043. https://doi.org/10.2147/ijn.S174072.

Hamid, Layth L. 2024. "*Allium sativum* extract mediate the biosynthesis of palladium nanoparticles as potential nanodrug for combating multidrug-resistant bacteria and wound healing." *Materials Chemistry and Physics* 321: 129507.

Hamman, Josias. 2008. "Composition and applications of *Aloe vera* leaf gel." *Molecules* 13 (8): 1599–1616.

Hashemi, Seyyed Abbas, Seyyed Abdollah Madani, and Saied Abediankenari. 2015. "The review on properties of *Aloe vera* in healing of cutaneous wounds." *BioMed Research International* 2015 (1): 714216.

Heinrich, Michael. 2015. "Ethnopharmacology: A short history of a multidisciplinary field of research." *Ethnopharmacology*. https://doi.org/10.1002/9781118930717.ch1. https://onlinelibrary.wiley.com/doi/abs/10.1002/9781118930717.ch1.

Heinrich, Michael, Joanne Barnes, Jose Prieto-Garcia, Simon Gibbons, and Elizabeth M Williamson. 2017. *Fundamentals of Pharmacognosy and Phytotherapy E-Book: Fundamentals of Pharmacognosy and Phytotherapy E-Book*. Elsevier Health Sciences.

Hewlings, Susan J, and Douglas S Kalman. 2017. "Curcumin: A review of its effects on human health." *Foods* 6 (10): 92. https://doi.org/10.3390/foods6100092

Hosseinzadeh, Hossein, Mohammad Hadi Modaghegh, and Zahra Saffari. 2009. "*Crocus sativus* L. (saffron) extract and its active constituents (crocin and safranal) on ischemia-reperfusion in rat skeletal muscle." *Evidence-Based Complementary and Alternative Medicine* 6 (3): 343–350.

Imran, Muhammad, Muhammad Nadeem, Muhammad Asif Khan, Sheraz Ahmed, Ali Imran, Rai Muhammad Amir, Muhammad Umair Arshad, Syed Amir Gilani, Farhan Saeed, and Abdur Rauf. 2017. "Curcumin and its allied analogues: Epigenetic and health perspectives – A review." *Food Reviews International* 34 (4): 321–345. https://doi.org/10.17221/584/2015-CJFS

Jacob, Daniel Etim, Sylvester Chibueze Izah, Imaobong Ufot Nelson, and Koko Sunday Daniel. 2024. "Indigenous knowledge and phytochemistry: Deciphering the healing power of herbal medicine." In *Herbal Medicine Phytochemistry: Applications and Trends*, 1953–2005. Springer.

Jafari-Sales, Abolfazl, Behboud Jafari, Javad Sayyahi, and Tahereh Zohoori-Bonab. 2015. "Evaluation of antibacterial activity of ethanolic extract of malva neglecta and *Althaea officinalis* L. on antibiotic-resistant strains of *Staphylococcus aureus*." *Journal of Biology and Today's World* 4 (2): 58–62.

Jagetia, Ganesh Chandra, and Bharat B Aggarwal. 2007. "Spicing up" of the immune system by curcumin." *Journal of Clinical Immunology* 27: 19–35.

Jagetia, Ganesh Chandra, and Golgod Krishnamurthy Rajanikant. 2012. "Acceleration of wound repair by curcumin in the excision wound of mice exposed to different doses of fractionated γ radiation." *International Wound Journal* 9 (1): 76–92.

Javadi, Behjat, Amirhossein Sahebkar, and Seyed Ahmad Emami. 2013. "A survey on saffron in major Islamic traditional medicine books." *Iranian Journal of Basic Medical Sciences* 16 (1): 1.

Javadi, Parisa, Mohammad Ali Derakhshan, Fatemeh Asghari, Sharmin Kharrazi, and Reza Faridi-Majidi. 2023. "Co-electrospun PCL-collagen nanofibers containing saffron-synthesized gold nanoparticles as an antioxidant wound dressing." *Proceedings of the Institution of Mechanical Engineers, Part H: Journal of Engineering in Medicine* 237 (11): 1318–1329.

Jettanacheawchankit, Suwimon, Siriruk Sasithanasate, Polkit Sangvanich, Wijit Banlunara, and Pasutha Thunyakitpisal. 2009. "Acemannan stimulates gingival fibroblast proliferation; expressions of keratinocyte growth factor-1, vascular endothelial growth factor, and type I collagen; and wound healing." *Journal of Pharmacological Sciences* 109 (4): 525–531.

Jikah, Agatha Ngukuran, Great Iruoghene Edo, Raghda S Makia, Emad Yousif, Tayser Sumer Gaaz, Endurance Fegor Isoje, Rapheal Ajiri Opiti, Patrick Othuke Akpoghelie, Ufuoma Augustina Igbuku, and Joseph Oghenewogaga Owheruo. 2024. "A review of the therapeutic potential of sulfur compounds in *Allium sativum*." *Measurement: Food 8*: 100195.

Jiménez-Medina, Eva, Angel Garcia-Lora, Laura Paco, Ignacio Algarra, Antonia Collado, and Federico Garrido. 2006. "A new extract of the plant *Calendula officinalis* produces a dual in vitro effect: Cytotoxic anti-tumor activity and lymphocyte activation." *BMC câncer* 6: 1–14.

Jiménez-Monreal, Ana Maria, Laura García-Diz, Magdalena Martínez-Tomé, Maria del Mar A. Mariscal, and Maria Angeles Murcia. 2009. "Influence of cooking methods on antioxidant activity of vegetables." *Journal of Food Science* 74 (3): H97–H103. https://doi.org/10.1111/j.1750-3841.2009.01091.x

Karole, Sarita, Sarika Shrivastava, Shefali Thomas, Bhawana Soni, Shifa Khan, Julekha Dubey, Shashi P Dubey, Nushrat Khan, and Deepak Kumar Jain. 2019. "Polyherbal formulation concept for synergic action: A review." *Journal of Drug Delivery & Therapeutics* 9 (2): 453–458.

Kartini, Kartini, Nina Wati, Rabbindra Gustav, Risa Wahyuni, Yosua Fernaldi Anggada, Risna Hidayani, Antoni Raharjo, Ridho Islamie, and Sulistyo Emantoko Dwi Putra. 2021. "Wound healing effects of *Plantago major* extract and its chemical compounds in hyperglycemic rats." *Food Bioscience* 41: 100937.

Khan, Anamika Devi, Varsha Rastogi, Pallavi Lavhale, and Jyoti Jain. 2022. "Novel approaches for herbal drug delivery in wound healing: A review." *Indian Journal of Pharmaceutical Sciences* 84 (2).

Kianitalaei, Alieh, Zohre Feyzabadi, Shokuhsadat Hamedi, and Mjjaper Qaraaty. 2019. "*Althaea officinalis* in traditional medicine and modern phytotherapy." *Journal of Advanced Pharmacy Education and Research* 9 (S2): 155.

Klein, Von. 1905. "The medical features of the Papyrus Ebers." *Journal of the American Medical Association* XLV (26): 1928–1935. https://doi.org/10.1001/jama.1905.52510260014001e.

Knipping, Karen, Elisabeth CAM van Esch, Selva C Wijering, Sicco van der Heide, Anthony E Dubois, and Johan Garssen. 2008. "In vitro and in vivo anti-allergic effects of *Arctium lappa* L." *Experimental Biology and Medicine* 233 (11): 1469–1477.

Kumar, Mohit, Amit Sharma, Tarun Parashar Rinka Juneja, Mohammad Arif, Aniruddh Dev Singh, and Ashok Kumar. 2025. "Biological evaluation of herbal preparations and formulation standardization: A literature study." *Journal of Neonatal Surgery* 14 (8s).

Kumar, Sunil, Bikarma Singh, and Vikas Bajpai. 2021. "*Andrographis paniculata* (Burm. f.) Nees: Traditional uses, phytochemistry, pharmacological properties and quality control/quality assurance." *Journal of Ethnopharmacology* 275: 114054.

Langer, Robert. 1990. "New methods of drug delivery." *Science* 249 (4976): 1527–1533.

Leach, Matthew J. 2008. "*Calendula officinalis* and wound healing: A systematic review." *Wounds: A Compendium of Clinical Research and Practice* 20 (8): 236–243.

Li, Cong, Zhi Luo, Hongqing Feng, and Zhou Li. 2025. "Polymers facilitating therapeutic efficacy and applications for traditional Chinese medicine." *Journal of Polymer Science* 63 (4): 839–863.

Li, Xiuling, Shunung Liang, Chee Hwee Tan, Shuwen Cao, Xiaoding Xu, Phei Er Saw, and Wei Tao. 2021. "Nanocarriers in the enhancement of therapeutic efficacy of natural drugs." *BIO Integration* 2 (2): 40.

Li, Zheng, Zhiyuan Zhang, Jie Ding, Yuanyuan Li, Guiyun Cao, Lihao Zhu, Yifei Bian, and Yuhong Liu. 2024. "Extraction, structure and bioactivities of polysaccharide from root of *Arctium lappa* L.: A review." *International Journal of Biological Macromolecules 265*: 131035.

Lin, Song-Chow, Chia-Hsien Lin, Chun-Ching Lin, Yun-Ho Lin, Chin-Fa Chen, I-Cheng Chen, and Li-Ya Wang. 2002. "Hepatoprotective effects of *Arctium lappa* Linne on liver injuries induced by chronic ethanol consumption and potentiated by carbon tetrachloride." *Journal of Biomedical Science* 9 (5): 401–409.

Lin, Tzu-Kai, Lily Zhong, and Juan Luis Santiago. 2018. "Anti-inflammatory and skin barrier repair effects of topical application of some plant oils." *International Journal of Molecular Sciences* 19 (1): 70.

Li Pomi, Federica, Vincenzo Papa, Francesco Borgia, Mario Vaccaro, Alessandro Allegra, Nicola Cicero, and Sebastiano Gangemi. 2023. "*Rosmarinus officinalis* and skin: Antioxidant activity and possible therapeutical role in cutaneous diseases." *Antioxidants* 12 (3): 680.

Maenthaisong, Ratree, Nathorn Chaiyakunapruk, Surachet Niruntraporn, and Chuenjid Kongkaew. 2007. "The efficacy of aloe vera used for burn wound healing: A systematic review." *Burns* 33 (6): 713–718.

Maimoona, Alya, Ismat Naeem, Zeb Saddiqe, and Khalid Jameel. 2011. "A review on biological, nutraceutical and clinical aspects of French maritime pine bark extract." *Journal of Ethnopharmacology* 133 (2): 261–277.

Maquart, François-Xavier, Loren Pickart, Maryvonne Laurent, Philippe Gillery, Jean-Claude Monboisse, and Jacques-Paul Borel. 1988. "Stimulation of collagen synthesis in fibroblast cultures by the tripeptide-copper complex glycyl-L-histidyl-L-lysine-Cu^{2+}." *FEBS Letters* 238 (2): 343–346.

Mehdizadeh, Roya, Mohammad-Reza Parizadeh, Ali-Reza Khooei, Soghra Mehri, and Hossein Hosseinzadeh. 2013. "Cardioprotective effect of saffron extract and safranal in isoproterenol-induced myocardial infarction in wistar rats." *Iranian Journal of Basic Medical Sciences* 16 (1): 56.

Miraj, Sepide, and Samira Alesaeidi. 2016. "A systematic review study of therapeutic effects of *Matricaria recuitta* chamomile (chamomile)." *Electronic Physician* 8 (9): 3024.

Mitra, Ashim K, and Ashim K Mitra. 2003. *Ophthalmic Drug Delivery Systems*. Marcel Dekker.

Mohanty, Chandana, and Sanjeeb K Sahoo. 2017. "Curcumin and its topical formulations for wound healing applications." *Drug Discovery Today* 22 (10): 1582–1592.

Moses, Rachael L, Thomas AK Prescott, Eduard Mas-Claret, Robert Steadman, Ryan Moseley, and Alastair J Sloan. 2023. "Evidence for natural products as alternative wound-healing therapies." *Biomolecules* 13 (3): 444.

Moshiri, Mohammad, Mohammad Vahabzadeh, and Hossein Hosseinzadeh. 2015. "Clinical applications of saffron (*Crocus sativus*) and its constituents: A review." *Drug Research* 65 (06): 287–295. https://doi.org/10.1055/s-0034-1375681

Nanda, Sanju, and Kumud Madan. 2021. "The role of safranal and saffron stigma extracts in oxidative stress, diseases and photoaging: A systematic review." *Heliyon* 7 (2).

Negi, Puran Sindh, Guddadarangavvanahally K. Jayaprakasha, Lakkakula Jagan Mohan Rao, and Krishnankutty K. Sakariah. 1999. "Antibacterial activity of turmeric oil: A byproduct from curcumin manufacture." *Journal of Agricultural and Food Chemistry* 47 (10): 4297–4300. https://doi.org/10.1021/jf990308d

Newman, David J, and Gordon M Cragg. 2020. "Natural products as sources of new drugs over the nearly four decades from 01/1981 to 09/2019." *Journal of Natural Products* 83 (3): 770–803.

Pagano, Cinzia, Maria Rachele Ceccarini, Marco Faieta, Alessandro di Michele, Francesca Blasi, Lina Cossignani, Tommaso Beccari, Eleonora Oliva, Paola Pittia, and Manuel Sergi. 2022. "Starch-based sustainable hydrogel loaded with *Crocus sativus* petals extract: A new product for wound care." *International Journal of Pharmaceutics* 625: 122067.

Parente, Leila Maria Leal, Ruy de Souza Lino Júnior, Leonice Manrique Faustino Tresvenzol, Marina Clare Vinaud, José Realino de Paula, and Neusa Margarida Paulo. 2012. "Wound healing and anti-inflammatory effect in animal models of *Calendula officinalis* L. growing in Brazil." *Evidence-Based Complementary and Alternative Medicine* 2012 (1): 375671.

Pathak, Deepika, and Avijit Mazumder. 2024. "A critical overview of challenging roles of medicinal plants in improvement of wound healing technology." *DARU Journal of Pharmaceutical Sciences* 32 (1): 379–419.

Patra, Jayanta Kumar, Gitishree Das, Leonardo Fernandes Fraceto, Estefania Vangelie Ramos Campos, Maria del Pilar Rodriguez-Torres, Laura Susana Acosta-Torres, Luis Armando Diaz-Torres, Renato Grillo, Mallappa Kumara Swamy, and Shivesh Sharma. 2018. "Nano based drug delivery systems: Recent developments and future prospects." *Journal of Nanobiotechnology* 16 (1): 33.

Posadzki, Paul. 2013. "Adverse effects of herbal medicines: An overview of systematic reviews." *Clinical Medicine* 13 (1): 7–12.

Poursamimi, Javad, Zhaleh Shariati-Sarabi, Jalil Tavakkol-Afshari, Seyed Ahmad Mohajeri, and Mojgan Mohammadi. 2020. "*Crocus Sativus* (Saffron): An immunoregulatory factor in the autoimmune and non-autoimmune diseases." *Iranian Journal of Allergy, Asthma and Immunology* 19 (S1): 27–42.

Predes, Fabricia S, Ana LTG Ruiz, João E Carvalho, Mary A Foglio, and Heidi Dolder. 2011. "Antioxidative and in vitro antiproliferative activity of *Arctium lappa* root extracts." *BMC Complementary and Alternative Medicine* 11 (1): 5.

Purcaru, Tomina, Diguță Camelia, and Florentina Matei. 2018. "Antimicrobial potential of Romanian spontaneous flora – A minireview." *Scientific Papers. Series B. Horticulture* 62.

Radha, Maharjan H., and Nampoothiri P. Laxmipriya. 2015. "Evaluation of biological properties and clinical effectiveness of Aloeávera: áAásystematic review." *Journal of Traditional and Complementary Medicine* 5 (1): 21–26.

Raghavan, Rahul, Sanith Cheriyamundath, and Joseph Madassery. 2018. "Exploring the mechanisms of cytotoxic and anti-inflammatory property of andrographolide and its derivatives." *Pharmacognosy Reviews* 12 (23).

Rahiman, Niloufar, Maryam Akaberi, Amirhossein Sahebkar, Seyed Ahmad Emami, and Zahra Tayarani-Najaran. 2018. "Protective effects of saffron and its active components against oxidative stress and apoptosis in endothelial cells." *Microvascular Research* 118: 82–89.

Ramasamy, Thamil Selvee, Ain Zubaidah Ayob, Hsu Hsu Lynn Myint, Sharmanee Thiagarajah, and Farahnaz Amini. 2015. "Targeting colorectal cancer stem cells using curcumin and curcumin analogues: Insights into the mechanism of the therapeutic efficacy." *Cancer Cell International* 15 (1): 15.

Ramli, Suzita, Wong Jun Xian, and Noor Azira Abd Mutalib. 2020. "A review: Antibacterial activities, antioxidant properties and toxicity profile of *Centella asiatica*." *Educatum Journal of Science, Mathematics and Technology* 7 (1): 39–47.

Ranzato, Elia, Simona Martinotti, and Bruno Burlando. 2011. "Wound healing properties of jojoba liquid wax: An in vitro study." *Journal of Ethnopharmacology* 134 (2): 443–449.

Rašković, Aleksandar, Isidora Milanović, Nebojša Pavlović, Tatjana Ćebović, Saša Vukmirović, and Momir Mikov. 2014. "Antioxidant activity of rosemary (*Rosmarinus officinalis* L.) essential oil and its hepatoprotective potential." *BMC Complementary and Alternative Medicine* 14 (1): 9.

Rates, Stela Maria Kuze. 2001. "Plants as source of drugs." *Toxicon* 39 (5): 603–613.

Reuter, Juliane, Andrea Jocher, Julia Stump, B Grossjohann, G. Franke, and Christoph M. Schempp. 2008. "Investigation of the anti-inflammatory potential of *Aloe vera* gel (97.5%) in the ultraviolet erythema test." *Skin Pharmacology and Physiology* 21 (2): 106–110.

Reynolds, T., and A. C. Dweck. 1999. "*Aloe vera* leaf gel: A review update." *Journal of Ethnopharmacology* 68 (1–3): 3–37. https://doi.org/10.1016/s0378-8741(99)00085-9.

Rezaei, Maryam, Zeynab Dadgar, Ali Noori-Zadeh, Seyed Alireza Mesbah-Namin, Iraj Pakzad, and Elham Davodian. 2015. "Evaluation of the antibacterial activity of the *Althaea officinalis* L. leaf extract and its wound healing potency in the rat model of excision wound creation." *Avicenna Journal of Phytomedicine* 5 (2): 105.

Russo, Daniela, Ludovica Lela, Nadia Benedetto, Immacolata Faraone, Gianluca Paternoster, Patricia Valentão, Luigi Milella, and Monica Carmosino. 2025. "Phytochemical composition and wound healing properties of *Echinacea angustifolia* DC. Root hydroalcoholic extract." *International Journal of Molecular Sciences* 26 (6): 2562.

Sadeghnia, Hamid R, Hamideh Shaterzadeh, Fatemeh Forouzanfar, and Hossein Hosseinzadeh. 2017. "Neuroprotective effect of safranal, an active ingredient of *Crocus sativus*, in a rat model of transient cerebral ischemia." *Folia Neuropathologica* 55 (3): 206–213.

Sahlabgi, Alaa, Dumitru Lupuliasa, Gabriela Stanciu, Simona Lupşor, Lavinia Lia Vlaia, Ramona Rotariu, Nicoleta Corina Predescu, Cristiana Rădulescu, Radu-Lucian Olteanu, and Sorina-Geanina Stănescu. 2025. "The development and comparative evaluation of rosemary hydroalcoholic macerate-based dermatocosmetic preparations: A study on antioxidant, antimicrobial, and anti-inflammatory properties." *Gels* 11 (3): 149.

Samarasinghe, WM Priyantha, Chandani Ranasinghe, K Hector Jayawardana, Seneviratnage Somaratne, and GM Kamal B Gunaherath. 2024. "A rare phytosterol, stigmast-5-en-3 β, 7 α, 22 α-triol and other secondary metabolites from *Leea indica* showing enhanced in vitro cell migration and proangiogenic activity." *Natural Product Research* 38 (19): 3466–3470.

Samuelsen, Anne Berit. 2000. "The traditional uses, chemical constituents and biological activities of *Plantago major* L. A review." *Journal of Ethnopharmacology* 71 (1–2): 1–21.

Sánchez, Maria, Elena González-Burgos, Irena Iglesias, and Maria Pilar Gómez-Serranillos. 2020. "Pharmacological update properties of *Aloe Vera* and its major active constituents." *Molecules* 25 (6):1324. https://doi.org/10.3390/molecules25061324.

Sari, Asti Arum, and Reza Rahmawati. 2023. "A bibliometric analysis of *Aloe vera* in wound healing." *Pharmacon: Jurnal Farmasi Indonesia* 20 (1): 36–49.

Shimada, Ken, Kentaro Ushijima, Chisato Suzuki, Michiko Horiguchi, Hitoshi Ando, Tomomi Akita, Mizuki Shimamura, Junki Fujii, Chikamasa Yamashita, and Akio Fujimura. 2018. "Pulmonary administration of curcumin inhibits B16F10 melanoma lung metastasis and invasion in mice." *Cancer Chemotherapy and Pharmacology* 82: 265–273.

Singer, Adam J, and Richard AF Clark. 1999. "Cutaneous wound healing." *New England Journal of Medicine* 341 (10): 738–746.

Soheilifar, Mohammad Hasan, Dara Dastan, Nastaran Masoudi-Khoram, Hoda Keshmiri Neghab, Sima Nobari, Seyed Mehdi Tabaie, and Razieh Amini. 2024. "In vitro and in vivo evaluation of the diabetic wound healing properties of Saffron (*Crocus Sativus* L.) petals." *Scientific Reports* 14 (1): 19373.

Stanisavljević, Ivana T, Saša S Stojičević, Dragan T Veličković, Miodrag L Lazić, and Vlada B Veljković. 2008. "Screening the antioxidant and antimicrobial properties of the extracts from plantain (*Plantago major* L.) leaves." *Separation Science and Technology* 43 (14): 3652–3662.

Surjushe, Amar, Resham Vasani, and Dattatray G. Saple. 2008. "*Aloe vera*: A short review." *Indian Journal of Dermatology* 53 (4): 163–166.

Taha, Shereen Sayed. 2022. "Chemical and biological studies on *Allium sativum* L. (1952–2020): A comprehensive review." *Journal of Advanced Biomedical and Pharmaceutical Sciences* 5 (1): 1–22.

Tanga, Mbappe. 2021. *Morphological Characteristics and Chemical Composition of s Skullcap (Scutellaria lateriflora L.) and Burdock (Arctium lappa L.) Cultivated under Different Conditions*. Cape Peninsula University of Technology.

Thakur, Lovely, Umang Ghodasra, Nilesh Patel, and Mahesh Dabhi. 2011. "Novel approaches for stability improvement in natural medicines." *Pharmacognosy Reviews* 5 (9): 48.

Tietel, Zipora, Shirin Kahremany, Guy Cohen, and Navit Ogen-Shtern. 2021. "Medicinal properties of jojoba (*Simmondsia chinensis*)." *Israel Journal of Plant Sciences* 68 (1–2): 38–47.

Tønnesen, Hanne Hjorth, Már Másson, and Thorsteinn Loftsson. 2002. "Studies of curcumin and curcuminoids. XXVII. Cyclodextrin complexation: Solubility, chemical and photochemical stability." *International Journal of Pharmaceutics* 244 (1–2): 127–135.

Triastuti, Asih, Dimas Adhi Pradana, Iwang Davi Setiawan, Nanang Fakhrudin, Setiawan Khoirul Himmi, Sitarina Widyarini, and Abdul Rohman. 2022. "In vivo anti-inflammatory activities of *Plantago major* extract and fractions and analysis of their phytochemical components using a high-resolution mass spectrometry." *Research in Pharmaceutical Sciences* 17 (6): 665–676.

Tümen, İbrahim, Esra Küpeli Akkol, Hakkı Taştan, Ipek Süntar, and Mehmet Kurtca. 2018. "Research on the antioxidant, wound healing, and anti-inflammatory activities and the phytochemical composition of maritime pine (*Pinus pinaster* Ait)." *Journal of Ethnopharmacology* 211: 235–246.

Vitale, Stefania, Sara Colanero, Martina Placidi, Giovanna Di Emidio, Carla Tatone, Fernanda Amicarelli, and Anna Maria D'Alessandro. 2022. "Phytochemistry and biological activity of medicinal plants in wound healing: An overview of current research." *Molecules* 27 (11): 3566.

Wichtl, Max. 2004. *Herbal Drugs and Phytopharmaceuticals: A Handbook for Practice on a Scientific Basis*. CRC Press.

Worakunphanich, Wiwan, Montarat Thavorncharoensap, Sitaporn Youngkong, Kunlawat Thadanipon, and Ammarin Thakkinstian. 2021. "Safety of *Andrographis paniculata*: A systematic review and meta-analysis." *Pharmacoepidemiology and Drug Safety* 30 (6): 727–739.

Wu, Feng, Dong Bian, Yong Xia, Zhi Gong, Qing Tan, Jian Chen, and Yong Dai. 2012. "Identification of major active ingredients responsible for burn wound healing of *Centella asiatica* Herbs." *Evidence-Based Complementary and Alternative Medicine* 2012: 848093. https://doi.org/10.1155/2012/848093.

Xu, Wenqian, Hongkang Zhu, Bin Hu, Yuliang Cheng, Yahui Guo, Weirong Yao, and He Qian. 2021. "Echinacea in hepatopathy: A review of its phytochemistry, pharmacology, and safety." *Phytomedicine* 87: 153572.

Xue, Taotao, Kaihua Ruan, Zhishu Tang, Jinao Duan, and Hongbo Xu. 2023. "Isolation, structural properties, and bioactivities of polysaccharides from *Althaea officinalis* Linn.: A review." *International Journal of Biological Macromolecules* 242: 125098.

Xue, Tao-Tao, Hong-Bo Xu, Zhi-Shu Tang, Jin-Ao Duana, Hong-Bo Liu, Xin-Bo Shi, and Zhong-Xing Song. 2021. "Progress in chemical compositions and pharmacological activities of *Althaea officinalis*." *Medicine Research* 5 (2): 210002.

Yadav, Ashutosh. 2021. "Traditional utilization and pharmacological properties of medicinal plants: A review." In *Traditional Utilization and Pharmacological Properties of Medicinal Plants*, 1–28. Apple Academic Press.

Zargari, A. 1991. *Medicinal Plants, Volume II*. Tehran University Publication.

Zeng, Xiaoyan, Zhenfeng Guan, Reyisha Yusufu, Siyi Li, Qi Wu, Yanping Feng, Tiantian Chen, Muxiang Luo, and Fenghuan Wei. 2024. "Multi-drug loaded electrospinning nanofibers promote healing with less scar formation by remodeling wound microenvironment via modulating TGF-β1/Smads/EN1 signal." *Chemical Engineering Journal* 499: 156154.

Zhang, Hui Sun, Ping Wang, Ying Han, and Xijun Wang. 2012. "Future perspectives of personalized medicine in traditional Chinese medicine: A systems biology approach." *Complementary Therapies in Medicine* 20 (1–2): 93–99.

Zhang, Qi. 2013. *WHO Traditional Medicine Strategy 2014–2023*. World Health Organization.

Zhang, Xiaorui. 2000. "General guidelines for methodologies on research and evaluation of traditional medicine." *World Health Organization* 1: 1–71.

Zhao, Chen, Hio-Tong Kam, Yan Chen, Guiyi Gong, Maggie Pui-Man Hoi, Krystyna Skalicka-Woźniak, Alberto Carlos Pires Dias, and Simon Ming-Yuen Lee. 2021. "Crocetin and its glycoside crocin, two bioactive constituents from *Crocus sativus* L.(saffron), differentially inhibit angiogenesis by inhibiting endothelial cytoskeleton organization and cell migration through VEGFR2/SRC/FAK and VEGFR2/MEK/ERK signaling pathways." *Frontiers in Pharmacology* 12: 675359.

Zhou, Xin, Yanling Guo, Kun Yang, Peng Liu, and Jun Wang. 2022. "The signaling pathways of traditional Chinese medicine in promoting diabetic wound healing." *Journal of Ethnopharmacology* 282: 114662.

8 Animal-Based Products in Wound Healing

Safeena Rashid[1], *Aarif Ali*[2]*, *Waseem Younis Khan*[1], *Farhat Jabeen*[3], *and Javeed Iqbal Ahmad Bhat*[2]

[1] Department of Biochemistry, School of Biological Sciences, University of Kashmir Srinagar, Jammu & Kashmir, India

[2] Division of Basic Sciences and Humanities, Faculty of Horticulture, SKUAST-K, Shalimar, Srinagar, Jammu & Kashmir, India

[3] Department of Clinical Biochemistry, School of Biological Sciences, University of Kashmir, Srinagar, Jammu & Kashmir, India

8.1 INTRODUCTION

Skin, the biggest and the first line of defense against the environment/diseases, is the vital organ of the body. In addition to protection, the main functions of the skin include regulation and sensation. Specifically, it guards against mechanical pressure impacts, lessens the effects of temperature fluctuations, fights microbial infections, restricts the effects of radiation, and prevents chemicals from entering. The skin is the largest organ in the body and plays a vital role in many physiological processes, including fluid balance, changes in peripheral circulation, and the regulation of body temperature (via perspiration and hair).

The skin is a defense system, an injury response, and a store of nutrients that help it maintain homeostasis (Otsuka et al., 2022; Shedoeva et al., 2019), protecting our body's interior organs, ligaments, muscles, and bones. Additionally, it contributes to the production of vitamin D, for which it acts as a reservoir. Damage to the skin's extensive network of nerve cells causes the affected areas to lose their ability to feel and transmit changes in the environment, such as heat, cold, touch, and pain. The loss of the skin's integrity and normal functioning, which can be brought on by mechanical, chemical, or physical factors, is referred to as a wound (Imran et al., 2015; Kumar et al., 2007). To ensure the organism's homeostasis, it is imperative to restore the skin's structure and functions as soon as possible when it is compromised. This way, the wound repair process can start just after the skin rupture (Muhlstadt et al., 2012). Injuries can impair a patient's independence, working ability, self-esteem, and general well-being. Acute wounds take seven to ten days to heal; however, chronic wounds may not take this long because of a sluggish and disorganized healing process. Numerous factors, such as underlying disease, infection, chronic inflammation, medication use, and oxidative stress brought on by aging, have been linked to the development of chronic wounds (Shaw and Martin, 2009). Recent wounds that have not yet moved through the various stages of healing are referred to as acute wounds. Surgical incisions, heat injuries, and abrasions are examples of wounds that can be either full-thickness scars that only harm the subcutaneous layer of skin (Szycher and Lee, 1992). Good circulation, immunological activity, and a moist atmosphere devoid

* Corresponding author: buttaarif31@gmail.com

DOI: 10.1201/9781003644958-8

of dead tissue or foreign objects are all characteristics of acute wound healing and they all support the regular function of the cells involved in the healing mechanism (Hosseinkhani et al., 2017). Furthermore, chronic wounds do not heal in the typical stages and are not promptly and efficiently treated (Li et al., 2007). Resistance to therapy is often seen, and the wound must be allowed to heal for a longer period since several environmental factors, including trauma, desiccation, and infection, can impede repair. Additionally, the presence of free oxygen radicals in wounds and host factors like impaired circulation, respiratory function, or immune deficiencies disrupt the healing process, which can result in additional tissue damage and postpone the usual proceeding phases of the healing process (Saini et al., 2016).

8.2 ANIMAL-BASED PRODUCTS IN WOUND HEALING

Wound care is crucial for human health since unhealed wounds and prolonged treatment times lead to higher economic expenditures, as well as social repercussions for medical facilities, physicians, patients, and their families (Young and McNaught, 2011). As a result, patients who receive inadequate wound care may experience additional issues, such as bleeding, infections, inflammation, and difficulties with tissue regeneration (Sharifi et al., 2020). The World Health Organization (WHO) estimates that about 80% of people worldwide treat their illnesses by traditional methods, with 85% of them using plant therapies (Ayaz et al., 2015). Throughout history, materials sourced from animals have been crucial in facilitating tissue repair, owing to their inherent compatibility with human biology. These products, obtained from different animal tissues, organs, and by-products, offer both structural and biochemical signals that aid in the healing process. Products derived from animals have historically been crucial in wound healing owing to their compatibility with biological systems, inherent bioactive properties, and capacity to promote tissue regeneration. Wound healing materials derived from animals, including collagen, gelatin, chitosan, honey, and fibrin, are commonly utilized in clinical practice. Their efficacy is attributed to characteristics such as biocompatibility, biodegradability, antimicrobial properties, and the capacity to enhance cell adhesion, migration, and proliferation. These natural biomaterials are available in various wound care formats, such as dressings, scaffolds, hydrogels, and grafts. Recent advancements in tissue engineering and biotechnology have broadened the use of animal-derived products, with innovative options like decellularized extracellular matrices, silk-based scaffolds, and marine collagen becoming increasingly popular. Nonetheless, the use of animal materials presents challenges, including potential immunogenic reactions, disease transmission risks, and ethical issues surrounding animal sourcing. Some of the important animal-derived products involved in wound healing are shown in Figure 8.1.

8.2.1 Honey

Honey is a sweet, viscous, and yellowish-brown sugar solution that bees and other insects collect from blossom nectar. The nectar is bio-synthesized, stored in the honeycomb to ripen, deposited, dehydrated, and stored for the bees (Ramli et al., 2018). It comprises 80–85% carbs, 15–17% water, 0.1–0.4% protein, and 0.2% ash, with trace amounts of organic acids, enzymes, vitamins, and minerals (Buba et al., 2013). In folklore, honey has long been used as a wound healing medicine. Pre-clinical research and clinical experiments have scientifically validated this traditional belief, showing that honey speeds up the repair of a variety of wounds such as burns, surgical wounds, and malignant wounds. According to *in vivo* research, lesions treated with honey typically healed more quickly than those left untreated. According to certain research, honey's effectiveness and efficiency in wound healing was more than that of hydrofiber dressings and hydrofiber silver (Khoo

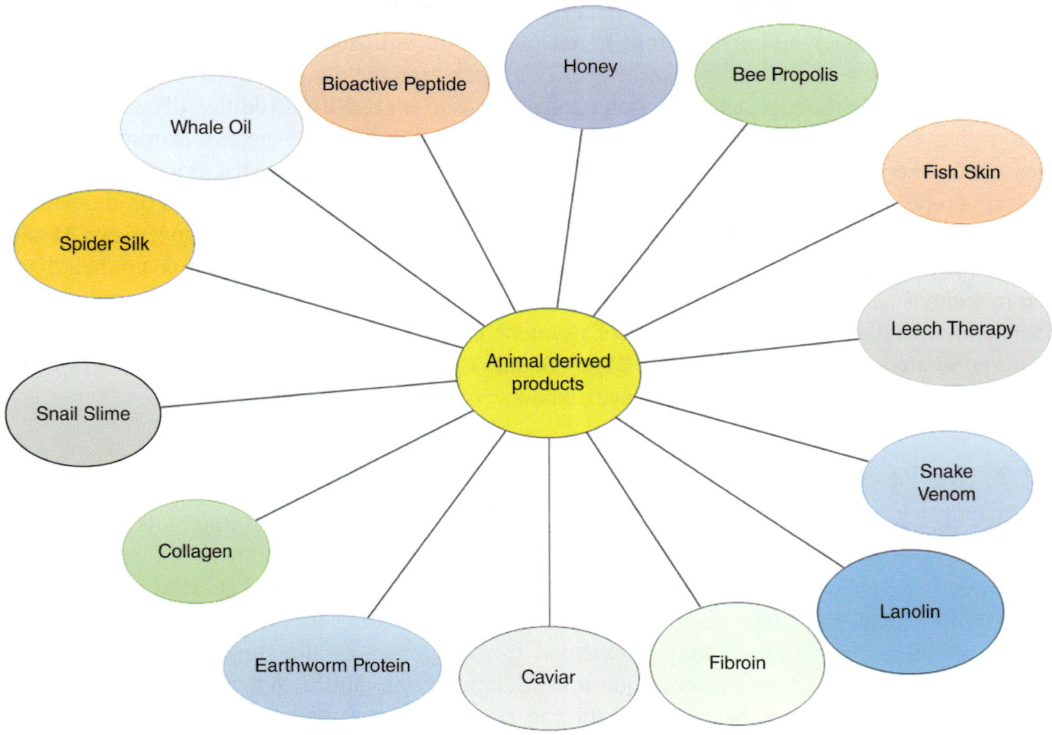

FIGURE 8.1 Important animal-derived products involved in wound healing.

et al., 2010), but it was on par with other common wound care treatments, including Intrasite Gel TM (Urai et al., 2012), Alpha® ointment (Paydar et al., 2016), and hydrocolloid dressing (Urai et al., 2012). It was also noted that the animals' wound sites had thinner scars as a result of the honey treatment (Urai et al., 2012).

Although the area of the wound decreased during the inflammatory phase, Nakajima and colleagues found that Japanese honey (*Acacia*) had few benefits because some wounds did not fully heal after 14 days (Nakajima et al., 2013). Another study was carried out the following year to look at the benefits of using hydrocolloid dressing and Japanese honey together on wound healing. The findings, which included bigger wounds, delayed re-epithelialization, and collagen deposition in the wounds, demonstrated that this involvement did not improve cutaneous wound healing (Mukai et al., 2015). To the best of our knowledge, no two honeys are same; their chemical compositions, biological characteristics, and therapeutic effectiveness vary according to the location, climate, and plant-based foods they are made from (Buba et al., 2013). In animal experiments, honey is administered differently in terms of dosage, frequency, duration, and method. Various differences could be explained by the different results found in various investigations. Furthermore, numerous human research using various wound types have confirmed the benefits of honey in wound healing. Ten patients with cutaneous wounds, ranging in age from 10 to 70, were included in the pilot trial. On day 7, honey reduced burning sensation and discoloration around the skin; on day 15, it reduced pain and swelling and showed positive effects on the development of healthy granulation (Vijaya and Nishteswar, 2012). After using the honey-coated bandage, the patients' wounds gradually shrank in size and became cleaner. Throughout the intervention period, exudation and wound discomfort did not significantly change (Lund-Nielsen et al., 2011). To evaluate the cosmetic results and wound healing qualities of honey in female patients undergoing elective plastic surgery (n = 52; ages 34–43), another randomized clinical trial was conducted. According to the results, using a honey dressing decreased the average scar width and decreased the risk of surgical problems such as severe erythema, dehiscence, and infection (Goharshenasan et al., 2016). Additionally, several studies have assessed honey's impact on burn injuries. Included were ten tangential excision patients, ages 2–60, who were dressed with Tualang honey. Tualang honey may have bactericidal and bacteriostatic properties, as evidenced by the considerable reduction in the bacterial load on the wound starting on day 6 (Nasir et al., 2010). In the second research, after 7 and 21 days, respectively, 84 and 100% of patients with superficial thermal burns (<40% TBSA) exhibited acceptable epithelialization in wounds treated with honey (Maghsoudi et al., 2011). The healing period for wounds treated with honey was likewise shorter than that of wounds treated with silver sulfadiazine in patients with first- and second-degree burns (<50% TBSA) (Baghel et al., 2009; Gupta et al., 2011). Overall, the published human research consistently revealed honey's beneficial effects.

8.2.2 Bee Propolis

Propolis "is a well-known natural material that bees collect from the bark" and buds of different trees and combine with pollen, wax, and bee enzymes. In their honeycombs, bees use propolis, sometimes called "bee glue," to smooth internal surfaces, fill gaps, and seal apertures. The Greek word "propolis" means "defense of the hive," with "pro" standing for "for" or "in defense" and "polis" meaning "city." The colony is shielded from illnesses by the sterile environment created by its antiseptic and antibacterial qualities (Salatino et al., 2005). When invaders, like other insects, die inside the hive, bees use propolis to embalm their remains to avoid decomposition (Toreti et al., 2013). Propolis has been the subject of several international investigations in recent decades, with numerous publications looking at its biological properties and chemical makeup. The composition of propolis, a complex mixture of different compounds, can vary greatly depending on the plant materials used and the period of collecting. Various factors including season, light exposure, altitude, and the availability of food and activity during collection might affect the composition of propolis (Mercan, 2006; Silici and Kutluca, 2005). Propolis' chemical composition and biological characteristics have been extensively studied; more than 300 chemical constituents have been found in propolis from different geographical areas (Bankova et al., 2000). Xu et al. (2009) state that flavonoids, phenolics, and aromatic chemicals are the main "chemical classes found in propolis."

Because of the antibacterial and anesthetic properties, propolis has been used in both orthodox and alternative medicine to cure wounds and ulcers. Because of its antiviral, antifungal, anti-inflammatory, anticancer, and antioxidant qualities, it has a wide range of medical uses (Rojczyk et al., 2020). The ancient Egyptians, Greeks, and Romans were pioneers in utilizing propolis for wound care and disinfection (Sforcin, 2007). Additionally, the ancient Egyptians incorporated propolis in the mummification process as a key preservative agent (Kuropatnicki et al., 2013). Roman soldiers utilized propolis as a first-aid remedy for battlefield injuries. During the Boer War, a formulation combining vaseline and propolis, known as propolisin vasogen, was employed for wound care.

Propolis is a significant natural resource for creating new pharmaceutical goods because it has been shown in recent tests to be safe and non-toxic for human ingestion. Wounds interrupt the continuity of the skin, which can occur due to burns, injuries, surgical cuts, or medical conditions. Microorganisms like bacteria and fungi can invade the wound area, leading to infections, hindering the healing process, and exacerbating local tissue damage. The properties of propolis, including its antimicrobial effects, ability to remodel skin tissue, and promotion of skin cell proliferation, have been associated with enhanced skin healing. Because of its advantages in wound healing, skin regeneration, purulent infection treatment, pain management, and local anesthetic, propolis has been used in traditional medicine. Propolis has shown remarkable therapeutic effectiveness across various types of wounds, including those related to diabetes, infections, surgical procedures, burns, and gastric ulcers (Chirumbolo, 2012). By reducing the inflammatory response, propolis can promote a more efficient healing process. In the early phases of wound healing, it increases the extracellular matrix (ECM) components; later, there is a noticeable decrease in these components. Additionally, propolis promotes "the expression of transforming growth factor-β, a multifunctional cytokine that is necessary for the early stages of wound healing such as inflammation, hemostasis" (Chirumbolo, 2012). Propolis is also helpful in treating burn wounds since it can reverse "the harmful effects of free radicals in the skin and tissues." By preventing the generation of reactive oxygen species (ROS), it prevents oxidative damage to biomolecules, slows down NF-κB activation, reduces eicosanoid synthesis, and lowers the expression of several inflammatory cytokines (Chirumbolo, 2012). Research conducted on diabetic rodents revealed that propolis significantly enhances the rate of wound healing (Martinotti and Ranzato, 2015). It aids in the re-epithelialization process and reduces inflammation by regulating the physiological levels of neutrophils and macrophages in the affected area, thereby preventing the chronic inflammation commonly seen in diabetes. Topical creams containing propolis demonstrated comparable outcomes (Martinotti and Ranzato, 2015).

8.2.3 Fish Skin

Fish skin can be utilized as an ingredient to cure burns and wounds because of the numerous benefits that have been described. Fish skin treatment reduces the need for painkillers and speeds up the healing of wounds (Ibrahim et al., 2020a, 2020b). Omega-3 polyunsaturated fatty acids (PUFAs), including eicosapentaenoic acid (EPA), docosahexaenoic acid (DHA), and collagen I and III, are all found in fish skin. Both EPA and DHA have antioxidant and antibacterial properties. The inadequate ability of injured tendon tissues to regenerate is one of the clinical problems. It may be possible to combine stem cells with acellular pig and cow ECM. However, using fish skin can be more effective for the reasons listed, including accessibility, lower biological risk, and religious considerations. This problem was resolved by Liu et al. (2022) using decellularized tilapia fish skin (DTFS). In addition to increasing tendon-derived stem cell (TDSC) migration and tonic differentiation, the crosslinker-modified acellular skin sample did not exhibit toxicity. The rat Achilles tendon defect model *in vivo* study's findings demonstrated that the DTFS containing cultivated cells can aid in tendon tissue regeneration. Thus, DTFS using TDSCs is a viable treatment strategy for tendon tissue regeneration (Liu et al., 2022). Since, the Food and Drug Administration (FDA) approved fish skin for use as wound dressings, it has become a common skin substitute in therapeutic settings. Furthermore, fish skin cellular and/or tissue-based products (CTPs) have shown promising outcomes in treating venous leg ulcers and other acute and chronic wounds (Fiakos et al., 2020). Fish skin CTPs have been shown to have antiviral and antibacterial properties and are a productive tool for speeding wound repair when compared to other animal-derived skin transplant materials. For example, in an acute wound scenario, fish skin grafts heal more quickly than dehydrated amnion chorionic membrane and pig intestine submucosa. It has been shown that decellularized fish skin CTPs are an appropriate option for people who might be more likely to experience hypersensitive reactions to foreign substances (Fiakos et al., 2020; Patel and Lantis, 2019). Ideally, the CTP will create a structural matrix that promotes epidermal and dermal regeneration. The treatment of diabetic foot ulcers (DFUs) is one of the frequent uses for CTPs today. Studies have also demonstrated that acellular dermal matrix (ADM) has the ability to treat acute wounds, including burn wounds (Lima-Junior et al., 2019). Burns rank as the sixth most common cause of non-fatal pediatric injuries. The WHO states that low- and middle-income countries currently have a child burn mortality rate that is more than seven times greater than that of high-income countries, with a much higher frequency.

In addition to the patient's age, general health, conditions the extent, severity, and source of the burn all affect how quickly a burn wound heals. In addition to reducing the chances of infection and fluid loss, early wound closure also reduces mortality and hospital stay duration. However, each patient's needs should be taken into consideration while making therapy recommendations. Numerous research have suggested using fish skin because it possesses nearly all of these qualities (Costa et al., 2019; Júnior et al., 2020). ADM contributes significantly to the healing of wounds on mouse skin by producing biological components involved in the healing process. Because ADM contains peptides with antimicrobial qualities, it may be utilized to heal acute wounds.

> With an even greater concentration of type I collagen, it has been shown that the morphological structure and non-infectious microbiota of tilapia skin are similar to those of human skin. These factors support its therapeutic application as a biomaterial for burn treatment.
>
> (Costa et al., 2019; Júnior et al., 2020)

An open-label, monocentric, randomized phase II pilot study was conducted on 30 children aged 2–12 who had superficial partial thickness burns. The skin of tilapia was handed to the kid's epithelialization (Júnior et al., 2020). The findings demonstrated the effectiveness and affordability of using tilapia skin to treat children's superficial partial-thickness burns. The skin of tilapia was observed to stick to the wound bed quite nicely epithelialization (Júnior et al., 2020). The patient was released following ten days of epithelialization. The results showed that fish skin may adhere to the wound bed and function effectively as a flexible xenograft. "Successful burn therapy was demonstrated by its non-toxicity, ability to act as a barrier against pathogens, and 10-day epithelialization" (Júnior et al., 2020). A 23-year-old male patient's right upper limb was repaired from a superficial partial thickness burn, while his left upper limb was repaired from a deep partial thickness burn, demonstrating the advantages of using a fish skin graft. In this instance, two sites experienced full re-epithelialization 12 and 17 days following therapy, respectively (Lima-Junior et al., 2019). NexoBrid was used to coat burns following enzymatic debridement, depending on the severity of the injury. Significant pain relief, hydration retention, and quicker wound healing were the outcomes of using fish skin dressings. Furthermore, enhanced skin elasticity, thickness, and pigmentation were shown to produce better functional and aesthetic outcomes. Patients who had fish skin treatment reported less discomfort and itching than those who had split-thickness skin graft (STSG) treatment. Above all, compared to wounds treated with Supratel, wounds treated with fish skin had more pliable skin and produced significantly more sebum (Wallner et al., 2022).

8.2.4 Leech Therapy

The mechanical and metabolic features of leeches make them one of the most natural and efficient methods for healing wounds. The application of medicinal leech therapy to post-traumatic wounds can increase tissue salvage in flap surgery or damaged tissue replantation. Additionally, it can be used for persistent lesions that don't heal, such as pressure ulcers, DFUs, and venous leg ulcers (Ghods et al., 2019). It is preferred because medical leech therapy is inexpensive, quick, efficient, simple, and useful.

Leeches have a dorso-ventral flattened body. The pieces resemble a drawer on the front and back. There is a set number of 34 segments in their bodies. They have a total of 300 teeth in their mouths, including floating teeth in each jaw and their bites are Y-shaped. Despite high chlorine content, leeches perish because they are highly sensitive to abrupt changes in their living environment, including temperature and chlorine levels. Leeches draw oxygen from the body's surface and consume it, either as air oxygen or as dissolved oxygen in water (Aloto and Eticha, 2018; Ayhan and Mollahaliloglu, 2018). Only roughly 15 of the over 800 leech species found worldwide can be categorized as medical leeches. The genus *Hirudo* in the family Hirudinea is often home to the leeches utilized in medical leech therapy. These leeches are recognized as therapeutic and non-toxic. Sterile leeches that have never been used on another person are among the most crucial considerations when selecting leeches for medicinal applications. Furthermore, the only leeches that can be utilized for treatment are those that belong to the medical leech category.

Leeche's therapeutic modes of action have been variably interpreted from antiquity to the present. In leech therapy, the leech will bite you once and then suck blood for 20–45 minutes as you bond. The primary therapeutic advantages were thought to be derived from phlebotomy, which involves feeding leeches their host's blood. "The bioactive substances in leech saliva that are released into the host's bloodstream during sucking," however, have also been shown to have therapeutic benefits (Abdullah et al., 2012). Leeches use their suction cups, which are found in the head and tail parts of their bodies, to hold onto their prey. The chemicals that the leech emits when it starts feeding cause the arteries to widen, making the blood and fluids more fluid. Following the leech bite, it promotes blood flow, protects itself (antimicrobial activity), prevents detection (analgesic and anti-inflammatory effects), inhibits adhesion, aggregation, and coagulation (inhibition of platelet function and anticoagulant effect), and develops a suction pathway (ECM degradation) (Sig et al., 2017).

The leeches leave the attachment region after the feeding session because they are completely satisfied. Leeches' digestive tract contains endoenzymes that protect blood. *Pseudomonas hirudinis* and *Aeromonas hydrophila* are examples of endosymbiotic bacteria that aid in the digestion of blood after feeding and stop it from decomposing (Abdualkader et al., 2013). Even though more than 100 bioactive substances with different molecular weights are found in leech saliva in the form of protein and peptide structures, certain of these substances have been identified as playing a major active role. The effects of hirudotherapy on flap survival and wound healing in rats and mice have been studied in animals. According to these studies, leeches promote wound healing and flap salvage while reducing necrosis by increasing blood flow in the bite site. In one study, the researchers used albino Wistar rats to examine the effects of topical phenytoin (PHT) treatment and leech therapy on the healing of incisional skin wounds. Rats in the medical leech (*Hyacinthus orientalis*) treatment group experienced a noticeably quicker rate of wound healing than rats in the topical PHT treatment group (Darestani et al., 2014). In another study, the researchers used a mouse model to examine the efficacy of ischemic preconditioning and hirudotherapy in flaps that had been subjected to ischemia for an extended period of time. The flap survival rate rose to an average of 88%, and the rate of necrosis reduced satisfactorily, particularly when the hirudotherapy group was contrasted with the control group (Moosavian et al., 2014). Although leech therapy is typically considered safe, it may carry risks such as bleeding, infection, and anemia. This treatment might not be appropriate for every patient, especially those who have bleeding disorders. Further research is required to establish the best practices for leech therapy, investigate the possible advantages of various leech species, and evaluate the treatment's long-term efficacy.

8.2.5 Snake Venom

There are over 100,000 recognized venomous animal species worldwide. Each of these species has the ability to produce venoms that often contain over 100 different molecules. Animal venoms are used effectively for defense and hunting. Venoms are not composed of isolated toxins but of complex chemical mixtures of pharmacologically active components, including proteins, peptides, and enzymes with specific biological activities, as well as some non-protein compounds such as carbohydrates, lipids, metal ions, and other yet unidentified substances. Nevertheless, venom toxicity is usually associated with a few toxins present in the venom. Bites or stings from certain venomous animals can cause acute envenomation that can lead to death. The administration of an antivenom is the most effective treatment for envenomation according to WHO recommendations. To optimize the effectiveness of these treatments, it is essential to have a better knowledge of the composition of the venom. Venoms are emitted and transported by well-developed exocrine venom glands, integrated with a delivery system, which have several crucial functions for these animals, including the capture, elimination, and digestion of their prey. They can also serve as a defense mechanism against potential enemies. An intriguing and newly developing field of biomedical research is the function of snake venom in wound healing. Even though snake venom is often thought to be dangerous, several of its bioactive ingredients have demonstrated encouraging therapeutic benefits for tissue regeneration and repair. Snake venom contains a wide array of biologically active molecules such as the following:

- Disintegrins
- Metalloproteinases (SVMPs)
- Phospholipases A_2 (PLA_2s)
- Growth factor-like peptides
- L-amino acid oxidases (LAAOs)

These components can modulate inflammation, cell migration, angiogenesis, and tissue remodeling all of which are essential for wound treatment. In recent decades, studies have revealed that snake venoms also contain a large amount of non-enzymatic proteins. In contrast to the enzymatic proteins found in snake venoms, non-enzymatic proteins generally play a role in immobilizing prey (Chen et al., 2018). Typically, they interact with specific membrane receptors, ion channels, or plasma proteins, thereby disrupting the physiological processes of the prey, leading to neurotoxic and cardiotoxic effects (McCleary and Manjunatha Kini, 2013). They are classified into different protein families based on their amino acid sequences and structure. Some of the well-established non-enzymatic protein families in snake venoms include the following: (i) cysteine-rich secretory proteins (CRISPs) or helveprins; (ii) snaclecs (C-type lectins and related proteins); (iii) proteinase inhibitors; (iv) nerve growth factors; (v) bradykinin-potentiating peptides; (vi) natriuretic peptides; (vii) three-finger toxins; (viii) sarafotoxins; (ix) cobra venom factors; (x) vascular endothelial growth factors; (xi) vespryns; (xii) waprins; and (xiii) veficolins.

8.2.5.1 Mechanisms Involved in Wound Healing by Snake Venom

Following processes are involved in wound healing by snake venom as mentioned below:

1. *Inflammatory Phase Modulation*: As a result of venom's L-amino acid oxidase (LAAO), hydrogen peroxide (H_2O_2) functions as a pro-inflammatory and antibacterial signal. By producing metabolites of arachidonic acid that helps in the recruitment of immune cells, PLA_2 enzymes also serve to modulate inflammation (Salama et al., 2024).
2. *Angiogenesis (Formation of New Blood Vessels)*: Vascular endothelial growth factor (VEGF) like proteins found in snake venom imitate genuine VEGF and encourage capillary development into injured tissue. For tissue regeneration, this improves the delivery of oxygen and nutrients (Moura-da-Silva and Baldo, 2012).
3. *Cell Migration and Proliferation*: Disintegrins facilitate keratinocyte and fibroblast migration, which is necessary for wound closure, by attaching to integrins on cell surfaces. ECM can be broken down by *Snake Venom Metalloproteinases* (SVMPs), creating room for new cell development and promoting re-epithelialization (Swenson et al., 2007).
4. *ECM Remodeling and Tissue Repair*: SVMPs facilitate the deposition of new tissue by breaking down damaged collagen and matrix proteins. By balancing matrix synthesis and degradation, they also aid in scar reconstruction (Fox and Serrano, 2009).
5. *Antibacterial and Antioxidant Effects*: The broad-spectrum antibacterial qualities of PLA2s and LAAOs lower the danger of infection during the healing process. Additionally, their slight stimulation of oxidative stress causes cells to respond adaptively, improving survival and repair.

8.2.6 Lanolin

Since ancient times, lanolin—a complex blend of fatty acids, hydrocarbons, sterols, aliphatic alcohols, and high molecular weight esters—has been utilized for its moisturizing qualities. Beginning in 1882, crude lanolin was purified into lanolin wax, which was then processed into a variety of derivatives. This process is still ongoing today, with newer, more purified anhydrous lanolins.

The fat from sheep's wool is used to make it. Lanolin, also known as wool wax, wool yolk, or wool grease, is an oily substance that sheep's sebaceous glands secrete under their skin. An agricultural compound derived from sheep's wool is called lanolin. It is golden to pale brown in color, according to research conducted 2000 years ago. It is creamy or greasy in texture (Wang et al., 2020). Lanolin comes in two varieties; hydrous lanolin contains between 25 and 30% water. Anhydrous lanolin has a mild odor and a melting point between 38°C and 42°C. These components lubricate the skin's surface, giving it a smooth and silky appearance. Lanolin mixes well with other ingredients in cosmetic and personal care products and aids in the formation of emulsions. There are a number of technical uses for lanolin, including the pharmaceutical industry, cosmetics (particularly lanolin derivatives), personal care items, and industrial sectors. Only topical applications are permitted for the lanolin (Nozimoto et al., 2024). In water, it may absorb more than 200% of its own weight. Emollient, barrier-repairing, anti-inflammatory, antimicrobial, and skin-protective properties are all present in lanolin. It has a lengthy history of beneficial use, is safe to use topically, and has a hydrating effect. Lanolin creates an air-permeable barrier when it is applied to wounded skin, which promotes moist wound healing. Additionally, when applied to an eczematous hand, it creates an air-permeable barrier and encourages moist wound healing. According to Yuksel et al. (2021), it has antibacterial, anti-inflammatory, skin-protective, and barrier-repairing properties. A statistically significant improvement in the epithelization rate, dermal thickness, and an increase in the number of cells in the dermis were observed, which were attributed to the exclusive effect of lanolin cream (Chvapil et al., 1988). A topical preparation combining lanolin, petrolatum, and hyaluronic acid (HA) was developed and evaluated *in vivo* (Figueira et al., 2022). The HA-containing preparation revealed an improvement in the healing process, with a reduced wound area and an increased collagen content at day 14, compared to the ointment without HA (Figueira et al., 2022). A comparative study demonstrated that human-processed lanolin (HPA lanolin) had a notable effect on reducing pain and promoting healing of nipple trauma associated with breastfeeding (Neto et al., 2018).

Lanolin's occlusive and emollient properties make it an important asset in wound healing, especially for superficial wounds and dry skin conditions. Its effectiveness is confirmed by clinical research, especially when combined with other substances such as HA. Nevertheless, it is prudent to remain vigilant regarding possible allergic reactions, and this should not replace professional medical care for serious or infected wounds.

8.2.7 Fibroin

A ubiquitous protein polymer, silk, is made by different insects like bees, spiders, and silkworms. The main sequences, structural characteristics, and fiber sequences of silk fibers from different species vary. Two varieties of silkworms—the domesticated mulberry and the wild non-mulberry—produce the majority of silk. The *Bombyx mori* family of farmed silkworms produces mulberry silk. The main sources of wild variety silk are Eri (*Philosamia ricini/Samia ricini*), Tussar (*Antheraea mylitta*), and Muga (*Anthrraea asana/assamensis*) (Gupta et al., 2021). Compared to silkworm silk, spider silk is more mechanically strong. However, domestic *B. mori* filaments are more commonly used in the commercial silk industry due to the scarcity of spider silks (Murugesh, 2017). Only silk from silkworms has historically been produced in huge quantities, and it has recently been exploited as a biomaterial due to its durability and availability (Vepari and Kaplan, 2007). The two proteins present in *B. mori* silkworm silk are fibroin and sericin. About 70% of the silk's weight is made up of the former, which is found inside the thread, while the remaining 30% is made up of the thin layer that envelops the interior thread. It also contains fat/wax (0.8–1%) and color/ash (1–1.4%) (Belhaj et al., 2012). "The middle and anterior glands of the silkworm release sericin, whereas the posterior gland produces fibroin." The US FDA has approved the use of silk, a natural protein polymer, for medical purposes. The outermost silk sericin, which can trigger an immunological response when paired with fibroin, is extracted from mulberry silk to create silk fibroin (Altman et al., 2003; Wang et al., 2014). The two exocrine silk glands (located on both sides of the body) concurrently link the spinnerets together with sericin as the larvae spin by releasing two extremely thin (10 μm diameter) fibroin double strands through them. According to Inoue et al. (2000), the presence of air makes the protein fiber stronger and more resilient.

Fibrin, which makes up the majority of silk, is the inner core and provides mechanical strength, while sericin acts as the glue-like outer layer. Each silk fiber has two sericin-coated silk filaments (Koh et al., 2015). Silk fibers are thought to be composed of nanofibers, which are the building blocks of silk and have a diameter of 3.5 nm. These nanobrils, which are 20–200 nm in diameter, entwine and interact strongly with one another to produce larger microbril units (Ling et al., 2018). The arrangement of micro- and nanofibrils is parallel to the silk filaments. Silk fibroin, a key structural protein of silk, is composed of polypeptide chains with molecular weights between 200 and 350 kDa. The main structural component of silk fiber is an H–L complex, which is composed of a heavy (H-) and light (L-) chain polypeptide connected by a single disulfide bond at the C-terminus of the H-chain. Additionally, the structure of silk fiber is present and helps to its overall integrity through the glycoprotein P25, which is noncovalently attached to the H–L-chains. The fibrous qualities of silk fibers are mostly due to the H-chain. According to Yu et al. (2017), glycine (46%), serine (12%), and alanine (30%) are the most prevalent amino acids in this chain. Most of the acidic amino acids, including valine, leucine, and isoleucine, are located in the nonfibrous L-chains (Shimura, 1983).

The skin protects the human body from the environment, dehydration, and pathogenic agents. According to Singer and Clark (1999), wound healing is a complicated process that involves interactions with several cells and matrices as well as several overlapping phases, such as inflammation, the formation of new tissue, and tissue remodeling that all take place simultaneously. Inflammation, which happens soon after an injury and can last for up to two days, is the first stage of wound healing. The immune system, inflammatory pathways, and coagulation cascade must all be activated in order to stop infection, eliminate dying and dead tissues, and stop more fluid and blood losses. According to Gonzalez et al. (2016), "phagocytosis and the generation of many cytokines and growth factors are two crucial functions of inflammatory cells such as neutrophils and macrophages in the healing process." New tissue development, which is associated with collagen/matrix deposition, re-epithelialization, angiogenesis, and wound contraction, is the second stage of wound healing (Gonzalez et al., 2016; Midwood et al., 2004). "Rebuilding the extracellular matrix and epidermis occurs in conjunction with the remodelling or maturation phase of the wound healing process, which can continue for a year or more, depending on the extent of the injury" (Gurtner et al., 2008; Midwood et al., 2004).

8.2.7.1 Different Forms of Silk Fibroin

8.2.7.1.1 Silk Fibroin Solution

The silk fibroin solution is made by reducing the silk cocoon in an aqueous solution of 0.02M Na_2CO_3 for 40 minutes of boiling at 95°C, and then using distilled water to remove the glue-like sericin proteins (Chung et al., 2015; Lee et al., 2014; Martinez-Mora et al., 2012). Next, degummed silk fibroin is dissolved in a molar ratio of 1:2:8 using water, ethanol, and calcium chloride for 50 minutes at 98°C. To create silk fibroin solutions, this solution is filtered over a dialysis membrane for three days. According to Park et al. (2018), silk fibroin solution dramatically accelerated healing both *in vitro* and *in vivo*. Furthermore, rabbit corneal epithelial lesions have been treated with soluble silk protein given as an eye drop to promote healing (Abdel-Naby et al., 2017).

8.2.7.1.2 Nanofibrous Scaffolds from Silk Fibroin

An interesting kind of skin substitute composed of silk fibrin is called a silk fibrin scaffold. According to Lee et al. (2014) and Gholipourmalekabadi et al. (2015), electrospinning produces a pseudo-three-dimensional structure that facilitates cell attachment and proliferation. Two-dimensional scaffolds created with electrospinning technology are crucial to the creation of skin substitute goods that contain all the ingredients required for efficient skin regeneration and wound healing (Ige et al., 2012).

8.2.7.1.3 Hydrogels from Silk Fibroin

Hydrogels are three-dimensional networks of polymers having a high swelling ratio in an aqueous solution. Because of their generally good biocompatibility, hydrogels are usually generated from naturally occurring polymers like chitosan (Murakami et al., 2010), alginate (Bouhadir et al., 2001), collagen (Niiyama and Kuroyanagi, 2014), hyaluronic acid (Burdick and Prestwich, 2011), and silk fibroin (Ju et al., 2014; Lee et al., 2017). Numerous biomedical applications, such as wound healing, have made extensive use of hydrogels (Drury and Mooney, 2003). In addition to promoting cell adhesion, growth, migration, proliferation, and ECM formation, silk fiber has the potential to increase the mechanical strength of hydrogels derived from other natural polymers (Kapoor and Kundu, 2016).

8.2.7.1.4 Silk Fibroin Sponges/Blend Films

Silk fibroin mixes, such as sponges and films, can improve wound healing. Regenerated silk fiber solutions have also been used to create porous sponges, which are crucial materials for tissue engineering (Liu et al., 2008; Shen et al., 2010). "By providing a network of linked pores with a wide surface area inside a predefined three-dimensional space, sponge scaffolds do facilitate tissue development and cell adhesion." According to Lee et al. (2017), the collagen hybrid sponge made of silk and duck feet may be used as a dermal replacement for full-thickness skin defects. In order to treat DFUs, Liu et al. (2019) employed a freeze-drying technique to produce a silkfibroin sponge that would allow neurotensin-loaded gelatin microspheres to be released gradually.

8.2.7.2 Silk Fibroin's Mechanism of Wound Healing

According to Infanger et al. (2019), silk fibroin promotes wound healing by activating the NF-κB signaling pathway. Through complex signaling networks, NF-κB controls a range of cell activities, including as adhesion, proliferation, inflammation, and the elimination of ROS. Accordingly, it is thought that NF-κB signaling is important for the healing of corneal epithelial wounds as well as many other wounds (Wang et al., 2013). Toll-like receptors (TLRs) and tumor necrosis factor receptor (TNFR), two significant NF-κB mediators, were expressed greater in cells activated by silk fibroin. They also promoted the expression of many different proteins, such as cyclin, fibronectin, VEGF, vimentin, transforming growth factor (TGF), interleukin-1b (IL-1b), interleukin-10 (IL-10), and epidermal growth factor (EGF), which are regulated by NF-κB and govern wound healing. According to Park et al. (2018), silk fiber enhanced wound healing by increasing NF-κB signaling (Infanger et al., 2019). Silk fiber offered protection "in a rat model of burn injury" by obstructing the apoptotic pathway (Aykac et al., 2018). Numerous intricate cellular signaling pathways, such as Wnt and Notch signaling (Shi et al., 2015), TGF beta (TGFβ) signaling (Cheng et al., 2016; Shi et al., 2015), mitogen-activated proteinkinase (MAPK) signaling (Chen et al., 2016), and AKT/mTOR signaling (Xing et al., 2015), are intricately coordinated during the wound healing process. MAPK signaling (Thuraisingam et al., 2010) and AKT/mTOR signaling (Xing et al., 2015) are two biological pathways implicated in wound healing.

8.2.8 Caviar

A luxury meal, caviar is derived from the eggs of sturgeon (Acipenseridae family) and contains a variety of biomolecules and nutrients. Many studies have been published that look at the lipid composition of caviar since it was proposed that the lipids from fish eggs are essential nutrients for humans and that different lipid classes have different roles in animals. These include the caviar's high PUFAs, DHA, and EPA,which have been associated with better skin health (Kendall et al., 2017; Yoshida et al., 2016).

Caviar is a popular, rich-value seafood product that is very abundant in nutrients, particularly fatty acids, when compared to other fish. Since DHA and EPA are the primary constituents of phospholipids, the caviar had high levels of PUFAs, even though the composition of fatty acids can vary depending on factors like species, habitats, and cultivation conditions (Abbas and Hrachya, 2015; Caprino et al., 2008; Mol and Turan, 2008; Shin et al., 2010). Numerous studies have shown that PUFAs can cure cardiovascular and inflammatory illnesses as well as skin conditions similar to atopic dermatitis when consumed through diet (Kiecolt-Glaser et al., 2012; Yoshida et al., 2016). Compounds from marine organisms have been shown in several recent studies to have anti-aging properties for the skin. In human dermal fibroblasts (HDFs), for instance, caviar extract from beluga sturgeon combined with selenium and CoQ10 has been found to enhance collagen type I expression (Marotta et al., 2012). Additionally, fatty acids found in caviar have been shown to influence adipocytes' production of adiponectin, which is predicted to benefit skin (Song et al., 2017). These findings demonstrate that caviar extract influences the synthesis of adiponectin, which enhances skin health. Furthermore, in differentiated adipocytes, caviar extract and DHA enhanced the expression of adiponectin, which may reduce skin photoaging by inhibiting the expression of matrix metalloproteins-1 (MMP-1).

8.2.9 Earthworm Protein

Earthworms (*Lumbrucis terrestris*) are decomposers in the global ecosystem and are members of the annelida phylum in the animalia kingdom. Eighteen amino acids, fatty acids, microelements, lumbritin, lumbrofebrin, terrestrolum brolysin, purine, choline, cholesterin, and vitamins are all present in the coelomic fluid that fills the segmented earthworm's body cavity (Zhang et al., 2006). According to Balamurugan et al. (2008), indigenous people all over the world, but especially in Asia, including China, India, Myanmar, Korea, and Vietnam, have long extracted and used physiologically active substances from earthworms. According to earlier research, earthworms have antiulceral (Prakash et al., 2007), antioxidative (Balamurugan et al., 2008), hepatoprotective (Balamurugan, 2007), anti-inflammatory (Balamurugan et al., 2007), antimicrobial, anticancer (Cooper et al., 2004), antiapoptotic (Han et al., 2014; Jones et al., 2016), anticoagulation, and fibrinolytic properties (Mihara et al., 1991). The speed at which wounds heal may be accelerated by all of these characteristics and modes of action.

Earthworms are often used as a model to study wound healing and the effects of spidroin protein, as well as UV-A and UV-C light (Abd El-Aziz et al., 2022). Furthermore, earthworm tissue extract has been found to have the remarkable ability to enhance wound healing in mice (Ponesakki et al., 2023). This is done by promoting collagen synthesis, fibroblast proliferation, blood capillary formation, and reducing inflammation (Deng et al., 2018), and similarly, earthworm coelomic fluid accelerates wound healing potential in rabbits (Singh et al., 2015).

In ancient China, physicians used discarded earthworms to treat a variety of common ailments such as burns, arthritis, itching, and inflammation (Hrzenjak et al., 1998). With the advancement of modern science, several bioactive components with medical functions have been extracted from earthworms, including G-90. Recent research by Yang et al. (2017) demonstrated that an extract from the regenerative tissue of the head of *Eisenia fetida*, designated as "G-90," is capable of activating the proliferative phase of the acute wound healing process in both *in vivo* and *in vitro* settings. Grdisa et al. (2001) demonstrated that G-90 was able to promote cell proliferation and migration by increasing the expression of growth factors associated with the wound healing process. The earthworm *Eudrilus eugeniae* is a fascinating species that uses its segmented clitellum to ensure effective regeneration. If the clitellum segments are not intact, the amputated body segments cannot regenerate (Christyraj et al., 2020). The clitellum and the factors it release play a fundamental role in determining survival, regeneration, and differentiation capabilities (Paul et al., 2022). Proteins derived from tail-amputated *E. fetida* showed positive effects on wound healing (Li et al., 2023). These proteins stimulated the expression of basic fibroblast growth factor (bFGF) and activated the Ras/Raf/MEK/ERK signaling pathway, which promoted cell migration and wound closure (Li et al., 2023). Proteomic research has revealed specific proteins in the ES2 fraction of earthworm extract (EE) that promoted the proliferation of fibroblasts and keratinocytes (Yang et al., 2017). In addition, these proteins activated the expression of bFGF and VEGF, which are crucial for tissue regeneration during the wound healing process (Yang et al., 2017). A systematic review and meta-analysis involving different animal models (mice, rats, rabbits) found that EE significantly increased wound healing rates, decreased healing time, and improved tissue regeneration compared to control groups (He et al., 2021). In deep second-degree burn models, treatment with EE resulted in faster wound healing, reduced edema, increased collagen content, and improved histological findings, indicating its potential as a therapeutic agent for burns (He et al., 2021).

In summary, earthworm-derived proteins could become an important natural therapy in wound care, especially for burns, diabetic ulcers, and chronic wounds, pending further clinical validation.

8.2.10 Collagen

The collagen being the most prevelant protein is essential for preserving the skin's physiological function and structural integrity (Chhum et al., 2016; Qi et al., 2017). Collagen has been accepted as a prime focus in skin tissue engineering because of its wonderful qualities, which include minimal antigenicity, great biocompatibility, and exceptional bioactivity (Mondal et al., 2018). Over the past few decades, remarkable progress has been made in the development of innovative collagen dressings for the treatment of chronic wounds (Bohn et al., 2016). Collagen wound dressings can be classified as films, sponges, hydrogels, nanofibers, and powders based on the various scaffold material formulations (Gaspar-Pintiliescu et al., 2019; Rastogi et al., 2009).

8.2.10.1 Collagen Films

These offer a physical barrier to infection and a protective layer; collagen films have been used extensively as wound dressings (Chattopadhyay and Raines, 2014). Collagen films made by freeze-drying or wind-drying are pliable and flexible, which improves their adherence to the wound and makes them comfortable to wear. Furthermore, because they are thin and transparent, it is easy to observe the wound and track the healing process more precisely (Abdelrahman and Newton, 2011). The size, kind, and location of the wound may affect how long the film dressings need to be changed, and they can be kept for up to seven days (Weller et al., 2020). Collagen films are effective for superficial wounds and abrasions, but because of their poor fluid absorption capacity, wound fluid accumulates beneath the dressing, allowing exudate to infiltrate and foreign microorganisms to penetrate the wound surface. For larger wounds, they are therefore inconvenient. Additionally, for an effective collagen film seal, the skin surrounding the wound needs to be intact.

8.2.10.2 Collagen Sponges

Collagen sponges are powerful wound dressings that can effectively absorb blood and wound exudate due to their highly interconnected and porous structures (Cheng et al., 2017; Sudharsan et al., 2013). In addition to promoting hemostasis and healing and lowering the risk of skin immersion, collagen sponges have the potential to permeabilize water vapor and air, which can create a humid environment and have a heat-insulating effect (Chen et al., 2019; Huang et al., 2020). Collagen sponges are a type of porous scaffold material that is created by freeze-drying and has a high pore diameter, good mechanical strength, and good plasticity. By altering the collagen content and the freeze-drying process, the porosity of collagen sponges can be changed (Chattopadhyay and Raines, 2014). Wounds with a lot of exudate, wounds that are partially or completely thick, and transplant donor sites can all benefit from the usage of collagen sponges. As soon as the sponge dressings are saturated with absorbed exudate, they must be changed; this can happen every day or just once or twice a week (Rahmani and Mooney, 2018).

8.2.10.3 Collagen Hydrogels

The perfect wet environment for healing dry wounds is provided by collagen hydrogels, which are cross-linked scaffolding materials that contain 70–90% water. The semi-permeable nature of collagen hydrogels to gas and liquid facilitates the autolysis debridement of necrotic tissue and provides a moist environment for the wound (Rahmani and Mooney, 2018). Collagen hydrogels' soft texture makes it possible to cut them to better fit the wound, and their transparency makes it possible to see the wound without taking off the dressing (Rahmani and Mooney, 2018). Furthermore, their ineffective performance leaves no residue or harm when replaced, and their moderate cooling action can greatly lessen surgical pain and inflammation (Raafat et al., 2018). To meet the demands of various skin injuries, collagen can undergo chemical and physical cross-linking to create hydrogels with a range of mechanical strengths (Sarrigiannidis et al., 2021). Hydrogel dressings should normally be changed every one to three days, depending on the wounds' level of hydration (Thirumaleshwar et al., 2012).

8.2.10.4 Collagen Nanofibers

Collagen nanofibers are electrospun wound dressings that successfully prevent excessive exudate accumulation and wound dryness because of their medium water vapor permeability. Collagen nanomaterials are 3D biomaterials with a size range of 1–100 nm that can be made using a number of techniques, such as melt-blowing, self-assembly, template synthesis, phase separation, and electrospinning (Lo and Fauzi, 2021). Due to its affordability and ease of use, electrospinning technology is the most widely used technique for creating collagen nanomaterials. It can also be used to create collagen nanofibers (Lannutti et al., 2007). Strongly promoting gas diffusion, cell proliferation, adhesion, migration, and wound healing are collagen nanofibers with a specific surface area and high porosity (Mbese et al., 2021).

8.2.10.5 Collagen Powders

Collagen powders are ground-up wound dressings that have a greater surface area for interaction with the wound and provide a better covering. Collagen powders can be employed as a drug delivery method because of their tiny particle size and large specific surface area, which increase their activity and speed up the release of adsorbed medicines (Segalova et al., 1981). In the meanwhile, secondary wounds such as pressure sores, diabetic ulcers, venous stasis ulcers, second-degree burns, and post-radiation dermatitis can be effectively treated with commercial collagen powders made from bovine cartilage (Kallis and Friedman, 2018). Collagen dressings are essential for halting bleeding, reducing inflammation, encouraging angiogenesis, and accelerating tissue repair. Collagen wound dressings have been shown in clinical investigations to effectively heal various chronic wounds, significantly lowering patient suffering and significantly reducing treatment duration. Collagen dressings come in a variety of forms, each with pros and cons that should be carefully considered based on the wound's condition.

8.2.11 Snail Slime

The discovery of natural adhesion phenomena and mechanisms has led to the development of a new generation of tissue adhesives in recent decades. Snails are terrestrial mollusks that release mucus when they are agitated or crawling. The mucus secreted by snails allows them to keep consistent attachment to the "smooth surfaces of trees and rocks during their crawl" (Denny, 1984). Over 2000 years ago, burns, abscesses, and other lesions were treated using land snail mucus (Greistorfer et al., 2017). The natural sticking behavior and vague bioactivity of snail mucus suggest that it could be a natural adhesive for wound healing. Snail slime, also known as snail mucin, has garnered significant interest in the fields of dermatology and wound healing because of its abundant bioactive compounds. It is rich in glycoproteins, hyaluronic acid, antimicrobial peptides, glycolic acid, allantoin, and antioxidants, all of which play a vital role in skin regeneration and repair. Snail mucin helps to decrease inflammation at wound locations by regulating cytokine production and preventing excessive immune reactions that could hinder healing. It promotes fibroblast activity, leading to enhanced collagen deposition, which is crucial for the remodeling of the ECM and the tensile strength of healed tissue. Snail slime demonstrates antimicrobial characteristics against different bacterial strains, thereby lowering the likelihood of wound infections. Hyaluronic acid and glycoproteins found in snail slime assist in moisture retention, creating a moist wound environment that encourages re-epithelialization. Additionally, snail mucin can boost angiogenesis, aiding in the formation of new blood vessels essential for delivering nutrients to healing tissues.

A research investigation into the extract of *Helix aspersa* (a widely recognized garden snail) revealed that it notably sped up the healing process of wounds and diminished inflammation in a rat model of burns (Tsoutsos et al., 2009). The mucus of the *H. aspersa* Muller snail is composed of glycolic acid and lactic acid, which are present at an estimated concentration of 0.1%. Additionally, it contains allantoin, which makes up 0.5% of the mucus weight, along with collagen and elastin (0.3%), glycosaminoglycans (GAG), lectin, antimicrobial peptides (AMPs), and protease inhibitors (ranging from 1.3 to 1.8%), as well as mucins (Greistorfer et al., 2017; Gubitosa et al., 2020). This snail filtrate possesses therapeutic properties that are applicable in both cosmetic and medical fields (Greistorfer et al., 2017; Gubitosa et al., 2020). In the realm of cosmetics, it promotes the production of collagen, elastin, and other skin elements that aid in the repair of photoaging signs (Greistorfer et al., 2017; Gubitosa et al., 2020). The application of snail mucin gel on the skin of rats resulted in enhanced angiogenesis, indicated by an increased count of new blood vessels in the treated areas, which implies better oxygenation and nutrient supply to the tissue (Herman et al., 2024). The mucus produced by *H. aspersa* might influence a range of skin processes in a multifaceted manner by encouraging fibroblast proliferation (Trapella et al., 2018), stimulating healing and regenerative processes (Aouji et al., 2023), and protecting against transepidermal water loss (TEWL). Studies indicate that the application of snail mucus gel on the skin can boost angiogenesis during the wound healing process in rats (Rosanto et al., 2021). The glycosaminoglycan obtained from snail mucus significantly accelerated the healing of full-thickness wounds in diabetic mice by alleviating inflammation and dermal edema, while also promoting angiogenesis (Wu et al., 2020). The hydrogel that includes snail glycosaminoglycan and methacrylated gelatin enhances diabetic wound healing through the suppression of inflammatory cytokines and the polarization of macrophages (Zhou et al., 2023). The ophthalmic solution (GlicoPro®) obtained from snail mucus revealed anti-inflammatory and analgesic properties, in addition to inducing regeneration and bio-adhesivity in corneal cells (Mencucci et al., 2021). It may have a vital role in the treatment of dry eye disease (Mencucci et al., 2021). Furthermore, snail mucus has exhibited therapeutic promise against a range of cancers, including breast cancer (El-Ouar et al., 2013), colorectal adenocarcinoma (Georgieva et al., 2020), prostate cancer (Smith and Odero-Marah, 2012), melanoma (Ellijimi et al., 2018), and erythromyeloid leukemia (Stabili et al., 2015).

Snail slime exhibits potential advantages for wound healing, providing anti-inflammatory, antimicrobial, and regenerative effects. While initial research and pre-clinical studies back its effectiveness, further high-quality clinical trials involving humans are necessary to standardize its use in medical settings.nic acid, glycolic signs.,

8.2.12 Spider Silk

Spider silk was first used as a wound treatment in ancient Roman medicine. In a thorough analysis, Newman and Newman discussed the historical uses of spider silk in medicine. Numerous uses were in dermatology, including as a treatment for skin lesions and as a styptic. But these early uses were empirical rather than grounded on fundamental research. The remarkable mechanical qualities, biocompatibility, and potential for wound healing of spider silk proteins—especially spidroins—have attracted a lot of interest in biomedical research. These proteins, which are mostly produced from the silk glands of spiders, are excellent candidates for tissue engineering and regenerative medicine applications due to their exceptional strength, flexibility, and biodegradability (Baoyong et al., 2010).

The bioactive motifs found in spider silk proteins, including fibronectin, can be functionalized to improve cell adhesion, migration, and proliferation. For example, it has been demonstrated that recombinant spider silk proteins linked with cell-binding domains promote the proliferation of different human cell types, such as fibroblasts and keratinocytes, hence aiding in tissue regeneration (Lee et al., 2022). Spider silk scaffolds have been shown to promote collagen deposition and the formation of new blood vessels, both of which are essential for effective wound healing. In rat models, spider silk protein-functionalized scaffolds produced more growth factors, including fibroblast growth factor (bFGF) and hydroxyproline, indicating better collagen synthesis and tissue remodeling (Baoyong et al., 2010). The natural antibacterial qualities of spider silk proteins aid in preventing wound infections. A safer healing environment can be achieved by topically applying spidroin-containing gels, which have demonstrated inhibitory effects against clinical bacterial isolates, including methicillin-resistant *Staphylococcus aureus* (MRSA) (Chouhan et al., 2019). The proteins in spider silk are biodegradable, which lowers the chance of chronic inflammation by ensuring that they do not remain in the body. According to Abd Ellah et al. (2019), their biocompatibility facilitates the material's integration with native tissues, reducing unfavorable immune responses and encouraging the body's natural healing processes.

Deep second-degree burn wounds can now be covered with dressings made of spider silk, which promotes skin regeneration and speeds up healing. According to Chouhan et al. (2018), these dressings promote cellular activity and increase the expression of growth factors that are necessary for tissue repair. Bioadhesives made from injectable spider silk have been developed for quick closure of surgical wounds. These adhesives have advantages over conventional suturing techniques due to their better bonding properties and biocompatibility (Brankovic et al., 2024). To treat full-thickness burn wounds, functionalized silk scaffolds have been used as readily available grafts. These grafts increase healing results by encouraging collagen deposition, vascularization, and re-epithelialization (Baoyong et al., 2010). Optimizing the manufacturing of recombinant spider silk proteins to make them more affordable and scalable for therapeutic applications is the main goal of ongoing research. The qualities of biomaterials derived from spider silk could be improved by developments in genetic engineering and synthetic biology, which could result in more efficacious treatments for a range of wounds and tissue ailments.

8.2.13 Whale Oil

Whale oil, particularly cod liver oil, has long been used to aid in wound healing due to its richness of bioactive compounds, including vitamins and omega-3 fatty acids. Although there are few direct research on the wound healing mechanisms of whale oil, comparable marine oils and their constituents can provide valuable insights. It has been demonstrated that applying cod liver oil ointment topically speeds up the neovascularization and epithelialization of experimental wounds. Wounds treated with a 25% cod liver oil ointment epithelialized considerably more quickly (mean 8.9 days) than controls (mean 13.9 days) in a study using hairless mice. Furthermore, the lesions treated with cod liver oil showed faster neovascularization, suggesting an enhanced blood supply that is necessary for tissue regeneration (Terkelsen et al., 2000). The anti-inflammatory qualities of cod liver oil might be part of its ability to heal wounds. By regulating the inflammatory response, the oil's constituents can lessen excessive inflammation, which can obstruct healing. Cod liver oil balances pro-inflammatory and anti-inflammatory mediators, promoting the resolution phase of inflammation and facilitating the shift to tissue repair and regeneration (Silva et al., 2018). Collagen synthesis is known to be influenced by omega-3 fatty acids found in cod liver oil, including DHA and EPA. By influencing the activity of fibroblasts, the cells that make collagen, these fatty acids can promote the production of ECM and maintain structural integrity while wounds heal. Because oxidative stress damages cells and tissues, it can impede the healing process. Antioxidants included in cod liver oil can counteract free radicals, minimizing oxidative damage and creating a healing environment.

To promote healing and provide bioactive compounds to the site of injury, cod liver oil has been made into ointments that may be applied directly to wounds. Cod liver oil is occasionally used to wound dressings in order to combine the dressing's protective qualities with its ability to promote healing. By supplying vital nutrients that aid in general tissue regeneration, oral use of cod liver oil may promote systemic healing processes.

Although there is little concrete proof that whale oil aids in wound healing, comparable marine oils like cod liver oil have shown promise. Their omega-3 fatty acid and vitamin content, which reduce inflammation, encourage collagen formation, and improve tissue healing, is mostly responsible for these effects. To completely grasp the potential of whale oil in wound healing applications, more research is required (Terkelsen et al., 2000).

8.2.14 Bioactive Peptides

It has been demonstrated that bioactive peptides facilitate tissue repair, lower inflammation, and quicken the regenerative processes necessary for wound closure, making them essential for wound healing. These peptides can be obtained from different sources like endogenous proteins, dietary proteins, and ECM proteins.

Pro-inflammatory cytokines including TNF-α, IL-6, and IL-1β are essential for the initial phases of wound healing and can be modulated by bioactive peptides. Collagen-derived peptides, for example, offer anti-inflammatory properties that lessen excessive inflammation, which may otherwise hinder tissue repair. Specific peptides facilitate the activation and recruitment of macrophages, which are essential for eliminating infections and debris and promoting the subsequent phases of healing. By encouraging the release of VEGF, bioactive peptides can promote angiogenesis. For the regenerative process to proceed, angiogenesis is essential for supplying oxygen and nutrients to the mending tissue. Fibroblasts are in charge of making collagen and the ECM. In order to close wounds and regenerate tissue, bioactive peptides promote fibroblast migration, proliferation, and differentiation. A variety of peptides, including those derived from collagen, encourage the synthesis of collagen, a crucial part of the ECM that provides the new tissue with strength and structure. Additionally, the collagen matrix aids in the anchoring of newly formed blood vessels. Cells such as fibroblasts, endothelial cells, and keratinocytes can migrate and proliferate to the wound site in response to signals from ECM protein-derived peptides like collagen or fibronectin. Keratinocytes must traverse the wound bed during the epithelialization process in order to restore the skin barrier. Cells such as fibroblasts, endothelial cells, and keratinocytes can migrate and proliferate to the wound site in response to signals from ECM protein-derived peptides like collagen or fibronectin. Keratinocytes must traverse the wound bed during the epithelialization process in order to restore the skin barrier. Certain bioactive peptides, particularly those made from collagen or milk, have antibacterial qualities that serve to keep the wound from becoming infected and create a more conducive healing environment.

8.2.14.1 Mechanisms of Action of Bioactive Peptides

Bioactive peptides exert their effects on wound healing through several pathways.

8.2.14.1.1 Receptor-Mediated Signaling

Numerous bioactive peptides work by binding to cell surface receptors such as integrins and G-protein coupled receptors (GPCRs). Collagen-derived peptides, such Pro-Hyp (proline-hydroxyproline), for instance, attach to certain fibroblast receptors to initiate intracellular signaling cascades that cause cell migration, proliferation, and collagen synthesis.

8.2.14.1.2 Gene Expression Regulation

Certain transcription factors, including NF-kB, AP-1, and Smad proteins, which regulate the expression of genes involved in inflammation, cell proliferation, ECM synthesis, and tissue remodeling, can be activated or inhibited by bioactive peptides.

8.2.14.1.3 Modulation of Matrix Metalloproteinases (MMPs)

Parts of the ECM are broken down by MMPs. Their control is essential to ensuring appropriate tissue remodeling during wound healing. MMP activity is modulated by certain bioactive peptides to promote tissue remodeling and avoid excessive matrix destruction that can hinder healing.

8.2.14.1.4 Interaction with Growth Factors

Bioactive peptides can influence growth factors like VEGF, TGF-β, and EGF, which are critical for angiogenesis, ECM deposition, and cell proliferation.

8.3 IMPORTANCE OF ANIMAL-DERIVED COMPOUNDS IN WOUND CARE

Wounds have been treated with natural substances for thousands of years. Numerous plants and animals contain natural chemicals that are widely accessible and can be used to treat wounds. Using traditional Chinese and Indian remedies, they have demonstrated efficacy in healing. Some of the animal-based compounds used in wound care are given below.

8.3.1 Natural Compounds with Anti-Inflammation Properties

One of the key steps in the healing of wounds is the inflammatory response. Hemostasis, the elimination of pathogens, injured tissues, and wound cleaning are all impacted by inflammation and anti-inflammation (Reinke and Sorg, 2012). However, a prolonged period of inflammation will impact the healing process of wounds and result in a pathological condition (Landen et al., 2016).

8.3.1.1 Bilirubin

Bilirubin is the animal-based compound targeting inflammatory response. It is the final result of heme catabolism in mammals; bilirubin is a reddish-orange substance that is crucial for cellular defense (Stocker et al., 1987). A study assessed that wound healing process in rat skin after treating it with bilirubin, based on their hypothesis that it would aid in wound healing (Kant et al., 2017). Bilirubin treatment of the wound resulted in a rise in IL-10 expression and a decrease in pro-inflammatory markers such as TNF-α and ICAM-1. Rats who received treatment also showed increase in hydroxyproline, glucosamine, and wound contraction. The impact of bilirubin on growth factors, cytokines, and angiogenesis during wound healing in diabetic rats was also investigated (Ram et al., 2016).

8.3.2 Natural Compounds with Antibacterial Properties

Natural substances that possess antibacterial qualities are crucial in the process of wound healing, as they help to prevent or minimize infections that could otherwise hinder recovery and lead to complications. There are two ways to explain a compound's antibacterial activity: either it suppresses antibacterial resistance or it inhibits the formation of essential bacterial components (Khameneh et al., 2019).

8.3.2.1 Honey

The oldest documented use of bee honey to treat wounds dates back thousands of years, to an ancient Egyptian trauma manual from between 2600 and 2200 BCE (Majtan, 2011; Zubair and Aziz, 2015). According to Kumar and Kothari (2021) and Manyi-Loh et al (2011), "honey is a concentrated aqueous solution of inverted sugars with a pH of 3.6 that contains 40% fructose, 40% glucose, 20% water, enzymes, vitamins, and minerals." The natural enzyme glucose oxidase generates hydrogen peroxide, which is what gives most ordinary honey its antibacterial properties. Highly reactive free radicals are created during the breakdown of hydrogen peroxide, and these radicals react with the bacteria to destroy them (Manyi-Loh et al., 2011). However, due to their low pH and high content of supersaturated sugar, a number of other "non-peroxide" types of honey, including Manuka and jelly bush, have antibacterial qualities of their own (Mandal and Mandal, 2011). In particular, Atrott and Henle proposed that methylglyoxal, a special antibacterial component that is only accountable for the honey's unique antibacterial activity, is present in considerable amounts in Manuka honey (Atrott and Henle, 2009).

8.3.2.2 Propolis

Propolis was used by the Egyptians, Persians, and Romans. Propolis can be obtained from honey bees, tree buds, poplar, willow, elm, alder, birch, beech, and other plant sources (Kuropatnicki et al., 2013). Propolis contains almost 300 different chemical compounds, including polyphenols, phenolic aldehydes, amino acids, steroids, and other chemicals (Huang et al., 2014). The primary components of propolis include flavonoids, glycerides, phenylpropanoids, and cinnamic acids and their esters (Czyzewska et al., 2015; Uzel et al., 2005). The flavonoids, esters, and aromatic acids in propolis are primarily responsible for its antibacterial properties against Gram-positive bacteria (Uzel et al., 2005).

8.3.3 Natural Compounds with Collagen Promotion Properties

Collagen is the most prevalent protein in the body. Collagen draws fibroblasts and promotes the production of new collagen in the wound bed during the healing process. Collagen dressing technique promotes angiogenesis, autolytic debridement, and re-epithelialization while also aiding in the development of new tissues. Therefore, substances that might encourage the production of collagen are crucial to the healing process.

8.3.3.1 β-Glucans

β-glucans are glucose polymers found in fungi, grains, and yeast. These substances fall under the category of biological response modifiers, according Bohn and BeMiller (1995). Numerous studies have demonstrated that soluble and particulate β-glucans improve immune function by having anti-infective, anticancer, and immunomodulatory effects (Browder et al., 1990; Pretus et al., 1991). β-glucans improve wound healing by encouraging macrophage infiltration, which in turn stimulates tissue granulation, collagen deposition, and re-epithelialization. Because of its high stability and resistance to wound proteases, β-glycon wound cure is a great wound healing agent (Majtan and Jesenak, 2018).

8.3.3.2 Amino Acids and Peptides

Natural procollagen compounds that consist of "amino acids and peptides for wound healing" can also be found in animals (bees, snails, fish) in addition to traditional therapeutic plants. Honey's low pH contribute to the establishment and maintenance of ideal circumstances for fibroblasts, which need an acidic environment to perform tasks, including organizing and migrating collagen (Mellin et al., 1992). According to a study, amino acids present in *Mytilus galloprovincialis* and *Rapana venosa* promote dermal and epidermal neoformation to speed up the healing of skin wounds (Badiu et al., 2010).

The mechanism underlying the increased benefits of these amino acids for wound healing was proposed by Shearer et al. (1997) to be closely related to the differential control "of macrophage arginine metabolism, in which TGF-β1 may play a crucial coregulatory function." Furthermore, the terrestrial snail *Cryptozona bistrialis*' bioactive peptide promotes the migration of NIH/3T3 murine fibroblast cells *in vitro*. Additionally, *in vivo* experiments on Wistar albino rats with and without diabetes shown "that the Crypto-zona bistrialis-peptide was effective in promoting wound healing" (Ulagesan et al., 2018). The substantial increase in collagen content brought about by the improved movement of fibroblasts and epithelial cells to the site of injury is thought to be the cause of the increased wound contraction. However, the precise chemical formula was not displayed by the extract chemicals derived from animal sources.

8.4 CONCLUSION

Effective wound healing is a complex biological process that is vital to restoring tissue integrity and improving patient quality of life. In recent years, animal-derived natural compounds have emerged as promising agents in wound care due to their intrinsic **antimicrobial**, **anti-inflammatory**, **immunomodulatory**, and **tissue-regenerative** properties. These bioactive substances, including honey, silk fibroin, snail slime, etc., work at various levels of the healing cascade, from reducing inflammation and oxidative stress to stimulating fibroblast proliferation, angiogenesis, and ECM remodeling. Unlike many synthetic drugs, animal-based compounds offer **biocompatibility** and **multifunctionality**, often acting simultaneously on microbial control, immune modulation, and tissue regeneration. Their capacity to **support wound closure, regulate exudate**, and **prevent infection** makes them particularly valuable in an era of rising antimicrobial resistance and chronic wound burden. However, despite the therapeutic promise, challenges remain. The **standardization of biological materials**, potential **immunogenicity**, variability in **compound composition based on source and environment,** and **lack of large-scale clinical validation** are critical barriers to widespread adoption. Moreover, the **mechanistic understanding of how these animal-derived agents influence cellular and molecular events** in wound healing remains incomplete. In conclusion, animal-derived natural products hold immense potential in modern wound management. With sustained research, innovation, and clinical integration, these biomolecules may significantly advance the efficacy and safety of wound care therapies, addressing current limitations and improving patient outcomes across diverse wound types.

GLOSSARY

Angiogenesis: The process by which new blood vessels develop from pre-existing ones is known as angiogenesis.

Animal product: Products derived from animals, including both edible and non-edible products, are referred to as animal products.

Antioxidant: Antioxidants are molecules that neutralize free radicals, which are very reactive molecules that can harm cells and tissues, thereby protecting the body from oxidative stress.

Collagen: Collagen is a structural protein essential to the body because it gives different tissues and organs strength, suppleness, and support.

Fibroin: Fibroin is the primary structural protein found in silk fibers that silkworms and certain other insects (such as spiders) produce.

Healing: The process by which the body restores damaged tissues and reaches a normal, functional condition is referred to as healing. The body attempts to repair the skin (or other afflicted tissues) and stop infection while wounds heal.

Hematostatis: The process by which the body stops and prevents bleeding, or blood loss, following an injury is known as hemostasis.

Inflammation: It is the body's normal, defensive reaction to damage, infection, or dangerous stimuli (such as irritants or toxins).

Mucus: Mucus is a viscous fluid produced by numerous mucous membranes throughout the body, particularly in the digestive, reproductive, and respiratory systems, to keep organs and tissues healthy and functioning properly.

Venom: Venom is a poisonous chemical produced by some creatures, like snakes, spiders, scorpions, and some fish or amphibians mostly for defense or to catch prey.

Wound: An injury to the body that usually involves a break in the skin or other body tissues is called a wound. Numerous things, including physical trauma, scrapes, abrasions, burns, and even internal injuries, can cause it. Minor scrapes to more serious injuries that may necessitate medical attention are all examples of wounds.

REFERENCES

Abbas Esmaeil M., and Hovhannisyan G. Hrachya. "Fatty acid composition of caviar and liver from cultured great sturgeon (*Huso huso*)." *International Food Research Journal* 22, no. 3 (2015):1083–1089.

Abd El-Aziz, Fatma El-Zahraa A., Helal F. Hetta, Basma N. Abdelhamid, and Noura H. Abd Ellah. "Antibacterial and wound-healing potential of PLGA/spidroin nanoparticles: A study on earthworms as a human skin model." *Nanomedicine* 17, no. 6 (2022): 353–365.

Abd Ellah, Noura H., Fatma El-Zahraa A. Abd El-Aziz, Sara A. Abouelmagd, Basma N. Abd El-Hamid, and Helal F. Hetta. "Spidroin in carbopol-based gel promotes wound healing in earthworm's skin model." *Drug Development Research* 80, no. 8 (2019): 1051–1061.

Abdel-Naby, Waleed, Brigette Cole, Aihong Liu, Jingbo Liu, Pengxia Wan, Ryan Schreiner, David W. Infanger, Nicholas B. Paulson, Brian D. Lawrence, and Mark I. Rosenblatt. "Treatment with solubilized silk-derived protein (SDP) enhances rabbit corneal epithelial wound healing." *PLoS One* 12, no. 11 (2017): e0188154.

Abdelrahman, Tarig, and Heather Newton. "Wound dressings: Principles and practice." *Surgery (oxford)* 29, no. 10 (2011): 491–495.

Abdualkader, Abdualrahman M., Ahmed M. Ghawi, Mohamed Alaama, Mohd. Awang, and Ahmed Merzouk. "Leech therapeutic applications." *Indian Journal of Pharmaceutical Sciences* 75, no. 2(2013): 127–137.

Abdullah, Swaid, Latief M. Dar, Adil Rashid, and Anita Tewari. "Hirudotherapy/leech therapy: Applications and indications in surgery." *Archives of Clinical and Experimental Surgery* 1, no. 3 (2012): 172–180.

Aloto, Desta, and Eyob Eticha. "Leeches: A review on their pathogenic and beneficial effects." *Journal of Veterinary Science & Technology* 9, no. 01 (2018).

Altman, Gregory H., Frank Diaz, Caroline Jakuba, Tara Calabro, Rebecca L. Horan, Jingsong Chen, Helen Lu, John Richmond, and David L. Kaplan. "Silk-based biomaterials." *Biomaterials* 24, no. 3 (2003): 401–416.

Aouji, Marouane, Amine Rkhaila, Bouchra Bouhaddioui, Malak Zirari, Hala Harifi, Youness Taboz, Lalla Aicha Lrhorfi, and Rachid Bengueddour. "Chemical composition, mineral profile, anti-bacterial, and wound healing properties of snail slime of *Helix aspersa* Müller." *BioMedicine* 13, no. 4 (2023): 10.

Atrott, Julia and Thomas Henle. "Methylglyoxal in manuka honey–correlation with antibacterial properties." *Czech Journal of Food Sciences* 27, no. Special Issue 1 (2009): S163–S165.

Ayaz, Muhammad, Fazal Subhan, Jawad Ahmed, Arif-ullah Khan, Farhat Ullah, Ihsan Ullah, Gowhar Ali, Nawazish-i-Husain Syed, and Sajid Hussain. "Sertraline enhances the activity of antimicrobial agents against pathogens of clinical relevance." *Journal of Biological Research-Thessaloniki* 22 (2015): 1–8.

Ayhan, Hüseyin, and Salih Mollahaliloğlu. "Tıbbi sülük tedavisi: Hirudoterapi." *Ankara Medical Journal* 18, no. 1 (2018): 141–148.

Aykac, Asli, Buse Karanlik, and Ahmet Ozer Sehirli. "Protective effect of silk fibroin in burn injury in rat model." *Gene* 641 (2018): 287–291.

Badiu, Diana L., Rafael Luque, Elena Dumitrescu, Anca Craciun, and Danut Dinca. "Amino acids from *Mytilus galloprovincialis* (L.) and *Rapana venosa* molluscs accelerate skin wounds healing via enhancement of dermal and epidermal neoformation." *The Protein Journal* 29 (2010): 81–92.

Baghel, Praveen Singh, S. Shukla, Raj Kumar Mathur, and R. Randa. "A comparative study to evaluate the effect of honey dressing and silver sulfadiazene dressing on wound healing in burn patients." *Indian Journal of Plastic Surgery* 42, no. 02(2009): 176–181.

Balamurugan, Manoharan. "Restoration of histoarchitecture in the paracetamol-induced liver damaged rat by earthworm extract, Lampito mauritii (Kinberg)" *European Review for Medical and Pharmacological Sciences* 11, no. 6(2007): 407.

Balamurugan, Manoharan, Kandasamy Parthasarathi, Edwin L. Cooper, and Latchumanan S. Ranganathan. "Earthworm paste (Lampito mauritii, Kinberg) alters inflammatory, oxidative, haematological and serum biochemical indices of inflamed rat." *European Review for Medical and Pharmacological Sciences* 11, no. 2(2007): 77.

Balamurugan, Mariappan, Kasi Parthasarathi, Lalpet Souri Ranganathan, and Edwin L. Cooper. "Hypothetical mode of action of earthworm extract with hepatoprotective and antioxidant properties." *Journal of Zhejiang University Science B* 9 (2008): 141–147.

Bankova, Vassya S., Solange L. de Castro, and Maria C. Marcucci. "Propolis: Recent advances in chemistry and plant origin." *Apidologie* 31, no. 1 (2000): 3–15.

Baoyong, Lu, Zheng Jian, Chen Denglong, and Li Min. "Evaluation of a new type of wound dressing made from recombinant spider silk protein using rat models." *Burns* 36, no. 6 (2010): 891–896.

Belhaj Khalifa, Imen, Neji Ladhari, and Mustapha Touay. "Application of sericin to modify textile supports." *Journal of the Textile Institute* 103, no. 4(2012): 370–377.

Bhat, Gajanan, ed. *Structure and Properties of High-Performance Fibers*. Woodhead Publishing, 2016.

Bohn, Gregory, Brock Liden, Gregory Schultz, Qingping Yang, and Daniel J. Gibson. "Ovine-based collagen matrix dressing: Next-generation collagen dressing for wound care." *Advances in Wound Care* 5, no. 1 (2016): 1–10.

Bohn, John A., and James N. BeMiller. "(1→ 3)-β-d-Glucans as biological response modifiers: A review of structure-functional activity relationships." *Carbohydrate Polymers* 28, no. 1 (1995): 3–14.

Bouhadir, Kamal H., Kuen Yong Lee, Eben Alsberg, Kelly L. Damm, Kenneth W. Anderson, and David J. Mooney. "Degradation of partially oxidized alginate and its potential application for tissue engineering." *Biotechnology Progress* 17, no. 5 (2001): 945–950.

Brankovic, Marija, Fatima Zivic, Nenad Grujovic, Ivan Stojadinovic, Strahinja Milenkovic, and Nikola Kotorcevic. "Review of spider silk applications in biomedical and tissue engineering." *Biomimetics* 9, no. 3 (2024): 169.

Browder, William., David Williams, Henry Pretus, Giorgio Olivero, Francesco Enrichens, Patrizio Mao, and Allessandro Franchello. "Beneficial effect of enhanced macrophage function in the trauma patient." *Annals of Surgery* 211, no. 5 (1990): 605.

Buba, Fatimah, Abubakar Gidado, and Aliyu Shugaba. "Analysis of biochemical composition of honey samples from North-East Nigeria." *Biochemistry and Analytical Biochemistry* 2, no. 3 (2013): 139.

Burdick, Jason A., and Glenn D. Prestwich. "Hyaluronic acid hydrogels for biomedical applications." *Advanced Materials* 23, no. 12 (2011): H41–H56.

Caprino, Fabio, Vittorio Maria Moretti, Federica Bellagamba, Giovanni Mario Turchini, Maria Letizia Busetto, Ivan Giani, Maria Antonietta Paleari, and Mario Pazzaglia. "Fatty acid composition and volatile compounds of caviar from farmed white sturgeon (*Acipenser transmontanus*)." *Analytica chimica acta* 617, no. 1–2 (2008): 139–147.

Chattopadhyay, Sayani, and Ronald T. Raines. "Collagen-based biomaterials for wound healing." *Biopolymers* 101, no. 8 (2014): 821–833.

Chen, Jing, Yu Chen, Yajie Chen, Zicheng Yang, Bo You, Ye Chun Ruan, and Yizhi Peng. "Epidermal CFTR suppresses MAPK/NF-κB to promote cutaneous wound healing." *Cellular Physiology and Biochemistry* 39, no. 6 (2016): 2262–2274.

Chen, Na, Siqi Xu, Yuhan Zhang, and Feng Wang. "Animal protein toxins: Origins and therapeutic applications." *Biophysics Reports* 4, no. 5 (2018): 233–242.

Chen, Zelin, Ziwen Wang, Taotao Jin, Gufang Shen, Yu Wang, Xu Tan, Yibo Gan et al. "Fibrogenic fibroblast-selective near-infrared phototherapy to control scarring." *Theranostics* 9, no. 23 (2019): 6797.

Cheng, Fang, Yue Shen, Ponnuswamy Mohanasundaram, Michelle Lindström, Johanna Ivaska, Tor Ny, and John E. Eriksson. "Vimentin coordinates fibroblast proliferation and keratinocyte differentiation in wound healing via TGF-β–Slug signaling." *Proceedings of the National Academy of Sciences* 113, no. 30 (2016): E4320–E4327.

Cheng, Xiaochen, Ziyu Shao, Chengbo Li, Lejun Yu, Mazhar Ali Raja, and Chenguang Liu. "Isolation, characterization and evaluation of collagen from jellyfish *Rhopilema esculentum* Kishinouye for use in hemostatic applications." *PLoS One* 12, no. 1 (2017): e0169731.

Chhum, Panharith, Hongtao Yu, Bo An, Brian R. Doyon, Yu-Shan Lin, and Barbara Brodsky. "Consequences of glycine mutations in the fibronectin-binding sequence of collagen." *Journal of Biological Chemistry* 291, no. 53 (2016): 27073–27086.

Chirumbolo, Salvatore. "Flavonoids in propolis acting on mast cell-mediated wound healing." *Inflammopharmacology* 20 (2012): 99–101.

Chouhan, Dimple, Tshewuzo-U. Lohe, Naresh Thatikonda, V. G. M. Naidu, My Hedhammar, and Biman B. Mandal. "Silkworm silk scaffolds functionalized with recombinant spider silk containing a fibronectin motif promotes healing of full-thickness burn wounds." *ACS Biomaterials Science & Engineering* 5, no. 9(2019): 4634–4645.

Chouhan, Dimple, Naresh Thatikonda, Linnea Nileback, Mona Widhe, My Hedhammar, and Biman B. Mandal. "Recombinant spider silk functionalized silkworm silk matrices as potential bioactive wound dressings and skin grafts." *ACS Applied Materials & Interfaces* 10, no. 28 (2018): 23560–23572.

Christyraj, Jackson Durairaj Selvan, Ananthaselvam Azhagesan, Mijithra Ganesan, Karthikeyan Subbiah Nadar Chelladurai, Vennila Devi Paulraj, and Johnson Retnaraj Samuel Selvan Christyraj. "Understanding the role of the clitellum in the regeneration events of the earthworm *Eudrilus eugeniae*." *Cells Tissues Organs* 208, no. 3–4 (2020): 134–141.

Chung, Eun-Jae, Hyung Woo Ju, Hyun Jung Park, and Chan Hum Park. "Three-layered scaffolds for artificial esophagus using poly (ϵ-caprolactone) nanofibers and silk fibroin: An experimental study in a rat model." *Journal of Biomedical Materials Research Part A* 103, no. 6 (2015): 2057–2065.

Chvapil, Milos, John A. Gaines, and Thomas Gilman. "Lanolin and epidermal growth factor in healing of partial-thickness pig wounds." *The Journal of Burn Care & Rehabilitation* 9, no. 3 (1988): 279–284.

Cooper, Edwin L., Binggen Ru, and Ning Weng. "Earthworms: Sources of antimicrobial and anticancer molecules." *Complementary and Alternative Approaches to Biomedicine* 546 (2004): 359–389.

Costa, Bruno Almeida, Edmar Maciel Lima Júnior, Manoel Odorico de Moraes Filho, Francisco Vagnaldo Fechine, Maria Elisabete Amaral de Moraes, Francisco Raimundo Silva Júnior, Maria Flaviane Araújo do Nascimento Soares, and Marina Becker Sales Rocha. "Use of tilapia skin as a xenograft for pediatric burn treatment: A case report." *Journal of Burn Care & Research* 40, no. 5 (2019): 714–717.

Czyzewska, Urszula, Joanna Konończuk, Joanna Teul, Paweł Drągowski, Renata Pawlak-Morka, Arkadiusz Surażyński, and Wojciech Miltyk. "Verification of chemical composition of commercially available propolis extracts by gas chromatography–mass spectrometry analysis." *Journal of Medicinal Food* 18, no. 5 (2015): 584–591.

Darestani, Kaveh Darabi, Sayid Mahdi Mirghazanfari, Katayoun Gohari Moghaddam, and Somayeh Hejazi. "Leech therapy for linear incisional skin-wound healing in rats." *Journal of Acupuncture and Meridian Studies* 7, no. 4 (2014): 194–201.

Deng, Zhen-han, Jian-jian Yin, Wei Luo, Ronak Naveenchandra Kotian, Shan-shan Gao, Zi-qing Yi, Wen-feng Xiao, Wen-ping Li, and Yu-sheng Li. "The effect of earthworm extract on promoting skin wound healing." *Bioscience Reports* 38, no. 2 (2018): BSR20171366.

Denny, Mark W. "Mechanical properties of pedal mucus and their consequences for gastropod structure and performance." *American Zoologist* 24, no. 1 (1984): 23–36.

Drury, Jeanie L., and David J. Mooney. "Hydrogels for tissue engineering: Scaffold design variables and applications." *Biomaterials* 24, no. 24 (2003): 4337–4351.

Ellijimi, Chedli, Manel Ben Hammouda, Houcemeddine Othman, Wassim Moslah, Jed Jebali, Hazem Ben Mabrouk, Maram Morjen et al. "*Helix aspersa* maxima mucus exhibits antimelanogenic and antitumoral effects against melanoma cells." *Biomedicine & Pharmacotherapy* 101 (2018): 871–880.

El-Ouar, Ibtissem, Cornelia Braicu, Dalila Naimi, Alexandru Irimie, and Ioana Berindan-Neagoe. "Anti tumour effect of aqueous extract from *Helix aspersa*." *Pharmacognosy Magazine* 13, no. 50 (2013): 281–285.

Fiakos, Gabriella, Zeming Kuang, and Evan Lo. "Improved skin regeneration with acellular fish skin grafts." *Engineered Regeneration* 1 (2020): 95–101.

Figueira, Thiago Gomes, Francisco Vieira Dos Santos, and Sérgio Akinobu Yoshioka. "Development, characterization and in vivo evaluation of the ointment containing hyaluronic acid for potential wound healing applications." *Journal of Biomaterials Science, Polymer Edition* 33, no. 12 (2022): 1511–1530.

Fox Jay W and Solange M T Serrano. "Timeline of key events in snake venom metalloproteinase research." *Journal of Proteomics* 72, no. 2 (2009): 200–209.

Gaspar-Pintiliescu, Alexandra, Ana-Maria Stanciuc, and Oana Craciunescu. "Natural composite dressings based on collagen, gelatin and plant bioactive compounds for wound healing: A review." *International Journal of Biological Macromolecules* 138 (2019): 854–865.

Georgieva, Ani, Katerina Todorova, Ivan Iliev, Valeriya Dilcheva, Ivelin Vladov, Svetlozara Petkova, Reneta Toshkova, Lyudmila Velkova, Aleksandar Dolashki, and Pavlina Dolashka. "Hemocyanins from *Helix* and *Rapana* snails exhibit in vitro antitumor effects in human colorectal adenocarcinoma." *Biomedicines* 8, no. 7 (2020): 194.

Ghods, Roshanak, Mojtaba Abdi, and Matineh Pourrahimi. "Leech therapy indications: A scoping review." *Traditional Medicine Research* 4, no. 3 (2019): 118.

Gholipourmalekabadi, Mazaher, Masoud Mozafari, Mojgan Bandehpour, Mohammad Salehi, Marzieh Sameni, Hector Hugo Caicedo, Ahmad Mehdipour, Hatef Ghasemi Hamidabadi, Ali Samadikuchaksaraei, and Hossein Ghanbarian. "Optimization of nanofibrous silk fibroin scaffold as a delivery system for bone marrow adherent cells: In vitro and in vivo studies." *Biotechnology and Applied Biochemistry* 62, no. 6 (2015): 785–794.

Goharshenasan, Peiman, Shahideh Amini, Ali Atria, Hamidreza Abtahi, and Ghasemali Khorasani. "Topical application of honey on surgical wounds: A randomized clinical trial." *Forschende Komplementärmedizin/Research in Complementary Medicine* 23, no. 1 (2016): 12–15.

Gonzalez, Ana Cristina de Oliveira, Tila Fortuna Costa, Zilton de Araújo Andrade, and Alena Ribeiro Alves Peixoto Medrado. "Wound healing – A literature review." *Anais brasileiros de dermatologia* 91, no. 5 (2016): 614–620.

Grdisa, Mira, Maja Popovic, and Terezija Hrzenjak. "Glycolipoprotein extract (G-90) from earthworm *Eisenia foetida* exerts some antioxidative activity." *Comparative Biochemistry and Physiology Part A: Molecular & Integrative Physiology* 128, no. 4 (2001): 821–825.

Greistorfer, Sophie, Waltraud Klepal, Norbert Cyran, Andreas Gugumuck, Livia Rudoll, Johannes Suppan, and Janek von Byern. "Snail mucus – Glandular origin and composition in *Helix pomatia*." *Zoology* 122 (2017): 126–138.

Gubitosa, Jennifer, Vito Rizzi, Paola Fini, Anna Laurenzana, Gabriella Fibbi, Clara Veiga-Villauriz, Fiorenza Fanelli et al. "Biomolecules from snail mucus (*Helix aspersa*) conjugated gold nanoparticles, exhibiting potential wound healing and anti-inflammatory activity." *Soft Matter* 16, no. 48 (2020): 10876–10888.

Gupta, Sharda, Hussam Alrabaiah, Marquette Christophe, Mohammad Rahimi-Gorji, Sohail Nadeem, and Arindam Bit. "Evaluation of silk-based bioink during pre and post 3D bioprinting: A review." *Journal of Biomedical Materials Research Part B: Applied Biomaterials* 109, no. 2 (2021): 279–293.

Gupta, Shilpi Singh, Onkar Singh, Praveen Singh Bhagel, Sonia Moses, Sumit Shukla, and Raj Kumar Mathur. "Honey dressing versus silver sulfadiazene dressing for wound healing in burn patients: A retrospective study." *Journal of Cutaneous and Aesthetic Surgery* 4, no. 3 (2011): 183–187.

Gurtner, Geoffrey C., Sabine Werner, Yann Barrandon, and Michael T. Longaker. "Wound repair and regeneration." *Nature* 453, no. 7193 (2008): 314–321.

Han, Chien-Kuo, Wei-Wen Kuo, Chia-Yao Shen, Tung-Sheng Chen, Peiying Pai, Chang-Hai Tsai, Feng-Yueh Lo, Da-Tong Ju, and Chih-Yang Huang. "Dilong prevents the high-KCl cardioplegic solution administration-induced apoptosis in H9c2 cardiomyoblast cells mediated by MEK." *The American Journal of Chinese Medicine* 42, no. 06 (2014): 1507–1519.

He, Miao, Wen-Qing Xie, Guo Cheng, Wen-Ping Li, Deng-Jie Yu, Hong-Fu Jin, Zhen-Han Deng, and Yu-Sheng Li. "The therapeutic effects of earthworm extract on deep second-degree burn wound healing." *Annals of Palliative Medicine* 10, no. 3 (2021): 2869879–2862879.

Herman, Anna, Patrycja Wińska, Małgorzata Białek, and Andrzej P. Herman. "Biological properties of the mucus and eggs of *Helix aspersa* Müller as a potential cosmetic and pharmaceutical raw material: A preliminary study." *International Journal of Molecular Sciences* 25, no. 18 (2024): 9958.

Hosseinkhani, Ayda, Maryam Falahatzadeh, Elahe Raoofi, and Mohammad M. Zarshenas. "An evidence-based review on wound healing herbal remedies from reports of traditional Persian medicine." *Journal of Evidence-Based Complementary & Alternative Medicine* 22, no. 2 (2017): 334–343.

Hrzenjak, Terezija, Maja Popović, Tamara Božić, Mira Grdisa, Dorde Kobrehel, and Ljerka Tiška-Rudman. "Fibrinolytic and anticoagulative activities from the earthworm *Eisenia foetida*." *Comparative Biochemistry and Physiology Part B: Biochemistry and Molecular Biology* 119, no. 4 (1998): 825–832.

Huang, Shuai, Cui-Ping Zhang, Kai Wang, George Q. Li, and Fu-Liang Hu. "Recent advances in the chemical composition of propolis." *Molecules* 19, no. 12 (2014): 19610–19632.

Huang, Ying, Xin Zhao, Zhiyi Zhang, Yongping Liang, Zhanhai Yin, Baojun Chen, Lang Bai, Yong Han, and Baolin Guo. "Degradable gelatin-based IPN cryogel hemostat for rapidly stopping deep noncompressible hemorrhage and simultaneously improving wound healing." *Chemistry of Materials* 32, no. 15 (2020): 6595–6610.

Ibrahim, Ahmed, Dalia Hassan, Noura Kelany, Saber Kotb, and Mahmoud Soliman. "Validation of three different sterilization methods of Tilapia skin dressing: Impact on microbiological enumeration and collagen content." *Frontiers in Veterinary Science* 7 (2020a): 597751.

Ibrahim, Ahmed, Mahmoud Soliman, Saber Kotb, and Magda M. Ali. "Evaluation of fish skin as a biological dressing for metacarpal wounds in donkeys." *BMC Veterinary Research* 16 (2020b): 1–10.

Ige, Oladeji O., Lasisi E. Umoru, and Sunday Aribo. "Natural products: A minefield of biomaterials." *International Scholarly Research Notices* 2012, no. 1 (2012): 983062.

Imran Hina, Mansoor Ahmad, Atiq-ur, Yaqeen Zahra Rahman, Tehmina Sohail, Nudrat Fatima, Wasif Iqbal, and Shafay Yaqeen Syed. "Evaluation of wound healing effects between *Salvadora persica* ointment and Solcoseryl jelly in animal model." *Pakistan Journal of Pharmaceutical Sciences.* 28, no. 5 (2015): 1777–1780.

Infanger, David W., Waleed Abdel-Naby, Jinniece J. Kalal, Nicholas B. Paulson, Yue Bai, and Brian D. Lawrence. "Silk-derived protein-4 (SDP-4) inhibits nuclear factor kappa B (NF-κB) inflammatory signaling that underlies dry eye disease (DED)." *Investigative Ophthalmology & Visual Science* 60, no. 9 (2019): 2820–2820.

Inoue, Satoshi, Kazunori Tanaka, Fumio Arisaka, Sumiko Kimura, Kohei Ohtomo, and Shigeki Mizuno. "Silk fibroin of *Bombyx mori* is secreted, assembling a high molecular mass elementary unit consisting of H-chain, L-chain, and P25, with a 6: 6: 1 molar ratio." *Journal of Biological Chemistry* 275, no. 51 (2000): 40517–40528.

Jones, Hefin R., Calum T. Robb, Mauro Perretti, and Adriano G. Rossi. "The role of neutrophils in inflammation resolution." In *Seminars in Immunology*, vol. 28, no. 2, pp. 137–145. Academic Press, 2016.

Ju, Hyung Woo, Ok Joo Lee, Bo Mi Moon, Faheem A. Sheikh, Jung Min Lee, Jung-Ho Kim, Hyun Jung Park et al. "Silk fibroin based hydrogel for regeneration of burn induced wounds." *Tissue Engineering and Regenerative Medicine* 11 (2014): 203–210.

Júnior, Edmar Maciel Lima, Manoel Odorico de Moraes Filho, Bruno Almeida Costa, Ana Paula Negreiros Nunes Alves, Maria Elisabete Amaral de Moraes, Alex Marques do Nascimento Uchôa, Camila Barroso Martins et al. "Lyophilised tilapia skin as a xenograft for superficial partial thickness burns: A novel preparation and storage technique." *Journal of Wound Care* 29, no. 10 (2020): 598–602.

Kallis, Penelope J., and Adam J. Friedman. "Collagen powder in wound healing." *Journal of Drugs in Dermatology: JDD* 17, no. 4 (2018): 403–408.

Kant, Vinay, Babu Lal Jangir, Ankit Nigam, Vinod Kumar, and Sonu Sharma. "Dose regulated cutaneous wound healing potential of quercetin in male rats." *Wound Medicine* 19 (2017): 82–87.

Kapoor, Sonia, and Subhas C. Kundu. "Silk protein-based hydrogels: Promising advanced materials for biomedical applications." *Acta Biomaterialia* 31 (2016): 17–32.

Kendall, Alexandra C., Magdalena Kiezel-Tsugunova, Luke C. Brownbridge, John L. Harwood, and Anna Nicolaou. "Lipid functions in skin: Differential effects of n-3 polyunsaturated fatty acids on cutaneous ceramides, in a human skin organ culture model." *Biochimica et Biophysica Acta (BBA)-Biomembranes* 1859, no. 9 (2017): 1679–1689.

Khameneh, Bahman, Milad Iranshahy, Vahid Soheili, and Bibi Sedigheh Fazly Bazzaz. "Review on plant antimicrobials: A mechanistic viewpoint." *Antimicrobial Resistance & Infection Control* 8 (2019): 1–28.

Khoo, Yan-Teng, Ahmad Sukari Halim, Kirnpal-Kaur B. Singh, and Noor-Ayunie Mohamad. "Wound contraction effects and antibacterial properties of *Tualang* honey on full-thickness burn wounds in rats in comparison to hydrofibre." *BMC Complementary and Alternative Medicine* 10 (2010): 1–8.

Kiecolt-Glaser, Janice K., Martha A. Belury, Rebecca Andridge, William B. Malarkey, Beom Seuk Hwang, and Ronald Glaser. "Omega-3 supplementation lowers inflammation in healthy middle-aged and older adults: A randomized controlled trial." *Brain, Behavior, and Immunity* 26, no. 6 (2012): 988–995.

Koh, Leng-Duei, Yuan Cheng, Choon-Peng Teng, Yin-Win Khin, Xian-Jun Loh, Si-Yin Tee, Michelle Low et al. "Structures, mechanical properties and applications of silk fibroin materials." *Progress in Polymer Science* 46 (2015): 86–110.

Kumar, Binu, M. Vijayakumar, R. Govindarajan, and P. Pushpangadan. "Ethnopharmacological approaches to wound healing—exploring medicinal plants of India." *Journal of Ethnopharmacology* 114, no. 2(2007): 103–113.

Kumar P, and V Kothari. (eds.). *Wound Healing Research: Current Trends and Future Directions*. Springer Singapore; 2021. doi: 10.1007/978-981-16-2677-7.

Kuropatnicki, Andrzej K., Ewelina Szliszka, and Wojciech Krol. "Historical aspects of propolis research in modern times." *Evidence-Based Complementary and Alternative Medicine* 2013, no. 1 (2013): 964149.

Landen, Ning Xu, Dongqing Li, and Mona Ståhle. "Transition from inflammation to proliferation: A critical step during wound healing." *Cellular and Molecular Life Sciences* 73 (2016): 3861–3885.

Lannutti, Joseph, Darrell Reneker, Tea Ma, David Tomasko, and Darrell Farson. "Electrospinning for tissue engineering scaffolds." *Materials Science and Engineering: C* 27, no. 3(2007): 504–509.

Lee, Jung Min, Md Tipu Sultan, Soon Hee Kim, Vijay Kumar, Yeung Kyu Yeon, Ok Joo Lee, and Chan Hum Park. "Artificial auricular cartilage using silk fibroin and polyvinyl alcohol hydrogel." *International Journal of Molecular Sciences* 18, no. 8 (2017): 1707.

Lee, Ok Joo, Hyung Woo Ju, Ji Heui Kim, Jung Min Lee, Chang Seok Ki, Jung-Ho Kim, Bo Mi Moon, Hyun Jung Park, Faheem A. Sheikh, and Chan Hum Park. "Development of artificial dermis using 3D electrospun silk fibroin nanofiber matrix." *Journal of Biomedical Nanotechnology* 10, no. 7 (2014): 1294–1303.

Lee, Woong-Jin, Kyoungjoo Cho, Aaron-Youngjae Kim, and Gyung-Whan Kim. "Injectable click fibroin bioadhesive derived from spider silk for accelerating wound closure and healing bone fracture." *Materials* 15, no. 15 (2022): 5269.

Li, Jie, Juan Chen, and Robert Kirsner. "Pathophysiology of acute wound healing." *Clinics in Dermatology* 25, no. 1 (2007): 9–18.

Li, Tianyi, Yujie Sun, Jiaqi Wang, Chenning Zhang, and Yikun Sun. "Promoted skin wound healing by tail-amputated eisenia foetida proteins via the Ras/Raf/MEK/ERK signaling pathway." *ACS Omega* 8, no. 15 (2023): 13935–13943.

Lima-Junior, Edmar Maciel, Manoel Odorico de Moraes Filho, Bruno Almeida Costa, Francisco Vagnaldo Fechine, Maria Elisabete Amaral de Moraes, Francisco Raimundo Silva-Junior, Maria Flaviane Araújo do Nascimento Soares, Marina Becker Sales Rocha, and Cybele Maria Philopimin Leontsinis. "Innovative treatment using tilapia skin as a xenograft for partial thickness burns after a gunpowder explosion." *Journal of Surgical Case Reports* 2019, no. 6 (2019): rjz181.

Ling, Shengjie, Wenshuai Chen, Yimin Fan, Ke Zheng, Kai Jin, Haipeng Yu, Markus J. Buehler, and David L. Kaplan. "Biopolymer nanofibrils: Structure, modeling, preparation, and applications." *Progress in Polymer Science* 85 (2018): 1–56.

Liu, Haifeng, Hongbin Fan, Yue Wang, Siew Lok Toh, and James CH Goh. "The interaction between a combined knitted silk scaffold and microporous silk sponge with human mesenchymal stem cells for ligament tissue engineering." *Biomaterials* 29, no. 6 (2008): 662–674.

Liu, Jianghui, Liwei Yan, Wei Yang, Yong Lan, Qiyu Zhu, Hongjie Xu, Canbin Zheng, and Rui Guo. "Controlled-release neurotensin-loaded silk fibroin dressings improve wound healing in diabetic rat model." *Bioactive Materials* 4 (2019): 151–159.

Liu, Zhe, Ming-Zhao Yu, Hao Peng, Ruo-Tao Liu, Thou Lim, Chang-Qing Zhang, Zhen-Zhong Zhu, and Xiao-Juan Wei. "Decellularized tilapia fish skin: A novel candidate for tendon tissue engineering." *Materials Today Bio* 17 (2022): 100488.

Lo, Samantha, and Mh Busra Fauzi. "Current update of collagen nanomaterials—Fabrication, characterisation and its applications: A review." *Pharmaceutics* 13, no. 3 (2021): 316.

Lund-Nielsen, Betina, Lis Adamsen, Hans Jørn Kolmos, Mikael Rørth, Anders Tolver, and Finn Gottrup. "The effect of honey-coated bandages compared with silver-coated bandages on treatment of malignant wounds—A randomized study." *Wound Repair and Regeneration* 19, no. 6 (2011): 664–670.

Maghsoudi, Hadi, Farshad Salehi, Mohammad K. Khosrowshahi, Mohammad Baghaei, Mohammad Nasirzadeh, and Reza Shams. "Comparison between topical honey and mafenide acetate in treatment of burn wounds." *Annals of burns and fire disasters* 24, no. 3(2011): 132.

Majtan, Juraj. "Methylglyoxal—A potential risk factor of manuka honey in healing of diabetic ulcers." *Evidence-Based Complementary and Alternative Medicine* 2011, no. 1 (2011): 295494.

Majtan, Juraj, and Milos Jesenak. "β-Glucans: Multi-functional modulator of wound healing." *Molecules* 23, no. 4 (2018): 806.

Mandal, Manisha Deb, and Shyamapada Mandal. "Honey: Its medicinal property and antibacterial activity." *Asian Pacific Journal of Tropical Biomedicine* 1, no. 2 (2011): 154–160.

Manyi-Loh, Christy E., Anna M. Clarke, and Roland N. Ndip. "An overview of honey: Therapeutic properties and contribution in nutrition and human health." *African Journal of Microbiology Research* 5, no. 8 (2011): 844–852.

Marotta, Francesco, Ascanio Polimeni, Umberto Solimene, Aldo Lorenzetti, Emilio Minelli, Shalini Jain, Reza Rastmanesh, Sonia Sedriep, and Vincenzo Soresi. "Beneficial modulation from a high-purity caviar-derived homogenate on chronological skin aging." *Rejuvenation Research* 15, no. 2 (2012): 174–177.

Martinez-Mora Celia, Mrowiec Anna, García-Vizcaíno Eva María, Alcaraz Antonia, Cenis José Luis, Nicolás Francisco José. Fibroin and sericin from *Bombyx mori* silk stimulate cell migration through upregulation and phosphorylation of c-Jun. *PLoS One* 7, no. 7 (2012):e42271

Martinotti, Simona, and Elia Ranzato. "Propolis: A new frontier for wound healing?" *Burns & Trauma* 3 (2015): 1–7.

Mbese, Zintle, Sibusiso Alven, and Blessing Atim Aderibigbe. "Collagen-based nanofibers for skin regeneration and wound dressing applications." *Polymers* 13, no. 24 (2021): 4368.

McCleary, Ryan JR, and R. Manjunatha Kini. "Non-enzymatic proteins from snake venoms: A gold mine of pharmacological tools and drug leads." *Toxicon* 62 (2013): 56–74.

Mellin, Theodore N., Robert J. Mennie, Doreen E. Cashen, John J. Ronan, Joanna Capparella, Mary Lou James, Jerry Disalvo et al. "Acidic fibroblast growth factor accelerates dermal wound healing." *Growth Factors* 7, no. 1 (1992): 1–14.

Mencucci, Rita, Giovanni Strazzabosco, Virginia Cristofori, Andrea Alogna, Daria Bortolotti, Roberta Gafà, Michela Cennamo et al. "GlicoPro, novel standardized and sterile snail mucus extract for multi-modulative ocular formulations: New perspective in dry eye disease management." *Pharmaceutics* 13, no. 12 (2021): 2139.

Mercan, N. "Antimicrobial activity and chemical compositions of Turkish propolis from different regions." *African Journal of Biotechnology* 5, no. 11 (2006).

Midwood, Kim S., Leyla Valenick Williams, and Jean E. Schwarzbauer. "Tissue repair and the dynamics of the extracellular matrix." *The International Journal of Biochemistry & Cell Biology* 36, no. 6 (2004): 1031–1037.

Mihara, Hisashi, Hiroyuki Sumi, Tomoyuki Yoneta, Hideaki MizuMOTO, Ryuzo Ikeda, Masao Seiki, and Masugi Maruyama. "A novel fibrinolytic enzyme extracted from the earthworm, *Lumbricus rubellus*." *The Japanese Journal of Physiology* 41, no. 3 (1991): 461–472.

Mol, Suhendan, and Sabahat Turan. "Comparison of proximate, fatty acid and amino acid compositions of various types of fish roes." *International Journal of Food Properties* 11, no. 3 (2008): 669–677.

Mondal MIH, (ed.). *Cellulose-Based Superabsorbent Hydrogels; Polymers and Polymeric Composites: A Reference Series*. Cham, 1643–166. Springer International Publishing; 2018.

Moosavian, Hamid Reza, Sayid Mahdi Mirghazanfari, and Katayoun Gohari Moghaddam. "Effect of ischemia preconditioning and leech therapy on cutaneous pedicle flaps subjected to prolonged ischemia in a mouse model." *Aesthetic Plastic Surgery* 38 (2014): 1024–1029.

Moura-da-Silva, Ana M., and Cristiani Baldo. "Jararhagin, a hemorrhagic snake venom metalloproteinase from *Bothrops jararaca*." *Toxicon* 60, no. 3 (2012): 280–289.

Muhlstadt, Michael, Christoph Thomé, and Claudia Kunte. "Rapid wound healing of scalp wounds devoid of periosteum with milling of the outer table and split-thickness skin grafting." *British Journal of Dermatology* 167, no. 2(2012): 343–347.

Mukai, Kanae, Miki Koike, Saki Nakamura, Yuka Kawaguchi, Fumika Katagiri, Saki Nojiri, Yuki Yamada et al. "Evaluation of the effects of a combination of Japanese honey and hydrocolloid dressing on cutaneous wound healing in male mice." *Evidence-Based Complementary and Alternative Medicine* 2015, no. 1 (2015): 910605.

Murakami, Kaoru, Hiroshi Aoki, Shingo Nakamura, Shin-ichiro Nakamura, Megumi Takikawa, Motoaki Hanzawa, Satoko Kishimoto et al. "Hydrogel blends of chitin/chitosan, fucoidan and alginate as healing-impaired wound dressings." *Biomaterials* 31, no. 1 (2010): 83–90.

Murugesh Babu K. *Silk: Processing, Properties and Applications*. Woodhead Publishing; 2013.

Nakajima, Yukari, Yuki Nakano, Sono Fuwano, Natsumi Hayashi, Yukiho Hiratoko, Ayaka Kinoshita, Megumi Miyahara et al. "Effects of three types of Japanese honey on full-thickness wound in mice." *Evidence-Based Complementary and Alternative Medicine* 2013, no. 1 (2013): 504537.

Nasir, Nur-Azida Mohd, Ahmad Sukari Halim, Kirnpal-Kaur Banga Singh, Ananda Aravazhi Dorai, and Mehru-Nisha Muhammad Haneef. "Antibacterial properties of tualang honey and its effect in burn wound management: A comparative study." *BMC Complementary and Alternative Medicine* 10 (2010): 1–7.

Neto, Corintio Mariani, Rosemeire Sartori de Albuquerque, Sonia Cristina de Souza, Renata Oliveira Giesta, Andrea Penha Spinola Fernandes, and Bárbara Mondin. "Comparative study of the use of HPA lanolin and breast milk for treating pain associated with nipple trauma." *Revista Brasileira de Ginecologia e Obstetrícia/RBGO Gynecology and Obstetrics* 40, no. 11 (2018): 664–672.

Niiyama, Hayato, and Yoshimitsu Kuroyanagi. "Development of novel wound dressing composed of hyaluronic acid and collagen sponge containing epidermal growth factor and vitamin C derivative." *Journal of Artificial Organs* 17 (2014): 81–87.

Nozimoto, Ingrid Naomi Pires, Beatriz Aparecida da Silva, Melissa Diniz Bandeira, Ana Paula da Silva, Sandra Kalil Bussadori, Elaine Marcílio Santos, and Ana Luiza Cabrera Martimbianco. "Nonpharmacological interventions for treating breastfeeding nipple pain: Systematic review and meta-analysis." *Breastfeeding Medicine* 19, no. 8 (2024): 599–611.

Otsuka, Takayoshi, Ho-Man Kan, and Cato T. Laurencin. "Regenerative engineering approaches to scar-free skin regeneration." *Regenerative Engineering and Translational Medicine* 8, (2022): 1–23.

Park, Ye Ri, Md Tipu Sultan, Hyun Jung Park, Jung Min Lee, Hyung Woo Ju, Ok Joo Lee, Dong Jin Lee, David L. Kaplan, and Chan Hum Park. "NF-κB signaling is key in the wound healing processes of silk fibroin." *Acta Biomaterialia* 67 (2018): 183–195.

Patel, Munir, and John C. Lantis II. "Fish skin acellular dermal matrix: Potential in the treatment of chronic wounds." *Chronic Wound Care Management and Research* (2019): 59–70.

Park Ye Ri, Md Tipu Sultan, Hyun Jung Park, Jung Min Lee, Hyung Woo Ju, Ok Joo Lee, Dong Jin Lee, David L. Kaplan and Chan Hum Park. "NF-κB signaling is key in the wound healing processes of silk fibroin." *Acta Biomaterialia* 67, (2018): 183–195.

Paul, Sayan, Subburathinam Balakrishnan, Arun Arumugaperumal, Saranya Lathakumari, Sandhya Soman Syamala, Vijithkumar Vijayan, Selvan Christyraj Jackson Durairaj, Vaithilingaraja Arumugaswami, and Sudhakar Sivasubramaniam. "Importance of clitellar tissue in the regeneration ability of earthworm *Eudrilus eugeniae*." *Functional & Integrative Genomics* 22, no. 4 (2022): 1–32.

Paydar, Shahram, Majid Akrami, Amirreza Dehghanian, Roshanak Alavi Moghadam, Mohsen Heidarpour, Amir Bahari Khoob, and Behnam Dalfardi. "A comparison of the effects of alpha and medical-grade honey ointments on cutaneous wound healing in rats." *Journal of Pharmaceutics* 2016, no. 1 (2016): 9613908.

Ponesakki, Vasanthakumar, Sandhya Soman Syamala, Sayan Paul, Arun Arumugaperumal, Padmashree Selvakumar, Vijithkumar Vijayan, A. P. M. S. Raj, and Sudhakar Sivasubramaniam. "Activation of mouse ear lobe tissue regeneration by metabolites of earthworm." *J. Exp. Zool* 26, no. 1(2023): 1191–1205.

Prakash, Manikandan, Manoharan Balamurugan, Kandasamy Parthasarathi, Gopal Gunasekaran, Edwin L Cooper and Latchumanan S. Ranganathan. "Anti-ulceral and anti-oxidative properties of" earthworm paste" of Lampito mauritii (Kinberg) on Rattus Norvegicus." *European Review for Medical & Pharmacological Sciences* 11, no. 1 (2007).

Pretus, Howard A., Howard. E. Ensley, Robert B. McNamee, Edwin L. Jones, I. William Browder, and D. L. Williams. "Isolation, physicochemical characterization and preclinical efficacy evaluation of soluble scleroglucan" *The Journal of Pharmacology and Experimental Therapeutics* 257, no. 1(1991): 500–510.

Qi, Yu, Hui Wang, Kai Wei, Ya Yang, Ru-Yue Zheng, Ick Soo Kim, and Ke-Qin Zhang. "A review of structure construction of silk fibroin biomaterials from single structures to multi-level structures." *International Journal of Molecular Sciences* 18, no. 3 (2017): 237.

Qin, Zhen, Robert M. Balimunkwe, and Tian Quan. "Age-related reduction of dermal fibroblast size upregulates multiple matrix metalloproteinases as observed in aged human skin in vivo." *British Journal of Dermatology* 177, no. 5(2017): 1337–1348.

Raafat, Amany I., Naeem M. El-Sawy, Nagwa A. Badawy, Eglal A. Mousa, and Asmaa M. Mohamed. "Radiation fabrication of Xanthan-based wound dressing hydrogels embedded ZnO nanoparticles: In vitro evaluation." *International Journal of Biological Macromolecules* 118 (2018): 1892–1902.

Rahmani, Sahar, and David J. Mooney. "Tissue-engineered wound dressings for diabetic foot ulcers." *The Diabetic Foot: Medical and Surgical Management* (2018): 247–256.

Ram, Mahendra, Vishakha Singh, Sanjay Kumawat, Vinay Kant, Surendra Kumar Tandan, and Dinesh Kumar. "Bilirubin modulated cytokines, growth factors and angiogenesis to improve cutaneous wound healing process in diabetic rats." *International Immunopharmacology* 30 (2016): 137–149.

Ramli, Nur Zuliani, Kok-Yong Chin, Khairul Anwar Zarkasi, and Fairus Ahmad. "A review on the protective effects of honey against metabolic syndrome." *Nutrients* 10, no. 8 (2018): 1009.

Rastogi, Sanjay, Mancy Modi, and Brijesh Sathian. "The efficacy of collagen membrane as a biodegradable wound dressing material for surgical defects of oral mucosa: A prospective study." *Journal of Oral and Maxillofacial Surgery* 67, no. 8 (2009): 1600–1606.

Reinke, Johannes M., and Heiko Sorg. "Wound repair and regeneration." *European Surgical Research* 49, no. 1(2012): 35–43.

Rojczyk, Ewa, Agnieszka Klama-Baryła, Wojciech Łabuś, Katarzyna Wilemska-Kucharzewska, and Marek Kucharzewski. "Historical and modern research on propolis and its application in wound healing and other fields of medicine and contributions by Polish studies." *Journal of Ethnopharmacology* 262 (2020): 113159.

Rosanto, Yosaphat Bayu, Cahya Yustisia Hasan, Rahardjo Rahardjo, and Tri Wahyu Pangestiningsih. "Effect of snail mucus on angiogenesis during wound healing." *F1000Research* 10 (2021): 181.

Saini, Sapna, Anju Dhiman, and Sanju Nanda. "Traditional Indian medicinal plants with potential wound healing activity: A review." *International Journal of Pharmaceutical Sciences and Research* 7, no. 5 (2016): 1809.

Salama, Walaa H., Ahmed A. Abd-Rabou, Roqaya I. Bassuiny, Amr E. El Hakim, and Yasser E. Shahein. "Exploration of antimicrobial and anticancer activities of L-amino acid oxidase from Egyptian *Naja haje* venom." *Toxicon* 242 (2024): 107708.

Salatino, Antonio, Érica Weinstein Teixeira, Giuseppina Negri, and Dejair Message. "Origin and chemical variation of Brazilian propolis." *Evidence-Based Complementary and Alternative Medicine* 2, no. 1 (2005): 33–38.

Sarrigiannidis, Stylianos O., José M. Rey, Oana Dobre, Cristina González-García, Matthew J. Dalby, and Manuel Salmeron-Sanchez. "A tough act to follow: Collagen hydrogel modifications to improve mechanical and growth factor loading capabilities." *Materials Today Bio* 10 (2021): 100098.

Segalova, Natalia E., Anna M. Dubinskaya, and Lidiya A. Ivanova. "Preparation of finely divided collagen powder by low-emperature grinding." *Pharmaceutical Chemistry Journal* 15, no. 7(1981): 523–527.

Sforcin, Jose M. "Propolis and the immune system: a review." *Journal of ethnopharmacology* 113, no. 1(2007): 1–14.

Sharifi, Shahriar, Mohammad Javad Hajipour, Lisa Gould, and Morteza Mahmoudi. "Nanomedicine in healing chronic wounds: Opportunities and challenges." *Molecular Pharmaceutics* 18, no. 2 (2020): 550–575.

Shaw, Tanya J., and Paul Martin. "Wound repair at a glance." *Journal of Cell Science* 122, no. 18 (2009): 3209–3213.

Shearer, Jeffry D., John R. Richards, Charles D. Mills, and Michael D. Caldwell. "Differential regulation of macrophage arginine metabolism: A proposed role in wound healing." *American Journal of Physiology-Endocrinology and Metabolism* 272, no. 2 (1997): E181–E190.

Shedoeva, Aleksandra, David Leavesley, Zee Upton, and Chen Fan. "Wound healing and the use of medicinal plants." *Evidence-Based Complementary and Alternative Medicine* 2019, no. 1 (2019): 2684108.

Shen, Weiliang, Xiao Chen, Jialin Chen, Zi Yin, Boon Chin Heng, Weishan Chen, and Hong-Wei Ouyang. "The effect of incorporation of exogenous stromal cell-derived factor-1 alpha within a knitted silk-collagen sponge scaffold on tendon regeneration." *Biomaterials* 31, no. 28 (2010): 7239–7249.

Shi, Yan, Bin, Shu, Ronghua Yang, Yingbin Xu, Bangrong Xing, Jian Liu, Lei Chen et al. "Wnt and Notch signaling pathway involved in wound healing by targeting c-Myc and Hes1 separately." *Stem Cell Research & Therapy* 6 (2015): 1–13.

Shimura, Kazuhiko. "The physiology and biology of spinning in Bombyx mori. IV. Chemical composition and biosynthesis of silk proteins." (1983): 455-461.

Shin, Joong-Han, Alexandra CM Oliveira, and Barbara A. Rasco. "Quality attributes and microbial storage stability of caviar from cultivated white sturgeon (*Acipenser transmontanus*)." *Journal of Food Science* 75, no. 1 (2010): C43–C48.

Sig, Ali K., Mustafa Guney, Aylin Uskudar Guclu, and Erkan Ozmen. "Medicinal leech therapy—An overall perspective." *Integrative Medicine Research* 6, no. 4 (2017): 337–343.

Silici, Sibel, and Semiramis Kutluca. "Chemical composition and antibacterial activity of propolis collected by three different races of honeybees in the same region." *Journal of Ethnopharmacology* 99, no. 1 (2005): 69–73.

Silva, Jessica R., Beatriz Burger, Carolina MC Kühl, Thamiris Candreva, Mariah BP Dos Anjos, and Hosana G. Rodrigues. "Wound healing and omega-6 fatty acids: From inflammation to repair." *Mediators of Inflammation* 2018, no. 1 (2018): 2503950.

Singer, Adam J., and Richard AF Clark. "Cutaneous wound healing." *New England Journal of Medicine* 341, no. 10 (1999): 738–746.

Singh, Ran V., Hari P. Aithal, Prakash Kinjavdekar, Abhijit M. Pawde, Karam P. Singh, Mudasir B. Gugjoo, and Amarpal. "Evaluation of the healing potential of earthworm coelomic fluid in full thickness skin wounds in rabbits." *Veterinarski arhiv* 85, no. 2 (2015): 189–200.

Smith, Brian N., and Virginia A, Odero-Marah. "The role of Snail in prostate cancer." *Cell adhesion & migration* 6 no. 5 (2012): 433–441.

Song, Jia, Cheng Li, Yushan Lv, Yi Zhang, William Kwame Amakye, and Limei Mao. "DHA increases adiponectin expression more effectively than EPA at relative low concentrations by regulating PPARγ and its phosphorylation at Ser273 in 3T3-L1 adipocytes." *Nutrition & Metabolism* 14 (2017): 1–11.

Stabili, Loredana, Roberto Schirosi, Maria Giovanna Parisi, Stefano Piraino, and Matteo Cammarata. "The mucus of *Actinia equina* (Anthozoa, Cnidaria): An unexplored resource for potential applicative purposes." *Marine Drugs* 13, no. 8 (2015): 5276–5296.

Stocker, Roland, Yorihiro Yamamoto, Antony F. McDonagh, Alexander N. Glazer, and Bruce N. Ames. "Bilirubin is an antioxidant of possible physiological importance." *Science* 235, no. 4792 (1987): 1043–1046.

Sudharsan, S., P. Seedevi, R. Saravanan, P. Ramasamy, S. Kumar Vasanthakumar, Vairamani, A. Srinivasan, and A. Shanmugam. "Isolation, characterization and molecular weight determination of collagen from marine sponge Spirastrella inconstans (Dendy)." *African Journal of Biotechnology* 12, no. 5(2013): 504–511.

Swenson Stephen D, Swapnika Ramu and Francis S. Markland. "Anti-angiogenesis and RGD-containing snake venom disintegrins." *Current Pharmaceutical Design* 13, no. 28 (2007): 2860–2871.

Szycher, Michael, and Steven James Lee. "Modern wound dressings: A systematic approach to wound healing." *Journal of Biomaterials Applications* 7, no. 2 (1992): 142–213.

Terkelsen, Louise Holmen, Anni Eskild-Jensen, Hanne Kjeldsen, John Howard Barker, and Vibeke Elisabeth Hjortdal. "Topical application of cod liver oil ointment accelerates wound healing: An experimental study in wounds in the ears of hairless mice." *Scandinavian Journal of Plastic and Reconstructive Surgery and Hand Surgery* 34, no. 1 (2000): 15–20.

Thirumaleshwar, Shailesh, Parthasarthi K Kulkarni, and Devegowda V Gowda. "Liposomal hydrogels: A novel drug delivery system for wound dressing." *Current Drug Therapy* 7, no. 3 (2012): 212–218.

Thuraisingam, Thusanth, Yong Zhong Xu, Kalyn Eadie, Mitra Heravi, Marie-Cristine Guiot, Rony Greemberg, Matthias Gaestel, and Danuta Radzioch. "MAPKAPK-2 signaling is critical for cutaneous wound healing." *Journal of Investigative Dermatology* 130, no. 1 (2010): 278–286.

Toreti, Viviane Cristina, Helia Harumi Sato, Glaucia Maria Pastore, and Yong Kun Park. "Recent progress of propolis for its biological and chemical compositions and its botanical origin." *Evidence-Based Complementary and Alternative Medicine* 2013, no. 1 (2013): 69

Trapella, Claudio, Roberta Rizzo, Stefania Gallo, Andrea Alogna, Daria Bortolotti, Fabio Casciano, Giorgio Zauli, Paola Secchiero, and Rebecca Voltan. "HelixComplex snail mucus exhibits pro-survival, proliferative and pro-migration effects on mammalian fibroblasts." *Scientific Reports* 8, no. 1 (2018): 17665.

Tsoutsos, Dimosthenis, Despoina Kakagia, and Konstantinos Tamparopoulos. "The efficacy of *Helix aspersa* Müller extract in the healing of partial thickness burns: A novel treatment for open burn management protocols." *Journal of Dermatological Treatment* 20, no. 4 (2009): 219–222.

Ulagesan, Selvakumari, Kamatchi Sankaranarayanan, and Amutha Kuppusamy. "Functional characterisation of bioactive peptide derived from terrestrial snail *Cryptozona bistrialis* and its wound-healing property in normal and diabetic-induced Wistar albino rats." *International Wound Journal* 15, no. 3 (2018): 350–362.

Urai, Tamae, Kanae Mukai, Junko Sugama, and Toshio Nakatani. "Effectiveness of Indonesian honey on the acceleration of cutaneous wound healing: An experimental study in mice." PhD diss., Kanazawa University, 2012.

Uzel, Atac, Özant Önçağ, Dilşah Çoğulu, and Ömür Gençay. "Chemical compositions and antimicrobial activities of four different Anatolian propolis samples." *Microbiological Research* 160, no. 2 (2005): 189–195.

Vepari, Charu, and David L. Kaplan. "Silk as a biomaterial." *Progress in Polymer Science* 32, no. 8–9 (2007): 991–1007.

Vijaya, Kumari K., and K. Nishteswar. "Wound healing activity of honey: A pilot study." *AYU (An International Quarterly Journal of Research in Ayurveda)* 33, no. 3 (2012): 374–377.

Wallner, Christoph, Jana Holtermann, Marius Drysch, Sonja Schmidt, Felix Reinkemeier, Johannes Maximilian Wagner, Mehran Dadras et al. "The use of intact fish skin as a novel treatment method for deep dermal burns following enzymatic debridement: A retrospective case-control study." *European Burn Journal* 3, no. 1 (2022): 43–55.

Wang, Ling, Xiaolin Wu, Ting Shi, and Luo Lu. "Epidermal growth factor (EGF)-induced corneal epithelial wound healing through nuclear factor κB subtype-regulated CCCTC binding factor (CTCF) activation." *Journal of Biological Chemistry* 288, no. 34 (2013): 24363–24371.

Wang, Xiaosan, Wang Xiaohan, Yang Chen, Wenhua Jin, Qingzhe Jin, and Xingguo Wang. "Enrichment of branched chain fatty acids from lanolin via urea complexation for infant formula use." *LWT* 117 (2020): 108627.

Wang, Zheng, Yeshun Zhang, Jinxiang Zhang, Lei Huang, Jia Liu, Yongkui Li, Guozheng Zhang, Subhas C. Kundu, and Lin Wang. "Exploring natural silk protein sericin for regenerative medicine: An injectable, photoluminescent, cell-adhesive 3D hydrogel." *Scientific Reports* 4, no. 1 (2014): 7064.

Weller, Carolina D., Victoria Team, and Geoffrey Sussman. "First-line interactive wound dressing update: A comprehensive review of the evidence." *Frontiers in Pharmacology* 11 (2020): 155.

Wu, Yuebo, Zhipeng Zhou, Lan Luo, Maixian Tao, Xiao Chang, Lian Yang, Xingyong Huang, Lin Hu, and Mingyi Wu. "A non-anticoagulant heparin-like snail glycosaminoglycan promotes healing of diabetic wound." *Carbohydrate Polymers* 247 (2020): 116682.

Xing Wei, Wei Guo, Cun-Hua Zou, Ting-Ting Fu, Xiang-Yun Li, Ming Zhu, Jun-Hua Qi, Jiao Song, Chen-Hui Dong, Zhuang Li, Yong Xiao, Pei-Song Yuan, Hong Huang and Xiang Xu. "Acemannan accelerates cell proliferation and skin wound healing through AKT/mTOR signaling pathway." *Journal of Dermatological Science* 79, no. 2 (2015): 101–109.

Xu, Yuanjun, Liping Luo, Bin Chen, and Yuxin Fu. "Recent development of chemical components in propolis." *Frontiers of Biology in China* 4 (2009): 385–391.

Yang, Yuwei, Haicong Hu, Wenqi Wang, Xiaojie Duan, Shilin Luo, Xiongfei Wang, and Yikun Sun. "The identification of functional proteins from amputated lumbricus *Eisenia fetida* on the wound healing process." *Biomedicine & Pharmacotherapy* 95 (2017): 1469–1478.

Yoshida, Shinya, Koji Yasutomo, and Toshiyuki Watanabe. "Treatment with DHA/EPA ameliorates atopic dermatitis-like skin disease by blocking LTB4 production." *The Journal of Medical Investigation* 63, no. 3.4 (2016): 187–191.

Young, Alistair, and Clare-Ellen McNaught. "The physiology of wound healing." *Surgery (Oxford)* 29, no. 10 (2011): 475–479.

Yuksel, Yasemin Topal, Line Brok Nørreslet, and Jacob P. Thyssen. "Allergic contact dermatitis in patients with atopic dermatitis." *Current Dermatology Reports* 10 (2021): 67–76.

Zhang, Mei, Xu Li, Yi Liu, Feng Ye, and Genquan Qiu. "Effects of extract of dilong (pheretima) on the scalded skin in rats." *Journal of Traditional Chinese Medicine= Chung i tsa Chih Ying wen pan* 26, no. 1 (2006): 68–71.

Zhou, Zhipeng, Tuo Deng, Maixian Tao, Lisha Lin, Luyun Sun, Xuemei Song, and Dongxiu Gao et al. "Snail-inspired AFG/GelMA hydrogel accelerates diabetic wound healing via inflammatory cytokines suppression and macrophage polarization." *Biomaterials* 299 (2023): 122141.

Zubair, Raheel, and Nasir Aziz. "As smooth as honey—The historical use of honey as topical medication." *JAMA Dermatology* 151, no. 10 (2015): 1102–1102.

9 Therapeutic Approaches to Wound Healing

Amna Tazheer[1], *Huma Hameed*[1*], *Tehreem Mushtaq*[1], *Mahtab Ahmad Khan*[1], *Muhammad Zaman*[1], *Ana Cláudia Paiva-Santos*[2,3], *Ali Irfan*[4], and *Tooba Mehboob*[1]

[1] Faculty of Pharmaceutical Sciences, University of Central Punjab (UCP), Lahore, Pakistan

[2] Department of Pharmaceutical Technology, Faculty of Pharmacy of the University of Coimbra, University of Coimbra, Coimbra, Portugal

[3] REQUIMTE/LAQV, Group of Pharmaceutical Technology, Faculty of Pharmacy of the University of Coimbra, University of Coimbra, Coimbra, Portugal

[4] The Green Institute of Chemical, Biomedical and Environmental Sciences (GICBES), Lahore, Pakistan

9.1 CONVENTIONAL APPROACHES FOR WOUND HEALING

Skin constitutes 16% of the overall body weight, covering an average area of 1.85 m^2, and is recognized as the largest organ in the human body. Besides its crucial role in maintaining homeostasis, the skin serves as a protective barrier against external elements, helping to prevent infections and fluid loss. Consequently, maintaining the integrity of the skin is essential for overall health. Skin injuries or wounds resulting from trauma, burns, chronic illnesses, and surgical procedures can lead to disability and emotional distress, presenting significant challenges for healthcare systems worldwide. Skin wounds are categorized into acute (healing) and chronic (non-healing) wounds, depending on their effects and underlying causes. Minor acute wounds can be repaired and typically heal through the skin's inherent self-healing capabilities. The skin's healing process is facilitated by various dynamically coordinated cellular responses, remodeling of the extracellular matrix (ECM), and the involvement of growth factors (Aragona et al., 2017). This healing process aims to restore both the functional and anatomical integrity of the skin. In contrast, larger skin injuries and chronic wounds tend to heal slowly, making them challenging to treat due to their vulnerability to infections and fluid loss. Even with rigorous investigations and numerous strategies aimed at improving skin wound healing, the treatment of large and chronic wounds still represents an unmet therapeutic need, primarily due to the complexities involved in assessment and management of wound care. At present, achieving effective and high-quality wound healing presents significant challenges for both clinicians and researchers, leading to substantial costs. This is evidenced by the expanding global market for wound care products, which was approximately USD 12 billion in 2020 and is projected to grow to USD 18.7 billion

* Corresponding author: huma.hameed@ucp.edu.pk

DOI: 10.1201/9781003644958-9

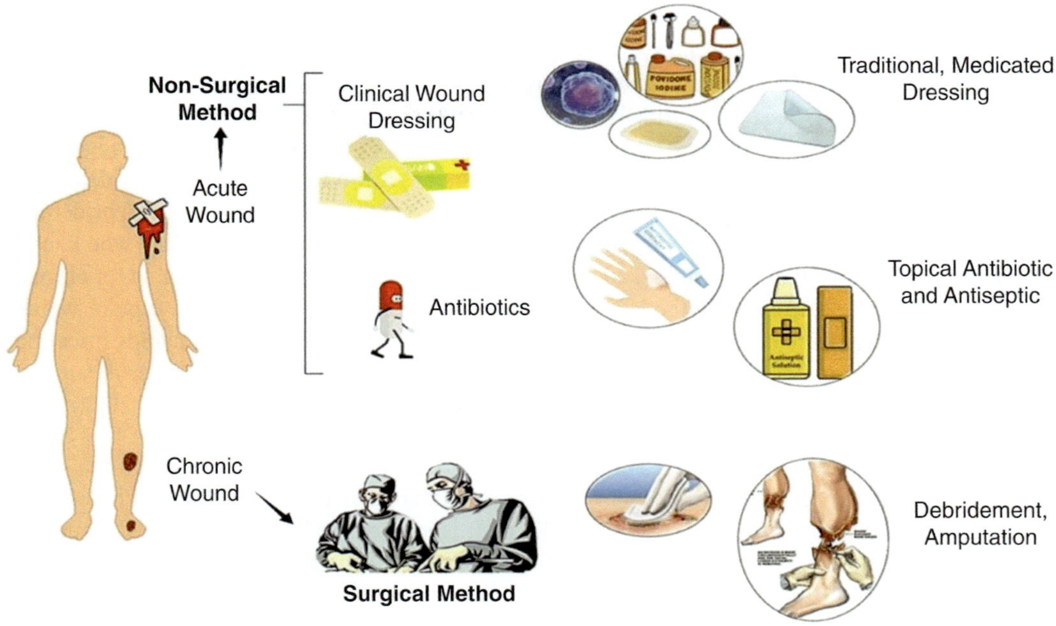

FIGURE 9.1 Conventional approaches in wound healing. The figure describes the conventional wound healing processes that are classified mainly into surgical and non-surgical procedures. Non-surgical procedures are commonly used in acute wounds and comprise clinical dressing of the wound, medicated or traditional dressings, topical antibiotics, and antiseptic solutions and many more choices for preventing infection and promoting natural healing. Conversely, surgical intervention is reserved for chronic or severe wounds, involving debridement, amputation, etc. to eliminate necrotic tissue and encourage healing. These established methods are the basis of clinical wound management. (Image generated through BioRender Software.)

by 2027 (Sen, 2021). Typically, skin wound treatments are categorized as either "Conventional" or "Regenerative".

Conventional wound healing techniques are those traditional methods employed to treat and care for wounds. Alternative strategies for wound care often prioritize the maintenance of a sterile and hydrated environment to shield the wound, reduce the risk of infection, and facilitate the natural healing process. They are all categorically divided into two broad types: surgical and non-surgical as shown in Figure 9.1. Each technique is important in facilitating tissue repair and avoiding complications.

9.1.1 Surgical Procedures in Wound Healing

Surgical methods of wound healing are normally used when the wound is big, deep, or complex and needs invasive procedures to ensure a good recovery (Tavelli et al., 2023). Surgical methods deal with physically maneuvering the tissue to regain its integrity and function. Some common surgical procedures involve debridement, flap surgery, suturing, and stapling and skin substitutes and dermal matrices. These surgical methods are important in treating wounds that cannot heal on their own or need more sophisticated treatments to achieve quicker recovery, reduce complications, and enhance cosmetic results.

9.1.1.1 Debridement

Debridement is a removal of necrotic, infectious, or dead tissue from lesions by surgery that will help them heal faster and more effectively (Mayer et al., 2024; Tognetti et al., 2021). It can be achieved through one of several debridement modalities, sharp (scalpel), mechanical (irrigation or gauze), enzymatic (surface enzymes), autolytic (moist retainers), or biological (sterile maggot therapy) (Nowak et al., 2022). Sharp debridement is fast but needs a good professional. Debridement of the mechanical variety may be hurtful and actually damage healthy tissue. Enzymatic debridement is selective but slow, whereas autolytic debridement is gentle and suitable for patients who cannot tolerate more vigorous methods. Biological debridement is highly selective and suitable for chronic wounds (Tran et al., 2023). All methods have special indications depending on the type of wound and condition of the patient. Effective and timely debridement increases healing, avoids infection, and prepares the wound bed for additional treatment. It is usually the initial step prior to surgical closure or grafting. Effective wound care starts with proper debridement (Dayya et al., 2022).

9.1.1.2 Skin Grafting

Skin grafting is the process of taking healthy skin from a donor site and transplanting it to a wound site that cannot heal naturally (Ozhathil et al., 2021). It is applied for burns, ulcers, and traumatic wounds. Two forms are prevalent: split-thickness grafts, which consist of the epidermis and some of the dermis, and full-thickness grafts, which contain both layers of skin and are utilized in areas where they will be visible for better cosmetic appearance (Schlottmann et al., 2021). Mesh grafts can expand and drain, best for extensive wounds. Autografts, which are taken from the patient's own body, minimize rejection risk. Allografts and xenografts are temporary from donors or animals (Schlottmann et al., 2021). Proper wound bed preparation, vascular supply, and infection control are the keys to successful grafting. Immobilization after surgery helps the graft adhere and survive (Stone et al., 2021). Grafts speed healing, minimize fluid loss, and guard underlying tissues. Under appropriate care, grafts greatly enhance wound healing and appearance (Yoon et al., 2022).

9.1.1.3 Flap Surgery

Flap surgery is a method in which live tissue and its blood supply are moved to close a wound (Thomas et al., 2021; Windisch et al., 2022). It is employed for compound wounds with bare bones, tendons, or joints that are more than a skin graft (Hemmingsson et al., 2022). Flaps may be local (next to the area), regional (not next to but proximal to the area), or free (from a distant area with a need for microvascular anastomosis). It provides improved contour, longevity, and functional recovery (Inouye et al., 2021). Transferred tissue could be skin, muscle, fat, or fascia. Survival of the flap depends on vascular supply; hence, careful planning before surgery is a must. Post-operative follow-up involves looking for complications such as necrosis or infection (Tan et al., 2021). Flaps offer early and permanent cover for large and deep wounds. They allow for both structural and aesthetic restoration. Flap surgery is a gold standard for reconstructive wound care (Rogoń et al., 2024).

9.1.1.4 Suturing and Stapling

Suturing and stapling are basic methods employed to close wounds and promote healing. Sutures may be absorbable, which dissolve with time, or non-absorbable, which need to be removed after healing. Staples are metal clips employed for quick closure, especially in surgical wounds and high-tension sites (Chandrasekaran et al., 2023). Technique requires alignment of wound edges with minimal tension to minimize scarring and dehiscence. Different types of stitching, such as interrupted or continuous sutures, are chosen according to the location and type of wound. Sterile technique is crucial to avoid infection during surgery (Van de Kuit et al., 2022). Early removal of non-absorbable materials provides the best cosmetic outcome. These techniques enhance tissue approximation, reduce contamination, and accelerate healing (Balomenos et al., 2023). Advanced adhesive strips or tissue glues can be employed as adjuncts to improve outcomes. Suturing and stapling are effective and necessary maneuvers in wound management (Chandrasekaran et al., 2024). Skin substitutes and dermal matrices are bioengineered materials that facilitate wound healing by replicating the structure of natural skin (Sierra-Sánchez et al., 2021). They are especially useful in situations where autografts are not possible because there are insufficient donor sites (Bay et al., 2021). They are scaffolds that enhance cell migration, angiogenesis, and dermal regeneration. They assist in lowering the risk of infection and inflammation and hastening healing in acute and chronic wounds (Tognetti et al., 2021). Skin substitutes are either temporary or permanent and are frequently utilized in the treatment of burns and diabetic foot ulcers. Their successful use is based on appropriate wound bed preparation and aseptic conditions. They provide enhanced cosmetic results and lower the incidence of multiple operations (Holl et al., 2021). This strategy involves the convergence of biotechnology and clinical science for improved healing of wounds. With growing innovation, skin substitutes are increasingly being used in wound care (Palackic et al., 2022). Surgical wound healing procedures are described in Table 9.1.

9.1.2 Non-Surgical Therapies in Wound Healing

Non-surgical wound healing directs attention toward ensuring and improving natural repair processes within the body without resorting to invasive methods (Lei et al., 2024). The strategies for non-surgical wound healing center on optimizing the wound environment, tissue regeneration, and complication prevention, including infection (Dixon and Edmonds, 2021). Dressings like hydrocolloids, foams, and hydrogels provide the wound with an optimal level of moisture to speed up healing, as well as to protect it against external pathogens. Topical therapies, such as antimicrobial ointments, growth factors, and enzymatic creams, aid in the healing process through bacterial load reduction and tissue regeneration stimulation. Compression is usually utilized in chronic venous ulcers for the promotion of circulation and swelling reduction, while offloading methods are essential in pressure ulcer prevention, especially in diabetics (Heyens et al., 2021). More advanced non-surgical treatments such as electrical stimulation, ultrasound, and nutritional supplementation have also demonstrated positive impacts in wound healing through increased cell migration, decreased inflammation, and tissue repair. Such non-surgical approaches are especially useful in the treatment of chronic or non-healing wounds, offering viable alternatives to, or adjunctive therapy with, surgical intervention (Mamun et al., 2024b).

9.1.2.1 Wound Dressings

Current wound dressings are a key part of acute and chronic wound management in providing a moist healing environment. They facilitate optimal temperature and pH, exudate absorption, and prevention of the wound from external infection (Liang et al., 2021). Hydrocolloid dressings facilitate autolytic debridement by encapsulating wound fluids. Foam dressings are extremely absorbent and applied to heavy exudate wounds (Laurano et al., 2022). Hydrogel dressings moisten dry or necrotic tissue, facilitating gentle debridement. Seaweed-derived alginate dressings support hemostasis and are appropriate for bleeding wounds (Brumberg et al., 2021). Transparent films facilitate oxygen exchange with visibility of the wound area. Antimicrobial dressings, including those with silver or iodine, are used to prevent infection. Dressing selection is based on the type of wound, size, depth, and amount of exudate. Frequent dressing changes and correct technique optimize the healing of wounds. These dressings have a very important function in non-surgical wound care (Wang et al., 2022).

9.1.2.2 Topical Drugs

Topical drugs are applied directly onto wounds to facilitate healing by providing antimicrobial, anti-inflammatory, or regenerative mediators (Derwin et al., 2022). Examples include silver sulfadiazine for burns, mupirocin for bacterial infections, and collagenase ointment for enzymatic debridement. Growth factors such as platelet-derived growth factor (PDGF) may be applied topically to induce cell proliferation and angiogenesis (Cancio, 2021). Topical agents decrease bacterial load, modulate inflammation, and induce granulation tissue formation. Their efficacy relies on proper diagnosis and wound status. Inadequate use or misuse may create resistance or hypersensitivity

TABLE 9.1
Surgical Procedures in Wound Healing

Surgical Procedure	Mechanism of Action	Advantages	Disadvantages	Current Application	Marketed Product	Future Perspectives	Reference
Debridement	Removal of necrotic tissue stimulates wound healing by resetting the wound bed.	Promotes healing Reduces infection risk	Painful May remove healthy tissue	Chronic wounds, ulcers, burn care	Versajet Hydrosurgery, Debrisoft Pad	Advancements in enzymatic and biological debridement agents.	Claes et al. (2023) and Misra et al. (2023)
Skin grafting	Grafts provide immediate wound coverage and integrate with the host tissue	Immediate coverage Prevents infection and fluid loss	Requires donor site Scarring possible	Burns, trauma, post-tumor excision wounds	Integra, Epicel, AlloDerm	Development of stem cell-based and gene-enhanced grafts for better integration and function	Braza and Fahrenkopf (2023) and Prabowo et al. (2022)
Flap surgery	Transfers vascularized tissue to the wound site to support healing and function	Vascularized tissue lowers infection risk Restores function.	Complex procedure Longer recovery	Large trauma wounds, reconstructive surgery	DIEP flap systems, free flap devices	Robotic-assisted flap surgery and regenerative tissue flaps for enhanced precision and outcomes	Rogoń et al. (2024) and Shaikh et al. (2021)
Suturing and stapling	Physically approximates wound edges to support tissue regeneration and closure	Fast closure Reduces infection risk.	Scarring Potential infection	Surgical incisions, trauma lacerations	Ethicon sutures, Covidien staplers	Smart sutures with sensors and antimicrobial properties are in development	Chandrasekaran et al. (2024) and Al-Ajlouni et al. (2023)
Skin substitutes	Acts as a scaffold for cell growth and re-epithelialization.	Promotes healing Protects wound.	Expensive Risk of poor integration.	Burns, diabetic ulcers, venous leg ulcers	Apligraf, Dermagraft, Biobrane	Use of stem cells and 3D bioprinting to create fully functional skin equivalents	Vecin and Kirsner (2023) and Potekaev et al. (2021)

(De Francesco et al., 2022). Topicals are combined with dressings for extended release in some of them. The sterile application avoids contamination. As used properly, topical agents hasten healing and enhance clinical outcome. They play a significant role in non-surgical wound management strategies (Wong et al., 2022).

9.1.2.3 Antimicrobials and Antiseptics

Antimicrobials and antiseptics prevent and control infection of wounds by lowering the microbial load (Cwajda-Białasik et al., 2022). Povidone-iodine and chlorhexidine are frequently used antiseptics for cleaning surgical areas or wounds. Silver dressings exert long-term antimicrobial action and are particularly helpful in chronic wounds (Maillard et al., 2021). Topical antibiotics such as mupirocin and neomycin have specific activity against bacteria. Effective as they are, excessive use of antiseptics harms healthy tissue and prolongs healing (Alves et al., 2021). Thus, they are generally prescribed for a short duration or in infected wounds. Topical antimicrobial resistance is rare but must still be watched out for. These drugs also aid in the control of biofilm-related infections. They need to be applied with care under sterile conditions. In conclusion, antiseptics and antimicrobials are essential in wound healing as they control infection-related complications (Daeschlein, 2013).

9.1.2.4 Compression Therapy

Compression therapy is mostly applied in the treatment of venous leg ulcers and chronic venous insufficiency. It entails controlled pressure application to enhance venous return, minimize edema, and increase tissue oxygenation (Nickles et al., 2023). Elastic bandages, compression stockings, and multi-layer compression systems are some of the common equipment used. Compression decreases inflammation, enhances lymphatic drainage, and promotes ulcer healing (Patton et al., 2023). Technique is critical to prevent excessive pressure, which may hinder circulation. It is contraindicated in peripheral arterial disease patients except when vascular status has been evaluated (Shi et al., 2021). Compliance with therapy is a major factor in outcomes since chronic use is necessary. It also prevents recurrence of ulcer. Patient education and follow-up are necessary for long-term success. It is a non-invasive, cost-effective method of treating chronic wounds.

9.1.2.5 Offloading Methods

Offloading methods are used to reduce pressure on certain areas to enable wounds to heal, particularly in diabetic foot ulcers (Bus et al., 2024). Total contact casts (TCC) are the gold standard and evenly distribute weight over the foot. Removable devices such as controlled ankle movement (CAM) walkers and custom orthotics provide flexibility and frequent wound examination. Specialized footwear and insoles are used to prevent recurrence of ulcers (Tran and Haley, 2021). Shear forces and microtrauma are decreased by offloading, which can exacerbate wounds. Pressure mapping may direct the most beneficial offloading. Compliance is paramount; patient education greatly enhances compliance. Inappropriate use may cause delayed healing or new pressure points (Tai et al., 2024). Offloading is usually used in conjunction with dressings or debridement for best effect. It is a useful, non-surgical intervention to aid healing in pressure-induced and neuropathic wounds.

9.1.2.6 Hypertonic Saline and Sugar Pastes

Hypertonic saline and sugar pastes are effective but conventional agents for wound care in infected or necrotic wounds. Their high osmolarity extracts fluid from the wound, lowering edema and bacterial burden (de Cabrera et al., 1988). Dehydration of microorganisms occurs by dressings of hypertonic saline, reducing the viability of the environment for bacteria to grow. Sugar pastes, commonly sucrose or honey solutions, encourage autolytic debridement and granulation tissue growth. They also decrease odor and exudate in highly draining wounds. They are particularly effective in resource-constrained environments owing to their cost-effectiveness (Biswas et al., 2010). Although straightforward, they necessitate frequent dressing changes and vigilant observation. Their effectiveness can be augmented when combined with contemporary wound dressings. The agents are particularly useful in controlling chronic ulcers or infected wounds. Even though they are conventional, they are gaining renewed use for their cost-effectiveness and antibacterial action (Emily, 2023).

9.1.2.7 Honey

Honey has been used in wound care for centuries due to its natural antimicrobial, anti-inflammatory, and healing properties. It promotes autolytic debridement, maintains a moist wound environment, and stimulates tissue regeneration. Manuka honey is particularly noted for its high antibacterial potency (Vaghardoost et al., 2018). Herbal treatments like aloe vera, turmeric, neem, and calendula are also applied topically for their anti-inflammatory, antioxidant, and antimicrobial effects. These agents enhance healing, soothe the wound, and reduce the risk of infection. Many herbal extracts promote collagen synthesis and angiogenesis. However, product standardization and quality control remain challenges (Naseri et al., 2021). These natural remedies are widely accepted in traditional medicine and are now being integrated into evidence-based care. They are especially popular for patients seeking natural or complementary therapies. When used appropriately, honey and herbs can significantly support wound healing (Chotchoungchatchai et al., 2020).

9.1.2.8 Electrical Stimulation Therapy

Electrical stimulation therapy (EST) is utilized to treat the wound using low-level electrical currents to facilitate healing. It mimics the electrical signals of the body, increasing cell migration, angiogenesis, and collagen synthesis. EST stimulated fibroblast activity and improved oxygenation of the wound bed (Luo et al., 2021). It reduces bacterial load and inflammation and also facilitates quicker recovery. High-voltage pulsed current (HVPC) and low-intensity direct current are some of the various forms. Therapy is commonly delivered via electrodes arranged circumferentially around the wound site (Ashrafi et al., 2017). EST is particularly beneficial with pressure ulcers, diabetic ulcers, and venous leg ulcers. Therapy is usually delivered several times a week under professional supervision. While usually safe, patient selection and device calibration are necessary. Electrical stimulation is a new, non-invasive modality in chronic wound care (Verdes et al., 2022).

9.1.2.9 Ultrasound Therapy

Ultrasound therapy employs sound waves to stimulate tissue repair and wound healing (Palanisamy et al., 2022). Low-frequency ultrasound, either contact or non-contact, supports debridement and bacterial decrease. It enhances blood flow, increases cell permeability, and aids in protein synthesis (Mahran, 2023). The therapy also minimizes inflammation and edema by stimulating lymphatic drainage. Ultrasound can break biofilms, rendering chronic wounds more amenable to other therapies. It is especially useful for diabetic and venous ulcers that are hard to heal (Wakabayashi et al., 2020). Non-contact machines minimize patient discomfort and are simple to use. Treatments are brief and given several times a week. If added to standard therapy, ultrasound therapy can make healing much faster. It is an effective and safe adjunct to the treatment of hard-to-heal wounds (Hess et al., 2003).

9.1.2.10 Nutrition

Nutritional support is vital for wound healing, as inadequate nutrition can impair or delay the entire process. Proteins are required for immune function, tissue repair, and the synthesis of collagen. Essential micronutrients such as vitamin C, zinc, and vitamin A are needed for immune response, angiogenesis, and epithelialization (Saeg et al., 2021). Lack of these micronutrients may result in the failure of wound closure, infection, or dehiscence. Metabolic needs are higher in chronic wounds or burns, where specific nutritional programs are necessary. Oral supplements or enteral nutrition may be required for patients with marked deficits (Sriram et al., 2016). Hydration also facilitates cell metabolism and transport of nutrients. Regular nutritional evaluation needs to be included in wound care, particularly in elderly or hospitalized populations. Individualized planning is facilitated by registered dietitians. A holistic strategy integrating nutrition with wound therapy enhances outcomes significantly (Casey, 2003). Non-Surgical wound healing procedures are described in Table 9.2. The differences between surgical and non-surgical approaches are shown in Table 9.3.

TABLE 9.2
INon-Surgical Therapies in Wound Healing

Non-Surgical Procedure	Mechanism of Action	Advantages	Disadvantages	Current Application	Marketed Product	Future Prespective	Reference
Wound dressings	Maintains a moist environment, reduces desiccation, and promotes autolytic debridement	Protects wound Promotes faster healing	Requires frequent changes Can be expensive	Optimal dressing selection for different wound types	Chronic wounds, pressure ulcers, surgical sites	DuoDERM, Allevyn, Tegaderm	Brumberg et al. (2021)
Topical agents	Direct antimicrobial, anti-inflammatory, or regenerative effects depending on the agent used	Reduces infection Stimulates tissue regeneration	Irritation or allergy risks Limited efficacy for some wounds	Resistance development Formulation stability	Minor burns, infected wounds, chronic ulcers	Silver sulfadiazine, Regranex, Iodosorb	Alves et al. (2021), Cancio (2021), and De Francesco et al. (2022)
Compression therapy	Enhances venous return, reduces edema, and improves tissue perfusion	Reduces edema; Improves circulation	Uncomfortable; Improper use may damage skin	Patient compliance Risk of ischemia if applied incorrectly	Venous leg ulcers, lymphedema.	Coban, Jobst, Profore	Shi et al., (2021)
Hypertonic saline and sugar pastes	Osmotic action draws fluid and toxins from wound, reduces bacterial load, and supports healing	Cost-effective Antimicrobial Reduces edema	Discomfort Frequent changes Skin irritation	Patient discomfort Formulation standardization	Infected wounds, mild to moderate exudative wounds	Mesalt, Sugar-Dome	Haesler (2023) and de Cabrera et al. (1988)
Honey and herbal treatments	Natural antimicrobial and anti-inflammatory activity Promotes autolytic debridement and healing	Natural; antimicrobial Stimulates regeneration	Expensive Allergic reactions possible Limited data for deep wounds	Standardization of herbal products Regulatory barriers	Minor burns, pressure sores, infected wounds	Medihoney, WoundCare+, Manuka Wound Gel	Vaghardoost et al. (2018)
Electrical stimulation therapy	Low-voltage currents enhance cellular migration, angiogenesis, and collagen synthesis	Accelerates healing Improves blood flow and oxygenation	Requires equipment Not for pacemaker patients	Cost, training, and patient accessibility	Chronic wounds, diabetic ulcers	BioWave, HVPC devices	Ashrafi et al. (2017) and Hess et al. (2003)
Ultrasound therapy	Microvibrations increase circulation, reduce inflammation, and stimulate cellular activity and protein synthesis	Non-invasive Enhances regeneration	Not for deep or infected wounds Requires skilled use	Equipment cost Treatment duration	Chronic wounds, pressure ulcers	MIST Therapy System, SONOCA 185	Palanisamy et al. (2022) and Mahran (2023)
Nutritional support	Provides essential nutrients (proteins, vitamins, zinc) required for collagen synthesis and immune function	Supports healing systemically Reduces infection risk	May not be sufficient alone Needs monitoring	Individualized plans Monitoring of deficiencies	Malnourished wound patients, elderly care	Juven, Arginaid, Glucerna	Casey (2003) and Cereda et al. (2024)

TABLE 9.3
Difference between Surgical and Non-Surgical

Aspect	Surgical Wound Healing	Non-Surgical Wound Healing	Reference
Approach	Involves direct intervention to remove damaged tissue and repair wounds through procedures like grafting and suturing	Focuses on promoting natural healing using external treatments without invasive surgery	Lei et al. (2024) and Mamun et al. (2024a)
Techniques	Skin grafting, flap surgery, debridement, suturing, skin substitutes	Dressings, topical agents, electrical stimulation, ultrasound therapy, compression therapy, offloading techniques	Mamun et al. (2024a), Liang et al. (2021), Luo et al. (2021), and Mayer et al. (2024)
Technology integration	Use of advanced tools like robotic systems for precise incision, laser technology for cutting, and 3D-printing for customized skin grafts	Introduction of nanotechnology, stem cell applications, bioelectronic devices, and smart wound dressings for controlled healing	Kolimi et al. (2022) and Arabpour et al. (2024)
Patient experience	Requires anesthesia, postoperative care, longer recovery, and potential discomfort	Less pain, minimal recovery time, non-invasive methods allow patients to engage in daily activities sooner	Yuan et al. (2022) and Dixon and Edmonds (2021)
Healing time	Faster healing for acute and complex wounds, but longer recovery period post-surgery due to surgical trauma	Slower healing, but the process supports the body's natural regeneration over time, especially for chronic wounds	Oryan et al. (2017) and Holl et al. (2021)
Complexity	High complexity due to the need for skilled surgeons, specialized operating rooms, and sophisticated equipment	Requires careful selection of treatments and regular monitoring, but overall, less complex than surgical interventions	Sahana and Rekha (2018)
Risk of complications	Higher risk of infections, scarring, wound dehiscence (wound opening), and complications from anesthesia	Lower risk of infection, though delayed healing or improper wound management can occur	Inouye et al. (2021) and Lei et al. (2024)
Cost	More expensive due to hospital stays, surgery, and post-operative care	Generally, less expensive for treatments like dressings and topical agents, although advanced therapies can be costly	Patel et al. (2019)
Effectiveness	Highly effective for large, deep, or traumatic wounds requiring immediate repair or coverage	Effective for chronic, non-healing wounds or those with limited surgical intervention options, like diabetic foot ulcers	Lei et al. (2024)
Innovation	Innovations include advanced skin substitutes, robotic surgery, 3D skin printing, and gene therapy for enhanced tissue repair	Non-invasive innovations such as stem cell therapy, nanoparticle drug delivery, bioelectric therapy, and smart bandages for controlled healing	Brumberg et al. (2021) and Lei et al. (2024)

9.2 INNOVATIVE STRATEGIES FOR WOUND HEALING

Innovative wound healing strategies include advanced, science-based methods to facilitate and speed up the healing process. These involve a range of contemporary techniques such as hydrogels, electrospun nanofibers, nanotherapeutics, stem cells, microfluidic systems, ECM, platelet-rich plasma (PRP), noncoding RNAs (ncRNAs), exosomes, negative pressure wound therapy (NPWT), and hyperbaric oxygen therapy (HBOT) as shown in Figure 9.2.

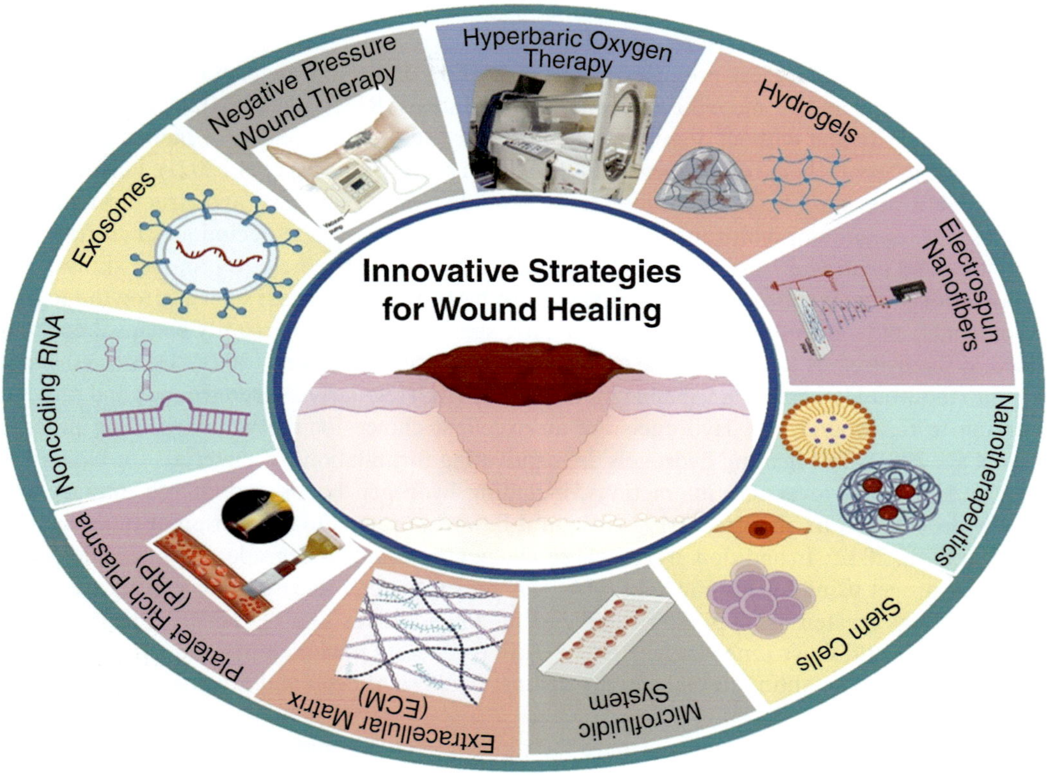

FIGURE 9.2 Innovative strategies for wound healing: The figure depicts a range of novel approaches to wound healing, each addressing particular aspects of tissue repair. Negative pressure wound therapy and hyperbaric oxygen therapy increase oxygenation and drainage to facilitate quicker healing. Hydrogels and electrospun nanofibers create moist environments and structural support that replicates the extracellular matrix. Nanotherapeutics, stem cells, and exosomes provide targeted regeneration and anti-inflammatory action. In addition, noncoding RNAs and platelet-rich plasma (PRP) control cellular behavior, whereas microfluidic systems and decellularized ECM facilitate accurate modeling and tissue integration. These methods together constitute a move toward personalized, regenerative wound therapy. (Image generated through BioRender Software.)

9.2.1 Hydrogels

Hydrogels are formulations containing a network of crosslinked polymeric structures. This three dimensional (3D) structure has the ability to retain high water content while remaining structurally intact after expanding (Gounden and Singh, 2024). At the molecular scale, the key characteristics that define the structure of hydrogel include the mesh size and the molecular weight of the polymer chains situated between the crosslinking points (Raghuwanshi and Garnier, 2019). Hydrogels can be crosslinked chemically (covalently) or physically (hydrogen bonding/entanglement) (Sanjanwala et al., 2022). Diffusion of water into a hydrogel is primarily responsible for its swelling (Rizwan et al., 2022). The rate of swelling is determined by the polymer content and crosslinking density (Kowalski et al., 2019). A large variety of hydrogels with varying ranges of characteristics were developed and studied for uses related to the wound healing process. The manufacturing process and crosslinking technique may also have an impact on structural factors, in addition to the chemical makeup and concentration of each polymer (Li et al., 2021). Several manufacturing techniques, such as casting, electrospinning, injectable, in situ 3D, bioprinting, 3D printing, and bioprinting, have been documented for use in wound healing applications.

They could exhibit various features, including structural, physical, rheological, disintegration, and water-retaining qualities (Norahan et al., 2023).

They have exhibited great promise in wound healing and tissue regeneration. They offer a moist environment and promote the autolysis of wound scabs (Ribeiro et al., 2024). Hydrogels can be removed easily and without disrupting the newly created tissue because of their weak adherence, which prevents them from adhering to the wound bed (Brumberg et al., 2021). Furthermore, various additives could be added to hydrogels to modify their biological and structural characteristics. The creation of multifunctional hydrogels through a variety of techniques has given the area of wound healing a fresh perspective and made it possible for them to promote skin regeneration effectively (Wang et al., 2022). A variety of bioactive substances, such as growth factors, antioxidants, and anti-inflammatory and antibiotic substances could be readily encapsulated and added to the hydrogels to improve the healing process (Kharaziha et al., 2021). Injecting hydrogels into the wounded area is a valuable option to help fill irregularly shaped wounds and provide adhesion to the wound bed. Hydrogels are an excellent choice for the wound healing process (Yao et al., 2021). Self-healing hydrogels are captivating formulations of materials for biological applications. Using reversible interactions, including hydrogen bonds and charge interactions, these hydrogels can fix their recurring structural flaws and return to their original functionality (Bertsch et al., 2022). In addition, hydrogels can be nontoxic and biodegrade. For instance, a self-healing hydrogel using hyaluronic acid (HA) was made that can release antibacterial agents on demand. It was created using prefabricated casting and additives like PDGF-BB (PDGFBB) (Li et al., 2023). In addition to encouraging *in vivo* skin regeneration with no irritation, the biological data demonstrated an antimicrobial activity against *Escherichia coli* and *Staphylococcus aureus in vitro* (Shahzadi et al., 2023).

Although multifunctional hydrogels have their merits, they encounter problems in wound healing treatments (Maleki et al., 2021). More research must be conducted on higher-level hydrogels with infection-sensing capabilities and intelligent drug-release systems (Seidi et al., 2021). Biomacromolecules such as gums, keratin, zein, and deoxyribonucleic acid (DNA) can be added to improve their mechanical strength, self-healing, antibacterial, and ECM-mimic properties (Asadi et al., 2021). Combining cell-laden hydrogels and stem cell-derived exosomes can hasten recovery and make individualized therapies possible (Li et al., 2022).

9.2.2 Electrospun Nanofibers

Traditional dressings have been unable to address the growing global need for wound care and treatment because of their low antibacterial properties and many other problems. The electrospinning technology is gaining popularity among researchers because of its simplicity and versatility. An advanced wound dressing, the electrospun nanofiber membrane is thought to have distinctive morphological and functional activities which are related to those of an ECM (Zheng et al., 2025). They have excellent promise for delivering and encapsulating active chemicals that can improve wound healing mechanisms (Gao et al., 2021). Membranes made of electrospun nanofibers have versatility, microporosity, and a significant surface-to-volume ratio making them appropriate for a variety of biomedical uses, such as drug administration, scaffolds for tissue engineering (Parham et al., 2020).

The electrospinning technique, which is used to prepare superfine fibers, has evolved over the years. When provided with voltage, the electrospinning fluid changes shape and eventually develops into a Taylor cone. As the jet expands, it bends and whips about. As the solvent evaporates, the jet narrows to a nanoscale scale before solidifying and settling onto the collector as nanofibers. By this principle, the size and morphology of nanofibers can be changed by adjusting formulation characteristics, process variables, and atmospheric variables during the electrospinning process (Ahmadi Bonakdar and Rodrigue, 2024; Xue et al., 2019). Electrospinning technology creates nanofibers that resemble ECM have small pore sizes and are highly porous. As a result, it has gained a lot of interest and is being utilized to make specialized nanofibers for biomedical and various other purposes (Ekrami et al., 2023). This technique has progressively transformed into multiple-fluid electrospinning, double-fluid electrospinning, and single-fluid electrospinning (Du et al., 2022).

Nanofiber dressings for wounds developed via electrospinning offer various benefits. First, it is a perfect environment for cell growth, migration, adhesion, and differentiation since its framework and functional activities are the same as those of the normal ECM (Gao et al., 2021). Second, the electrospinning polymer matrix may combine the long-term reliability of synthetic polymers with the compatibility of natural polymers (Zha et al., 2020). Additionally, a wide range of physiologically active substances, including antimicrobial medications, inorganic nanoparticles, hormones, and others may be loaded optimally into the nanofiber membrane due to its vast surface area and porous structure (Nayl et al., 2022). The rate and duration at which the drug becomes available to exert its therapeutic effect can be adjusted by modifying the fiber structure, thus encouraging successful wound healing. Though natural and synthetic polymers have many beneficial characteristics, their ability to aid wound healing is limited. Electrospinning nanofibers can be made from a variety of polymers and electrospinning processes, each with its own set of chemical and physical features (Elango et al., 2023). For the achievement of desired properties, it is critical to select the appropriate multicomponent polymers and apply them using electrospinning technology. PCL (Aldana and Abraham, 2017; Yoshimoto et al., 2003), poly[lactic acid-co-glycolic acid] (PLAGA) (Kumbar et al., 2008), PVA (Cho et al., 2012), PLA (Eren Boncu and Ozdemir, 2022), and PLCL (Daristotle et al., 2021) are some of the most widely used synthetic polymers. Natural polymers exhibit higher responsive and compatible characteristics but have lower mechanical qualities. Most often used natural polymers for the preparation used in healing diabetic foot ulcers include chitosan (CS) (Mirbagheri et al., 2023), collagen (Lin et al., 2012), gelatin (GT) (Massoumi et al., 2019), zein (Fernandez et al., 2009), and elastin (Kitazono et al., 2004).

Dressings loaded with drug factors are made from nanofiber materials, which allow them to perform their functions during different phases of wound healing and aid in wound repair (Gul et al., 2021). Electrospun nanofiber dressings with drug factors primarily include pharmaceuticals or growth factors with antibacterial and hemostasis capabilities in the electrospun solution, improving the efficacy of wound excipients and aiding the healing process (Cheng et al., 2015; Yang et al., 2020). However, the electrospun nanofiber dressings effect should be thoroughly investigated in several clinical studies to confirm its future therapeutic application.

9.2.3 Nanotherapeutics

Nanotechnologies are classified into nanotherapeutic-based drug delivery systems and nanocarrier-based drug delivery systems. There are two categories of nanocarrier-based DDS: lipid-based nanoformulations (such as liposomes) and polymer nanoformulations (including polymeric NPs and inorganic NPs) (Alshawwa et al., 2022; Kahraman et al., 2017). These nanotechnology-based methods deliver targeted medications in a desired, adjustable pattern and at the level required for the target site. They may also serve as transporters for bioactive compounds used to heal serious wounds. Several forms of nanoparticles have been documented in the studies for the healing process, which can be divided into four major stages: hemostasis, inflammatory, proliferative, and remodeling. Mechanisms including blood clotting, platelet aggregation, and vasoconstriction take place during the hemostasis phase. Polymeric nanoparticles, zinc oxide (ZnO) NPs, and cerium oxide (CeO_2) NPs have been shown to assist this phase by promoting immediate clot formation and stabilizing the wound surface (Barroso et al., 2020; Salimi and Mohammadipanah, 2021).

During the inflammatory phase, neutrophils and macrophages move to the wounded site, causing phagocytosis and emit several growth factors and cytokines. Nanomaterials such as polymeric NPs, carbon-based nanomaterials, silver nanoparticles, gold nanoparticles, ceramic NPs, and liposomes have been shown to increase immune cell activity and regulate inflammation, resulting in faster wound healing (Barroso et al., 2020; Mordorski and Prow, 2016). Fibroblast proliferation, production of collagen, re-epithelization, and angiogenesis are all characteristics of the proliferative phase. Copper nanoparticles, micelles, liposomes, iron oxide NPs, and ceramic NPs all play important roles in this phase by stimulating cellular development, aiding tissue regeneration, and increasing vascularization at the wound site (Mordorski and Prow, 2016; Nandhini et al., 2024). Finally, the last phase of wound healing (remodeling phase) is distinguished by the substitution of collagen III by fibroblasts with collagen I, the remodeling of the ECM, and scar tissue development. Nanomaterials such as iron oxide nanoparticles, and polymeric nano scaffolds aid in wound reorganization, collagen organization, and scar reduction, allowing for more effective tissue repair (Jansman and Hosta-Rigau, 2019). These nanomaterials, with their new physicochemical features, have the potential to improve the healing process.

The application of nanomaterials in the process of wound healing is extending rapidly, and clinical research is being done on nanotherapy-based wound healing therapies (Hamdan et al., 2017; Naderi et al., 2018). The physiochemical characteristics of nanomaterials, such as their large surface area, high surface area-to-volume ratios, and nano size, are the main causes of their growing utilization (Mihai et al., 2019). In wound management, two kinds of nanomaterials are commonly used: (1) nanomaterials having inherent qualities that promote wound healing, and (2) nanoparticles that serve as vehicles for therapeutic agent delivery (Naskar and Kim, 2020). Silver (Ag) (Naskar et al., 2020b), gold (Au) (Naskar et al., 2018), and zinc oxide (ZnO) (Naskar et al., 2020a) are the most studied metallic nanoparticles because of their good physiochemical properties and antimicrobial activity. Nonmetallic nanoparticles have been considered as potential wound healing agents. Carbon-based has shown significant potential as nanomedicine in a wide range of biomedical applications, including bioimaging, tissue regeneration, and controlled pharmaceutical products (Zhou et al., 2010). In the past, chitosan was the main ingredient used in wound healing applications due to its antibacterial, biocompatible, and biodegradable qualities (Matica et al., 2019). For wound healing applications, chitosan-associated nanocomposites outperform chitosan. For wound healing applications, antibiotics have been effectively delivered to their target area via a variety of nanomaterials that possess biocompatible and biodegradable properties, such as peptide nanostructures, polymeric nanoparticles, and liposomes (Cheow and Hadinoto, 2011; Chereddy et al., 2016; Hadinoto et al., 2013; Li et al., 2013; Mao et al., 2018; Mugabe et al., 2005).

9.2.4 Stem Cells

The capacity of stem cells to self-renew and differentiate into a variety of cell types sets them apart. As a potential treatment for wound healing, they have been shown to produce growth factors and cytokines that promote immunomodulation and regeneration (Farabi et al., 2024).The potential of different types of stem cell for improving wound healing has been examined in several animal models and clinical studies. All three germ layers (ectoderm, mesoderm, and endoderm) can be formed by pluripotent cells called embryonic stem cells (ESCs) (Yamanaka, 2020). ESCs can develop into a range of cell types, including mature cells like hematopoietic stem cells, specialized immune cells, and epidermal cells that are involved in wound healing. ESCs were successfully developed into functional keratinocytes *in vitro* producing an epidermal layer that might act as a temporary substitute in patients waiting for autologous grafts (Riha et al., 2021). Induced pluripotent stem cells (iPSCs) present a prospective therapeutic substitute for ESCs as well while avoiding significant ethical difficulties associated with their usage. iPSCs, like ESCs, can differentiate into cell types useful in wound healing therapy, including keratinocytes, fibroblasts, and epithelial stem cells (Gorecka et al., 2019). The possibility for malignant transformation or the formation of teratomas is a critical consideration in the clinical usage of ESCs and iPSCs. Multipotent stem cells, or MSCs, may be obtained from a range of organs (Soliman et al., 2021). Because of their pro-regenerative and immunomodulatory abilities as well as their capacity to reach injured sites, they may aid in the healing process. Originally, bone marrow—which is still a commonly utilized tissue source for cell separation—was used to extract MSCs (Yamaguchi et al., 2024). Compared to cell control or saline therapy, bone marrow-derived (BM)-MSCs demonstrated considerable advantages in a variety of animal wound healing models, comprising incisional, excisional, and diabetic chronic wounds (Banu et al., 2024). By increasing levels of vascular endothelial growth factor (VEGF) and hepatocyte growth factor (HGF), BM-MSCs have been shown to stimulate angiogenesis and fibroblast and keratinocyte migration (Ma et al., 2023). Additionally, MSCs may cause wounds to become inflamed, which would aid in the healing process. Human umbilical cord MSCs accelerated the healing of rat wounds by reducing the number of inflammatory cells and pro-inflammatory cytokines such as tumor necrosis factor-alpha (TNF) and interleukin-1 (IL-1) (Yang et al., 2024).

The effectiveness of stem cell-based treatment for wound healing is being investigated in ongoing clinical studies. MSCs produced from adipose tissue and allogenic bone marrow are being researched as potential therapies for second-degree burns, skin ulcers, and wounds that heal abnormally. Furthermore, autologous stromal vascular fraction transplantation is being investigated for its ability to stimulate tissue regeneration in chronic venous stasis ulcers (Cai et al., 2023). An increasing amount of research indicates that the administration of stem cells, namely MSCs, may be able to significantly increase the rate and caliber of wound healing.

Recent advances in stem cell therapy have significantly boosted wound healing techniques. Scaffold engineering, using biomimetic matrices and bioactive polymers, optimizes the stem cell microenvironment and improves regeneration potential. 3D bioprinting enables targeted stem cell delivery, enabling complex skin grafting and vascularized tissue engineering (Ho et al., 2023; Manita et al., 2021). Hypoxic preconditioning is known to promote angiogenesis and stem cell survival, allowing for better integration into ischemic tissues (Garcia et al., 2022; Lee et al., 2016).

9.2.5 Microfluidic Systems

Systems that employ micropipes to manage or dispose of tiny fluids—which may be used to create homogeneous droplets in environments at the micron level—are referred to as microfluidic technology (Rosellini and Cascone, 2023). Continuous liquids are split into minute droplets using this approach, which uses a combination of shear forces and surface tension. The size of microdroplets, which are usually micron or nanoscale in size, depends on the channel diameter and flow velocity of the different phases. T-connected, co-fluidic, and flow-focused channels are only a few of the various channel forms found in microfluidic devices, sometimes referred to as microfluidic chips. Under shear loads, the continuous phase squeezes the scattered phase as it reaches the flow-focused channel from both sides, creating discrete droplets. By utilizing a flow-focusing tube, hydrogel microspheres were produced that were both sticky and injectable (Chen et al., 2022).

At the orifice of the conical capillary, droplets are formed continuously, allowing the dispersed phase to blend with the continuous phase in a co-flow channel. The continuous phase, which includes surfactants, moves in the same direction. Lu Fan et al. used this method with microfluidic technology to produce a microcarrier with photo-responsive drug delivery microspheres (PDDMs). In a T-junction channel, the perpendicular movement of both phases creates shear forces, resulting in flow fluctuations and droplet formation. The main applications for this rather simple technique are DNA analysis and cell screening. In a T-junction channel device, microfluidic technology was used to create polydimethylsiloxane (PDMS) microspheres (Zhang et al., 2022). Microfluidic technology offers several benefits over conventional technology, such as cheap cost, tiny droplet volume, and great size controllability (Debski et al., 2018; Han et al., 2022). To maintain the uniqueness of microdroplets, customized microfluidic devices are designed according to the structure of microspheres and their functional requirements.

Conventional wound repair assays, like the scratch assay and cell exclusion assay, have been translated and enhanced by modern advances in microfluidics and lab-on-a-chip technologies. This is in line with the global trend toward automation, miniaturization, and the incorporation of cell-based diagnostics. These tiny cell analysis systems offer more precise spatial and temporal control over a variety of dynamic microenvironmental factors, including oxygen gradients, shear stress, and metabolic processes, at the molecular, cellular, and tissue levels. This results in more realistic *in vitro* models that closely resemble the wound's *in vivo* microenvironment (Shabestani Monfared et al., 2021).

To improve traditional wound healing assays, a variety of lab-on-a-chip and microfluidics devices have been developed. These devices may find application in fundamental research, drug discovery, and diagnostics. The primary goal of microfluidic chip technology is to enable automated and miniature bioassays (Joshi, 2016). Scalable, high-throughput cell-based analytical assays are made possible by the small volume requirements of miniature microfluidic devices (Gao et al., 2016). These technologies can be used to develop innovative customized devices, especially for pharmacological and diagnostic screening, thanks to current plastic mass manufacturing processes (Huh, 2015). Human body responses to certain treatments are often not replicated in preclinical studies using animal models. Furthermore, a patient's reaction to therapy may differ from another's because of lifestyle and genetic variables.

Therefore, it is possible to save significant costs and demonstrate precise forecasts throughout the initial stages of medication testing by using a human cell-based intermediate *in vitro* model (Felder et al., 2019; Gao et al., 2016). The creation of inexpensive, disposable microdevices lays the groundwork for parallel, sophisticated analysis. Because silicone-based polymers are so biocompatible, they are frequently employed in the construction of microdevices. Additionally, it is said to be rather easy to manufacture fluid channels using cutting, molding, and hot embossing (Joshi, 2016). PDMS is a silicone polymer that is widely used in microdevice fabrication. It is optically transparent and gas permeable, which enables efficient CO_2 and O_2 exchange and makes it suitable for use in cell culture (Shabestani Monfared et al., 2021). The microdevice system's transparency allows for the imaging and tracking of cells and fluids.

9.2.6 Extracellular Matrix (ECM)

The ECM is an acellular, protein-rich matrix required for structural support and cellular adhesion. Previous research has shown that the ECM modulates the healing of wounds by influencing cellular pathways of signaling. The ECM has direct effects on cell activity, such as adhesion, proliferation, migration, and survival (Diller and Tabor, 2022). Cells working on the ECM induce extracellular protease production, which in turn modulates growth factor bioavailability (Vijayan et al., 2021). The two most common proteins in the native ECM are collagen and elastin. Collagen is the most common protein identified in the ECM. On the other hand, depending on age, gender, and anatomical location, elastin makes up 0.69–7.9% of the skin's ECM. Laminin, fibronectin, vitronectin, and tenascin are among the other proteins present in the ECM, while proteoglycans and glycosaminoglycans (GAGs) comprise the remaining 3–5% of the ECM (Diller and Tabor, 2022). Proteins in the ECM play an important role in wound healing by providing structural support, regulating cellular communication, and stimulating tissue restoration. Collagen preserves tissue integrity and promotes epidermal and dermal differentiation, whereas elastin maintains elasticity. Fibronectin promotes platelet adhesion, leukocyte migration, and elastin deposition, which all help with wound healing. Laminin and vitronectin help in cellular signaling, hemostasis, and adhesion, which aids in tissue repair. Tenascin also promotes cell migration and proliferation, which accelerates the wound process of healing (Potekaev et al., 2021). ECM is a crucial structural element of tissues. The "wound healing cascade" is a sequence of healing processes that can be triggered by either acute or chronic damage to the ECM. Hemostasis, the first cessation point of blood flow and blood clotting, is soon followed by the stage of inflammation, the granulation-rich proliferative stage, and the remodeling phase which results in collagen I tissue. These procedures overlap. Cellular signaling pathways, many of which include the ECM, are responsible for all of this (Diller and Tabor, 2022).

ECM-produced biomaterials are a basic but promising strategy for supporting diabetic wound healing. Biomaterials composed of only a few ECM components fall short of capturing the complexity of the ECM. To more closely mimic the structure, intricacy, and usefulness of the original ECM, new approaches will probably be needed. Biomaterials generated from ECM are produced using three primary techniques. A biomaterial is then created by combining decellularized tissues, *in vitro*-generated ECM, and purified ECM components. Isolated ECM constituents including collagen, fibrin, HA, and elastin are used to create biomaterials. Collagen and HA are two examples of components that can be combined to create ECM-derived biomaterials. ECM biomaterials formed from decellularized tissue are produced from cadavers or animals. Biomaterials facilitate wound healing by angiogenesis, immunomodulation, and antibacterial action (Hosty et al., 2024).

9.2.7 Platelet-rich Plasma (PRP)

Numerous medical applications, such as non-surgical regeneration of tissues, wound closure, and rejuvenation, use biological preparations based on plasma proteins and autologous PRP formulation. PRP and protein-based technologies perform a variety of systemic and local cellular and molecular functions; their application has produced encouraging patient outcomes, that are carried out sequentially to restore the proper functioning and integrity of damaged tissues.

In 1954, Kingsley et al. adopted the term "platelet-rich plasma" to characterize the standardization of platelet concentrate preparations for transfusions (Kingsley, 1954). Following that, an autologous development of a "platelet-fibrinogen-thrombin mixture" reinforced fibrin polymerization in corneal wounds by including a high concentration of platelets in fibrin glue (Rosenthal et al., 1978). In later years, Knighton et al. used a blood-derived material called "platelet-derived wound healing factors" (Knighton et al., 1986) to describe the function of platelets in wound healing procedures. Marx et al., on the other hand, first used the term PRP in regenerative medicine in 1998, pointing out that platelet growth factors can promote tissue repair (Marx, 2001). The denser fibrin-based PRP was among the several platelets concentrate formulations that were eventually referred to by the label PRP. The release of platelet components from lysosomal granules, as well as the activities of several platelet adhesion molecules, are required for PRP-induced repair of tissues, regeneration, and wound rehabilitation. Furthermore, platelet physiological functions including angiogenesis, immunomodulation, cell signaling, and metabolism are impacted by the presence or absence of certain leukocytes, and these factors in turn impact tissue healing and functional tissue restoration (Everts et al., 2020).

PRP contains a variety of proteins that are required throughout various stages of wound healing. Von Willebrand factor, fibrinogen, fibronectin, vitronectin, and thrombospondin-1 (TSP-1), among other factors, regulate hemostasis to stabilize the clot and commence repair. Cell migration (chemotaxis) is facilitated by PDGF, HGF, IL-8, and SDF-1, which direct critical cells to the wound site. IL-1β, VEGF, TGF-β, and angiopoietins promote angiogenesis and inflammation by forming new blood vessels and regenerating tissues. IGF-1, TGF-β, MMPs, and BMPs play key roles in ECM formation and remodeling, providing necessary structural support for healing. Many of these proteins have various functions in different biological pathways, demonstrating the complexities of PRP in wound repair (Chaudhary et al., 2022).

Modern low-volume ultrafiltration devices with tiny membrane pore sizes can yield a protein-rich product that can be used either by itself or in combination with high-definition PRP formulations to make a consolidated multicellular protein-rich platelet concentrate (PR-PRP) preparation. Other studies have documented the synergistic effects of PDGF, non-PDGF, and PR-PRP with more diverse growth factors and other important non-platelet derivatives than PRP alone (Everts et al., 2024). When PR-PRPs are active, they function as a biological matrix that preserves and improves connections between its molecular and cellular components (Everts et al., 2024). Additionally, the versatility of PR-PRP technology makes it possible to create patient-specific formulations for both non-surgical and surgical uses. PRP therapies are justified by the fact that precise delivery of concentrated platelets into injured and degraded tissues stimulates physiological pathways, which in turn triggers tissue regeneration by releasing a number of PDGF, cytokines, molecules, and sticky proteins. Additionally, local tissue regulators promote tissue healing through cell signaling, paracrine, autocrine, and intracrine pathways after PRP injection.

9.2.8 Noncoding RNAs

NcRNAs serve an important function in regulating certain biological responses. It is accomplished by influencing the activities of a diverse set of targets. NcRNAs are categorized into two categories according to the length of their nucleotide sequences: short NcRNAs (nucleotide units <200) and long NcRNAs (more than 200 nucleotide units). In this section, the roles of NcRNAs, such as circular RNAs (circRNAs), long NcRNAs (lncRNAs), and microRNAs (miRNAs), in wound healing will be discussed (Aghayants et al., 2024; Hombach and Kretz, 2016). Their role will be explained in this unique process to grasp a better concept of the complex mechanism of the healing procedure.

MiRNAs regulate the processes of wound healing by modulating important signaling pathways. They inhibit nuclear factor kappa beta (NF-kB) and activate phosphoinositide 3-kinase/protein kinase B (PI3K/AKT) to facilitate tissue repair. Signal transducer and activator of transcription 3 (STAT3) induces M2 macrophage polarization through dual-specificity tyrosine-phosphorylation regulated kinase 1A/STAT3 (DYRK1A/STAT3) and Janus kinase 2/STAT3 (JAK2/STAT3) signaling. VEGF signaling through hypoxia-inducible factor 1-alpha (HIF1A), AKT/endothelial nitric oxide synthase (AKT/eNOS), and p38 mitogen-activated protein kinase (p38MAPK) is involved in the promotion of blood vessel formation. The Wnt/β-catenin pathway is responsible for promoting cellular migration and growth for the process of re-epithelialization. Matrix metalloproteinase-9 (MMP-9) and matrix metalloproteinase-1 (MMP-1) are involved in tissue remodeling (Jiang et al., 2022).

LncRNAs include H19, metastasis-associated lung adenocarcinoma transcript 1 (MALAT1), growth arrest-specific transcript 5 (GAS5), HOX transcript antisense intergenic RNA (HOTAIR), antisense ncRNA in the ink4 locus (ANRIL), and KLF3 antisense RNA 1 (KLF3-AS1). H19 aids in all stages of the healing process (Aghayants et al., 2024). It results in inflammation by activating the peroxisome proliferator-activated receptor gamma/STAT3 (PPARγ/STAT3) pathway. H19 controls PI3K-AKT signaling in angiogenesis. It also promotes migration and cellular growth of keratinocytes, while inhibiting pyroptosis via reducing NLR family pyrin domain containing 3 (NLRP3) activation. MALAT1 stimulates anti-inflammatory responses by targeting miR-1914-3p and activating milk fat globule-EGF factor 8 (MFGE8). MALAT1 also regulates transforming growth factor beta 1 (TGFb1)-induced epithelial-to-mesenchymal transition (EMT) in keratinocytes. GAS5 stimulates STAT1 to activate M1 macrophages, whereas its inhibition favors M2 macrophage polarization and wound healing. It also stimulates GAS5 inflammatory signaling pathways. HOTAIR, ANRIL, and KLF3-AS1 are principally engaged in boosting the extension of blood vessels from the current vasculature (Aghayants et al., 2024). Circular RNAs (circRNAs) are closed-loop molecules lacking poly-A tails which regulate cell growth, metabolic processes, cell death, and inflammatory processes. They act as transcriptional regulators, miRNA sponges, and clinical biomarkers, and interact with RNA-binding proteins. Circ-Snhg11 is involved in diabetic wound healing, and research shows its expression in bone marrow mesenchymal stem cell-derived exosomes (BMSC-Exos) enhances angiogenesis and the healing process. Future wound healing research might benefit greatly from the development of ncRNA-based biomarkers because of their unique expression. Furthermore, modulating miRNAs may be a more successful technique as compared to typical medications targeting single proteins, because a single miRNA can influence a whole functional network (Herter and Xu Landén, 2017). As a result, studying the functions of miRNAs in wound regeneration could result in new treatment strategies related to wound healing that have previously been complicated to treat.

9.2.9 Exosomes

Exosomes are extracellular vesicles with diameters ranging from 30 to 150 nm. These may consist of DNA, RNA, lipids, and proteins. Once exosomes have been taken up by cells, their internal material is carried to the targeted cell cytoplasm, causing changes in the recipients' functional condition. Exosomes are proven to participate in intercellular communication (Li and Wu, 2022).

Exosomes obtained through various sources have the ability to affect the activities of wound healing cells, promoting neovascular development (Xu et al., 2020), promoting the deposition of collagen (Wang et al., 2021), and inhibiting the process of inflammation (Ti et al., 2015), all of which accelerate healing. Exosomes affect the targeted tissues by the following mechanism: (a) Surface receptors on exosomes interact directly with target cell surface receptors; (b) Exosome surface protein fragments connect with surface receptors on target cells; (c) After merging with the targeted cell membrane, the exosome membrane emits its contents into the cytoplasm of the target cell; and (d) Exosomes are phagocytosed by receptor cells (Barros et al., 2018).

Exosomes improve the activity of wound healing cells by modifying macrophage polarization and facilitating the transition from pro-inflammatory M1 to anti-inflammatory M2 macrophages, thus speeding up wound healing. They also improve intercellular communication between endothelial cells, which promotes angiogenesis, collagen formation, and inflammation reduction (Li and Wu, 2022). Exosomes can help macrophages phenotypically polarize from M1 to M2, decrease inflammatory mechanisms via boosting the rate of M2/M1 polarization, upregulating anti-inflammatory factor production, and dramatically enhance wound healing rate. Exosomes made from human circulating fibrocytes have been shown to speed wound healing in diabetic mice *in vivo* and to stimulate diabetic keratinocyte proliferation and migration *in vitro*. Exosomes decrease inflammation by stimulating the PTEN/AKT signaling pathway, which promotes macrophage phenotypic alterations. Exosomes stimulate fibroblast proliferation and migration by stimulating various signaling pathways. It helps keratinocytes proliferate and migrate. Proteins are the primary components of exosomes. Guo and colleagues found that PRP-Exos capsules containing bFGF, PDGF, VEGF, and TGF-β improve wound healing (Beanes et al., 2003; Losi et al., 2013; Martino et al., 2011). Exosomes made from human urine-derived stem cells (USC-Exos) exhibit significant amounts of deleted in malignant brain tumors 1 (DMBT1) protein and increase EC angiogenic activity, hence improving angiogenesis and wound closure in diabetes-induced rat models (Chen et al., 2018).

Exosomes have a high potential for promoting the wound healing process. The injected exosomes' duration at the site was constrained, nevertheless, because the movement of body fluids prevented them from being added at specific sites. To organize exosomes and promote wound healing, hydrogels can be used as scaffolding. Since then, we have learned that chitosan hydrogel has several uses. The ability of hydrogel enables it to cover the site of injury, providing a suitable environment, providing optimal conditions for cell growth. Additionally, in diabetic mice, it raises the rate at which wounds close, thanks to placental MSC-derived exosomes. Further the present use of exosomes in combination with different hydrogels has demonstrated enhanced stability, sustained release, and improved therapeutic efficacy in wound healing applications (Li and Wu, 2022).

Smart-engineered naturally produced exosomes with certain inherited/conjugated surface proteins and bioactive compounds can efficiently interface with target cells and be delivered to specified organs (Joorabloo and Liu, 2023). These nano-vesicles are sophisticated delivery methods due to their inherent stability, low immunogenicity, and high drug loading and targeting capabilities (Joorabloo and Liu, 2023). They are less likely to be detected by immune cells and eliminated from the body. This prolonged circulation can also improve distribution to certain tissues. Small compounds, genes, proteins, and incorporated nanoparticles are some of the therapeutic cargos loaded into exosomes for wound healing (Li and Wu, 2022).

9.2.10 Negative Pressure Wound Therapy

NPWT, formally referred to as vacuum-assisted wound closure (VAC) therapy, exerts pressure on the area where the wound occurred. A suction pump, a fluid collection device, a polyurethane foam sponge, and a semi-occlusive adhesive cover are included in the system. The foam has pores ranging from 400 to 600 μm; it can be cut to meet any wound's size and geometry. The foam cells' open structure guarantees that negative pressure is evenly distributed throughout the whole wound surface (Normandin et al., 2021). Different types of sponges have been produced to satisfy the needs of a wider range of wounds. For example, sponges can be divided into two different subgroups: the WHITE FOAM (3M, KCI) dressing, which is less adhesive and hence best suited for more delicate wounds or underlying tissues, and the GRANUFOAM (3M, KCI) dressing, which is appropriate for both acute and chronic wounds. The chosen foam is then applied to the skin using an adhesive membrane and coupled to a suction pump that produces negative pressures ranging from −50 to −200 mm Hg. The pressure can be applied either continuously or occasionally. VAC therapy is beneficial through four major modes of action, such as macro deformation, fluid drainage, wound stability, and micro deformation; other mechanisms include metabolic alterations, perfusion alteration, and bacterial load reduction (Orgill et al., 2009; Venturi et al., 2005).

To promote the healing process, NPWT with Instillation (NPWTi) alternates topical solution application with the conventional VAC system. This technique allows for the adjustment of parameters such as solution type, dwell time, and pressure duration to meet a variety of therapeutic objectives (Kanapathy et al., 2020). NPWTi-d has demonstrated considerable benefits in enhancing granulation tissue production while decreasing bacterial burden, making it an excellent adjuvant therapy for complex wounds. However, the optimal parameters and solution types have yet to be determined (Faust et al., 2021). NPWT provides a variety of dressing and cleansing choices, including GRANUFOAM, WHITEFOAM, and the VERAFLO cleanse dressing. The decision is based on parameters such as wound geometry, location, and infectious state. The VERAFLO wash dressing is very useful for NPWTi-d since it evenly distributes topical solutions and removes thick exudates, making it an excellent choice (Normandin et al., 2021). Closed incision negative pressure therapy (CPT) is a technique for promoting healing and reducing problems in closed surgical wounds. It entails providing negative pressure to the closed wound to reduce tension, edema, and fluid collection, thereby improving perfusion and lowering the risk of infection and dehiscence. Studies have indicated that ciNPT can considerably reduce surgical site problems, such as infections and wound dehiscence, as compared to conventional dressings (Hsiao et al., 2022). Additionally, it has been linked to shorter hospital stays and less discomfort. NPWT has been extensively examined and proven to improve wound healing in various therapeutic settings, including diabetic ulcers, pressure ulcers, chronic wounds, and skin grafts. Studies have shown that NPWT results in higher wound closure rates, fewer amputations, and shorter treatment times than traditional therapy. It also improves skin graft survival and lowers the need for multiple grafts (Hsiao et al., 2022; Zaver and Kankanalu, 2023). NPWT is useful in addressing complicated wounds because it reduces infection and inflammation. Overall, NPWT improves wound care outcomes.

NPWT is regarded as a safe method that rarely causes issues. However, various problems of varied severity have been recorded, with the majority of them attributed to poor technique or inappropriate patient selection (Normandin et al., 2021). In other words, the majority of adverse events associated with VAC therapy are preventable. Toxic shock syndrome and enteric fistula are serious consequences of VAC therapy, which can be avoided with proper wound care and the use of protective interfaces. Blood loss might also cause hemodynamic instability.

9.2.11 Hyperbaric Oxygen Therapy

Pure oxygen is therapeutically inhaled in a chamber with a pressure of 0.2–0.3 MPa as part of hyperbaric oxygen treatment (HBOT). HBOT's causes a significant increase in reactive oxygen species (ROS) and nitrogen radical (RNS) production and is responsible for both its therapeutic and detrimental effects on specific organs, by redirecting the breakdown of hypoxia inducible factor (HIF-1) to pathways that are not affected by oxygen levels. It increases HIF-1 concentration. The healing process is accelerated by elevated levels of HIF-1, which promotes the synthesis of several growth factors. HBOT promotes the release of heat shock proteins (HSP); it functions as a guide for HIF-1 (Růžička et al., 2021). It improves the efficiency of some antibiotics, promotes fibroblast proliferation and collagen formation, and inhibits the action of proteolytic enzymes. It boosts the formation of ROS and RNS while also promoting antioxidant enzyme activity.

Furthermore, it may have a positive or negative impact on many tissues. For instance, long-term exposure to excessive oxygen levels (much lower level as compared to the level in HBOT) causes pulmonary edema, chronic pulmonary fibrosis, emphysema, inflammation, collagen deposition, and diffuse alveolar damage in the lungs. According to Dejmek et al. (2018), lung fibroblasts exposed to HBOT showed reduced mitochondrial oxygen consumption and decreased viability at the cellular level. The possible antibacterial effects of HBOT have lately been investigated, and the spike in oxygen content is potentially harmful to anaerobic microbes, namely clostridium strains. HBOT stimulates the generation of ROS and RNS; they cause irreparable damage by oxidizing the pathogen's proteins and lipids when they spontaneously react with its organelles. Furthermore, by temporarily inhibiting signaling pathways of inflammation and the synthesis of pro-inflammatory cytokines, such as those from macrophages, HBOT enhances defense against germs. HBOT has been shown to reduce the release of IL-1, IL-6, IL-8, and TNFα by circulating monocytes, simultaneously increasing the secretion of the beneficial cytokine IL-10.

As shown in the neutrophil model using HLA 60 cells, HBOT also encourages leukocyte phagocytosis (Almzaiel et al., 2013). According to the same study, following HBOT, neutrophil-like cells' respiratory burst activity increased. HBOT improves the efficacy of several antibiotics, including cefazolin, ciprofloxacin, tobramycin, vancomycin, and linezolid. It has been demonstrated that HBOT increases fibroblast proliferation in human dermal fibroblast cells, but not at the start of treatment but rather on day 7. It stimulates growth factor release and neovascularization in human keratinocyte cultures (Hollander et al., 2000). All of these events, when combined, may help to accelerate wound healing using HBOT. These innovative approaches of wound healing are summarized in Table 9.4.

9.3 CONCLUSION

Wound healing is a dynamic and multifaceted process that includes many physiological pathways and cell interactions. Conventional surgical treatments are still irreplaceable for the treatment of severe or complex wounds, providing accurate tissue repair and quick closure. Nevertheless, they are usually associated with invasiveness and increased risks of complications. Non-surgical treatments, such as dressings, topical treatments, and physical modalities, offer necessary assistance to natural regeneration and are particularly useful in chronic wound management. New advances like hydrogels, electrospun nanofibers, nanotherapeutics, stem cells, exosomes, and microfluidic systems

TABLE 9.4
Innovative Strategies for Wound Healing

Innovative Strategy	Mechanism of Action	Advantages	Challenges	Current Applications	Marketed Products	Future Perspectives	References
Hydrogels	Hydrogel active compounds accelerate the healing process, increase granulation and collagen synthesis, and promote more efficient re-epithelialization Fibroblast proliferation, and angiogenesis angiogenesis, and collagen deposition	Capable of replenishing water and maintaining a moist atmosphere, simple and convenient to replace, provides a layer of protection, simulates the natural extracellular matrix (ECM), and maintains a humid environment to promote wound recovery and cutaneous tissue regeneration	Mechanical stability, biological characteristics, limited number of polymers, limited absorption ability and is not ideal for wounds with heavy exudate, safety, and efficiency concerns	Burn wounds, diabetic ulcers, post-surgical wounds	Suprasorb® G, Neoheal® Hydrogel, Simpurity™ Hydrogel, NU-GEL™, ActivHeal®	Creating multipurpose hydrogels and new dressings with exosomes produced from stem cells. Commercial potential and clinical translation potential of the hydrogels that were created for wound recovery Developing hydrogels for scarless tissue regeneration. Creating multipurpose hydrogels and new dressings with exosomes produced from stem cells Marketability and clinical translation potential of the hydrogels produced for wound care	Bolivar-Monsalve et al. (2022), Ying et al. (2020), and Asadi et al. (2021)
Electrospun nanofibers	Their microstructure closely resembles the human extracellular matrix (ECM), providing an optimal environment for adhesion, and cell proliferation	High porosity, biomimetic structure, adjustable characteristics, large surface area, and adaptable fiber shape and surface function	It lacks the required cell guiding capabilities, has uneven fiber characteristics, requires extensive clinical studies to guarantee the ultimate clinical use, and is complicated to manufacture	Pressure sore, diabetic foot ulcer, surgical wound, skin grafting, wound healing	FastClot/ SurgiClot®, TuneCoat™™, BIOcular™™, Nanotrix®, ReDura™, Rivelin®	Antimicrobial nanofibers, bioactive nano scaffolds	Gao et al. (2021) and Parham et al. (2020)

Nanotherapeutics	Targeted therapy, sustained release, penetration into deeper layers	Increased bioavailability, reduced toxicity, controlled release	Regulatory hurdles, clinical translation, stability issues, complex manufacturing, cost, and toxicity of nanomaterials are the principal concerns	Wound infection control, anti-inflammatory treatment, burn wounds, diabetic wounds	Acticoat™, PolyMem® dressings	Increasing targeting effectiveness and site specificity for more successful wound healing applications based on the nanomaterial utilized, focusing on Smart nanoparticles, AI-driven nanoformulations	Hamdan et al. (2017), Naderi et al. (2018), and Mihai et al. (2019)
Stem cells	Modulate immune response, increase fibroblasts and keratinocytes migration, enhance angiogenesis, and modulate the process of inflammation	Potential for complete tissue regeneration, long-lasting effects	Ethical concerns, risk of tumorigenesis, high cost	Used in both acute and long-term injuries, such as burns, diabetic foot ulcers, vein lumen ulcers, PUs	Grafix® S2RM®, ACTEX™	Gene-edited stem cells, bioengineered stem cell therapies	Banu et al. (2024), Ma et al. (2023), and Yang et al. (2024)
Microfluidic systems	Simulates wound environment, real-time analysis	High precision, reduced reagent use, personalized medicine	Limited clinical translation, and fabrication challenges	Wound-on-a-chip models, drug testing	....	AI-integrated microfluidic platforms	Joshi (2016), Gao et al. (2016), and Huh (2015)
Extracellular matrix (ECM)	Provides scaffold, modulates cell signaling, promotes healing	Biocompatible, supports cell migration, natural healing	Costly extraction, immune rejection concerns	Decellularized ECM grafts, ECM-based biomaterials	Xcellistem, Novosorb, Xcellistem, Promogran, MatriDerm, AlloDerm, Pelnac, Acell Wound Powder	Bioengineered ECM with growth factors	Potekaev et al. (2021) and Diller and Tabor (2022)

(Continued)

TABLE 9.4 *(Continued)*
Innovative Strategies for Wound Healing

Innovative Strategy	Mechanism of Action	Advantages	Challenges	Current Applications	Marketed products	Future Perspectives	References
Platelet-rich plasma (PRP)	Enhances cell proliferation, angiogenesis, and collagen synthesis	Autologous, reduces inflammation, easy to obtain	Variability in PRP formulations, standardization issues	Used in pressure ulcers (bedsores), diabetic ulcers, leg ulcers, skin transplants, first and second-degree thermal burns, cuts, cuts, and wounds caused by surgery	AutoloGel™, Aurix™, RegenKit®-Wound Gel	PRP nanocarriers and the application of PRP products with biomaterials in tissue engineering employing 3D bioprinting	Everts et al. (2020) and Chaudhary et al. (2022)
Noncoding RNAs	Modulate inflammation, enhance cell migration and proliferation	Precision medicine potential, gene-based regulation	Delivery challenges, off-target effects, overexpression, complexity of noncoding RNA	Gene therapy for chronic wounds	....	ncRNA-loaded biomaterials, CRISPR-based regulation	Aghayants et al. (2024) and Hombach and Kretz (2016)
Exosomes	Transfer bioactive molecules, regulate immune response	Natural carriers, low immunogenicity, regenerative potential	Isolation complexity, low yield, stability concerns, immunosuppression, and tumorigenesis during the treatment process	Stem cell-derived exosome therapy, wound regeneration	---------------	Engineered exosomes with targeted therapy	Xu et al. (2020), Wang et al. (2021), and Ti et al. (2015)
Negative pressure wound therapy (NPWT)	Drainage of fluids, stabilization of the wound environment, improves blood circulation, lowers microbial stress and improves tissue granulation	Reduces wound size, and improves healing speed	Cost, pain during dressing changes, device dependency	Post-surgical wounds, traumatic injuries	RENASYS TOUCH PICO VAC Dressing Medela Wound VAC Smith & Nephew PICO 7	NPWT with bioactive dressings, automated NPWT systems	Hsiao et al. (2022) and Zaver and Kankanalu (2023)

Hyperbaric oxygen therapy (HBOT)	Enhances oxygen delivery, promotes angiogenesis, reduces infections, increases fibroblast proliferation	Non-invasive, enhances antibiotic effects, accelerates healing	High cost, multiple sessions required, contraindications	Diabetic ulcers, radiation-induced wounds	Sechrist Model 4000, Sechrist Model, OxyHealth Vitaeris 320, OxyHealth Vitaeris 240, BARA-MED® 200, Summit to Sea Hyperbaric Chambers	Personalized HBOT, AI-driven oxygen therapy	Hollander et al. (2000) and Růžička et al. (2021)
Platelet-rich plasma (PRP)	Enhances cell proliferation, angiogenesis, and collagen synthesis	Autologous, reduces inflammation, easy to obtain	Variability in PRP formulations, standardization issues,	Used in pressure ulcers (bedsores), diabetic ulcers, leg ulcers, skin transplants, first and second-degree thermal burns, cuts, cuts, and wounds caused by surgery	AutoloGel™, Aurix™, RegenKit®-Wound Gel	PRP nanocarriers and the application of PRP products with biomaterials in tissue engineering employing 3D bioprinting	Everts et al. (2020) and Chaudhary et al. (2022)

exhibit high potential in revolutionizing wound care as a more targeted, efficient, and customized practice. These new approaches replicate the ECM, provide targeted therapies, and control inflammation and angiogenesis efficiently. The integration of conventional methods with intelligent, regenerative, and low-invasive therapies facilitates a complete and patient-oriented strategy for wound healing. The future directions are toward the integration of bioengineered materials, nanomedicine, and advanced diagnostics to improve therapeutic outcomes and enable scar-free, accelerated healing.

GLOSSARY

AKT/eNOS pathway: A signaling pathway that enhances nitric oxide production, encouraging blood flow and tissue repair.

Angiopoietins: A group of growth factors involved in the regulation of blood vessel maturation, stability, and remodeling during healing.

ANRIL (antisense noncoding RNA in the INK4 locus): A lncRNA that modulates inflammation and cell cycle activity during wound repair.

Compression therapy: Utilization of therapeutic pressure to enhance blood flow and minimize swelling in wound areas.

Debridement: The body's natural process of breaking down dead tissue with the use of moist-retentive dressings.

Electrical stimulation therapy (EST): Application of electrical currents as a catalyst for wound healing and tissue restoration.

Electrospun nanofiber: Thin fibers produced through electrospinning employed as scaffolds for tissue regeneration support.

Exosomes: Nano-vesicles secreted by cells that play a role in cell signaling and repair of the tissue.

Extracellular matrix (ECM): A protein and polysaccharide structural network of proteins and polysaccharides that maintains cell growth and tissue regeneration.

Fibrinogen: A plasma protein that is transformed into fibrin to create a blood clot in the process of wound healing.

Fibronectin: An extracellular matrix protein that facilitates cell adhesion, migration, and tissue repair.

Flap surgery: A method involving tissue with a blood supply transferred from one location to another to close a wound.

GAS5 (growth arrest-specific 5): A control lncRNA that suppresses cell proliferation and controls inflammatory responses in wound healing.

GRANUFOAM (3M, KCI) dressing: A reticulated open-cell foam that is used in NPWT to promote granulation tissue formation and contraction of the wound.

HIF1A (hypoxia-inducible factor 1-alpha): A transcription factor that activates genes involved in response to low oxygen, promoting angiogenesis and wound healing.

HOTAIR (HOX transcript antisense RNA): A long noncoding RNA involved in the modulation of gene expression and wound healing and tissue regeneration.

Hydrogel: A polymer network with high water content that creates a moist environment that favors healing of wounds.

Hyperbaric oxygen therapy (HBOT): The therapeutic application of 100% oxygen at higher pressure to enhance oxygen delivery to injured tissue.

Hypoxia-inducible factor 1 (HIF-1): A transcription factor induced in low oxygen to control genes concerned with angiogenesis and cell survival during wound healing.

IL-1β (interleukin-1 beta): A cytokine with pro-inflammatory effects that is involved in the initiation of the wound healing response.

JAK2/STAT3 signaling: An important signaling pathway induced by cytokines that controls cell proliferation, survival, and immune responses during wound healing.

MALAT1 (metastasis-associated lung adenocarcinoma transcript 1): A long noncoding RNA that is involved in the promotion of cell migration, angiogenesis, and tissue repair.

Microfluidic systems: Devices that handle small volumes of fluids for drug testing and analysis of cell behavior.

MMP-1 (matrix metalloproteinase-1): An enzyme that degrades collagen to enable cell migration and tissue repair.

MMP-9 (matrix metalloproteinase-9): An enzyme that degrades extracellular matrix components to help in tissue remodeling during wound repair.

Nanoparticles (NPs): Extremely small particles employed for targeted drug delivery in nanotherapeutics.

Nanotherapies: Application of nanoscale materials or systems for delivering targeted treatment in wound healing.

Negative pressure wound therapy (NPWT): A method involving vacuum pressure to drain fluids and facilitate healing.

Noncoding RNAs (ncRNAs): RNA molecules that control gene expression without being translated into proteins.

Non-surgical procedure: A treatment process that does not include cutting into the body or excising tissue.

p38MAPK (p38 mitogen-activated protein kinase): A kinase that is engaged in cellular responses to stress and inflammation in wound healing.

Platelet-rich plasma (PRP): A concentration of a patient's own platelets applied to enhance healing and regeneration.

PPARγ/STAT3 pathway: A signaling pathway that regulates inflammation, cell differentiation, and tissue regeneration in wound healing.

Reactive oxygen and nitrogen species (ROS and RNS): Chemically reactive molecules that serve double functions in wound healing by serving as signaling agents for tissue repair and also adding to oxidative stress.

Skin grafting: Transferring skin from one part of the body to another to cover and heal wounds.

Skin substitutes: Human-made substances that replicate natural skin utilized to support wound closure and regeneration.

Stem cells: Cells capable of differentiation into other cell types and facilitating tissue regeneration.

Topical drugs: Pharmaceuticals directly applied to the skin to address localized wound states.

Ultrasound therapy: Application of sound waves to facilitate tissue healing and inflammatory reduction in wounds.

VEGF (vascular endothelial growth factor): A growth factor that induces the growth of new blood vessels (angiogenesis).

VERAFLO therapy: A negative pressure wound therapy with instillation that provides topical solutions into the wound to facilitate cleansing and healing.

Von Willebrand factor (vWF): A blood glycoprotein that participates in platelet adhesion and clot formation at sites of wounds.

White foam (3M, KCI) dressing: A hydrophilic polyurethane foam that is utilized in negative pressure wound therapy (NPWT) to enhance moist wound healing and exudate management.

REFERENCES

Aghayants, Sis, Jinjin Zhu, Jing Yu, Rui Tao, Sicheng Li, Shengzhi Zhou, Yunhua Zhou, and Zhanyong Zhu. 2024. "The emerging modulators of non-coding RNAs in diabetic wound healing." *Frontiers in Endocrinology* 15: 1465975.

Ahmadi Bonakdar, Mahboubeh, and Denis Rodrigue. 2024. "Electrospinning: Processes, structures, and materials." *Macromol* 4 (1): 58–103.

Al-Ajlouni, Jihad M, Mohammed S Alisi, Yazan S Hammad, Ahmed A Alsousi, Haya O Karameh, Sanad H Kawasmi, Rwand H Aladwan, Reem A Awawdeh, Yahya A Almazaraa, and Freih O Abu Hassan. 2023. "Staples versus sutures wound closure in hip and knee arthroplasty: A prospective cohort study." *Journal of Wound Care* 32 (2): 98–103.

Aldana, Ana A, and Gustavo A Abraham. 2017. "Current advances in electrospun gelatin-based scaffolds for tissue engineering applications." *International Journal of Pharmaceutics* 523 (2): 441–453.

Almzaiel, Anwar J, Richard Billington, Gary Smerdon, and A John Moody. 2013. "Effects of hyperbaric oxygen treatment on antimicrobial function and apoptosis of differentiated HL-60 (neutrophil-like) cells." *Life Sciences* 93 (2–3): 125–131.

Alshawwa, Samar Zuhair, Abeer Ahmed Kassem, Ragwa Mohamed Farid, Shaimaa Khamis Mostafa, and Gihan Salah Labib. 2022. "Nanocarrier drug delivery systems: Characterization, limitations, future perspectives and implementation of artificial intelligence." *Pharmaceutics* 14 (4): 883.

Alves, Paulo J, Ruben T Barreto, Brigitte M Barrois, Luc G Gryson, Sylvie Meaume, and Stan J Monstrey. 2021. "Update on the role of antiseptics in the management of chronic wounds with critical colonisation and/or biofilm." *International Wound Journal* 18 (3): 342–358.

Arabpour, Zahra, Fatemeh. Abedi, Mohammad Salehi, Seyed Mohammad Baharnoori, Mohammad Soleimani, and Ali Reza Djalilian. 2024. "Hydrogel-based skin regeneration." *International Journal of Molecular Sciences* 25 (4). https://doi.org/10.3390/ijms25041982

Aragona, Mariaceleste, Sophie Dekoninck, Steffen Rulands, Sandrine Lenglez, Guilhem Mascré, Benjamin D. Simons, and Cédric Blanpain. 2017. "Defining stem cell dynamics and migration during wound healing in mouse skin epidermis." *Nature Communications* 8 (1): 14684.

Asadi, Nahideh, Hamidreza Pazoki-Toroudi, Azizeh Rahmani Del Bakhshayesh, Abolfazl Akbarzadeh, Soodabeh Davaran, and Nasim Annabi. 2021. "Multifunctional hydrogels for wound healing: Special focus on biomacromolecular based hydrogels." *International Journal of Biological Macromolecules* 170: 728–750.

Ashrafi, M., T. Alonso-Rasgado, M. Baguneid, and A. Bayat. 2017. "The efficacy of electrical stimulation in lower extremity cutaneous wound healing: A systematic review." *Experimental* Dermatology 26 (2): 171–178. doi: 10.1111/exd.13179.

Balomenos, Dimitrios B, Pagona G Gouletsou, and Apostolos D Galatos. 2023. "Evaluation of incisional wound healing in dogs after closure with Staples or tissue glue and comparison to intradermal suture pattern." *Animals* 13 (3): 426.

Banu, S Amitha, Abhijit M Pawde, Khan Sharun, E Kalaiselvan, Shivaraju Shivaramu, Karikalan Mathesh, Vikas Chandra, Rohit Kumar, Swapan Kumar Maiti, and Med Ram Verma. 2024. "Evaluation of bone marrow-derived mesenchymal stem cells with eggshell membrane for full-thickness wound healing in a rabbit model." *Cell and Tissue Banking* 25 (2): 493–508.

Barros, Francisco M, Fatima Carneiro, Jose C Machado, and Sónia A Melo. 2018. "Exosomes and immune response in cancer: Friends or foes?" *Frontiers in Immunology* 9: 730.

Barroso, Andreia, Henrique Mestre, Andreia Ascenso, Sandra Simões, and Catarina Reis. 2020. "Nanomaterials in wound healing: From material sciences to wound healing applications." *Nano Select* 1 (5): 443–460.

Bay, Caroline, Zachary Chizmar, Edward M Reece, Jessie Z Yu, Julian Winocour, Joshua Vorstenbosch, and Sebastian Winocour. 2021. "Comparison of skin substitutes for acute and chronic wound management." *Seminars in Plastic Surgery* 35 (3): 171–180.

Beanes, Steven R, Catherine Dang, Chia Soo, and Kang Ting. 2003. "Skin repair and scar formation: The central role of TGF-β." *Expert Reviews in Molecular Medicine* 5 (8): 1–22.

Bertsch, Pascal, Mani Diba, David J Mooney, and Sander CG Leeuwenburgh. 2022. "Self-healing injectable hydrogels for tissue regeneration." *Chemical Reviews* 123 (2): 834–873.

Biswas, Anindita, Monika Bharara, Catherine Hurst, Robert Gruessner, David Armstrong, and Howard Rilo. 2010. "Use of sugar on the healing of diabetic ulcers: A review." *Journal of Diabetes Science and Technology* 4 (5): 1139–45. https://doi.org/10.1177/193229681000400512

Bolívar-Monsalve, Edna Johana, Carlos Fernando Ceballos-González, Carolina Chávez-Madero, Brenda Guadalupe de la Cruz-Rivas, Silvana Velasquez Marin, Shirley Mora-Godínez, Luisa María Reyes-Cortés, Ali Khademhosseini, Paul S Weiss, and Mohamadmahdi Samandari. 2022. "One-step bioprinting of multi-channel hydrogel filaments using chaotic advection: Fabrication of pre-vascularized muscle-like tissues." *Advanced Healthcare Materials* 11 (24): 2200448.

Braza, Matthew E, and Matthew P Fahrenkopf. 2023. "Split-thickness skin grafts." In *StatPearls [Internet].* StatPearls Publishing.

Brumberg, Valentin, Tatiana Astrelina, Tatiana Malivanova, and Alexander Samoilov. 2021. "Modern wound dressings: Hydrogel dressings." *Biomedicines* 9 (9): 1235.

Bus, Sicco A., David G. Armstrong, Ryan T. Crews, Chris Gooday, Gustav Jarl, Kirsten Kirketerp-Møller, Vijay Viswanathan, and Peter A. Lazzarini. 2024. "Guidelines on offloading foot ulcers in persons with diabetes (IWGDF 2023 update)." *Diabetes/Metabolism Research and Reviews* 40 (3): e3647. https://doi.org/10.1002/dmrr.3647

Cai, Yuan, Fang Zhang, Jingwei Feng, Bihua Wu, Hai Li, Shune Xiao, Zairong Feng Lu, Chengliang Wei, and Deng. 2023. "Long-term follow-up and exploration of the mechanism of stromal vascular fraction gel in chronic wounds." *Stem Cell Research & Therapy* 14 (1): 163.

Cancio, Leopoldo C. 2021. "Topical antimicrobial agents for burn wound care: History and current status." *Surgical Infections* 22 (1): 3–11.

Casey, Gillian. 2003. "Nutritional support in wound healing." *Nursing Standard* 17 (23): 55–8. https://doi.org/10.7748/ns.17.23.55.s57

Cereda, Emanuele, Nicola Veronese, and Roberto Caccialanza. 2024. "Nutritional therapy in chronic wound management for older adults." *Current Opinion in Clinical Nutrition & Metabolic Care* 27 (1): 3–8. https://doi.org/10.1097/mco.0000000000000990

Chandrasekaran, Malarvizhi, Parthasarathy Ramadoss, Lalithkumar Loganathan, Gowtham Arunachalam, E. G. Prasad, K. V. S. Kishore, and K. V. S. Janardhan. 2024. "A comparative study of skin sutures versus skin staplers in emergency abdominal surgeries in tertiary care hospital." *International Journal of Academic Medicine and Pharmacy* 6 (3): 608–610.

Chandrasekaran, Nissanth, Shri Vishnu Manokaran, Dr Gandhamsetty Sai Ganesh, Harish Babu Jayakaran, and Yeshwanth Subash. 2023. "Comparison of surgical site infection in post-op patients for whom wound closure is done using staplers and sutures." *International Journal of Orthopaedics* 9 (1): 33–36.

Chaudhary, Preeti Kumari, Sanggu Kim, and Soochong Kim. 2022. "An insight into recent advances on platelet function in health and disease." *International Journal of Molecular Sciences* 23 (11): 6022.

Chen, Chun-Yuan, Shan-Shan Rao, Lu Ren, Xiong-Ke Hu, Yi-Juan Tan, Yin Hu, Juan Luo, Yi-Wei Liu, Hao Yin, and Jie Huang, 2018. "Exosomal DMBT1 from human urine-derived stem cells facilitates diabetic wound repair by promoting angiogenesis." *Theranostics* 8 (6): 1607.

Chen, Kaili, Fei Wang, Rui Ding, Zhengwei Cai, Tianyuan Zou, Andi Zhang, Dongye Guo, Bin Ye, Wenguo Cui, and Mingliang Xiang. 2022. "Adhesive and injectable hydrogel microspheres for inner ear treatment." *Small* 18 (36): 2106591.

Cheng, Feng, Jing Gao, Lu Wang, and Xingyou Hu. 2015. "Composite chitosan/poly (ethylene oxide) electrospun nanofibrous mats as novel wound dressing matrixes for the controlled release of drugs." *Journal of Applied Polymer Science* 132 (24).

Cheow, Wean Sin, and Kunn Hadinoto. 2011. "Factors affecting drug encapsulation and stability of lipid–polymer hybrid nanoparticles." *Colloids and Surfaces B: Biointerfaces* 85 (2): 214–220.

Chereddy, Kiran Kumar, Gaëlle Vandermeulen, and Véronique Préat. 2016. "PLGA based drug delivery systems: Promising carriers for wound healing activity." *Wound Repair and Regeneration* 24 (2): 223–236.

Cho, Daehwan, Anil N Netravali, and Yong Lak Joo. 2012. "Mechanical properties and biodegradability of electrospun soy protein Isolate/PVA hybrid nanofibers." *Polymer Degradation and Stability* 97 (5): 747–754.

Chotchoungchatchai, S., O. Krairit, P. Tragulpiankit, and S. Prathanturarug. 2020. "The efficacy of honey and a Thai Herbal Oil preparation in the treatment of pressure ulcers based on Thai traditional medicine wound diagnosis versus standard practice: An open-label randomized controlled trial." *Contemporary Clinical Trials Communications* 17: 100538. doi: 10.1016/j.conctc.2020.100538.

Claes, Karel EY, Ignace De Decker, Stan Monstrey, Yaron Shoham, Tom Vyncke, Bernard Depypere, Edward De Wolf, Fien Decuypere, Bernd Lannau, and Henk Hoeksema. 2023. "Helpful hints in deciding what and when to operate after enzymatic debridement." *Burns* 49 (1): 80–90.

Cwajda-Białasik, Justyna, Paulina Mościcka, and Maria T Szewczyk. 2022. "Antiseptics and antimicrobials for the treatment and management of chronic wounds: A systematic review of clinical trials." *Advances in Dermatology and Allergology/Postępy Dermatologii i Alergologii* 39 (1): 141–151.

Daeschlein, G. 2013. "Antimicrobial and antiseptic strategies in wound management." *International Wound Journal* 10 (Suppl 1): 9–14. doi: 10.1111/iwj.12175

Daristotle, John L, Metecan Erdi, Lung W Lau, Shadden T Zaki, Priya Srinivasan, Manogna Balabhadrapatruni, Omar B Ayyub, Anthony D Sandler, and Peter Kofinas. 2021. "Biodegradable, tissue adhesive polyester blends for safe, complete wound healing." *ACS Biomaterials Science & Engineering* 7 (8): 3908–3916.

Dayya, David, Owen J O'Neill, Tania B Huedo-Medina, Nusrat Habib, Joanna Moore, and Kartik Iyer. 2022. "Debridement of diabetic foot ulcers." *Advances in Wound Care* 11 (12): 666–686.

Debski, Pawel R, Karolina Sklodowska, Jacek A Michalski, Piotr M Korczyk, Miroslaw Dolata, and Slawomir Jakiela. 2018. "Continuous recirculation of microdroplets in a closed loop tailored for screening of bacteria cultures." *Micromachines* 9 (9): 469.

de Cabrera, María A., Toshio Okamoto, and Carlos Cabrera-Peralta. 1988. "[Influence of pre- and post-operative treatment with saline glucose solution on oral surgery wound healing. Histological study in rats]." *Rev Faculdade Odontol Lins* 1 (1): 4–11.

De Francesco, Francesco, Michele Riccio, and Shiro Jimi. 2022. "Contribution of topical agents such as hyaluronic acid and silver sulfadiazine to wound healing and management of bacterial biofilm." *Medicina* 58 (6): 835.

Dejmek, Jiří, Michaela Kohoutová, Martina Kripnerová, Markéta Čedíková, Zdeněk Tůma, Václav Babuška, Lukáš Bolek, and Jana Kuncová. 2018. "Repeated exposure to hyperbaric hyperoxia affects mitochondrial functions of the lung fibroblasts." *Physiological Research* 67.

Derwin, Rosemarie, Declan Patton, Pinar Avsar, Helen Strapp, and Zena Moore. 2022. "The impact of topical agents and dressing on pH and temperature on wound healing: A systematic, narrative review." *International Wound Journal* 19 (6): 1397–1408.

Diller, Robert B, and Aaron J Tabor. 2022. "The role of the extracellular matrix (ECM) in wound healing: A review." *Biomimetics* 7 (3): 87.

Dixon, Danielle, and Michael Edmonds. 2021. "Managing diabetic foot ulcers: Pharmacotherapy for wound healing." *Drugs* 81 (1): 29–56.

Du, Yutong, Xinyi Zhang, Ping Liu, Deng-Guang Yu, and Ruiliang Ge. 2022. "Electrospun nanofiber-based glucose sensors for glucose detection." *Frontiers in Chemistry* 10: 944428.

Ekrami, Elena, Mahvash Khodabandeh Shahraky, Matin Mahmoudifard, Mona Sadat Mirtaleb, and Parvin Shariati. 2023. "Biomedical applications of electrospun nanofibers in industrial world: A review." *International Journal of Polymeric Materials and Polymeric Biomaterials* 72 (7): 561–575.

Elango, Jeevithan, Camilo Zamora-Ledezma, and José Eduardo Maté-Sánchez de Val. 2023. "Natural vs synthetic polymers: How do they communicate with cells for skin regeneration—A review." *Journal of Composites Science* 7 (9): 385.

Eren Boncu, Tugba, and Nurten Ozdemir. 2022. "Electrospinning of ampicillin trihydrate loaded electrospun PLA nanofibers I: Effect of polymer concentration and PCL addition on its morphology, drug delivery and mechanical properties." *International Journal of Polymeric Materials and Polymeric Biomaterials* 71 (9): 669–676.

Everts, Peter, Kentaro Onishi, Prathap Jayaram, José Fábio Lana, and Kenneth Mautner. 2020. "Platelet-rich plasma: New performance understandings and therapeutic considerations in 2020." *International Journal of Molecular Sciences* 21 (20): 7794.

Everts, Peter A, José Fábio Lana, Robert W Alexander, Ignacio Dallo, Elizaveta Kon, Mary A Ambach, André van Zundert, and Luga Podesta. 2024. "Profound properties of protein-rich, platelet-rich plasma matrices as novel, multi-purpose biological platforms in tissue repair, regeneration, and wound healing." *International Journal of Molecular Sciences* 25 (14): 7914.

Farabi, Banu, Katie Roster, Rahim Hirani, Katharine Tepper, Mehmet Fatih Atak, and Bijan Safai. 2024. "The efficacy of stem cells in wound healing: A systematic review." *International Journal of Molecular Sciences* 25 (5): 3006.

Faust, Elizabeth, Jude L Opoku-Agyeman, and Amir B Behnam. 2021. "Use of negative-pressure wound therapy with instillation And dwell time: An overview." *Plastic and Reconstructive Surgery* 147 (1S-1): 16S–26S.

Felder, Marcel, Bettina Trueeb, Andreas Oliver Stucki, Sarah Borcard, Janick Daniel Stucki, Bruno Schnyder, Thomas Geiser, and Olivier Thierry Guenat. 2019. "Impaired wound healing of alveolar lung epithelial cells in a breathing lung-on-a-chip." *Frontiers in Bioengineering and Biotechnology* 7: 3.

Fernandez, Avelina, Sergio Torres-Giner, and Jose Maria Lagaron. 2009. "Novel route to stabilization of bioactive antioxidants by encapsulation in electrospun fibers of zein prolamine." *Food Hydrocolloids* 23 (5): 1427–1432.

Gao, AnXiu, YunLi Tian, ZhuanZhuan Shi, and Ling Yu. 2016. "A cost-effective microdevice bridges microfluidic and conventional in vitro scratch/wound-healing assay for personalized therapy validation." *BioChip Journal* 10: 56–64.

Gao, Chen, Liyuan Zhang, Juan Wang, Miao Jin, Qianqian Tang, Zhongrong Chen, Yue Cheng, Runhuai Yang, and Gang Zhao. 2021. "Electrospun nanofibers promote wound healing: Theories, techniques, and perspectives." *Journal of Materials Chemistry B* 9 (14): 3106–3130.

Garcia, John P, Francisco R Avila, Ricardo A Torres, Karla C Maita, Abdullah S Eldaly, Brian D Rinker, Abba C Zubair, Forte J Antonio, and Rachel Sarabia-Estrada. 2022. "Hypoxia-preconditioning of human adipose-derived stem cells enhances cellular proliferation and angiogenesis: A systematic review." *Journal of Clinical and Translational Research* 8 (1): 61.

Gorecka, Jolanta, Valentyna Kostiuk, Arash Fereydooni, Luis Gonzalez, Jiesi Luo, Biraja Dash, Toshihiko Isaji, Shun Ono, Shirley Liu, and Shin Rong Lee. 2019. "The potential and limitations of induced pluripotent stem cells to achieve wound healing." *Stem Cell Research & Therapy* 10: 1–10.

Gounden, Varshan, and Moganavelli Singh. 2024. "Hydrogels and wound healing: Current and future prospects." *Gels* 10 (1): 43.

Gul, Aysegul, Izabela Gallus, Akshat Tegginamath, Jiri Maryska, and Fatma Yalcinkaya. 2021. "Electrospun antibacterial nanomaterials for wound dressings applications." *Membranes* 11 (12): 908.

Hadinoto, Kunn, Ajitha Sundaresan, and Wean Sin Cheow. 2013. "Lipid–polymer hybrid nanoparticles as a new generation therapeutic delivery platform: A review." *European Journal of Pharmaceutics and Biopharmaceutics* 85 (3): 427–443.

Haesler, Emily. 2023. "WHAM evidence summary: Sugar dressing for wound healing and treating wound infection in resource limited settings." *WCET Journal - Portuguese Edition* 43 (2). https://doi.org/10.33235/wcet.43.2.35-40.

Hamdan, Suzana, Irena Pastar, Stefan Drakulich, Emre Dikici, Marjana Tomic-Canic, Sapna Deo, and Sylvia Daunert. 2017. "Nanotechnology-driven therapeutic interventions in wound healing: Potential uses and applications." *ACS central Science* 3 (3): 163–175.

Han, Ji Hwan, Chul Min Kim, Tae-Hyun Kim, Songwan Jin, and Gyu Man Kim. 2022. "Development of in situ microfluidic system for preparation of controlled porous microsphere for tissue engineering." *Pharmaceutics* 14 (11): 2345.

Hemmingsson, Oskar, Felix Binnermark, Christoffer Odensten, Martin Rutegård, Karl A Franklin, and Markku M Haapamäki. 2022. "Excision and suture in the midline versus Karydakis flap surgery for pilonidal sinus: Randomized clinical trial." *BJS Open* 6 (2): zrac007.

Herter, Eva K, and Ning Xu Landén. 2017. "Non-coding RNAs: New players in skin wound healing." *Advances in Wound Care* 6 (3): 93–107.

Hess, Charles L., Mary A. Howard, and Christopher E. Attinger. 2003. "A review of mechanical adjuncts in wound healing: Hydrotherapy, ultrasound, negative pressure therapy, hyperbaric oxygen, and electrostimulation." *Annals of Plastic Surgery* 51 (2): 210–218. doi: 10.1097/01.Sap.0000058513.10033.6b

Heyens, Leen JM, Dana Busschots, Ger H Koek, Geert Robaeys, and Sven Francque. 2021. "Liver fibrosis in non-alcoholic fatty liver disease: From liver biopsy to non-invasive biomarkers in diagnosis and treatment." *Frontiers in Medicine* 8: 615978.

Ho, Jasmine, Dominic Yue, Umber Cheema, Henry C Hsia, and Alan Dardik. 2023. "Innovations in stem cell therapy for diabetic wound healing." *Advances in Wound Care* 12 (11): 626–643.

Holl, Jordan, Cezary Kowalewski, Zbigniew Zimek, Piotr Fiedor, Artur Kaminski, Tomasz Oldak, Marcin Moniuszko, and Andrzej Eljaszewicz. 2021. "Chronic diabetic wounds and their treatment with skin substitutes." *Cells* 10 (3): 655.

Hollander, Dirk A, Mohssen Y Hakimi, Andre Hartmann, Kerstin Wilhelm, and Joachim Windolf. 2000. "The influence of hyperbaric oxygenation (HBO) on proliferation and differentiation of human keratinocyte cultures in vitro." *Cell and Tissue Banking* 1: 261–269.

Hombach, Sonja, and Markus Kretz. 2016. "Non-coding RNAs: Classification, biology and functioning." *Non-Coding RNAs in Colorectal Cancer* 937: 3–17.

Hosty, Louise, Thomas Heatherington, Fabio Quondamatteo, and Shane Browne. 2024. "Extracellular matrix-inspired biomaterials for wound healing." *Molecular Biology Reports* 51 (1): 830.

Hsiao, Hui-Yi, Wei-Chuan Hsieh, Frank Chun-Shin Chang, Jia-Wei Liu, Wen-Ling Kuo, David Chon-Fok Cheong, and Jung-Ju Huang. 2022. "The effect of negative pressure on wound healing and regeneration in closed incisions under high tension: Evidence from animal studies and clinical experience." *Journal of Clinical Medicine* 12 (1): 106.

Huh, Dongeun. 2015. "A human breathing lung-on-a-chip." *Annals of the American Thoracic Society* 12 (Supplement 1): S42–S44.

Inouye, Daniel, Sheng Zhou, Bhavishya Clark, Mark Swanson, Tamara Chambers, and Tamara N Chambers. 2021. "Two cases of impaired wound healing among patients with major head and neck free-flap reconstruction in the setting of COVID-19 infection." *Cureus* 13 (12).

Jansman, Michelle MT, and Leticia Hosta-Rigau. 2019. "Cerium-and iron-oxide-based nanozymes in tissue engineering and regenerative medicine." *Catalysts* 9 (8): 691.

Jiang, Yuanyuan, Xiang Xu, Long Xiao, Lihong Wang, and Sheng Qiang. 2022. "The role of microRNA in the inflammatory response of wound healing." *Frontiers in Immunology* 13: 852419.

Joorabloo, Alireza, and Tianqing Liu. 2023. "Engineering exosome-based biomimetic nanovehicles for wound healing." *Journal of Controlled Release* 356: 463–480.

Joshi, Preeti Nigam. 2016. "Cells and organs on chip—A revolutionary platform for." *Lab-on-a-Chip Fabrication and Application*: 77.

Kahraman, Emine, Sevgi Güngör, and Yıldız Özsoy. 2017. "Potential enhancement and targeting strategies of polymeric and lipid-based nanocarriers in dermal drug delivery." *Therapeutic Delivery* 8 (11): 967–985.

Kanapathy, Muholan, Angelos Mantelakis, Natasha Khan, Ibby Younis, and Afshin Mosahebi. 2020. "Clinical application and efficacy of negative pressure wound therapy with instillation and dwell time (NPWTi-d): A systematic review and meta-analysis." *International Wound Journal* 17 (6): 1948–1959.

Kharaziha, Mahshid, Avijit Baidya, and Nasim Annabi. 2021. "Rational design of immunomodulatory hydrogels for chronic wound healing." *Advanced Materials* 33 (39): 2100176.

Kingsley, CS. 1954. "Blood coagulation: Evidence of an antagonist to factor VI in platelet-rich human plasma." *Nature* 173 (4407): 723–724.

Kitazono, Eiichi, Hiroaki Kaneko, Takanori Miyoshi, and Keiichi Miyamoto. 2004. "Tissue engineering using nanofiber." *Journal of Synthetic Organic Chemistry, Japan* 62 (5): 514–519.

Knighton, david R, Kevin F Ciresi, Vance D Fiegel, Lorinda L Austin, and Ellen L Butler. 1986. "Classification and treatment of chronic nonhealing wounds." *Annals of Surgery* 204 (3): 322–330.

Kolimi, Prasad, Sravani Narala, Divya Nyavanandi, Ahmed A. A. Youssef, and Narendra Dudhipala. 2022. "Innovative treatment strategies to accelerate wound healing: Trajectory and recent advancements." *Cells* 11 (15). doi: 10.3390/cells11152439

Kowalski, Grzegorz, Karolina Kijowska, Mariusz Witczak, Łukasz Kuterasiński, and Marcin Łukasiewicz. 2019. "Synthesis and effect of structure on swelling properties of hydrogels based on high methylated pectin and acrylic polymers." *Polymers* 11 (1): 114.

Kumbar, Sangamesh G, Syam P Nukavarapu, Roshan James, Lakshmi S Nair, and Cato T Laurencin. 2008. "Electrospun poly (lactic acid-co-glycolic acid) scaffolds for skin tissue engineering." *Biomaterials* 29 (30): 4100–4107.

Laurano, Rossella, Monica Boffito, Gianluca Ciardelli, and Valeria Chiono. 2022. "Wound dressing products: A translational investigation from the bench to the market." *Engineered Regeneration* 3 (2): 182–200.

Lee, Dylan E, Nagi Ayoub, and Devendra K Agrawal. 2016. "Mesenchymal stem cells and cutaneous wound healing: Novel methods to increase cell delivery and therapeutic efficacy." *Stem Cell Research & Therapy* 7 1–8.

Lei, Yu, Ping Jiang, and Tian Tian. 2024. "Comparative analysis of surgical and non-surgical wound approaches in diabetic foot ulcer treatment: Meta-analysis and systematic review." *International Wound Journal* 21 (4): e14601.

Li, Chong, Xiaolin Zhang, Xinliang Huang, Xiaoying Wang, Guojian Liao, and Zhangbao Chen. 2013. "Preparation and characterization of flexible nanoliposomes loaded with daptomycin, a novel antibiotic, for topical skin therapy." *International Journal of Nanomedicine 8*: 1285–1292.

Li, Danyang, and Na Wu. 2022. "Mechanism and application of exosomes in the wound healing process in diabetes mellitus." *Diabetes Research and Clinical Practice* 187: 109882.

Li, Qirong, Dongxu Wang, Ziping Jiang, Rong Li, Tianyi Xue, Chao Lin, Yongzhi Deng, Ye Jin, and Baozhen Sun. 2022. "Advances of hydrogel combined with stem cells in promoting chronic wound healing." *Frontiers in Chemistry* 10: 1038839.

Li, Wenqi, Jing Hu, Cheng Chen, Xinyue Li, Honghua Zhang, Yanru Xin, Qingchang Tian, and Shuling Wang. 2023. "Emerging advances in hydrogel-based therapeutic strategies for tissue regeneration." *Regenerative Therapy* 24: 459–471.

Li, Yan, Juan Wang, Yan Wang, and Wenguo Cui. 2021. "Advanced electrospun hydrogel fibers for wound healing." *Composites Part B: Engineering* 223: 109101.

Liang, Yongping, Jiahui He, and Baolin Guo. 2021. "Functional hydrogels as wound dressing to enhance wound healing." *ACS Nano* 15 (8): 12687–12722.

Lin, Jiantao, Caihong Li, Yi Zhao, Jianchan Hu, and Li-Ming Zhang. 2012. "Co-electrospun nanofibrous membranes of collagen and zein for wound healing." *ACS Applied Materials & Interfaces* 4 (2): 1050–1057.

Losi, Paola, Enrica Briganti, Cesare Errico, Antonella Lisella, Elena Sanguinetti, Federica Chiellini, and Giorgio Soldani. 2013. "Fibrin-based scaffold incorporating VEGF-and bFGF-loaded nanoparticles stimulates wound healing in diabetic mice." *Acta Biomaterialia* 9 (8): 7814–7821.

Luo, R., J. Dai, J. Zhang, and Z. Li. 2021. "Accelerated skin wound healing by electrical stimulation." *Advanced Healthcare Materials* 10 (16): e2100557. doi: 10.1002/adhm.202100557.

Ma, Hui, Wing-Sum Siu, and Ping-Chung Leung. 2023. "The potential of MSC-based cell-free therapy in wound healing—A thorough literature review." *International Journal of Molecular Sciences* 24 (11): 9356.

Mahran, Hesham G. 2023. "Shockwave versus ultrasound therapy in the enhancement of Aloe vera in cutaneous wound healing." *Journal of Wound Care* 32 (Sup7a): cxxxix–cxlv. https://doi.org/10.12968/jowc.2023.32.Sup7a.cxxxix

Maillard, Jean-Yves, Günter Kampf, and Rose Cooper. 2021. "Antimicrobial stewardship of antiseptics that are pertinent to wounds: The need for a united approach." *JAC-Antimicrobial Resistance* 3 (1): dlab027.

Maleki, Aziz, Jiahui He, Shayesteh Bochani, Vahideh Nosrati, Mohammad-Ali Shahbazi, and Baolin Guo. 2021. "Multifunctional photoactive hydrogels for wound healing acceleration." *ACS Nano* 15 (12): 18895–18930.

Mamun, Abdul Aziz, Cheng Shao, Peng Geng, Shuai Wang, and Jun Xiao. 2024a. "Recent advances in molecular mechanisms of skin wound healing and its treatments." *Frontiers in Immunology* 15: 1395479. https://doi.org/10.3389/fimmu.2024.1395479

Mamun, Abdullah Al, Chuxiao Shao, Peiwu Geng, Shuanghu Wang, and Jian Xiao. 2024b. "Recent advances in molecular mechanisms of skin wound healing and its treatments." *Frontiers in Immunology* 15: 1395479.

Manita, Paula Gabriela, Itxaso Garcia-Orue, Edorta Santos-Vizcaino, Rosa Maria Hernandez, and Manoli Igartua. 2021. "3D bioprinting of functional skin substitutes: From current achievements to future goals." *Pharmaceuticals* 14 (4): 362.

Mao, Congyang, Yiming Xiang, Xiangmei Liu, Zhenduo Cui, Xianjin Yang, Zhaoyang Li, Shengli Zhu, Yufeng Zheng, Kelvin Wai Kwok Yeung, and Shuilin Wu. 2018. "Repeatable photodynamic therapy with triggered signaling pathways of fibroblast cell proliferation and differentiation to promote bacteria-accompanied wound healing." *ACS Nano* 12 (2): 1747–1759.

Martino, Mikaël M, Federico Tortelli, Mayumi Mochizuki, Stephanie Traub, Dror Ben-David, Gisela A Kuhn, Ralph Müller, Erella Livne, Sabine A Eming, and Jeffrey A Hubbell. 2011. "Engineering the growth factor microenvironment with fibronectin domains to promote wound and bone tissue healing." *Science Translational Medicine* 3 (100): 100ra89–100ra89.

Marx, Robert E. 2001. "Platelet-rich plasma (PRP): What is PRP and what is not PRP?" *Implant Dentistry* 10 (4): 225–228.

Massoumi, Hamed, Jhamak Nourmohammadi, Mehri Sadat Marvi, and Fathollah Moztarzadeh. 2019. "Comparative study of the properties of sericin-gelatin nanofibrous wound dressing containing halloysite nanotubes loaded with zinc and copper ions." *International Journal of Polymeric Materials and Polymeric Biomaterials* 68 (18): 1142–1153.

Matica, Mariana Adina, Finn Lillelund Aachmann, Anne Tøndervik, Håvard Sletta, and Vasile Ostafe. 2019. "Chitosan as a wound dressing starting material: Antimicrobial properties and mode of action." *International Journal of Molecular Sciences* 20 (23): 5889.

Mayer, Dieter O, William H Tettelbach, Guido Ciprandi, Fiona Downie, Jane Hampton, Heather Hodgson, Jose Luis Lazaro-Martinez, Astrid Probst, Greg Schultz, and Ewa Klara Stürmer. 2024. "Best practice for wound debridement." *Journal of Wound Care* 33 (Sup6b): S1–S32.

Mihai, Mara Madalina, Monica Beatrice Dima, Bogdan Dima, and Alina Maria Holban. 2019. "Nanomaterials for wound healing and infection control." *Materials* 12 (13): 2176.

Mirbagheri, Mahnaz Sadat, Sahar Akhavan-Mahdavi, Anwarul Hasan, Mohammad Saeed Kharazmi, and Seid Mahdi Jafari. 2023. "Chitosan-based electrospun nanofibers for diabetic foot ulcer management: Recent advances." *Carbohydrate Polymers* 313: 120512.

Misra, Priyanka, Rupali Kalsi, Sachit Anand Arora, Kumar Saurav Singh, Simoona Athar, Anchal Saini, and Anchal Saini Sr. 2023. "Effect of low-level laser therapy on early wound healing and levels of inflammatory mediators in gingival crevicular fluid following open flap debridement." *Cureus* 15 (2).

Mordorski, Breanne, and Tarl Prow. 2016. "Nanomaterials for wound healing." *Current Dermatology Reports* 5: 278–286.

Mugabe, Clement, Ali O Azghani, and Abdelwahab Omri. 2005. "Liposome-mediated gentamicin delivery: Development and activity against resistant strains of *Pseudomonas aeruginosa* isolated from cystic fibrosis patients." *Journal of Antimicrobial Chemotherapy* 55 (2): 269–271.

Naderi, Naghmeh, Dimitrios Karponis, Afshin Mosahebi, and Alexander M Seifalian. 2018. "Nanoparticles in wound healing; From hope to promise, from promise to routine." *Frontiers in Bioscience* 23: 1038–1059.

Nandhini, Jeyalakshmi, Elango Karthikeyan, and Selvaraj Rajeshkumar. 2024. "Nanomaterials for wound healing: Current status and futuristic frontier." *Biomedical Technology* 6: 26–45.

Naseri, S., M. Golpich, T. Roshancheshm, M. G. Joobeni, M. Khodayari, S. Noori, S. A. Zahed, S. Razzaghi, M. Shirzad, F. S. Salavat, and S. S. Dakhilpour. 2021. "The effect of henna and linseed herbal ointment blend on wound healing in rats with second-degree burns." *Burns* 47 (6): 1442–1450. doi: 10.1016/j.burns.2020.12.009

Naseri, Saeed, Mohammad Golpich, Tayebeh Roshancheshm, Mohammad Ghasem Joobeni, Mehdi Khodayari, Saeed Noori, Seyed Ali Zahed, Saeed Razzaghi, Mohammad Shirzad, Fatemeh Sadat Salavat, and Seyed Sadegh Dakhilpour. 2018. "Effect of bovine serum albumin immobilized Au–ZnO–graphene nanocomposite on human ovarian cancer cell." *Journal of Alloys and Compounds* 734: 66–74.

Naskar, Atanu, and Kwang-sun Kim. 2020. "Recent advances in nanomaterial-based wound-healing therapeutics." *Pharmaceutics* 12 (6): 499.

Naskar, Atanu, Sohee Lee, and Kwang-sun Kim. 2020a. "Antibacterial potential of Ni-doped zinc oxide nanostructure: Comparatively more effective against Gram-negative bacteria including multi-drug resistant strains." *RSC Advances* 10 (3): 1232–1242.

Naskar, Atanu, Sohee Lee, and Kwang-sun Kim. 2020b. "Easy one-pot low-temperature synthesized Ag-ZnO nanoparticles and their activity against clinical isolates of methicillin-resistant *Staphylococcus aureus*." *Frontiers in Bioengineering and Biotechnology* 8: 216.

Nayl, AbdElAziz A, Ahmed I Abd-Elhamid, Nasser S Awwad, Mohamed A Abdelgawad, Jinglei Wu, Xiumei Mo, Sobhi M Gomha, Ashraf A Aly, and Stefan Bräse. 2022. "Recent progress and potential biomedical applications of electrospun nanofibers in regeneration of tissues and organs." *Polymers* 14 (8): 1508.

Nickles, Michael A., William J. Ennis, Thomas F. O'Donnell, Jr., and Ian A. Altman. 2023. "Compression therapy in peripheral artery disease: A literature review." *Journal of Wound Care* 32 (Sup5): S25–s30. https://doi.org/10.12968/jowc.2023.32.Sup5.S25

Norahan, Mohammad Hadi, Sara Cristina Pedroza-González, Mónica Gabriela Sánchez-Salazar, Mario Moisés Álvarez, and Grissel Trujillo de Santiago. 2023. "Structural and biological engineering of 3D hydrogels for wound healing." *Bioactive Materials* 24: 197–235.

Normandin, Shanel, Tyler Safran, Sebastian Winocour, Carrie K Chu, Joshua Vorstenbosch, Amanda M Murphy, and Peter G Davison. 2021. "Negative pressure wound therapy: Mechanism of action and clinical applications." *Seminars in Plastic Surgery* 35 (3): 164–170.

Nowak, Marcela, Dorota Mehrholz, Wioletta Barańska-Rybak, and Roman Nowicki. 2022. "Wound debridement products and techniques: Clinical examples and literature review." *Advances in Dermatology and Allergology/Postępy Dermatologii i Alergologii* 39 (3):479–490.

Orgill, Dennis P, Ernest K Manders, Bauer E Sumpio, Raphael C Lee, Christopher E Attinger, Geoffrey C Gurtner, and H Paul Ehrlich. 2009. "The mechanisms of action of vacuum assisted closure: More to learn." *Surgery* 146 (1): 40–51.

Oryan, A., E. Alemzadeh, and A. Moshiri. 2017. "Burn wound healing: Present concepts, treatment strategies and future directions." *J Wound Care* 26 (1): 5–19. doi: 10.12968/jowc.2017.26.1.5

Ozhathil, Deepak K, Michael W Tay, Steven E Wolf, and Ludwik K Branski. 2021. "A narrative review of the history of skin grafting in burn care." *Medicina* 57 (4): 380.

Palackic, Alen, Robert P Duggan, Matthew S Campbell, Elliot Walters, Ludwik K Branski, Amina El Ayadi, and Steven E Wolf. 2022. "The role of skin substitutes in acute burn And reconstructive burn surgery: An updated comprehensive review." *Seminars in Plastic Surgery* 36 (1): 33–42.

Palanisamy, Prakash, Mahmoud Alam, Shuang Li, Simon K. H. Chow, and Yong-Ping Zheng. 2022. "Low-intensity pulsed ultrasound stimulation for bone fractures healing: A review." *Journal of Ultrasound in Medicine* 41 (3): 547–563. https://doi.org/10.1002/jum.15738

Parham, Shokoh, Anousheh Zargar Kharazi, Hamid Reza Bakhsheshi-Rad, Ahmad Fauzi Ismail, Hamid Ghayour, Hadi Nur, and Filippo Berto. 2020. "Electrospun nano-fibers for biomedical and tissue engineering applications: A comprehensive review." *Materials* 13 (9): 2153.

Patel, Shivani, Shweta Srivastava, Mahendra R. Singh, and Dinesh Singh. 2019. "Mechanistic insight into diabetic wounds: Pathogenesis, molecular targets and treatment strategies to pace wound healing." *Biomed Pharmacother* 112: 108615. https://doi.org/10.1016/j.biopha.2019.108615

Patton, Declan, Pelin Avsar, Alaa Sayeh, Alaa Budri, Thomas O'Connor, Seán Walsh, Louise Nugent, Declan Harkin, Niamh O'Brien, Joseph Cayce, Michelle Corcoran, María Gaztambide, and Zena Moore. 2023. "A meta-review of the impact of compression therapy on venous leg ulcer healing." *International Wound Journal* 20 (2): 430–447. doi: 10.1111/iwj.13891

Potekaev, Nikolai N, Olga B Borzykh, German V Medvedev, Denis V Pushkin, Marina M Petrova, Artem V Petrov, Diana V Dmitrenko, Elena I Karpova, Olga M Demina, and Natalia A Shnayder. 2021. "The role of extracellular matrix in skin wound healing." *Journal of Clinical Medicine* 10 (24): 5947.

Prabowo, Wahyu Haris, Awal Prasetyo, and Neni Susilaningsih. 2022. "The effect of multilevel doses of caffeine on tissue macrophage and blood lymphocyte count in autologous full thickness skin graft healing in Sprague Dawley rats." *Bioscientia Medicina: Journal of Biomedicine and Translational Research* 6 (5): 1697–1702.

Raghuwanshi, Vikram Singh, and Gil Garnier. 2019. "Characterisation of hydrogels: Linking the nano to the microscale." *Advances in Colloid and Interface Science* 274: 102044.

Ribeiro, Mariana, Marco Simões, Carla Vitorino, and Filipa Mascarenhas-Melo. 2024. "Hydrogels in cutaneous wound healing: Insights into characterization, properties, formulation and therapeutic potential." *Gels* 10 (3): 188.

Riha, Shaima Maliha, Manira Maarof, and Mh Busra Fauzi. 2021. "Synergistic effect of biomaterial and stem cell for skin tissue engineering in cutaneous wound healing: A concise review." *Polymers* 13 (10): 1546.

Rizwan, Muhammad, Syeda Rubina Gilani, Arjumand Iqbal Durrani, and Sobia Naseem. 2022. "Kinetic model studies of controlled nutrient release and swelling behavior of combo hydrogel using *Acer platanoides* cellulose." *Journal of the Taiwan Institute of Chemical Engineers* 131: 104137.

Rogoń, Ignacy, Agnieszka Rogoń, Mariusz Kaczmarek, Adam Bujnowski, Jerzy Wtorek, Filip Lachowski, and Jerzy Jankau. 2024. "Flap monitoring techniques: A review." *Journal of Clinical Medicine* 13 (18): 5467.

Rosellini, Elisabetta, and Maria Grazia Cascone. 2023. "Microfluidic fabrication of natural polymer-based scaffolds for tissue engineering applications: A review." *Biomimetics* 8 (1): 74.

Rosenthal, A Ralph, Peter R Egbert, Christina Harbury, Jerry L Hopkins, and Edward Rubenstein. 1978. "Use of platelet-fibrinogen-thrombin mixture to seal experimental penetrating corneal wounds." *Albrecht von Graefes Archiv für klinische und experimentelle Ophthalmologie* 207: 111–115.

Růžička, Jiří, Jiří Dejmek, Lukáš Bolek, Jiří Beneš, and Jitka Kuncová. 2021. "Hyperbaric oxygen influences chronic wound healing—A cellular level review." *Physiological Research* 70 (Suppl 3): S261.

Saeg, Farhad, Roberto Orazi, Gregory M. Bowers, and Jeffrey E. Janis. 2021. "Evidence-based nutritional interventions in wound care." *Plastic and Reconstructive Surgery* 148 (1): 226–238. https://doi.org/10.1097/prs.0000000000008061.

Sahana, Thimmapura G., and Parvathi D. Rekha. 2018. "Biopolymers: Applications in wound healing and skin tissue engineering." *Molecular Biology Reports* 45 (6): 2857–2867. https://doi.org/10.1007/s11033-018-4296-3

Salimi, Fatemeh, and Fatemeh Mohammadipanah. 2021. "Nanomaterials versus the microbial compounds with wound healing property." *Frontiers in Nanotechnology* 2: 584489.

Sanjanwala, Dhruv, Vaishali Londhe, Rashmi Trivedi, Smita Bonde, Sujata Sawarkar, Vinita Kale, and Vandana Patravale. 2022. "Polysaccharide-based hydrogels for drug delivery and wound management: A review." *Expert Opinion on Drug Delivery* 19 (12): 1664–1695.

Schlottmann, Frederik, Vesna Bucan, Peter M Vogt, and Nicco Krezdorn. 2021. "A short history of skin grafting in burns: From the gold standard of autologous skin grafting to the possibilities of allogeneic skin grafting with immunomodulatory approaches." *Medicina* 57 (3): 225.

Seidi, Khaled, Mohammad Hosein Ayoubi-Joshaghani, Mehdi Azizi, Tahereh Javaheri, Mehdi Jaymand, Effat Alizadeh, Thomas J Webster, Amirhossein Ahmadieh Yazdi, Mostafa Niazi, and Michael R Hamblin. 2021. "Bioinspired hydrogels build a bridge from bench to bedside." *Nano Today* 39: 101157.

Sen, Chandan K. 2021. "Human wound and its burden: Updated 2020 compendium of estimates." *Advances in Wound Care* 10 (5): 281–292.

Shabestani Monfared, Ghazal, Peter Ertl, and Mario Rothbauer. 2021. "Microfluidic and lab-on-a-chip systems for cutaneous wound healing studies." *Pharmaceutics* 13 (6): 793.

Shahzadi, Uzma, Rabia Zeeshan, Sobia Tabassum, Hina Khadim, Muhammad Arshad, Arsalan Ahmad Ansari, Sher Zaman Safi, Rana Intisar ul Haq, and Anila Asif. 2023. "Physico-chemical properties and in-vitro biocompatibility of thermo-sensitive hydrogel developed with enhanced antimicrobial activity for soft tissue engineering." *Polymers for Advanced Technologies* 34 (12): 3870–3884.

Shaikh, Muhammad Saad, Muhammad Sohail Zafar, Flavio Pisani, Mohid Abrar Lone, and Yasser Riaz Malik. 2021. "Critical features of periodontal flaps with regard to blood clot stability: A review." *Journal of Oral Biosciences* 63 (2): 111–119.

Shi, C., J. C. Dumville, N. Cullum, E. Connaughton, and G. Norman. 2021. "Compression bandages or stockings versus no compression for treating venous leg ulcers." *Cochrane Database Syst Rev* 7 (7): Cd013397. doi: 10.1002/14651858.CD013397.pub2

Sierra-Sánchez, Álvaro, Kevin H Kim, Gonzalo Blasco-Morente, and Salvador Arias-Santiago. 2021. "Cellular human tissue-engineered skin substitutes investigated for deep and difficult to heal injuries." *NPJ Regenerative Medicine* 6 (1): 35.

Soliman, Hesham, Marine Theret, Wilder Scott, Lesley Hill, Tully Michael Underhill, Boris Hinz, and Fabio MV Rossi. 2021. "Multipotent stromal cells: One name, multiple identities." *Cell Stem Cell* 28 (10): 1690–1707.

Sriram, K., R. A. Meguid, and M. M Meguid. 2016. "Nutritional support in adults with chyle leaks." *Nutrition* 32 (2): 281–286. doi: 10.1016/j.nut.2015.08.002

Stone, Randolph, Emily C Saathoff, David A Larson, John T Wall, Nathan A Wienandt, Skuli Magnusson, Hilmar Kjartansson, Shanmugasundaram Natesan, and Robert J Christy. 2021. "Accelerated wound closure of deep partial thickness burns with acellular fish skin graft." *International Journal of Molecular Sciences* 22 (4): 1590.

Tai, Ting-Yu, Kuan-Ju Lin, Hsin-Yi Chang, Yu-Cheng Wu, Chien-Uen Huang, Xiao-Yu Lin, Fang-Chun Tsai, Cheng-Sheng Tsai, Yu-Hsuan Chen, Fang-Yi Wang, and Shih-Ching Chang. 2024. "Early identification of delayed wound healing in complex diabetic foot ulcers treated with A dermal regeneration template: A novel clinical target and its risk factors." *International Journal of Surgery* 110 (2): 943–955. https://doi.org/10.1097/js9.0000000000000898

Tan, Yan Yu, Frank Liaw, Robert Warner, Simon Myers, and Ali Ghanem. 2021. "Enhanced recovery pathways for flap-based reconstruction: Systematic review and meta-analysis." *Aesthetic Plastic Surgery* 45 (5): 2096–2115.

Tavelli, Lorenzo, Shayan Barootchi, Martina Stefanini, Giovanni Zucchelli, William V Giannobile, and Hom-Lay Wang. 2023. "Wound healing dynamics, morbidity, and complications of palatal soft-tissue harvesting." *Periodontology 2000* 92 (1): 90–119.

Thomas, Deena Clare, Chong Li Tsu, Rose A Nain, Norkiah Arsat, Soong Shui Fun, and Nik Amin Sahid Nik Lah. 2021. "The role of debridement in wound bed preparation in chronic wound: A narrative review." *Annals of Medicine and Surgery* 71: 102876.

Ti, Dongdong, Haojie Hao, Chuan Tong, Jiejie Liu, Liang Dong, Jingxi Zheng, Yali Zhao, Huiling Liu, Xiaobing Fu, and Weidong Han. 2015. "LPS-preconditioned mesenchymal stromal cells modify macrophage polarization for resolution of chronic inflammation via exosome-shuttled let-7b." *Journal of Translational Medicine* 13 1–14.

Tognetti, Linda, Elisa Pianigiani, Francesca Ierardi, Guido Lorenzini, Donato Casella, Flavio Giulio Liso, Andrea De Pascalis, Elisa Cinotti, and Pietro Rubegni. 2021. "The use of human acellular dermal matrices in advanced wound healing and surgical procedures: State of the art." *Dermatologic Therapy* 34 (4): e14987.

Tran, David L, Ren-Wen Huang, Ernest S Chiu, Erin M Rajhathy, John H Gregory, Elizabeth A Ayello, and R Gary Sibbald. 2023. "Debridement: Technical considerations and treatment options for the interprofessional team." *Advances in Skin & Wound Care* 36 (4): 180–187.

Tran, Morica M., and Melanie N. Haley. 2021. "Does exercise improve healing of diabetic foot ulcers? A systematic review." *Journal of Foot and Ankle Research* 14 (1): 19. doi: 10.1186/s13047-021-00456-w

Vaghardoost, Reza, Seyed Gholam Mousavi Majd, Hamid Tebyanian, Hamid Babavalian, Leila Malaei, Mitra Niazi, and Ali Javdani. 2018. "The healing effect of sesame oil, camphor and honey on second degree burn wounds in rat." *World Journal of Plastic Surgery* 7 (1): 67–71.

Van de Kuit, A., R. J. Krishnan, W. H. Mallee, L. M. Goedhart, B. Lambert, J. N. Doornberg, T. M. J. S. Vervest, and J. Martin. 2022. "Surgical site infection after wound closure with staples versus sutures in elective knee and hip arthroplasty: A systematic review and meta-analysis." *Arthroplasty* 4 (1): 12.

Vecin, Nicole M, and Robert S Kirsner. 2023. "Skin substitutes as treatment for chronic wounds: Current and future directions." *Frontiers in Medicine* 10: 1154567.

Venturi, Mark L, Christopher E Attinger, Ali N Mesbahi, Christopher L Hess, and Katherine S Graw. 2005. "Mechanisms and clinical applications of the vacuum-assisted closure (VAC) device: A review." *American Journal of Clinical Dermatology* 6: 185–194.

Verdes, Mirela, Karin Mace, Lindsay Margetts, and Sarah Cartmell. 2022. "Status and challenges of electrical stimulation use in chronic wound healing." *Current Opinion in Biotechnology* 75: 102710. https://doi.org/10.1016/j.copbio.2022.102710

Vijayan, Amritha, C. K. Nanditha, and G. S. Vinod Kumar. 2021. "ECM-mimicking nanofibrous scaffold enriched with dual growth factor carrying nanoparticles for diabetic wound healing." *Nanoscale Advances* 3 (11): 3085–3092.

Wakabayashi, Naohiro, Atsushi Sakai, Hiroshi Takada, Takashi Hoshi, Hiroyuki Sano, Shigeyuki Ichinose, Hiroshi Suzuki, and Ryosuke Ogawa. 2020. "Noncontact phased-array ultrasound facilitates acute wound healing in mice." *Plastic and Reconstructive Surgery* 145 (2): 348e–359e. https://doi.org/10.1097/prs.0000000000006481

Wang, Jie, Hao Wu, Yixuan Peng, Yue Zhao, Youyou Qin, Yingbo Zhang, and Zhibo Xiao. 2021. "Hypoxia adipose stem cell-derived exosomes promote high-quality healing of diabetic wound involves activation of PI3K/Akt pathways." *Journal of Nanobiotechnology* 19: 1–13.

Wang, Lirong, Mengyun Zhou, Tailin Xu, and Xueji Zhang. 2022. "Multifunctional hydrogel as wound dressing for intelligent wound monitoring." *Chemical Engineering Journal* 433: 134625.

Windisch, Peter, Vincenzo Iorio-Siciliano, Daniel Palkovics, Luca Ramaglia, Andrea Blasi, and Anton Sculean. 2022. "The role of surgical flap design (minimally invasive flap vs. extended flap with papilla preservation) on the healing of intrabony defects treated with an enamel matrix derivative: A 12-month two-center randomized controlled clinical trial." *Clinical Oral Investigations* 26: 1811–1821.

Wong, Andrew Yew Wei, Bernard Soon Yang Ong, Ainsley Ryan Yan Bin Lee, Aaron Shengting Mai, Sathiyamoorthy Selvarajan, Satish R Lakshminarasappa, Sook Muay Tay, Bernard Ong, and Aaron S Mai. 2022. "Topical biological agents as adjuncts to improve wound healing in chronic diabetic wounds: A systematic review of clinical evidence and future directions." *Cureus* 14 (7).

Xu, Juan, Suwen Bai, Yadi Cao, Lei Liu, Yang Fang, Juan Du, Li Luo, Mingwei Chen, Bing Shen, and Qiu Zhang. 2020. "miRNA-221-3p in endothelial progenitor cell-derived exosomes accelerates skin wound healing in diabetic mice." *Diabetes, Metabolic Syndrome and Obesity* 13: 1259–1270.

Xue, Jiajia, Tong Wu, Yunqian Dai, and Younan Xia. 2019. "Electrospinning and electrospun nanofibers: Methods, materials, and applications." *Chemical Reviews* 119 (8): 5298–5415.

Yamaguchi, Natsuki, Eri Horio, Jukito Sonoda, Miu Yamagishi, Satomi Miyakawa, Fumihiro Murakami, Hideaki Hasegawa, Yasuhiro Katahira, Izuru Mizoguchi, and Yasuyuki Fujii. 2024. "Immortalization of mesenchymal stem cells for application in regenerative medicine and their potential risks of tumorigenesis." *International Journal of Molecular Sciences* 25 (24): 13562.

Yamanaka, Shinya. 2020. "Pluripotent stem cell-based cell therapy—Promise and challenges." *Cell Stem Cell* 27 (4): 523–531.

Yang, Qian, Jinfa Huang, Yixuan Liu, Qiqing Mai, Yuan Zhou, Lei Zhou, Lingling Zeng, and Kaixian Deng. 2024. "Human umbilical cord mesenchymal stem cells promote anti-inflammation and angiogenesis by targeting macrophages in a rat uterine scar model." *Stem Cell Reviews and Reports* 20 (6): 1555–1568.

Yang, Shuang, Xiaoming Li, Ping Liu, Maolan Zhang, Chao Wang, and Bo Zhang. 2020. "Multifunctional chitosan/polycaprolactone nanofiber scaffolds with varied dual-drug release for wound-healing applications." *ACS Biomaterials Science & Engineering* 6 (8): 4666–4676.

Yao, Yingxia, Andi Zhang, Congshan Yuan, Xiguang Chen, and Ya Liu. 2021. "Recent trends on burn wound care: Hydrogel dressings and scaffolds." *Biomaterials Science* 9 (13): 4523–4540.

Ying, Guoliang, Nan Jiang, Carolina Parra-Cantu, Guosheng Tang, Jingyi Zhang, Hongjun Wang, Shixuan Chen, Ning-Ping Huang, Jingwei Xie, and Yu Shrike Zhang. 2020. "Bioprinted injectable hierarchically porous gelatin methacryloyl hydrogel constructs with shape-memory properties." *Advanced Functional Materials* 30 (46): 2003740.

Yoon, Jaechul, Dogeon Yoon, Hyeongjin Lee, JiUn Lee, SeoYul Jo, Dohern Kym, Haejun Yim, Jun Hur, Wook Chun, and GeunHyung Kim. 2022. "Wound healing ability of acellular fish skin and bovine collagen grafts for split-thickness donor sites in burn patients: Characterization of acellular grafts and clinical application." *International Journal of Biological Macromolecules* 205: 452–461.

Yoshimoto, Hiroshi, Yong-Min Shin, Hiroshi Terai, and Joseph P. Vacanti. 2003. "A biodegradable nanofiber scaffold by electrospinning and its potential for bone tissue engineering." *Biomaterials* 24 (12): 2077–2082.

Yuan, Min, Kun Liu, Tao Jiang, Shuai Li, Jian Chen, Zhen Wu, Wei Li, Rui Tan, Wei Wei, Xiang Yang, Hao Dai, and Zhi Chen. 2022. "GelMA/PEGDA microneedles patch loaded with HUVECs-derived exosomes and Tazarotene promote diabetic wound healing." *Journal of Nanobiotechnology* 20 (1): 147. doi: 10.1186/s12951-022-01354-4

Zaver, Vasudev, and Pradeep Kankanalu. 2023. "Negative pressure wound therapy." In *StatPearls [Internet]*. StatPearls Publishing.

Zha, Fangwen, Wei Chen, Lifeng Zhang, and Demei Yu. 2020. "Electrospun natural polymer and its composite nanofibrous scaffolds for nerve tissue engineering." *Journal of Biomaterials Science, Polymer Edition* 31 (4): 519–548.

Zhang, Zhiguang, Fuqiang Chu, Xin Wang, Xu Zhou, and Guirong Xiong. 2022. "Microfluidic fabrication of a PDMS microlens for imaging tunability." *Langmuir* 38 (13): 4059–4064.

Zheng, Zesen, Huihui Zhang, Jiaxin Yang, Xiaoyang Liu, Lianglong Chen, Wenwen Li, Siqi Mi, Hai Zhou, Weihan Zheng, and Wanting Xue. 2025. "Recent advances in structural and functional design of electrospun nanofibers for wound healing." *Journal of Materials Chemistry B* 13 (18): 5226–5263.

Zhou, Zhiguo, Steve Joslin, Anthony Dellinger, Marion Ehrich, Brad Brooks, Qing Ren, Ulrich Rodeck, Robert Lenk, and Christopher L Kepley. 2010. "A novel class of compounds with cutaneous wound healing properties." *Journal of Biomedical Nanotechnology* 6 (5): 605–611.

10 Nutritional Support for Wound Healing

Zeemal Seemab Amin[1], *Sabiya Azam*[1], *Sami A. Al-Hussain*[2], *Yousef A. Bin Jardan*[3**] *Nusrat Shafiq*[4], *Bushra Shakoor*[4], *Ali Irfan*[5*], and *Magdi E. A. Zaki*[2***]

1 School of Biochemistry, Minhaj University Lahore, Quaid-e-Azam industrial State, Lahore, Pakistan

2 Department of Chemistry, College of Science, Imam Mohammad Ibn Saud Islamic University (IMSIU), Riyadh, Saudi Arabia

3 Department of Pharmaceutics, College of Pharmacy, King Saud University, Riyadh, Saudi Arabia. (Y.A.B.J)

4 Synthetic and Natural Products Discovery (SNPD) Laboratory, Department of Chemistry, Government College Women University Faisalabad, Pakistan. (N.S.)

5 Green Institute of Chemical, Biomedical and Environmental Sciences (GICBES), Lahore, Pakistan

10.1 OVERVIEW OF WOUND HEALING

Wound healing incarnates a tortuous and highly harmonized biological advancement linking a cascade of molecular and cellular crossing points to refurbish tissue integrity following grievance. This urbane mechanism is indispensable for preserving cutaneous barrier utility and mitigating the risk of infection (Schultz et al., 2011). This process is systematically divided into four distinct and overlying phases: hemostasis, inflammation, proliferation, and remodeling. The first stage of wound healing is hemostasis, which occurs immediately after disturbance (LaPelusa and Dave, 2025). This approach involves a series of tightly controlled life processes designed to lead to blockage of arteries and clot formation. Vasoconstriction, stimulated by vasoactive substances including adrenaline, norepinephrine, thromboxane A_2, prostaglandins, and serotonin, reduces blood flow into the damaged area. Platelets adhere to the exposed subendothelial collagen through von Willebrand factor (vWF), causing platelet activation and aggregation (Sabbah et al., 2024). Activated platelets release chemokines, cytokines, and growth factors (for instance, platelet-derived growth factor (PDGF) and transforming growth factor-β (TGF-β)) that attract additional cellular mediators necessary for repair. The extrinsic and intrinsic pathways of the coagulation cascade are activated, leading to the enzymatic transformation of fibrinogen into fibrin. It forms a provisional hemostatic plug alongside the

* Corresponding author: ybinjardan@ksu.edu.sa
** Corresponding author: raialiirfan@gmail.com
*** Corresponding author: mezaki@imamu.edu.sa

 DOI: 10.1201/9781003644958-10

aggregated platelets, preventing hemorrhage and creating a defensive environment against colonization. When the wound enters the inflammation phase, inflammatory cells infiltrate the wound site to clear pathogens and necrotic tissue (Scopelliti et al., 2022). As the first responders, neutrophils phagocytose bacteria secrete proteolytic enzymes to degrade cellular debris (Phillipson and Kubes, 2019). Thereafter, macrophages mediate the resolution of inflammation via the secretion of important growth factors (e.g., vascular endothelial growth factor (VEGF), fibroblast growth factor (FGF), TGF-β, allowing tissue repair and entering the proliferative stage. The proliferative phase is characterized by active reconstruction of the wound through multiple interdependent processes (Wilgus et al., 2013). Granulation tissue forms fibroblasts that infiltrate the wound bed and secrete the provisional extracellular matrix (ECM) (Diller and Tabor, 2022) characterized by type III collagen, fibronectin, and hyaluronic acid (HA), which renders structural support for regeneration of the tissue. Hypoxia-induced signaling (HIF-1α) stimulates endothelial cell proliferation, leading to the neovascularization of the wound, ensuring adequate perfusion and nutrient delivery (Han et al., 2022).

In collagen deposition and remodeling, fibroblasts undergo phenotypic differentiation into myofibroblasts, facilitating wound contraction while secreting collagen type I, which progressively enhances tensile strength (Mathew-Steiner et al., 2021). Keratinocytes proliferate and migrate from the wound margins, restoring the epidermal barrier through stratification and differentiation. The absolute phase of wound maturation is remodeling phase represents the ultimate stage of wound repair, a protracted process that may persist for months to years, during which the nascent tissue undergoes structural reinforcement and realignment (Muse et al., 2023). This phase is characterized by the progressive reorganization of collagen fibrils, transitioning from a disorganized matrix to a more structured, cross-linked architecture. Although the wound may regain approximately 80% of its original tensile strength, the healed tissue remains biomechanically inferior to uninjured skin due to the absence of fully restored dermal appendages (e.g., hair follicles, sweat glands) and persistent scar formation.

10.2 HEALING MECHANISM

Wound repair proceeds via distinct mechanisms, contingent upon the method of wound closure like in primary infection the wound edges are surgically repaired (via sutures, staples, or adhesive agents), facilitating rapid re-epithelialization and minimizing scar formation (Cotterell et al., 2024), universally employed in clean, surgically incised wounds with minimal tissue loss. Spontaneous healing is the secondary intention in which the wound is permitted to heal through natural granulation, contraction, and epithelialization, without primary closure. When primary closure is not appropriate for wounds that are contaminated, infected, or severely traumatized, this treatment is recommended (Tombulturk et al., 2022). Tertiary infection also known as "Delayed primary closure"; the wound is initially left open to manage infection or excessive exudate, followed by surgical closure once the wound bed is deemed viable. It is often utilized in high-risk surgical wounds or those requiring debridement prior to definitive repair (ElHawary et al., 2023). Long-term tissue reinforcement is the final phase that involves extensive ECM remodeling, where collagen fibrils undergo cross-linking and realignment under the influence of matrix metalloproteinases (MMPs) and tissue inhibitors of metalloproteinases (TIMPs) (Khalili-Tanha et al., 2025). This results in the gradual replacement of immature collagen type III with the more resilient collagen type I, ultimately optimizing scar strength and functionality.

10.3 DETERMINANTS OF WOUND HEALING EFFICIENCY

The trajectory and efficacy of wound repair are influenced by a multitude of intrinsic and extrinsic factors (Anderson and Hamm, 2014). Nutritional status consisting of proteins are essential for collagen synthesis and immune function. Micronutrients (e.g., vitamins A, C, zinc, and copper) play pivotal roles in enzymatic reactions and antioxidant defense. Vascular perfusion means adequate blood supply that ensures oxygen delivery, nutrient availability, and waste removal. Ischemia or peripheral vascular disease impairs fibroblast activity and angiogenesis. Biofilm formation and persistent infection prolong the inflammatory phase, delaying granulation and epithelialization. Systemic and comorbid factors, including diabetes mellitus (via hyperglycemia-induced microangiopathy), disrupt neutrophil function and collagen deposition (Tombulturk et al., 2022). Advanced age, obesity, and tobacco use compromise cellular proliferation and ECM remodeling. Corticosteroids suppress inflammation, while chemotherapeutic agents inhibit mitotic activity (Hodgens and Sharman, 2025). Negative pressure wound therapy (NPWT) and bioengineered skin substitutes may enhance healing in chronic wounds.

For these phases to progress well, adequate dietary assistance is necessary. Proteins, carbohydrates, and fats supply the energy and building blocks required for collagen formation, tissue repair, and immunological response (Barchitta et al., 2019). Proteins are essential macromolecules that play a crucial role in wound healing. They serve as building blocks for tissues, facilitate immune responses, and regulate biochemical pathways necessary for repair. Proteins contribute to wound healing through multiple mechanisms starting with structural support by using collagen, elastin, and keratin provide the scaffolding for new tissue formation (de Albuquerque et al., 2023). In natural wound healing, the body forms a provisional scaffold immediately after injury, primarily composed of fibrin and fibronectin from the clotting process. This temporary matrix traps platelets, growth factors, and immune cells, creating an initial framework for repair. Over time, fibroblasts migrate into the wound and begin secreting key ECM components such as collagen type III (Wang et al., 2020), HA, and proteoglycans, forming granulation tissue (Selmani et al., 2024). Angiogenesis ensures blood supply, while myofibroblasts help contract the wound. As healing progresses, the scaffold matures with collagen type III is gradually replaced by stronger collagen type I, and MMPs remodel the ECM into a more organized structure. However, the resulting scar tissue remains mechanically inferior to uninjured skin due to the lack of fully restored dermal structures.

Clot formation originates when fibrinogen polymerizes into fibrin, forming a stable clot to prevent bleeding (Kearney et al., 2022). When a blood vessel is injured, the body activates a fast and complex process called *hemostasis to* stop the bleeding. Steps involved in this mechanism are vascular spasm, the immediate response in which the blood vessel constricts to reduce blood flow to the area. Primary hemostasis is known as platelet plug formation in which platelets (tiny blood cells) quickly rush to the injury site. They stick to exposed collagen of the damaged vessel wall (helped by von Willebrand factor) (Patten and Wang, 2021). Platelets activate, change shape, release granules, and stick to each other forming a *temporary platelet plug.* Coagulation cascade is a domino effect where clotting factors (proteins in plasma) get activated. It converts prothrombin (factor II) into thrombin (factor IIa). Thrombin then converts into the soluble fibrinogen and then into fibrin, an insoluble. Fibrin forms a mesh over the platelet plug, stabilizing it into a strong clot. This final stable clot prevents further bleeding. Once the wound heals, the clot is no longer needed, so the body carefully breaks it down using fibrinolysis. The inactive enzyme plasminogen (trapped in the clot) is converted to plasmin by tissue plasminogen activator (tPA) or urokinase. Plasmin acts like molecular scissors and cuts down fibrin into smaller pieces called fibrin degradation products (FDPs), dissolving the clot. This clearing of the clot restores normal blood flow and ensures the vessel is clean (Liu et al., 2025). Table 10.1 illustrates the function and location of the proteins essential for wound healing.

Figure 10.1 depicts the three major phases of blood clot development in a damaged blood artery. When a blood vessel is injured, the injury to the vessel's lining causes the release of clotting factors, which initiates the hemostatic reaction. In the second stage, vasoconstriction occurs to decrease blood flow at the injury site, while platelets are activated and begin to collect, forming a temporary stopper known as the platelet plug. Furthermore, in the third step, fibrinogen is transformed into fibrin by the metabolic process of thrombin. Fibrin strands create a web around the platelet plug,

TABLE 10.1
Proteins Providing Structural Support to Tissues in the Body

Protein	Function	Location
Collagen I	Tensile strength	Dermis, mature scar
Collagen III	Early matrix support	Granulation tissue
Elastin	Tissue recoil	Blood vessels, skin
Fibronectin	Cell adhesion	Provisional matrix
Laminin	Basement membrane	Dermal-epidermal junction

resulting in the formation of a solid, insoluble blood clot that successfully closes the wound (Ouyang and Willner, 2025). The clotting cascade depicts the molecular transformation with clotting factors activating prothrombin, which subsequently transforms fibrinogen into fibrin, resulting in the development of a stable blood clot.

10.4 MACROMOLECULES AND THEIR ROLE IN WOUND HEALING

Wound healing is a complex biological process that relies heavily on the coordinated functions of four major classes of macromolecules: proteins, polysaccharides, nucleic acids, and lipids. Proteins serve as the primary pillars of repair, with structural proteins like collagen providing the ECM

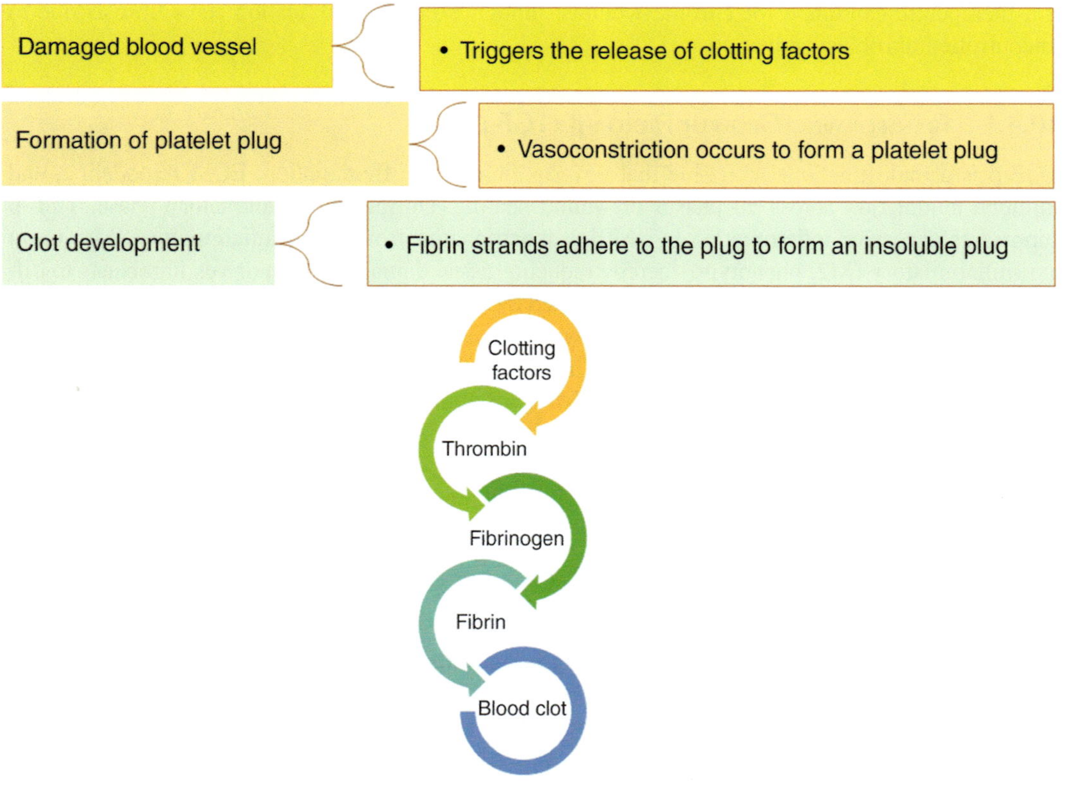

FIGURE 10.1 Blood clot formation in broken vessel.

framework, fibrinogen forming the initial clot scaffold, and growth factors such as PDGF and VEGF orchestrating cellular responses (Elieh-Ali-Komi et al., 2025).

Proteins serve as the architects of wound repair, performing multifaceted roles throughout the healing process. Structural proteins like collagen types I, III, and IV form the ECM scaffold, with type III predominating in early granulation tissue and type I providing strength to mature scars, while fibronectin and laminin facilitate cell adhesion and migration. The clotting process relies on fibrinogen polymerization into fibrin and vWF's platelet-anchoring function.

Signaling proteins including PDGF, TGF-β, and VEGF drive fibroblast activation and angiogenesis, with MMPs and TIMPs dynamically remodeling the matrix (Deng et al., 2024). Immune modulation occurs through cytokines (interleukin-6 (IL-6), tumor necrosis factor (TNF-α)) and immunoglobulins, with protein deficiencies leading to impaired healing treatable by recombinant proteins like rhPDGF (Knight et al., 2024).

PDGF is a critical signaling protein that initiates and sustains wound healing by coordinating cellular recruitment and ECM synthesis. Released primarily by platelets during clot formation, PDGF acts as a chemoattractant for neutrophils, macrophages, and fibroblasts, drawing these cells into the wound bed to clear debris and begin repair (Di Tolla et al., 2025). It binds to PDGF receptor-β (PDGFR-β) on fibroblasts and pericytes, activating intracellular pathways such as phosphatidylinositol 3-kinase (PI3K) and protein kinase B (AKT) and mitogen-activated protein kinase (MAPK) (Tavukcuoglu et al., 2024), which stimulates cell proliferation and migration. PDGF also primes macrophages to secrete additional growth factors, including VEGF and TGF-β (Panda and Mohanty, 2025) creating a self-amplifying loop that accelerates granulation tissue formation. By enhancing fibroblast activity, PDGF promotes the deposition of provisional ECM components like fibronectin and HA, which provide structural support for subsequent collagen synthesis. In preclinical models, exogenous PDGF-AA has been shown to reduce healing time by up to 30%, making it a therapeutic candidate for chronic wounds, though excessive application risks fibrosis due to uncontrolled fibroblast activation.

10.4.1 Transforming Growth Factor-β (TGF-β)

TGF-β is a multifunctional cytokine that governs fibroblast differentiation, ECM remodeling, and immune modulation across all phases of wound healing. During the inflammatory phase, TGF-β suppresses excessive inflammation by shifting macrophages from a pro-inflammatory (M1) to an anti-inflammatory (M2) phenotype, thereby reducing tissue damage. It then drives fibroblasts to differentiate into myofibroblasts—a process mediated by Mothers against decapentaplegic homolog 2/3 (Smad2/3) signaling—which are essential for wound contraction (Wong et al., 2025). TGF-β upregulates collagen types I and III, fibronectin, and TIMPs, stabilizing the ECM while suppressing MMPs to prevent premature breakdown (Zhao et al., 2024). Its ability to balance collagen synthesis and degradation is critical during the remodeling phase, where it ensures scar strength without excessive stiffness. However, dysregulated TGF-β signaling can lead to hypertrophic scarring or fibrosis, prompting research into targeted inhibitors (e.g., anti-TGF-β antibodies) or alternative isoforms like TGF-β3.

10.4.2 Vascular Endothelial Growth Factor (VEGF)

VEGF is the master regulator of angiogenesis, ensuring the formation of new blood vessels to deliver oxygen and nutrients to the healing wound. Hypoxia in the wound bed stabilizes HIF-1α, a transcription factor that upregulates VEGF expression in fibroblasts, macrophages, and keratinocytes (Shams et al., 2022). VEGF binds to its receptor VEGFR-2 on endothelial cells, triggering extracellular signal-regulated kinase (ERK) and focal adhesion kinase (FAK) pathways that stimulate cell proliferation, migration, and capillary sprouting (Xu et al., 2024). This process is vital during the proliferative phase, where nascent vessels integrate into the granulation tissue to support fibroblast and keratinocyte activity. VEGF also enhances vascular permeability, allowing plasma proteins and immune cells to infiltrate the wound. Its production is synergistically boosted by PDGF and TGF-β through macrophage-mediated paracrine signaling. Clinically, VEGF delivery via hydrogels or gene therapy has shown promise in treating ischemic wounds, though excessive angiogenesis can lead to dysfunctional vasculature (Xiaoming et al., 2024). Thus, precise spatiotemporal control of VEGF is crucial to ensure robust, functional neovascularization.

Under hypoxic conditions in granulation tissue, fibroblasts, macrophages, and keratinocytes upregulate HIF-1α, which stimulates the production of VEGF. VEGF acts on endothelial cells by binding to VEGF receptor 2 (VEGFR-2), initiating intracellular signaling cascades involving ERK and FAK. This leads to endothelial cell proliferation, migration, and capillary sprouting, thereby promoting angiogenesis. Newly formed nascent vessels contribute to granulation tissue formation and wound healing.

10.5 ROLE OF POLYSACCHARIDES IN WOUND HEALING

Polysaccharides function as hydration and signaling hubs, where HA creates a hydrated migration matrix and chondroitin sulfate organizes collagen fibrils (Gounden and Singh, 2024). Heparan sulfate stabilizes crucial growth factors, and HA fragments modulate inflammation through cluster of differentiation 44/toll-like receptor 4 (CD44/TLR4 pathways), making HA-based dressings valuable for burns and diabetic ulcers (Zhaot et al., 2025). Nucleic acids provide genetic instructions for repair proteins and fine-tune healing through microRNAs like miR-21 and miR-29 that regulate TGF-β signaling, while emerging nucleic acid therapies target MMP overexpression via small interfering RNA (siRNA) or deliver growth factor genes (Chylińska and Maciejczyk, 2025). When injury occurs, HA is cleaved into smaller fragments that bind to a CD44-TLR4 receptor complex on immune cells, triggering a protective anti-inflammatory response, that pretreatment with HA fragments in mice significantly reduces pro-inflammatory cytokines like IL-6 and TNFα after lipopolysaccharide (LPS) exposure, effectively protecting against sepsis. This protective effect is CD44-dependent, as mice lacking CD44 fail to suppress cytokine release or upregulate the critical protein, i.e., A20 or TNF-α, alpha-induced protein 3 (TNFAIP3).

10.5.1 Mechanistic Insights

HA fragments activate TLR4 signaling but diverge from LPS in their downstream effects. While LPS promotes pro-inflammatory cytokine storms via TLR4, HA–CD44 engagement upregulates A20, a key negative regulator of TLR signaling. A20 suppresses nuclear factor kappa B (NF-κB) activation, thereby limiting excessive inflammation. This process requires myeloid differentiation primary response protein 88 (MyD88) and TIR-domain-containing adaptor-inducing beta interferon (TRIF), adaptor proteins downstream of TLR4, confirming HA's reliance on canonical TLR4 signaling pathways. Importantly, siRNA-mediated knockdown of A20 abolishes HA's anti-inflammatory effects, demonstrating its pivotal role (Wilkinson and Hardman, 2020). HA-based biomaterials leverage this CD44/TLR4-A20 axis to modulate inflammation in chronic wounds and burns. By stabilizing growth factors and creating hydrated matrices, HA dressings enhance tissue regeneration while suppressing harmful immune responses. For diabetic ulcers, where TLR4 hyperactivation exacerbates inflammation, HA fragments offer a targeted strategy to restore balance by promoting A20-mediated tolerance.

CD44 interacts with TLRs, particularly TLR4, to modulate inflammation by regulating downstream signaling pathways and cytokine production. CD44, a transmembrane glycoprotein and major receptor for HA, plays a protective role in TLR-mediated inflammatory responses. When HA fragments bind to CD44, they form a complex with TLR4, which alters the typical pro-inflammatory signaling cascade. This interaction suppresses excessive inflammation by reducing NF-κB activation, a key transcription factor responsible for the production of pro-inflammatory cytokines like IL-6 and TNF-α.

High-molecular-weight HA (HMW-HA) starts to build up in the matrix of blood cells during hemostasis, while platelet transit to the wound premises starts the clot-forming process. HA interacts with leukocytes and lymphocytes CD44 receptor to promote immune cell activation when the healing process moves into the inflammatory phase. The production of pro-inflammatory cytokines including IL-6 and interleukin-8 (IL-8) is encouraged by this CD44 signaling cascade. At the same time, TLR2 signaling triggers macrophage activation in response to TGF-α and interleukin-1 beta (IL-1β), intensifying the inflammatory response. HA–CD44 complexes stimulate keratinocyte migration as well as expansion during the proliferative phase, which is necessary for re-epithelialization. Collagen is deposited by triggered fibroblasts, creating the supporting structure of granulation tissue. Additionally, HA and CDAL combine to produce complexes that improve matrix connectivity and promote tissue regeneration. Last, under the control of MMPs and CD44 signaling, fibroblasts undergo myofibroblast differentiation during the remodeling phase. These myofibroblasts play a crucial role in healing wounds and ECM remodeling, which helps to restore tissue integrity. All things considered, CD44–HA interactions coordinate vital cellular and molecular processes during wound healing, guaranteeing a smooth transition through every stage.

CD44 negatively regulates TLR-mediated inflammation by promoting an anti-inflammatory response. Like in macrophages lacking CD44, stimulation with TLR ligands such as LPS leads to significantly increased levels of pro-inflammatory cytokines compared to wild-type macrophages. CD44 dampens the inflammatory response by inhibiting common downstream signaling pathways shared by TLRs. Moreover, HA–CD44 engagement can induce cross-tolerance to LPS, wherein prior exposure to HA fragments reduces the inflammatory response to subsequent LPS stimulation. This phenomenon is mediated by alternate TLR4 signaling events that upregulate anti-inflammatory regulators like A20, which suppress NF-κB activation

CD44's cytoplasmic domain is crucial for its regulatory function in TLR signaling. It directly associates with TLR2 and potentially other TLRs to modulate their activity. For example, in osteoarthritis models, CD44 engagement with HA inhibits NF-κB nuclear translocation and reduces the expression of IL-1β and TNF-α in macrophages stimulated by TLR2 ligands.

10.5.2 CD44 Effect on the Production of Pro-Inflammatory Cytokines

The absence of CD44 significantly affects the production of pro-inflammatory cytokines by enhancing their expression, particularly during inflammatory responses mediated by TLRs. CD44, a transmembrane glycoprotein and receptor for HA, plays a crucial role in negatively regulating TLR-mediated inflammation. When CD44 is absent, this regulatory mechanism is impaired, leading to heightened inflammatory responses. Deficient CD44 macrophages produce elevated levels of pro-inflammatory cytokines such as TNF-α and IL-6 in response to bacterial stimuli like *Escherichia coli* or LPS. For example, in a model of *E. coli*-induced peritonitis, CD44 knockout (KO) mice exhibited significantly higher levels of TNF-α, IL-6, and chemokines such as macrophage inflammatory protein-2 (MIP-2) compared to wild-type mice. This suggests that CD44 normally acts to suppress excessive cytokine production during bacterial infections by modulating TLR signaling pathways (Jiang et al., 2011). Mechanistically, CD44 interacts with TLRs to inhibit downstream signaling cascades that activate NF-κB (Riehl et al., 2015), a transcription factor responsible for cytokine production. In the absence of CD44, this inhibition is lost, resulting in prolonged NF-κB activation and increased cytokine release (Lv et al., 2022). Additionally, CD44 deficiency has been associated with reduced expression of negative regulators of TLR signaling, such as A20 and suppressor of cytokine signaling 3 (SOCS3) (Feng et al., 2016), further amplifying the inflammatory response.

Carbohydrates stimulate fibroblast growth and enhance collagen production, which forms the framework for new tissue. Collagen synthesis is a key process during the proliferation phase of wound healing. They are integral part of glycoproteins, which contribute to wound structure and communication between cells. These molecules facilitate cell adhesion, migration, and proliferation—essential processes for epithelial repair (Nakakura et al., 2025). Carbohydrates regulate enzymes like hexokinase and citrate synthase, which are involved in wound-repair reactions. When HA binds to the cell surface CD44 receptor, a chain of events is triggered, which in turn trigger the NF-κB pathway. TNF-α, IL-1β, and IL-6 are important pro-inflammatory cytokines whose transcription and release are enhanced by this activation. During tissue damage and wound healing, these cytokines are essential for starting and maintaining the inflammatory response. The image emphasizes CD44 as a crucial mediator that connects intracellular inflammatory pathways to signals from the ECM.

10.6 ROLE OF LIPIDS IN WOUND HEALING

Lipids represent a structurally diverse class of hydrophobic or amphipathic organic molecules that serve fundamental roles in biological systems. Characterized by their low solubility in water and high solubility in nonpolar solvents, lipids fulfill critical functions in cellular architecture, energy homeostasis, and biochemical signaling (Wijesinghe, 2020).

10.6.1 Structural and Chemical Properties

Lipids are broadly categorized based on their chemical composition like *triglycerides, waxes, phospholipids, glycolipids, steroids, eicosanoids.* Triglycerides (TGs), or triacylglycerols, represent the primary storage form of lipids in biological systems, consisting of a glycerol backbone esterified with three fatty acid chains (Peng et al., 2025). These fatty acids may be saturated (e.g., palmitic acid) or unsaturated (e.g., oleic or linoleic acid), determining the physical and metabolic properties of the triglyceride. Biosynthesis occurs predominantly in the liver and adipose tissue through the Kennedy pathway (a crucial biosynthetic route for phospholipid production and plays a pivotal role in wound healing by supplying essential membrane components and signaling molecules (Holdaway et al., 2025). This metabolic pathway becomes particularly active in proliferating keratinocytes, fibroblasts, and endothelial cells at the wound site, where it generates phosphatidylcholine (PC) and phosphatidylethanolamine, the fundamental building blocks for cellular membranes during tissue regeneration (Choudhary et al., 2024).

During the inflammatory phase, Kennedy pathway-derived phospholipids serve as precursors for important mediators like platelet-activating factor while also contributing to macrophage polarization. As healing progresses to the proliferative stage, the pathway supports keratinocyte migration and fibroblast proliferation by providing the necessary lipids for membrane expansion and lamellipodia formation. The generated phospholipids also facilitate angiogenesis by maintaining endothelial cell integrity and participating in vascular tube formation (Qin et al., 2025). Notably, the Kennedy pathway exhibits phase-specific regulation throughout the healing process, with its activity dynamically adapting to meet the changing metabolic demands of different cell types during each healing stage, where glycerol-3-phosphate undergoes sequential acylation by fatty acyl-CoA esters via enzymes including glycerol-3-phosphate acyltransferase (GPAT), acylglycerol-phosphate acyltransferase (AGPAT), and diacylglycerol acyltransferase (DGAT), ultimately forming triglycerides under the regulation of insulin, which promotes lipogenesis, and glucagon/epinephrine, which inhibit it (Lee et al., 2018).

During energy demand, triglycerides are hydrolyzed by hormone-sensitive lipase (HSL) (Althaher, 2022) and adipose triglyceride lipase (ATGL) into free fatty acids and glycerol, which enter β-oxidation for ATP production or ketogenesis during prolonged fasting (Awad et al., 2024). Physiologically, triglycerides serve as the most energy-dense macromolecules (9 kcal/g), stored in adipose tissue for long-term energy reserves, thermal insulation, and organ protection (Murai, 2015). Additionally, their oxidation generates metabolic water, crucial for survival in arid environments. Clinically, elevated triglyceride levels (hypertriglyceridemia) are associated with atherosclerosis, acute pancreatitis, and metabolic syndrome (Bashir et al., 2024), while deficiencies in essential fatty acids (e.g., ω-3 and ω-6) impair skin integrity and wound healing (McDaniel et al., 2008). Dietary triglycerides vary in composition, with saturated fats (e.g., animal fats) linked to cardiovascular risk and unsaturated fats (e.g., olive oil, fish oil) offering cardioprotective benefits. Triglycerides constitute <30% of daily caloric intake, emphasizing the importance of balanced consumption to maintain metabolic health. Understanding triglyceride metabolism is thus pivotal for addressing obesity, diabetes, and lipid disorders (Williams, 2020), with therapeutic strategies targeting pathways such as PPAR-γ activation or lipoprotein lipase modulation (Saiki et al., 2024). Lipids play a multifaceted and indispensable role in wound healing, contributing to structural repair, energy metabolism, and the regulation of inflammatory and regenerative processes (Jerab et al., 2025). As fundamental components of cell membranes, phospholipids, sphingolipids, and ceramides maintain the structural integrity of keratinocytes and fibroblasts, ensuring their proper function during tissue regeneration. These lipids are particularly critical in restoring the epidermal barrier (Kumar et al., 2015), as they replenish the lipid matrix of the stratum corneum, preventing transepidermal water loss and shielding the wound from pathogenic invasion.

Beyond their structural role, lipids serve as a vital energy source, fueling cellular activities essential for wound repair. Free fatty acids derived from triglyceride breakdown provide substrates for ATP production, supporting the high metabolic demands of proliferating fibroblasts and collagen-producing cells. Notably, polyunsaturated fatty acids (PUFAs), including omega-3 (eicosapentaenoic acid, docosahexaenoic acid) and omega-6 (arachidonic acid, linoleic acid), modulate immune responses and enhance tissue regeneration. Omega-3 PUFAs, for instance, exert anti-inflammatory effects by competing with arachidonic acid for enzymatic conversion, thereby reducing the production of pro-inflammatory eicosanoids while promoting the synthesis of specialized pro-resolving mediators (SPMs) such as resolvins and protectins. Lipid-mediated signaling is central to the coordination of wound healing phases. Eicosanoids—including prostaglandins (e.g., PGE2) and leukotrienes (e.g., LTB4)—orchestrate early inflammatory responses by recruiting neutrophils and macrophages to the wound site. Conversely, SPMs actively resolve inflammation by enhancing macrophage phagocytosis of cellular debris and downregulating pro-inflammatory cytokine release. Additionally, sphingolipid derivatives like sphingosine-1-phosphate (S1P) regulate endothelial cell migration (Fan et al., 2021) and angiogenesis, facilitating the formation of new blood vessels necessary for nutrient delivery and tissue oxygenation. Arachidonic acid metabolites further influence vascular tone and permeability, optimizing the wound microenvironment for repair. The therapeutic application of lipids in wound care has gained significant attention (Torres et al., 2021), with lipid-based formulations demonstrating efficacy in managing both acute and chronic wounds. Natural lipid compounds, such as vegetable oils (e.g., olive oil, coconut oil) and honeybee-derived ceramides, exhibit antimicrobial and moisturizing properties, promoting epithelialization (Asokapandian et al., 2021). Advanced delivery systems, including lipid nanoparticles and bioactive dressings enriched with fatty acids (e.g., HA-lipid hybrids), enhance drug penetration and sustain anti-inflammatory effects. Moreover, lipidomics—the large-scale study of lipid pathways—has unveiled novel lipid biomarkers and therapeutic targets, such as the dysregulation of ceramide metabolism in diabetic wounds, offering new avenues for precision medicine in wound management (Han and Gross, 2022).

Lipids play a multifaceted role in wound healing, beginning with their structural contributions to cell membranes and skin integrity. Phospholipids, sphingolipids, and ceramides are essential components of cell membranes, maintaining the stability of keratinocytes and fibroblasts, which are critical for tissue repair and re-epithelialization. These lipids also restore the skin's protective barrier by rebuilding the lipid matrix of the stratum corneum, preventing microbial invasion and fluid loss (Starr et al., 2022). Energy provision is another key function, as fatty acids serve as substrates for ATP production, fueling the proliferation and migration of cells like fibroblasts and keratinocytes during tissue repair. Adequate dietary lipids are vital to sustain these energy-intensive processes. Lipids also act as signaling molecules that regulate inflammation. Bioactive lipids such as prostaglandins and leukotrienes initiate inflammation by recruiting immune cells like neutrophils to the wound site (Anand, 2020), while SPMs including resolvins and protectins, resolve inflammation by promoting macrophage activity and suppressing cytokine release. Dysregulation of these pathways can lead to chronic inflammation or fibrosis. Additionally, lipids like S1P drive angiogenesis by stimulating endothelial cell migration and capillary formation, while omega-3 and omega-6 fatty acids enhance collagen synthesis and ECM remodeling during tissue regeneration (Samson et al., 2021).

Therapeutically, lipids are leveraged in advanced wound care for their antimicrobial, anti-inflammatory, and regenerative properties (Fischer, 2020). Fatty acid-based dressings, such as those containing linoleic acid, accelerate re-epithelialization and reduce oxidative stress, while lipid nanoparticles deliver drugs or growth factors directly to the wound bed. Natural oils like virgin coconut oil promote healing by boosting fibroblast activity and reducing lipid peroxidation. Advances in lipidomics, such as liquid chromatography-mass spectrometry (LC–MS), enable researchers to map lipid signaling pathways across healing phases, offering insights into novel therapies for chronic wounds. In summary, lipids support wound healing through structural stability, energy provision, inflammation regulation, angiogenesis, and tissue remodeling, making them indispensable in both physiological repair and clinical interventions.

10.6.2 Lipids Support Tissue Repair and Re-Epithelialization

Structurally, lipids form the backbone of the plasma membrane, which is essential for maintaining cellular integrity and facilitating rapid repair after injury (Horsley, 2022). Phospholipids, sphingolipids, and ceramides contribute to membrane tension, rigidity, and fluidity, enabling resealing and remodeling of damaged membranes. These structural lipids also act as damage sensors, initiating early signaling responses that recruit repair machinery to the site of injury (Ghatak et al., 2023). For example, calcium influx following membrane disruption triggers lipid-mediated signaling pathways that coordinate vesicle fusion and delivery of new lipids to restore membrane integrity.

Lipids function as signaling molecules that regulate inflammation and cellular activity during wound healing. Bioactive lipids like prostaglandins and leukotrienes initiate inflammation by attracting immune cells such as macrophages to the wound site. SPMs derived from fatty acids help resolve inflammation by promoting macrophage activation and efferocytosis (Chiang and Serhan, 2020). This balance between pro-inflammatory and anti-inflammatory lipid signaling is critical for transitioning from the inflammatory phase to the proliferative phase of healing. Furthermore, fatty acids released through lipolysis in dermal adipocytes stimulate macrophage recruitment and contribute to revascularization and epithelial cell migration into the wound bed.

10.6.3 Metabolic Reprogramming

Lipids also play important role during tissue repair. Fatty acids serve as energy substrates for ATP production, fueling processes like fibroblast proliferation, collagen synthesis, and keratinocyte migration (Lee et al., 2020). Lipid metabolism influences cellular differentiation; for instance, adipocytes near wound edges undergo lipolysis and transdifferentiate into myofibroblasts that secrete growth factors and modify the ECM to support tissue regeneration. Lipidomic profiling has further revealed that specific lipid species, such as PCs and high-density lipoproteins (HDLs), are associated with faster re-epithelialization outcomes in burn injuries (Meikle et al., 2021).

Therapeutically, lipids have been incorporated into advanced wound care strategies due to their regenerative properties. Lipid-based formulations like fatty acid-enriched dressings enhance epithelialization by reducing oxidative stress and promoting fibroblast activity (Tenchov et al., 2021). Natural oils such as virgin coconut oil have shown efficacy in improving lipid peroxidation control while accelerating tissue repair. Advances in lipidomics continue to uncover new insights into lipid-mediated mechanisms in wound healing, offering potential for targeted therapies in chronic wounds or burns.

10.6.4 Resealing of the Plasma Membrane

Lipids play a crucial role in the resealing process of the plasma membrane by contributing both structurally and functionally. Structurally, lipids are the primary components of the plasma membrane, and their biophysical properties—such as tension, fluidity, and curvature—are essential for membrane repair (Ammendolia et al., 2021). After an injury, the plasma membrane undergoes rapid remodeling to restore its integrity. Lipids like phospholipids and sphingolipids dynamically rearrange to reduce membrane tension and facilitate resealing (Wiczew et al., 2021). For example, lateral movement, rotation, and flipping of lipids between membrane leaflets adjust lipid packing density, enabling the membrane to adapt to changes in tension and curvature at the wound site.

Functionally, lipids act as signaling molecules that coordinate the repair response (Jarc and Petan, 2020). Plasma membrane disruption triggers calcium influx from the extracellular space, which activates lipid-mediated signaling pathways. This includes the mobilization of lipid vesicles for exocytosis to patch the damaged area. Additionally, bioactive lipids like phosphatidylserine (PS) and S1P play signaling roles by recruiting repair machinery and promoting vesicle fusion at the injury site (Mierziak et al., 2021). Lipid rafts, specialized microdomains enriched with cholesterol and sphingolipids, aggregate proteins involved in repair processes and create localized platforms for efficient resealing.

Beyond individual lipid molecules, bulk lipid composition influences repair efficiency. For instance, membranes with higher levels of unsaturated fatty acids exhibit greater fluidity, which enhances vesicle fusion and membrane resealing. Conversely, rigid membranes with tightly packed saturated lipids may hinder repair kinetics. These structural and signaling roles underscore the importance of lipids in maintaining plasma membrane integrity and cellular homeostasis after injury.

10.6.5 Alteration of Membrane Tension and Rigidity

Lipids influence membrane tension and rigidity during the repair process by dynamically altering their structural organization and biophysical properties (Santinho et al., 2021). The plasma membrane is composed of various lipids, including phospholipids, sphingolipids, and cholesterol, which collectively determine its fluidity, curvature, and mechanical stability. Upon injury, the biophysical properties of the membrane are disrupted, requiring lipids to adapt and facilitate resealing (Renne and Ernst, 2023).

Lipids play a fundamental role in maintaining membrane integrity and facilitating repair through dynamic regulation of membrane tension and rigidity. Following injury, the lateral mobility, rotational diffusion, and trans-bilayer movement of lipids enable rapid adaptation to mechanical stress. A key mechanism involves the redistribution of phospholipids such as PS, which undergoes ATP-dependent translocation from the inner to outer leaflet via scramblases, generating localized outward curvature that reduces membrane tension and promotes resealing. This process is complemented by the presence of non-bilayer-forming lipids like phosphatidylethanolamine (PE), whose intrinsic negative curvature enhances membrane flexibility and facilitates topological remodeling at damage sites.

The mechanical properties of membranes are further modulated through lipid-mediated alterations in packing density. Membrane fluidity increases in damaged regions through the incorporation of unsaturated fatty acids (e.g., oleic acid), which introduce kinks in acyl chains and reduce intermolecular cohesion, enabling vesicle fusion for patch formation. Conversely, saturated fatty acids and cholesterol enhance rigidity by promoting tighter lipid packing and stabilizing microdomains, providing structural reinforcement to withstand mechanical stress. Experimental biophysical studies demonstrate that oleic acid can induce curvature frustration in bilayers, increasing bending rigidity while maintaining flexibility—a critical balance for effective repair.

Beyond their structural roles, lipids serve as essential signaling mediators that orchestrate repair processes. Calcium influx triggered by membrane injury activates lipid-dependent signaling cascades, including the recruitment of repair vesicles via synaptotagmin-mediated exocytosis.

Lipid rafts—membrane microdomains enriched in sphingolipids and cholesterol—function as organizational platforms that concentrate repair proteins (e.g., annexins, dysferlin) and facilitate vesicle docking and fusion. These coordinated structural and signaling functions highlight the sophisticated role of lipids in maintaining membrane homeostasis and ensuring efficient cellular repair.

10.6.6 Lipids as Damage Sensors in Repair

Lipids serve as primary sensors and initiators of plasma membrane repair through their unique capacity to detect structural disruptions and transduce biochemical signals following injury. As the fundamental structural constituents of cellular membranes, lipids are strategically positioned to respond immediately to mechanical damage through both biophysical and biochemical mechanisms. When membrane integrity is compromised, the resulting alterations in bilayer properties—including changes in fluidity, lateral tension, and phase separation—induce rapid lipid rearrangements that facilitate repair processes.

The initial response involves calcium-dependent lipid remodeling, where extracellular calcium influx through membrane breaches binds to anionic phospholipids such as PS. This interaction triggers multiple repair pathways: (1) PS externalization reduces local membrane tension through curvature generation, (2) calcium–PS complexes nucleate the assembly of repair scaffolds like annexins, and (3) synaptotagmin-mediated vesicle exocytosis is activated to patch the damaged area. Membrane microdomains, particularly lipid rafts enriched in cholesterol and sphingolipids, serve as organizational platforms that concentrate repair machinery and coordinate vesicle trafficking.

Concurrently, oxidative modification of membrane lipids generates potent damage signals. Oxidized phospholipids (e.g., oxPAPC) function as damage-associated molecular patterns (DAMPs) that activate pattern recognition receptors including TLR4 and nucleotide-binding domain, leucine rich–containing family, pyrin domain–containing-3 (NLRP3), bridging membrane repair with inflammatory responses. These oxidized lipid derivatives not only amplify local repair signaling but also recruit immune cells through chemotactic gradients, thereby integrating structural membrane restoration with broader tissue repair processes.

The dual functionality of lipids as both structural elements and signaling molecules enables a highly coordinated, multi-tiered response to membrane injury. This includes immediate physical stabilization through lipid redistribution and curvature modulation, intermediate-term repair through vesicle-mediated patching, and long-term tissue restoration through immunomodulatory signaling. The sophisticated integration of these lipid-mediated mechanisms underscores their central role in maintaining cellular homeostasis and orchestrating comprehensive wound healing responses.

10.7 MICRONUTRIENTS' ROLE IN WOUND HEALING

Essential to the human body, micronutrients play a major role in the wound healing process. These micronutrients, which consist of vitamins and minerals, are considered "micronutrients" because they are needed in tiny amounts, but they assist in the multiple stages of healing (Penny et al., 2022). Vitamin A is necessary for both the proliferation and differentiation of epithelial cells that are important in the step of re-epithelialization during the healing process (VanBuren and Everts, 2022). It also helps to regulate the inflammatory response by enhancing the migration and function of immune cells (e.g., neutrophils and macrophages) at the wound site. Deficiency of vitamin A is associated with impaired wound healing, which can lead to delayed closure of wounds and increased risk for infection (Hodge and Taylor, 2025). Similarly, vitamin C is a powerful antioxidant that protects your cells against oxidative stress and is responsible for collagen formation, one of the major structural proteins involved in the wound healing process. Vitamin C regulates the inflammatory response by stimulating immune cell migration and activity. It also plays a role in the process of angiogenesis, which is necessary for building new blood vessels to restore blood supply to the wound area. If there is a deficiency of vitamin C, this can lead to delayed wound healing, increased potential for infection, and chronic wound development (Kaźmierczak-Barańska et al., 2020). Vitamin E is an antioxidant that is soluble in lipids, and it is involved in protecting cell membranes and other cellular components from oxidative injury. It aids in the inflammatory response as well as moderating the immune system, both of which could affect the process of wound healing (Fatima et al., 2025). Vitamin E also has anti-inflammatory properties and could reduce the risk of excessive scarring or keloids (Wallert et al., 2021). Zinc is a vital mineral that is necessary for multiple cellular processes that occur during the wound healing process, including cell division, protein synthesis, and immune response regulation. Zinc is necessary for the correct function of enzymes involved in collagen synthesis, as well as the development of new blood vessels. Zinc deficiency can result in impaired wound healing, higher risk for infection, and delayed closure (Erfanian et al., 2024).

10.7.1 Amino Acids

Amino acids are particularly important for wound healing to provide the nutritional substrates required for the synthesis of proteins involved in tissue repair and regeneration. Some amino acids have been shown to be particularly important in the wound healing process including arginine, glutamine, and proline. Amino acids serve as the fundamental components of proteins, which are vital for tissue repair, immune function, and collagen production—all of which are important to healing from a wound (Erfanian et al., 2024). Arginine can function as a nitrogen donor in the synthesis of many amino acids required for wound healing. Arginine is also a precursor to nitric oxide (NO), which is believed to promote angiogenesis and vasodilation (Barbul et al., 1990). Glutamine is the most abundant plasma amino acid and is an important energy source for rapidly dividing cells, such as fibroblasts and leukocytes, and has a key role in immune function and collagen production (Yoo et al., 2020).

The phases of healing supported by amino acids.

1. *Hemostasis and Inflammation*: Amino acids also assist in immune cell function and in the production of cytokines to facilitate control of infections and to initiate repair of tissues (Ye et al., 2020).
2. *Proliferation Phase*: Essential for fibroblast proliferation, angiogenesis, and collagen matrix formation, amino acids are important for regeneration of skin, blood vessels, and connective tissue (Wang et al., 2022).
3. *Remodeling Phase*: Collagen is matured and aligned requiring a sustained supple of proline, lysine, and glycine. Supplements with specific nutrients (e.g., arginine, glutamine, vitamins, or minerals) may be beneficial in some individuals, especially for those with nutrient deficiencies and/or elevated nutritional needs (e.g., elderly, chronic disease, severe injury). Nutritional needs for wound healing will depend on many factors including the type and severity of the wound, the overall health of the individual and others. It is essential for healthcare professionals to assess the patients' nutrition status and nutritional support should be individualized to ensure the best potential healing outcomes (De Castro Bras and Frangogiannis, 2020).

10.7.1.1 Arginine

Arginine is an essential player in many processes associated with the wound healing cascade in a number of ways. The most important aspect is that arginine is the precursor to NO. NO is an excellent vasodilator that will improve perfusion and oxygen delivery to the wound, which is necessary for angiogenesis and faster wound healing (Chu and Delmore, 2021). Arginine is also a substrate for the enzyme arginase. Arginase is responsible for the synthesis of proline, the main amino acid of collagen. The formation and deposition of adequate collagen is a requirement for effective wound repair and a key component of nascent tissue. Arginine additionally modulates the immune response in wound healing. It can enhance the proliferation and activation of different immune cells, such as macrophage and lymphocyte populations needed for clearance of infection, debris removal, and initiation of repair. Arginine supplementation has also increased the antimicrobial effect against certain bacteria, especially Gram-negative bacteria, which may help improve wound healing or fight against wound infection, perhaps through disruption of bacterial arginine biosynthesis pathways (Yang et al., 2024). Finally, arginine can also have signaling functions, regulating many cellular processes in wound healing, like cell proliferation and migration.

Arginine can function as a nitrogen donor in the synthesis of many amino acids required for wound healing. Arginine is also a precursor to NO, which is believed to promote angiogenesis and vasodilation (Wu et al., 2021).

Glutamine is the most abundant plasma amino acid and is an important energy source for rapidly dividing cells, such as fibroblasts and leukocytes, and has a key role in immune function and collagen production.

Through multiple interrelated biochemical and physiological processes, arginine appears to play multiple roles in the wound healing process (Li et al., 2023).

1. *NO Production*: Arginine is the primary substrate for NO synthase (NOS), especially inducible NOS (iNOS), which produces NO and citrulline. NO creates a cytotoxic environment for acute wounds to mitigate disease by exerting an antibacterial effect during the inflammatory phase of healing (Almannai and El-Hattab, 2021). Moreover, NO enhances blood flow to the wound site that may deliver critical oxygen and nutrients necessary for tissue repair.
2. *Collagen Formation through the Arginase Pathway*: Arginine is converted to ornithine through arginase, which ultimately contributes to collagen synthesis through two pathways:
   - *Synthesis of Polyamines*: Ornithine is converted to polyamines (via ornithine decarboxylase), which is required for cell proliferation, angiogenesis, and tissue remodeling (Halloran et al., 2021).
   - *Synthesis of Proline/Glutamate*: Ornithine is converted to further proline and glutamate, which are precursors for collagen and ECM proteins. Supplementation of arginine significantly increases collagen deposition and wound-breaking strength in animal and human studies (Sedighi-Pirsaraei et al., 2024).
3. *Growth Hormone and Insulin-Like Growth Factor 1 (IGF-1)*: Arginine promotes IGF-1 release and growth hormone (GH). IGF-1 and GH stimulate:- Protein synthesis and fibroblast action- Improved nitrogen balance and hydroxyproline accrual, a marker of collagen synthesis, in humans (Oh et al., 2017).
4. *Inflammation Modulation*: For acute wounds, metabolism of arginine shifts from early metabolism of NO (i.e., anti-inflammatory, antimicrobial) to arginase-driven collagen synthesis later. For chronic wounds, arginine metabolism is dysregulated, and arginase activity decreased leads to poor collagen deposition. Arginase 1 (Arg1) deficiency correlates with sustained inflammation and delayed healing in preclinical studies (Gogoi et al., 2016).
5. *Clinical Evidence*: In human trials, supplementing 17–30 g/day of arginine increased IGF-1 levels, collagen synthesis, and lymphocyte activity in the elderly. In patients undergoing surgery, arginine supplementation led to decreased rates of wound dehiscence. Higher doses are safe and effective (up to 20 g/day), though research regarding timing and dose optimization continues. Figure 10.2 shows the biochemical path of arginine in the two primary stages of wound healing.

Depending on the state of healing, arginine, a crucial substrate, can be processed through two different routes. iNOS is the primary enzyme responsible for the metabolism of arginine during the inflammatory phase, resulting in the production of citrulline and NO. After being activated by the NO produced, macrophages release inflammatory cytokines such basic FGF (bFGF), TNF-α, and interleukin-1β (IL-1β). These molecules are essential for both pathogen removal and immunological response. The cellular breakdown of arginine shifts into the arginase route as the healing process moves into the proliferation phase, when it is converted into urea and ornithine. Proline and polyamines, that are necessary for the manufacture of collagen and cell division, respectively,

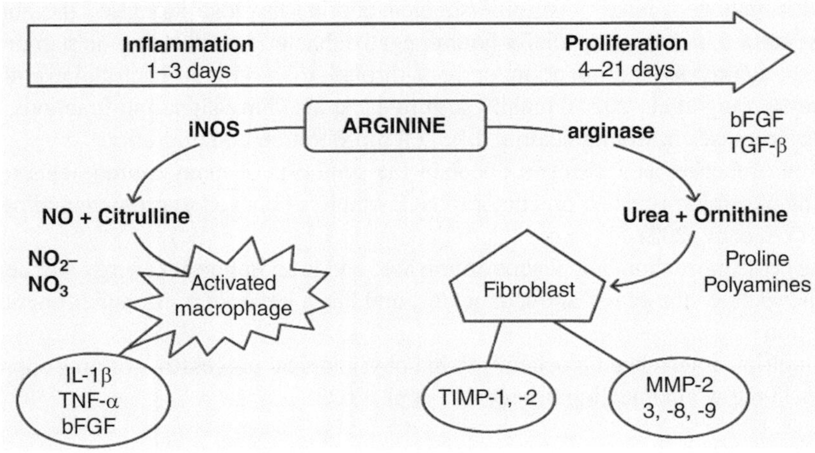

FIGURE 10.2 The figure depicts the biochemical trajectory of arginine over the two primary stages of wound healing, which are proliferation (4–21 days) and inflammation (1–3 days).

are synthesized from ornithine. These substances stimulate fibroblast activation, which results in the release of MMP-2, MMP-3, MMP-8, and MMP-9 and TIMP-1 and TIMP-2. Tissue regeneration and ECM reformation are controlled by this dynamic interaction. In general, the appropriate course of wound healing from inflammation to tissue repair depends on the balance between the iNOS and arginase pathways of arginine metabolism.

10.7.1.2 Proline

Proline has an important role in several stage of the wound healing process through several mechanism of action. It is an important amino acid in collagen synthesis and deposition and is a prerequisite for wound repair and new tissue formation. During wound healing, the endogenous biosynthesis of proline is important to ensure sufficient levels for collagen production (Linju and Rekha, 2023). The endogenous biosynthesis of proline can be increased to respond to the increases demand of collagen at the wound site. Peptides that contain proline, such as N-acetylated proline–glycine–proline (Ac-PGP), have been studied to reverse the non-healing of cutaneous wound and also promote neovascular by inducing migration and proliferation of endothelial progenitor cells (Zhu et al., 2023). In addition, proline can also protect cells against oxidative stress and cellular damage which is relevant to the survival and function of many of the cells involved in wound healing including fibroblasts and keratinocytes. Furthermore, proline can modulate the inflammatory response during wound healing. The availability and metabolism of proline can find influence in the activity of neutrophils and other immune cells which is essential for bringing carries of infection results in the initiation of the tissue repair process.

The following is a more elaborate description of the mechanisms:

1. *Collagen Synthesis and Stability*: Proline accounts for 23% of collagen (together with hydroxyproline) and establishes collagen's triple-helix structure (Karna et al., 2020).
2. *Hydroxylation*: Proline is hydroxylated to become hydroxyproline, which stabilizes the helical structure of collagen while improving tensile strength (Phang, 2023).
3. *Local Delivery*: During the earliest phases of healing, proline levels in the wound fluid are 50% higher than levels in plasma, demonstrating upregulated transport to the wound to enhance collagen production.
4. *Antioxidant Activity*: Proline decreases oxidative stress in wounds by increasing the activity of glutathione (GSH) peroxidase and superoxide dismutase. Reducing lipid peroxide levels and guarding the cells against free radicals (Ye et al., 2020).

5. *Wound Contraction and Epithelialization Topical Use*: In rat studies, 200 mg/day of micronized proline use improved epithelialization and wound contraction, faster healing than oral micronized proline (Gubbiveeranna et al., 2024). Systemic (intraperitoneal) use of proline increased mechanical strength in healed tissue probably due to collagen cross-linking.
6. *Synergy with Arginine and Nutrients*: Proline is biosynthesized from the arginine → ornithine → proline pathway. Collagen deposition is increased with arginine supplements, but it is not known whether proline is directly synthesized from arginine. Rigorous trials have shown that improved wound contraction in resistant wounds with the combination of proline, arginine, and vitamins A/C/E (Huynh et al., 2020).
7. *Clinical Evidence*: In trials, oral supplementation of proline and arginine reduced the wound area in humans by 38.4% over four weeks compared to controls. A total of 200 mg/day topically or systemically in rats, while human studies utilized 200 ml two times a day of nutrient-enriched formulas.

10.7.1.3 Glutamine

Glutamine is an essential amino acid that is an important part of multiple processes involved in the process of wound healing. Studies on glutamine supplementation have shown that it promotes wound healing primarily through increasing amounts of collagen following sustained injury (Shields, 2021). Mechanisms of action related to the positive effects of glutamine on wound healing can occur through glutamine serving as the main neuronal precursor for proline, which is the amino acid necessary for synthesis of collagen. Glutamine maintains adequate levels of GSH, an important antioxidant that protects cells when they are under oxidative stress and/or when they are depleted during the wound healing process (Wang et al., 2024). Glutamine helps modulate the inflammatory response by regulating the activity of immune cells, such as neutrophils, that are critical in removing infection and in initiating the wound repair process. Glutamine promotes the proliferation and migration of many different types of cells, such as fibroblasts and endothelial cells, to support tissue regeneration and angiogenesis. Glutamine supports the integrity of the intestinal epithelial barrier, which is important in preventing bacterial translocation, and thus decreases the risk of wound infection (Ren et al., 2020). Here is a comprehensive examination of its effects:

1. *NO Formation*: Glutamine indirectly aids in the production of NO by facilitating increases in levels of arginine and citrulline, the precursors of NO. NO plays an important role in the regulation of cell proliferation, collagen synthesis, and wound contraction during the proliferative phase of healing (Lopez-Sanchez et al., 2020).
2. *Antioxidant Role*: Glutamine assists in the production of GSH, an important antioxidant that decreases the oxidative stress from a wound. It serves to protect cells from degradation by free radicals and promotes tissue healing (Comino-Sanz et al., 2021).
3. *Gut Integrity and Pro-Inflammatory Cytokine Modulation*: Glutamine prevents intestinal atrophy and decreases the lactulose/mannitol ratio of excretion to prevent systemic inflammation associated with translocated bacteria. Glutamine decreases IL-6 and c-reactive protein (CRP), subsequently reducing total inflammation to promote a balanced healing environment (Sumbria et al., 2021).
4. *A Support Mechanism for Collagen Production*: Although glutamine itself is not a precursor to collagen, it does assist in collagen synthesis by facilitating energy for fibroblasts, epithelial cells, and immune cells. Responding to arginine and synergizing with it, to be metabolized into proline (a precursor of collagen). Rodent studies assessing hydroxyproline levels with and without glutamine show no statistically significant increases in the supplementation group with regard to hydroxyproline. The data suggest that if arginine is limited, glutamine may become more important (Stegen et al., 2020). Figure 10.3 shows the role of glutamine supplementation in the promotion of wound healing process.

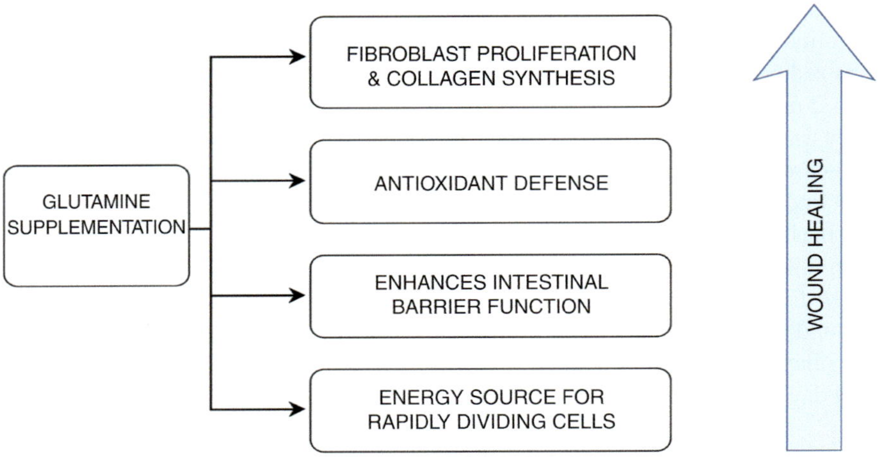

FIGURE 10.3 This demonstrates how glutamine supplementation promotes wound healing. A conditionally necessary amino acid, glutamine supports tissue regeneration and repair in a number of ways.

In order to build ECM and maintain the structural integrity of the healing tissue, it first stimulates the production of collagen and fibroblasts. Second, glutamine aids in the defensive mechanisms of antioxidants, lowering oxidative stress that may otherwise hinder the healing process. It also improves the function of the intestinal barrier, which is essential for avoiding infections and systemic inflammation that might slow healing. Finally, glutamine is a crucial energy source for cells that divide quickly, such immunological and epithelial cells, which speeds up the regeneration process. All of these activities work together to promote better and more effective wound healing. Table 10.2 is illustrating the essential micronutrients and their function in wound healing.

TABLE 10.2
Micronutrients' Role in Wound Healing

Category	Micronutrient	Role in Wound Healing
Amino acids	Arginine	Enhances immune function, promotes collagen deposition and wounds tensile strength
	Glutamine	Supports immune cells and intestinal health; helps reduce inflammation
	Proline and lysine	Precursors for collagen synthesis, essential for extracellular matrix formation
Vitamins	Vitamin A	Promotes epithelialization, fibroblast function, and immune response
	Vitamin C (ascorbic acid)	Crucial for collagen synthesis, antioxidant activity, and immune defense
	Vitamin E	Antioxidant that protects cell membranes; may help reduce oxidative stress at wound site
	Vitamin K	Essential for blood clotting during the initial phase of wound healing
	Vitamin D	Modulates inflammation and supports antimicrobial peptide production
Minerals	Zinc	Vital for DNA synthesis, cell proliferation, and immune function
	Iron	Necessary for oxygen transport and collagen cross-linking
	Copper	Required for collagen cross-linking and angiogenesis
	Magnesium	Supports protein and collagen synthesis; modulates inflammation
	Selenium	Antioxidant activity supports immune function and tissue repair

10.8 VITAMINS' ROLE IN WOUND HEALING

Vitamin C facilitates the formation and deposition of collagen which is important for the proper healing of wounds. Vitamin C increases the proliferation and survival of skin cells, including stem/progenitor cells leading to improved healing of wounds. Vitamin C alters the expression of gene and signaling pathways contributing to wound healing related to cell migration, angiogenesis, and inflammation (Bechara et al., 2022). Vitamin C acts as an antioxidant to decrease oxidative stress and maintain an environment conducive for wound healing (Dosedel et al., 2021). Vitamin C may stimulate NO production which is important for wound healing through improving blood flow and angiogenesis (Kashiouris et al., 2020). Vitamin C may stimulate the enzymes related to wound healing such as amylase and lipoxygenase. Vitamin C is shown to stimulate wound healing through topical application of vitamin C and its derivatives by stimulating collagen production and other mechanisms.

In conclusion, vitamin C serves several functions in wound healing, including promoting collagen synthesis, enhancing cell proliferation and survival, altering gene expression and signaling pathways, reducing oxidative stress, stimulating NO production, and possibly increasing the activity of the enzymatic players involved in wound healing.

Vitamin K promotes blood coagulation during hemostasis, which starts the healing process. B vitamins control TLR2 transmission and immune peptides, vitamin C boosts immunological function, vitamin E offers antioxidant protection, vitamin D promotes pro-inflammatory activities, and vitamin A functions as an anti-inflammatory agent during the inflammation phase. B vitamins encourage TGF-β function and lysyl oxidase efficiency for collagen cross-linking, vitamin E stimulates cell development, vitamin C helps avoid excessive scar formation, and vitamin A stabilizes collagen during proliferation. Vitamins A and C continue to help stabilize collagen and prevent scarring during the last remodeling phase, while Vitamin E increases the tensile strength of the wound. When taken together, these vitamins guarantee appropriate wound healing progression and effectiveness.

10.8.1 Vitamin D

Vitamin D promotes the proliferation and differentiation of epithelial cells, processes that are essential for healing wounds. Vitamin D is involved in the stimulation and repair of skin epithelial cells, such as alveolar type II (ATII) cells, to improve wound healing (Siregar and Hidayat, 2023). Vitamin D may decrease epithelial cell apoptosis, prevent more tissue damage and enhancing wound healing. Vitamin D may inhibit the epithelial-to-mesenchymal transition (EMT) process, which may impair wound healing (Zheng et al., 2020). Vitamin D possesses anti-inflammatory properties, which may lead to a favorable environment for wound healing by reducing excessive inflammation (Ao et al., 2021). Vitamin D may regulate gene expression and signaling pathways associated with wound healing, such as those involved with differentiation, cell proliferation and migration. Vitamin D may interact with, and modulate, pain signaling pathways which could affect the wound healing process. Vitamin D is needed for the barrier function of skin, which is critical for wound healing (Parke et al., 2021). Vitamin D deficiency may impair wound healing, and vitamin D status is an important consideration for optimizing wound healing. The molecular mechanisms by which vitamin D achieves its benefits in wound healing are complex and involve a variety of pathways, including those associated with the process of inflammation and many related to cell proliferation, differentiation, and signaling (van Schoor et al., 2024).

1. *Modulation of Immunity and Anti-Inflammatory Effects of Vitamin*
   - *D Cathelicidin (LL-37) Induction*: Vitamin D induces cathelicidin, an antimicrobial peptide that promotes innate immunity, modulates inflammation, and contributes to wound repair by enhancing both angiogenesis and re-epithelialization.
   - *Reduction of Inflammation*: Vitamin D reduces pro-inflammatory cytokines and enhances anti-inflammatory cytokines to create a favorable environment for healing.

2. *Skin Barrier Function and Re-Epithelialization*
   - *Vitamin D Receptors (VDR)*: The VDRs are expressed in skin cells and are important for activating, proliferating, and differentiating stem cells, which can promote barrier repair and re-epithelialization.
   - *Calcium Signaling*: Vitamin D and calcium act cooperatively to increase permeability barrier formation, which protects against fluid loss, and may also encourage wound closure (Bishop et al., 2021).
3. *Angiogenesis and Tissue Remodeling*
   - *Angiogenic Effects*: Vitamin D has an angiogenic effect that is critical for providing oxygen and nutrients to the wound space, facilitating tissue repair and remodeling (Kim et al., 2020).
   - *Collagen Synthesis*: Vitamin D might promote collagen synthesis indirectly by inducing growth factors (e.g., PDGF) that stimulate fibroblast activity and ECM synthesis (Shenoy et al., 2022).
4. *Glycemic Control for Diabetic Wounds*
   - **Supplementation** of vitamin D could stabilize blood glucose levels in patients with diabetes and indirectly promote wound healing from decreased metabolic stress (Halschou-Jensen et al., 2023).

10.8.2 Vitamin B

The use of vitamin B complex therapy consisting of vitamin B1, B6, and B12 may enhance neuroinflammation and regeneration of peripheral nerves after injury. Pantothenic acid (vitamin B5) has been shown to support surgical wound healing by enhancing skin barrier function and moisturization, though the exact molecular mechanisms of action are not well defined. Vitamin B complex therapy may be effective as an adjunctive therapy to routine surgical treatment of peripheral nerve injury, likely acting through modification of the inflammatory response and promoting nerve regeneration (Kalsi et al., 2024).

Vitamin B may support wound healing and tissue repair via its antioxidant properties as well as through regulation of cell signaling pathways associated with inflammation, and tissue regeneration. Vitamin B complex therapy may support the expression of growth factors and cytokines important for different phases of wound healing such as inflammation, proliferation, and remodeling (Saeg et al., 2021).

However, the specific mechanisms by which vitamin B treatments exert their effects on wound healing are not completely understood and merit further investigation. Further work is needed to define these pathways and signaling cascades involved in the processes of tissue repair and innervation.

1. *Energy Metabolism and Cellular Proliferation*
   - *Metabolic Support*: The B vitamins are key dietary determinants of energy production through the metabolism of carbohydrates, proteins, and fats to support cell proliferation and migration, which is necessary for closure of a wound (Hrubsa et al., 2022).
   - *Cell Growth*: B vitamins support the synthesis of nucleic acids and proteins, both of which are essential for tissue repair and growth as a wound heals (Peterson et al., 2020).
2. *Collagen Synthesis and Tissue Repair*
   - *Biotin (B7)*: Biotin plays a critical role in collagen synthesis by supporting enzyme function that produces collagen. A lack of biotin may negatively influence healing of a wound by impacting structural strength or collagen type (Kannan et al., 2024).
   - *Pantothenic Acid (B5)*: Pantothenic acid promotes the formation of proline, a key amino acid that supports collagen synthesis, to increase the strength and stabilization of the wound (Anastassakis, 2022).
3. *Inflammation Modulation and Immune Function*
   - *Thiamine (B1) and Riboflavin (B2)*: Both thiamine and riboflavin assist the body in maintaining healthy immune function, which is important for preventing infection and establishing a beneficial inflammatory response during the healing process.

- *Folate (B9) and Cobalamin (B12)*: Folate and cobalamin play essential roles in the synthesis and repair of DNA as cells turnover rapidly, which is especially important for closure of a wound (Boykin et al., 2020).

4. *Closure of Wounds and Strength of Tissue*
   - *Combinations of B Vitamins*: Certain combinations of B vitamins (e.g., B9 and B12; B3, B5, B6, and B10; B3, B5, and B7) have been shown to enhance wound closure rates due to positive effects on fibroblast migration and keratinocyte proliferation. Topical application of B vitamins may modulate inflammatory responses, enhance both fibroblasts and vascular populations, and improve collagen maturation in granulation tissues. Studies showed that B vitamins supplementation is capable of improving wound closure rates diabetic mice through improving microcirculation and decreasing homocysteine levels. Supplementation with vitamin B complex enhances periodontal healing by reducing pocket depth and enhancing collagen structure (Lyon et al., 2020).
5. *Evidence in Practice*
   - *Diabetic Wounds*: B vitamin supplementation has been shown to speed up wound closure in diabetic mice due to improved microcirculation and lower homocysteine levels (Bechara et al., 2021).
   - *Periodontal Healing*: Vitamin B complex supplementation enhances the healing of periodontal wounds, which is characterized by reduced pocket depth and altered collagen structure (El-Sayed et al., 2024).

10.9 MINERALS' ROLE IN WOUND HEALING

Inorganic ions and molecules can possibly impact the different stages of wound healing to include inflammation, proliferation, and remodeling. Mineral elements such as boron can interact with biological, metabolic, and molecular mechanisms responsible for altering the wound healing process. Mineral elements (e.g., those found in clay minerals) can be positively effective in wound healing and tissue regenerative medicine (Beyene et al., 2020). Physiological bone remodeling, which is essential for the repair of damaged bone, is a well-coordinated biological process involving the regulation of bone resorption and bone formation. Honey which contains vitamins, minerals, sugars, and amino acids may contain nutrients that aid in cellular support and wound repair. Minerals can play a role in the process of wound healing, and their effects and related molecular mechanisms can be useful, as in the case of curcumin. Complicated wound healing and bone healing concerns, such as unorganized fractures, can be directly associated with changes in bone mineral metabolism, especially in patients with chronic liver diseases. Bio alternatives can be used to facilitate the pro-osteogenic mechanisms associated with bone healing, which are important to repair damaged bone, based on supporting elements derived from recent studies (Grada and Phillips, 2022).

10.9.1 Boron

Nanoparticles containing boron can act as contaminants in LPS and peptidoglycan (components of bacteria) to stimulate the healing of wounds infected with bacteria (Korkmaz and Konca, 2020). Boron can be added to chitosan-based membranes or hydrogels to develop wound dressings that facilitate skin repair and the overall wound healing process (Zhang et al., 2023). The precise action mechanism of substantial effect of vascularized wound healing is missing but studies suggest that boron can provide beneficial impacts on the dynamic nature of wound healing in injured skin. Boron can promote the wound healing process by enhancing angiogenesis, the regeneration of new blood vessels, which is necessary for tissue regeneration. Boron-containing mesoporous bioactive glasses can release boron ions in a manner to offer a sustained delivery which may contribute to the potential of use for certain skin wound healing applications (Aslam et al., 2023). Modestly, at the molecular level, it is known that boron deficiency tolerance in plants involves mini transcriptional regulation and metabolic pathways to provide relatable insights into boron's role in wound healing.

Boron compounds have been shown to exhibit promising anticancer, antioxidant, and anti-inflammatory effects against glioblastoma, a form of brain cancer, which may also be applicable for the purpose of wound healing as well (Turkez et al., 2021). Boron-containing nano-hydroxyapatite composites could be beneficial for localized repair of bone injuries as boron is shown to modulate the molecular pathways for bone cell function and bone regeneration. Boron–nitrogen coordination bonds could also contribute to the self-healing properties of polyurethane elastomers which may also be relevant in the creation of wound healing materials. Here is the detail mechanism of how boron work in wound healing:

1. *Anti-Inflammatory Effects*
   - *Inflammation Modulation*: Boron modulates the inflammatory response by downregulating pro-inflammatory cytokines and may facilitate normal resolution of inflammation, as prolonged inflammation can hinder the healing process (Ri et al., 2023).
   - *Protease Inhibition*: Boron derivatives inhibit proteases (MMPs) present in the breakdown of the ECM, which is ultimately required to ensure the coordinated degradation and synthesis of the ECM occurs to allow for appropriate wound repair (Sedighi-Pirsaraei et al., 2024).
2. *Antimicrobial Activity*: Boron exhibits potent antimicrobial activity against a variety of pathogens commonly present in wounds, which reduces the risk of infection and may facilitate wound closure.
3. *Antioxidant Properties*: The antioxidant properties of boron protect against oxidative stress that is often present in chronic wounds and modulate reactive oxygen species (ROS) that could otherwise damage tissue and subsequently delay repair (Başaran et al., 2020).
4. *Cell Proliferation and Tissue Regeneration*
   - *Fibroblast Activity*: Boron has been shown to enhance fibroblast proliferation, migration, and turnover of the ECM to enhance the rate of wound closure while minimizing the formation of scars (Yigit et al., 2024).
   - *Collagen Formation*: Boron may indirectly mediate collagen formation through enhancing the expression of several collagen related proteins involved in wound contraction and ECM remodeling (i.e., collagen and TGF-β1) (Calabrese et al., 2023).
5. *Therapeutic Uses*
   - *Hydrogel Dressings*: Hydrogel dressings composed of boron have been created that can correspond with the various stages of wound healing, including visual assessment, anti-bacterial properties, and self-healing capacity (Soriano-Ursua et al., 2025).
6. *Clinical Evidence*: Studies have shown that boron can decrease time to recovery drastically in intensive care patients, signifying its validity as a therapeutic modality in wound healing.

10.9.2 Zinc

Zinc has an important role in the wound healing process through many mechanisms playing a role in different stages of the healing cascades. Below is more detail on its role:

1. *Immune Function and Inflammation*
   - Zinc is important in the normal development and function of immune cells such as T cells, B cells, neutrophils, macrophages, and natural killer cells. Zinc modifies the inflammatory response through the regulation of immune cell function and cytokine modulation.
   - Zinc reduces inflammation by enhancing the clearance of inflammatory mediators and necrotic tissue. This is essential for progression to the proliferative phase of healing (Mutlu et al., 2022).
2. *Coagulation and Hemostasis*
   Zinc contributes to activation and aggregation of platelets which are required to form a fibrin clot, the first stage of wound healing (Lebedeva et al., 2023).

3. *Tissue Repair and Re-Epithelialization*
   - Zinc is a cofactor for enzymes required in the synthesis of collagen. Collagen synthesis provides strength and integrity to the wound.
   - Zinc contributes to re-epithelialization by supporting keratinocyte migration and proliferation which are required for restoration of the epithelial barrier (Adjepong et al., 2020).
4. *Antioxidant Activity and Protection*
   - Zinc provides resistance to epithelial apoptosis by preventing ROS and bacterial toxins, potentially by mediating the antioxidant properties of metallothioneins (Chasapis et al., 2020).
   - Zinc has direct antimicrobial effects and inhibits bacterial growth and the risk of infection at the wound site.
5. *Angiogenesis and Tissue Remodeling*
   - Zinc aids in the process of angiogenesis, which is the development of new capillary blood vessels. This process is essential to supply oxygen and nutrients to the wound site while facilitating repair of the surrounding tissue (Tsuruoka et al., 2022).
   - Zinc acts as a modulator of MMPs involved in ECM remodeling and scar tissue formation during the final stages of the wound healing process (Hassan et al., 2023).
6. *Clinical Evidence*
   - *Benefits of Supplementation*: Zinc supplementation has been shown to benefit and improve outcomes in the treatment of patients who are zinc deficient, particularly chronic wounds such as leg ulcers. Topical zinc treatment is often more effective than oral treatments because of their localized nature (Duan et al., 2023).
   - *Consequences of Zinc Deficiency*: Zinc deficiency can contribute to delayed wound healing due to its potential implications on immune function, collagen synthesis, and tissue repair processes (Hussain et al., 2022).

10.10 HYDRATION'S ROLE IN WOUND HEALING

When healing a wound, hydration is vital for cellular function, nutrients, toxic byproduct removal, and overall environment for healing. The sections that follow discuss how hydration affects wound healing.

Key Roles of Hydration in Wound Healing

1. *Cellular Function and Skin Repair*: Water is a requirement for cellular processes that stimulate tissue repair. Skin that is hydrated is more supple, enabling cells to proliferate and migrate to close wounds. Sufficient hydration supports collagen synthesis, a fundamental step to forming new tissue matrices during tissue repair (Camilion et al., 2022).
2. *Nutrient and Oxygen Transport*: Water aids in the circulation of nutrients, oxygen, and growth factors to the wound area. Insufficient hydration retards the delivery of these substances, subsequently slowing or delaying the healing process. Growth factors that remain in a moist environment promote keratinocyte and fibroblast proliferation and facilitate repair of full-thickness tissue (Younis, 2020).
3. *Toxin Removal*: As wounds heal, waste products accumulate that need to be cleared. Hydration helps with renal function and creates lymphatic drainage, thus decreasing inflammation and swelling (edema) (Pedreros-Rosales et al., 2023).
4. *Moist Wound Healing*: Having a moist environment will promote re-epithelialization, decrease scarring, and minimize pain. Moisture promotes angiogenesis, collagen production and prevents new skin cells from drying out.
5. *Clinical Evidence Supporting Hydration:*
   - *Healing with Moisture Versus Dry*: Evidence indicates that wounds will heal faster in moist environments than when an open air, dry environment is used. Controlling hydration minimizes inflammatory responses and promotes regenerative healing without creating excess scarring (Nuutila and Eriksson, 2021).

- *Hydration beyond Normal Levels*: Hyper-hydration level has been shown to aid wound healing in a manner similar to moist environments without tissue damage or maceration. Modern dressing developed for anyone using it will ultimately help control and manage the moisture level that is most beneficial for wound healing by manage to prevent for example, maceration, by controlling the hydration level (Martins et al., 2023).

10.10.1 Phases of Wound Healing Influenced by Hydration

Hydration is a crucial necessity throughout every stage of wound healing. Systemic hydration guarantees adequate perfusion, immune assistance, and collagen restoration. Local moisture creates an optimal setting for cellular movement and matrix development. Effective management of hydration both orally and topically is vital for prompt and successful wound healing as shown in Figure 10.4.

1. *Hemostasis and Inflammation*: Upon injury to the skin, hydration helps form fibrin networks (scabs) and facilitates the movement of inflammatory cells to the injury location. Vessels dilate during an inflammatory response, which results in the exudation of fluid into the tissue, where it is taken up by the ECM (Guo et al., 2021).
2. *Proliferation Phase*: Hydration enhances ECM deposition and collagen formation, which results in granulation tissue with new blood vessels. A hydrated environment is also beneficial to nutrient and growth factor delivery and keratinocyte migration across the wound surface (Baron et al., 2020).
3. *Remodeling Phase*: Tissue remodeling reestablishes the skin barrier at the injury site. Maintaining hydration confirms optimal cellular activity during this phase (Kremer and Burkemper, 2024).

10.10.2 Hydration Strategies in Wound Care

1. *Drinking Water*: Ensuring systemic hydration recovers overall cellular function and supports wound healing from within.
2. *Moist Dressings*: Dressings like hydrogels, hydrocolloids, and alginates can provide a controlled moist environment to promote healing. NPWT systems can also optimize moisture balance while preventing contamination (Bold, 2020).
3. *Preventing Dehydration*: Hydration strategies are particularly beneficial for chronic wounds, diminishing desiccation and promoting contraction rates over time (Zhang et al., 2023).

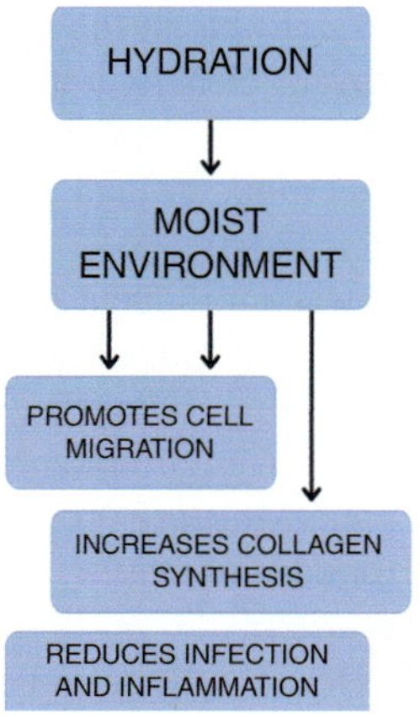

FIGURE 10.4 Effective management of hydration in wound healing.

10.11 NUTRITION ASSESSMENT AND GUIDELINES FOR VARIOUS WOUNDS

Nutrition is vital for wound healing. Malnutrition can impair immune function, collagen synthesis, and tissue repair mechanisms, which can take longer to heal and result in greater risk of complications. A thorough approach for nutrition assessment and intervention is necessary for optimal wound management, especially in regard to the different types of wounds (e.g., pressure ulcers, diabetic foot ulcers, surgical wounds, and trauma) (Sherf-Dagan et al., 2021).

1. *Nutrition Assessment in Wound Care*
   A comprehensive nutrition evaluation is an essential part of effective wound management. It aids in recognizing nutritional deficiencies that may hinder healing and directs specific interventions. This evaluation must be systematic, personalized, and continuous during the entire care process.
   a. *Initial Screening and Assessment*
      All wound patients should have a nutritional assessment. This is especially true for those with chronic or non-healing wounds.
      - *Key Components*: The key components include the following:
      - *Anthropometric Data*: Weight history, BMI, recent weight loss.
      - *Dietary Intake*: 24-hour recall, food regularity, capability to eat independently.
      - *Clinical Signs*: Emaciation, pallor, poor skin turgor, hair loss, mouth sores, edema.
      - *Laboratory Values*: Albumin, prealbumin, transferrin, total lymphocyte count (not diagnostic alone, but supportive).
      - *Comorbidities*: Diabetes, renal disease, cancer, cachexia, alcoholism.
   b. *Risk Factors for Malnutrition*
      - Older patients with poor mobility and/or patients with chronic illness.
      - Patients with large, exudative wounds or multiple wounds.
      - *Patients with Limited Nutrition*: Poor or no oral intake, dysphagia, or gastrointestinal problems.
2. *General Nutrition Guidelines for Wound Healing*
   Adequate nutrition is vital throughout every stage of wound healing. Lacking or having an imbalance of essential nutrients can hinder the healing process, elevate the risk of infection, and disrupt tissue repair. Tables 10.3 and 10.4 show the general nutritional recommendations for effective wound healing:
   a. *Macronutrients*
      Macronutrients such as proteins, carbohydrates, and fats are vital for the wound healing process. A sufficient and balanced consumption of these nutrients is essential to meet the metabolic needs of tissue repair, immune function, and collagen production. Table 10.3 shows the general nutritional recommendations of macronutrients for effective wound healing.

TABLE 10.3
Macronutrient's Nutritional Guidelines for Wound Healing

Nutrient	Role in Healing	Recommendations	Sources
Protein	Tissue repair, collagen, immune	1.25–1.5 g/kg/day (higher for severe wounds)	Meat, fish, eggs, dairy, legumes, nuts
Energy	Fuel for cell activity	30–35 kcal/kg/day (adjust for needs)	Whole grains, starchy veg, fats, oils
Carbohydrates	Prevent protein breakdown	50–60% of total energy	Whole grains, fruits, vegetables
Fat	Energy, cell membrane repair	20–35% of total energy	Oils, butter, nuts, seeds

TABLE 10.4
Micronutrient's Nutritional Guidelines for Wound Healing

Nutrient	Role in Healing	Recommendations	Sources
Vitamin C	Collagen synthesis, immunity	500 mg/day (supplement if deficient)	Citrus, berries, peppers, broccoli
Vitamin A	Epithelialization, immunity	700–900 mcg/day (supplement if deficient)	Liver, eggs, dairy, orange veg
Zinc	DNA synthesis, cell division	15–30 mg/day (short-term supplement if deficient)	Meat, shellfish, legumes, seeds
Iron	Oxygen transport, cell growth	As per deficiency	Red meat, beans, spinach
Arginine	Collagen, immune function	4.5–9 g/day (in some supplements)	Nuts, seeds, protein supplements
Glutamine	Cell proliferation	0.3–0.5 g/kg/day (in some supplements)	Protein-rich foods, supplements

b. *Micronutrients*
Micronutrients, though required in small amounts, are critical for every phase of wound healing from inflammation and immune response to tissue remodeling and epithelialization. Deficiencies in any of these vitamins or minerals can impair the healing process, increase infection risk, and delay recovery. Table 10.4 shows the general nutritional recommendations of micronutrients for effective wound healing.

c. *Hydration*
Acceptable fluid intake (minimum of 1.5 l/day unless contraindicated) is an important factor in skin turgor and cellular function.

3. *Nutrition for Specific Wound Types*
   a. *Pressure Ulcer*:
      - Increased fluid, protein (1.25–1.5 g/kg/day), energy, vitamins A, C, E, and zinc.
      - *Oral Supplements*: High-protein, high-calorie oral nutritional supplements have reduced incidence and promote healing.
      - *Documentation*: Monitor weight, intake, and regularly assess wound status (Yap and Holloway, 2021).
   b. *Diabetic Foot Ulcers*:
      - *Glycemic Control*: Important for healing; distribute carbohydrates equally to minimize blood sugar spikes.
      - *Protein*: 1.25–1.5 g/kg/day; complete/protein may be better than incomplete.
      - *Micronutrients*: Vitamin C, zinc, arginine, and glutamine supplementation may speed healing, especially with malnourished or dialysis patients.
      - Monitor blood glucose, renal function, and protein needs (Da Porto et al., 2022).
   c. *Surgical and Traumatic Wounds*:
      - *Protein and Energy*: Increased requirements to allow for tissue repair. A total of 60–100 g protein/day recommended.
      - *Micronutrients*: Stress vitamin C, vitamin A for immune support and wound healing. Zinc is useful as well due to its contribution to collagen synthesis.
      - *Oral Intake*: Encourage higher, but nutritious, density foods and consider supplements if intake is not adequate (Weimann et al., 2021).
4. *Nutritional Intervention Methods*
   a. *Dietary Recommendations*:
      - Eat a varied diet of foods from the five-food groups: grains, vegetables, fruits, milk or substitutes, sources of protein.
      - Have protein with every meal and snack.

TABLE 10.5
Different Wound Types and Their Key Nutrients for Healing

Wound Type	Protein (g/kg/day)	Key Micronutrients	Special Notes
Pressure ulcer	1.25–1.5	A, C, E, zinc	High-protein supplements, monitor exudate loss
Diabetic foot ulcer	1.25–1.5	C, zinc, arginine, glutamine	Glycemic control, renal function monitoring
Surgical/trauma	1.0–1.5	A, C, zinc	Emphasize protein, energy, and hydration
Chronic/non-healing	1.5+ (as tolerated)	All above	Consider specialized supplements, RD referral

- Eat fortified foods or add cheese, milk powder, or nut butters to meals, for additional protein and calories.
- Eat whole, fresh foods for phytonutrients and antioxidant properties (Basiri et al., 2022).

b. *Supplementation*:
- Oral nutritional supplements (e.g., shakes, bars) may be helpful for those with inadequate dietary protein intake.
- Multivitamin/mineral supplements may be useful for patients with inadequate intake or at risk for deficiencies.
- In general, targeted supplementation (zinc, vitamin C, arginine, glutamine) should be done only when clinical evidence of deficiency is present or in accordance with specific clinical guidelines.

c. *Special Considerations*:
- *Renal or Hepatic Impairment*: Should avoid adding excessive protein. Recommend reaching out to a nephrologist or hepatologist.
- *Dialysis*: Patients may need special amino acid supplementation, using the current knowledge of critical care nutrition. Monitoring is essential.
- *Diabetes*: Provide education on glycemic control, avoid added sugars from simple sugars, and monitor carbohydrate intake and/or careful planning.

d. *Monitoring and Follow-up*
- Reassessment regularly includes weight, wound healing progress, and dietary intake.
- *Plan Adaptations*: It should be based on healing time frame, lab values, and patient tolerance.
- *Patient Education*: Provide patients and caregivers the importance of nutrition in promoting healing of skin and tissue integrity, along with the tools to achieve greater nutrition needs and accomplish their goals.
- *Referrals*: Refer patient to a registered dietitian when complex cases arise or malnutrition is suspected. Table 10.5 shows the various types of wound and which nutrients for healing process.

10.12 EFFECTS OF MALNUTRITION ON WOUND HEALING

Malnutrition significantly impairs wound healing through multiple physiological mechanisms, increasing complications and delaying recovery. Optimal healing requires adequate protein, vitamins, minerals, and energy to support immune function, collagen synthesis, and tissue repair (Mahamoud et al., 2021).

Key Effects of Malnutrition on Wound Healing

1. *Delayed Inflammatory Phase*: Malnutrition prolongs the inflammatory stage by reducing immune cell function (e.g., neutrophils and lymphocytes) and impairing bacterial defense, increasing infection risk (Hannoodee and Nasuruddin, 2024).

2. *Reduced Collagen Synthesis and Tissue Strength*: Protein deficiency limits amino acids needed for collagen production, weakening wound tensile strength and increasing dehiscence risk. For example, protein–energy malnutrition (PEM) triggers muscle breakdown, diverting protein from healing to energy needs (Khatri et al., 2021).
3. *Increased Infection Rates*: Malnourished patients exhibit suppressed immunity (e.g., lower T-cell activity and antibody levels), raising susceptibility to infections. Studies link hypoalbuminemia (<3.5 g/dL) to a 28% delayed healing rate vs. 12% in well-nourished patients (Malone and Schultz, 2022).
4. *Impaired Progression to Healing Stages*: Nutrient deficiencies (e.g., vitamins A, C, zinc) disrupt fibroblast proliferation and angiogenesis, stalling wound progression. Chronic wounds in malnourished individuals often remain in the inflammatory phase due to insufficient growth factors.

10.13 CONCLUSION

Nutrition is an important aspect of wound healing. Nutrition involves the macronutrients and micronutrients contributing uniquely to tissue repair. Proteins are crucial for collagen formation, immune response, and cellular proliferation. Notably, protein or amino acids like arginine and glutamine have been shown to promote fibroblast activity and NO production. However, some studies suggest that protein requirements are based on wound type and clinical aspects of metabolism. Personalized nutrition approaches to wound healing may be warranted based on metabolic needs, the type of wound. Carbohydrates are the dominant source of energy for cellular functions such as phagocytosis and epithelialization, while lactate a byproduct of glucose stimulates collagen formation and angiogenesis. Lipids, particularly the omega-3 PUFAs, regulate inflammation and contribute to cell membrane integrity, but an excess of 80% of energy from fats has been linked to delayed healing as a result of oxidative stress. Interestingly, omega-3 diets resulted in decreased long-term wound strength in several animal studies, indicating excessive dietary fat in any way creates the potential to slow healing. Micronutrients serve as cofactors in important pathways; vitamin C is necessary for the cross-linking of collagen, while vitamin A facilitates epithelialization and mitigates the effects of corticosteroid-induced delays. Zinc is paramount for DNA synthesis and immune function, whilst iron is essential for carrying oxygen to cells that are proliferating. As a wound adjunct, curcumin may benefit the healing process, as it has been shown to have antioxidant and anti-inflammatory properties that reduce oxidative stress and improve collagen deposition. Products such as *Curcuma longa* in addition to other botanicals (*Centella asiatica*, *Aloe vera*) have been noted to promote fibroblast activity and hydration to promote wound healing. In conclusion, appropriate wound healing requires complex interactions of macro- and micronutrients, based on metabolic status and wound physiology. While proteins and carbohydrates provide the scaffolding and energy, micronutrients and bioactive compounds like curcumin adjust similarly detrimental inflammatory and regenerative processes. Personalized nutrition based on ongoing research remains key to minimizing problems and optimizing outcomes across varying populations.

GLOSSARY

Arginine: A semi-essential amino acid that supports collagen production, immune response, and the generation of nitric oxide.

Body mass index (BMI): An indicator of body fat, calculated from weight and height, can be utilized to assess malnutrition or obesity.

Collagen: A key structural protein that is produced during the proliferative stage of wound healing.

Dehydration: A condition characterized by a lack of fluids that negatively affects blood circulation, oxygen transport, and cellular activity in the wound area.

Glutamine: A crucial amino acid that supports immune function, promotes cell growth, and maintains gastrointestinal integrity during the healing process.

Macronutrients: Nutrients needed in significant quantities: proteins, carbohydrates, and fats each serving unique functions in the healing process.

Micronutrients: Vitamins and minerals that are vital in minor quantities for their enzymatic and structural contributions to tissue repair.

Moist wound healing: A principle of wound care that encourages quicker epithelialization and reduces scarring by keeping the local area hydrated.

REFERENCES

Adjepong, Dennis, Saira Jahangir, and Bilal Haider Malik. 2020. "The effect of zinc on post-neurosurgical wound healing: A review." *Cureus* (January). https://doi.org/10.7759/cureus.6770

Almannai, Mohammed, and Ayman W. El-Hattab. 2021. "Nitric oxide deficiency in mitochondrial disorders: The utility of arginine and citrulline." *Frontiers in Molecular Neuroscience* 14 (August). https://doi.org/10.3389/fnmol.2021.682780

Althaher, Arwa R. 2022. "An overview of hormone-sensitive lipase (HSL)." *The Scientific World Journal* 2022 (December): 1964684. https://doi.org/10.1155/2022/1964684

Ammendolia, Dustin A., William M. Bement, and John H. Brumell. 2021. "Plasma membrane integrity: Implications for health and disease." *BMC Biology* 19 (1): 71. https://doi.org/10.1186/s12915-021-00972-y

Anand, Paras K. 2020. "Lipids, inflammasomes, metabolism, and disease." *Immunological Reviews* 297 (1): 108–22. https://doi.org/10.1111/imr.12891

Anastassakis, Konstantinos. 2022. "Vit B5 (pantothenic acid)." In *Androgenetic Alopecia From A to Z : Vol. 2 Drugs, Herbs, Nutrition and Supplements*, edited by Konstantinos Anastassakis, 309–13. Cham: Springer International Publishing. https://doi.org/10.1007/978-3-031-08057-9_31

Anderson, Kristin, and Rose L. Hamm. 2014. "Factors that impair wound healing." *The Journal of the American College of Clinical Wound Specialists* 4 (4): 84–91. https://doi.org/10.1016/j.jccw.2014.03.001

Ao, Tomoka, Junichi Kikuta, and Masaru Ishii. 2021. "The effects of vitamin D on immune system and inflammatory diseases." *Biomolecules* 11 (11): 1624. https://doi.org/10.3390/biom11111624

Aslam, Awais Ali, Javeria Akram, Rana Adeel Mehmood, Arifa Mubarak, Amna Khatoon, Uzma Akbar, Sheikh Asrar Ahmad, and Muhammad Atif. 2023. "Boron-based bioactive glasses: Properties, processing, characterization and applications." *Ceramics International* 49 (12): 19595–605. https://doi.org/10.1016/j.ceramint.2023.03.164

Asokapandian, Sangamithra, Sreelakshmi S, Rajamanickam G. Lipids and oils: An overview. In: Gani A, Ashwar BA, editors. *Food biopolymers: structural, functional and nutraceutical properties*. Cham: Springer International Publishing; 2021. p. 389–411. doi:10.1007/978-3-030-27061-2_16.

Awad, Dominik, Pham Hong Anh Cao, Thomas L. Pulliam, Meredith Spradlin, Elavarasan Subramani, Tristen V. Tellman, Caroline F. Ribeiro, et al. 2024. "Adipose triglyceride lipase is a therapeutic target in advanced prostate cancer that promotes metabolic plasticity." *Cancer Research* 84 (5): 703–24. https://doi.org/10.1158/0008-5472.CAN-23-0555

Barbul, Arthur, Stella A. Lazarou, David T. Efron, Harold L. Wasserkrug, and Gregory Efron. 1990. "Arginine Enhances Wound Healing and Lymphocyte Immune Responses in Humans." *Surgery* 108 (2): 331–36; discussion 336–337. https:// doi.org/10.1016/0039-6060(90)90142-I

Barchitta, Martina, Andrea Maugeri, Giuliana Favara, Roberta Magnano San Lio, Giuseppe Evola, Antonella Agodi, and Guido Basile. 2019. "Nutrition and wound healing: An overview focusing on the beneficial effects of curcumin." *International Journal of Molecular Sciences* 20 (5): 1119. https://doi.org/10.3390/ijms20051119

Baron, Jens Malte, Martin Glatz, and Ehrhardt Proksch. 2020. "Optimal support of wound healing: New insights." *Dermatology* 236 (6): 593–600. https://doi.org/10.1159/000505291

Başaran Nurşen, Yalçın Duydu, Hatice Gül Anlar, Merve Bacanlı, Sevtap Aydin Dilsiz, Aylin Üstündağ, Can Özgür Yalçın, Tanja Schwerdtle, and Hermann M. Bolt. 2020. "Evaluation of oxidative stress and immune parameters of boron exposed males and females." *Food and Chemical Toxicology* 142 (August): 111488. https://doi.org/10.1016/j.fct.2020.111488

Bashir, Bilal, Jonathan Schofield, Paul Downie, Michael France, Darren M. Ashcroft, Alison K. Wright, Stefano Romeo, et al. 2024. "Beyond LDL-C: Unravelling the residual atherosclerotic cardiovascular disease risk landscape—Focus on hypertriglyceridaemia." *Frontiers in Cardiovascular Medicine* 11 (August): 1389106. https://doi.org/10.3389/fcvm.2024.1389106

Basiri, Raedeh, Maria Spicer, Cathy Levenson, Thomas Ledermann, Neda Akhavan, and Bahram Arjmandi. 2022. "Improving dietary intake of essential nutrients can ameliorate inflammation in patients with diabetic foot ulcers." *Nutrients* 14 (12): 2393. https://doi.org/10.3390/nu14122393

Bechara, Nada, Victoria M. Flood, and Jenny E. Gunton. 2022. "A systematic review on the role of vitamin C in tissue healing." *Antioxidants* 11 (8): 1605. https://doi.org/10.3390/antiox11081605

Bechara, Nada, Jenny E. Gunton, Victoria Flood, Tien-Ming Hng, and Clare McGloin. 2021. "Associations between nutrients and foot ulceration in diabetes: A systematic review." *Nutrients* 13 (8): 2576. https://doi.org/10.3390/nu13082576

Beyene, Robel T., Stephen Lentz Derryberry, and Adrian Barbul. 2020. "The effect of comorbidities on wound healing." *Surgical Clinics of North America* 100 (4): 695–705. https://doi.org/10.1016/j.suc.2020.05.002

Bishop, Emma L., Aiten Ismailova, Sarah Dimeloe, Martin Hewison, and John H White. 2021. "Vitamin D and immune regulation: Antibacterial, antiviral, anti-inflammatory." *JBMR Plus* 5 (1): e10405. https://doi.org/10.1002/jbm4.10405

Bold, Justine. 2020. "Supporting evidence-based practice in nutrition and hydration." *Wounds UK* 16 (2): 22–28.

Boykin, Joseph V., Glenn D. Hoke, Cassandra R. Driscoll, and Benita S. Dharmaraj. 2020. "High-dose folic acid and its effect on early stage diabetic foot ulcer wound healing." *Wound Repair and Regeneration* 28 (4): 517–25. https://doi.org/10.1111/wrr.12804

Calabrese, Edward, Peter Pressman, Evgenios Agathokleous, Gaurav Dhawan, Rachna Kapoor, and Vittorio Calabrese. 2023. "Boron enhances adaptive responses and biological performance via hormetic mechanisms." *Chemico-Biological Interactions* 376 (May): 110432. https://doi.org/10.1016/j.cbi.2023.110432

Camilion, Jose V., Siya Khanna, Sheela Anasseri, Coral Laney, and Harvey N. Mayrovitz. 2022. "Physiological, pathological, and circadian factors impacting skin hydration." *Cureus* (August). https://doi.org/10.7759/cureus.27666

Chasapis, Christos T., Panagoula-Stamatina A. Ntoupa, Chara A. Spiliopoulou, and Maria E. Stefanidou. 2020. "Recent aspects of the effects of zinc on human health." *Archives of Toxicology* 94 (5): 1443–60. https://doi.org/10.1007/s00204-020-02702-9

Chiang, Nan, and Charles N. Serhan. 2020. "Specialized pro-resolving mediator network: An update on production and actions." *Essays in Biochemistry* 64 (3): 443–62. https://doi.org/10.1042/EBC20200018

Choudhary, Vivek, Mrunal Choudhary, and Wendy B. Bollag. 2024. "Exploring skin wound healing models and the impact of natural lipids on the healing process." *International Journal of Molecular Sciences* 25 (7): 3790. https://doi.org/10.3390/ijms25073790

Chu, Andy S., and Barbara Delmore. 2021. "Arginine: What you need to know for pressure injury healing : Advances in skin & wound care." Accessed April 19, 2025. https://journals.lww.com/aswcjournal/fulltext/2021/12000/arginine__what_you_need_to_know_for_pressure.4.aspx

Chylińska, Natalia, and Mateusz Maciejczyk. 2025. "Hyaluronic acid and skin: Its role in aging and wound-healing processes." *Gels* 11 (4): 281. https://doi.org/10.3390/gels11040281

Comino-Sanz, Inés María, María Dolores López-Franco, Begoña Castro, and Pedro Luis Pancorbo-Hidalgo. 2021. "The role of antioxidants on wound healing: A review of the current evidence." *Journal of Clinical Medicine* 10 (16): 3558. https://doi.org/10.3390/jcm10163558

Cotterell, Asha, Michelle Griffin, Mauricio A Downer, Jennifer B Parker, Derrick Wan, and Michael T. Longaker. 2024. "Understanding wound healing in obesity." *World Journal of Experimental Medicine* 14 (1): 86898. https://doi.org/10.5493/wjem.v14.i1.86898

Da Porto, Andrea, Cesare Miranda, Gabriele Brosolo, Giorgio Zanette, Andrea Michelli, and Roberto Da Ros. 2022. "Nutritional supplementation on wound healing in diabetic foot: what is known and what is new?" *World Journal of Diabetes* 13 (11): 940–48. https://doi.org/10.4239/wjd.v13.i11.940

de Albuquerque, Priscilla Barbosa Sales, Natalie Emanuelle Ribeiro Rodrigues, Priscila Marcelino dos Santos Silva, Weslley Felix de Oliveira, Maria Tereza dos Santos Correia, and Luana Cassandra Breitenbach Barroso Coelho. 2023. "The use of proteins, lipids, and carbohydrates in the management of wounds." *Molecules* 28 (4): 1580. https://doi.org/10.3390/molecules28041580

De Castro Brás, Lisandra E., and Nikolaos G. Frangogiannis. 2020. "Extracellular matrix-derived peptides in tissue remodeling and fibrosis." *Matrix Biology* 91–92 (September): 176–87. https://doi.org/10.1016/j.matbio.2020.04.006

Deng, Ziqin, Tao Fan, Chu Xiao, He Tian, Yujia Zheng, Chunxiang Li, and Jie He. 2024. "TGF-β signaling in health, disease and therapeutics." *Signal Transduction and Targeted Therapy* 9 (1): 1–40. https://doi.org/10.1038/s41392-024-01764-w

Diller, Robert B., and Aaron J. Tabor. 2022. "The role of the extracellular matrix (ECM) in wound healing: A review." *Biomimetics* 7 (3): 87. https://doi.org/10.3390/biomimetics7030087

Di Tolla, Michele Francesco, Serena Romano, Pietro Vassetti, Domenico Perugini, Immacolata Filoso, Serena Cabaro, Giusy Ferraro, Francesco Oriente, Giuseppe Perruolo, and Flora Arvonio. 2025. "Platelet-derived growth factor as biomarker of clinical outcome for autologous platelet concentrate therapy in Grade I knee osteoarthritis." *Biologics: Targets and Therapy*: 137–47.

Doseděl, Martin, Eduard Jirkovský, Kateřina Macáková, Lenka Krčmová, Lenka Javorská, Jana Pourová, Laura Mercolini, et al. 2021. "Vitamin C—Sources, physiological role, kinetics, deficiency, use, toxicity, and determination." *Nutrients* 13 (2): 615. https://doi.org/10.3390/nu13020615

Maoping Duan, Tian Li, Bo Liu, Shuhua Yin, Jiachen Zang, Chenyan Lv, Guanghua Zhao, and Tuo Zhang. 2023. "Zinc nutrition and dietary zinc supplements." *Critical Reviews in Food Science and Nutrition* 63 (9): 1277–92. https://doi.org/10.1080/10408398.2021.1963664

ElHawary, Hassan, Jason Covone, Shafic Abdulkarim, and Jeffrey E. Janis. 2023. "Practical review on delayed primary closure: Basic science and clinical applications." *Plastic and Reconstructive Surgery – Global Open* 11 (8): e5172. https://doi.org/10.1097/GOX.0000000000005172

Elieh-Ali-Komi, Daniel, Marcus Maurer, and Frank Siebenhaar. 2025. "The impact of mast cells on the anatomy, cellular communication, and molecular immune network of lymph nodes." *Clinical Reviews in Allergy & Immunology* 68 (1): 35.

El-Sayed, Fawzy, Raluca Cosgarea, Anton Sculean, and Christof Doerfer. 2024. "Can vitamins improve periodontal wound Healing/Regeneration?" *Periodontology 2000* 94 (1): 539–602. https://doi.org/10.1111/prd.12513

Erfanian, Nafiseh, Mahsa Saffar, Sahba Khatami, Emad Yeganeh Khorasani, Saboura Nikpour, Mohammad Fereidouni, and Mohammad Hasan Namaei. 2024. "Modulating the Wnt signaling pathway in wound healing: The potential of L-carnosine, zinc sulfate, and B-complex vitamins cocktail." *Jundishapur Journal of Natural Pharmaceutical Products* 19 (2). https://doi.org/10.5812/jjnpp-143129

Fan, Xuehui, Lili Liu, Yue Shi, Fanghan Guo, Xiao He, Xiuli Zhao, Di Zhong, and Guozhong Li. 2021. "Recent advances of the function of sphingosine 1-phosphate (S1P) receptor S1P3." *Journal of Cellular Physiology* 236 (3): 1564–78.

Fatima, Ghizal, Andrej Dzupina, Abbas Ali Mahdi, Jan Fedacko, Aminat Magomedova, and Nasser Ghaly Yousif. 2025. "Role of vitamin-E as a vital nutrient in health and diseases." *Indian Journal of Clinical Biochemistry* (February). https://doi.org/10.1007/s12291-025-01309-0

Feng, Yi, Andrew J. Sanders, Fiona Ruge, Ceri-Ann Morris, Keith G. Harding, and Wen G. Jiang. 2016. "Expression of the SOCS family in human chronic wound tissues: Potential implications for SOCS in chronic wound healing." *International Journal of Molecular Medicine* 38 (5): 1349–58. https://doi.org/10.3892/ijmm.2016.2733

Fischer, Carol L. 2020. "Antimicrobial activity of host-derived lipids." *Antibiotics* 9 (2): 75. https://doi.org/10.3390/antibiotics9020075

Ghatak, Subhadip, Savita Khanna, Sashwati Roy, Mahesh Thirunavukkarasu, Seetur R. Pradeep, Brian C. Wulff, Mohamed S. El Masry, et al. 2023. "Driving adult tissue repair via re-engagement of a pathway required for fetal healing." *Molecular Therapy* 31 (2): 454–70. https://doi.org/10.1016/j.ymthe.2022.09.002

Gogoi, Mayuri, Akshay Datey, Keith T. Wilson, and Dipshikha Chakravortty. 2016. "Dual role of arginine metabolism in establishing pathogenesis." *Current Opinion in Microbiology* 29 (February): 43–48. https://doi.org/10.1016/j.mib.2015.10.005

Gounden, Varshan, and Moganavelli Singh. 2024. "Hydrogels and wound healing: Current and future prospects." *Gels* 10 (1): 43. https://doi.org/10.3390/gels10010043

Grada, Ayman, and Tania J. Phillips. 2022. "Nutrition and cutaneous wound healing." *Nutrition and the Skin, Part II* 40 (2): 103–13. https://doi.org/10.1016/j.clindermatol.2021.10.002

Gubbiveeranna, Vinod, G. T. Megha, C. G. Kusuma, H. Ravikumar, R. S. Thoyajakshi, S. Vijayakumar, S. N. Mathad, S. Nagaraju, Huda Wazzan, Anish Khan, Khalid A. Alzahrani, Asmaa M. Malash. 2024. "Effect of 'Procumbenase' a serine protease from Tridax procumbens aqueous extract on wound healing: A scar free healing of full thickness wounds." *International Journal of Biological Macromolecules* 273 (July): 133147. https://doi.org/10.1016/j.ijbiomac.2024.133147

Guo, Baolin, Ruonan Dong, Yongping Liang, and Meng Li. 2021. "Haemostatic materials for wound healing applications." *Nature Reviews Chemistry* 5 (11): 773–91. https://doi.org/10.1038/s41570-021-00323-z

Halloran, Katherine M, Claire Stenhouse, Guoyao Wu, and Fuller W. Bazer. 2021. "Arginine, agmatine, and polyamines: Key regulators of conceptus development in mammals." In *Amino Acids in Nutrition and Health: Amino Acids in Gene Expression, Metabolic Regulation, and Exercising Performance*, edited by Guoyao Wu, 85–105. Cham: Springer International Publishing. https://doi.org/10.1007/978-3-030-74180-8_6

Halschou-Jensen, Peter M., Jannie Sauer, Pierre Bouchelouche, Jesper Fabrin, Stig Brorson, and Søren Ohrt-Nissen. 2023. "Improved healing of diabetic foot ulcers after high-dose vitamin D: A randomized double-blinded clinical trial." *The International Journal of Lower Extremity Wounds* 22 (3): 466–74. https://doi.org/10.1177/15347346211020268

Han, Chen, May Barakat, and Luisa A. DiPietro. 2022. "Angiogenesis in wound repair: Too much of a good thing?" *Cold Spring Harbor Perspectives in Biology* 14 (10): a041225. https://doi.org/10.1101/cshperspect.a041225

Han, Xianlin, and Richard W. Gross. 2022. "The foundations and development of lipidomics." *Journal of Lipid Research* 63 (2): 100164. https://doi.org/10.1016/j.jlr.2021.100164

Hannoodee, Sally, and Dian N. Nasuruddin. 2024. "Acute inflammatory response." In *StatPearls [Internet]*. StatPearls Publishing. https://www.ncbi.nlm.nih.gov/sites/books/NBK556083/

Hassan, Nourhan, Thomas Krieg, Max Zinser, Kai Schröder, and Nadja Kroger. 2023. "An overview of scaffolds and biomaterials for skin expansion and soft tissue regeneration: Insights on zinc and magnesium as new potential key elements." *Polymers* 15 (19): 3854. https://doi.org/10.3390/polym15193854

Hodge, Christopher, and Christopher Taylor. 2025. "Vitamin A deficiency." In *StatPearls*. Treasure Island (FL): StatPearls Publishing. http://www.ncbi.nlm.nih.gov/books/NBK567744/

Hodgens, Alexander, and Tariq Sharman. 2025. "Corticosteroids." In *StatPearls*. Treasure Island (FL): StatPearls Publishing. http://www.ncbi.nlm.nih.gov/books/NBK554612/.

Holdaway, Courtney M., Kelly-Ann Leonard, Randal Nelson, Jelske Van Der Veen, Chinmayee Das, Russell Watts, Robin D. Clugston, Richard Lehner, and Rene L. Jacobs. 2025. "Alterations in phosphatidylethanolamine metabolism impacts hepatocellular lipid storage, energy homeostasis, and proliferation." *Biochimica et Biophysica Acta (BBA) – Molecular and Cell Biology of Lipids* 1870 (4): 159608. https://doi.org/10.1016/j.bbalip.2025.159608

Horsley, Valerie. 2022. "Adipocyte plasticity in tissue regeneration, repair, and disease." *Current Opinion in Genetics & Development* 76 (October): 101968. https://doi.org/10.1016/j.gde.2022.101968

Hrubša, Marcel, Tomáš Siatka, Iveta Nejmanová, Marie Vopršalová, Lenka Kujovská Krčmová, Kateřina Matoušová, Lenka Javorská, et al. n.d. "Biological properties of vitamins of the B-complex, Part 1: Vitamins B1, B2, B3, and B5."

Hussain, Azhar, Wenting Jiang, Xiukang Wang, Shumaila Shahid, Noreena Saba, Maqshoof Ahmad, Abubakar Dar, Syed Usama Masood, Muhammad Imran, and Adnan Mustafa. 2022. "Mechanistic impact of zinc deficiency in human development." *Frontiers in Nutrition* 9 (March). https://doi.org/10.3389/fnut.2022.717064

Huynh, Thi Yen Ly, Ilona Zareba, Weronika Baszanowska, Sylwia Lewoniewska, and Jerzy Palka. 2020. "Understanding the role of key amino acids in regulation of proline dehydrogenase/proline oxidase (Prodh/Pox)-dependent Apoptosis/Autophagy as an approach to targeted cancer therapy." *Molecular and Cellular Biochemistry* 466 (1–2): 35–44. https://doi.org/10.1007/s11010-020-03685-y

Eva Jarc, and Toni Petan. 2020. "A twist of FATe: Lipid droplets and inflammatory lipid mediators." *Biochimie*, Biogenesis and Fate of Lipid Droplets", guest edited by Dr Frédéric Carriere and Yonghua Li-Beisson, 169 (February): 69–87. https://doi.org/10.1016/j.biochi.2019.11.016

Jerab, Donia, Ferdinand Blangero, Paulo César Trindade da Costa, José Luiz de Brito Alves, Rym Kefi, Henda Jamoussi, Beatrice Morio, and Assia Eljaafari. 2025. "Beneficial effects of omega-3 fatty acids on obesity and related metabolic and chronic inflammatory diseases." *Nutrients* 17 (7): 1253. https://doi.org/10.3390/nu17071253

Jiang, Dianhua, Jiurong Liang, and Paul W. Noble. 2011. "Hyaluronan as an immune regulator in human diseases." *Physiological Reviews* 91 (1): 221–64. https://doi.org/10.1152/physrev.00052.2009

Kalsi, Rupali, Kala S. Bhushan, Mohit K. Mathur, Vineeta Gupta, Deepesh K. Gupta, and Sandeep K. Pandey. 2024. "To evaluate the effect of vitamin B complex on wound healing – A clinical and microbiological study." *Journal of Pharmacy and Bioallied Sciences* 16 (Suppl 1): S549. https://doi.org/10.4103/jpbs.jpbs_856_23

Kannan, Suganya, Jeyakumar Balakrishnan, and Prithiviraj Nagarajan. 2024. "Vitamin B7 (biotin) and its role in hair, skin and nail health." In *Hydrophilic Vitamins in Health and Disease*, edited by Anureet K. Shah, Paramjit S. Tappia, and Naranjan S. Dhalla, 233–52. Cham: Springer International Publishing. https://doi.org/10.1007/978-3-031-55474-2_10

Karna, Ewa, Lukasz Szoka, Thi Yen Ly Huynh, and Jerzy A. Palka. 2020. "Proline-dependent regulation of collagen metabolism." *Cellular and Molecular Life Sciences* 77 (10): 1911–18. https://doi.org/10.1007/s00018-019-03363-3

Kashiouris, Markos G., Michael L' Heureux, Casey A. Cable, Bernard J. Fisher, Stefan W. Leichtle, and Alpha A. Fowler. 2020. "The emerging role of vitamin C as a treatment for sepsis." *Nutrients* 12 (2): 292. https://doi.org/10.3390/nu12020292

Kaźmierczak-Barańska, Julia, Karolina Boguszewska, Angelika Adamus-Grabicka, and Bolesław T. Karwowski. 2020. "Two faces of vitamin C—Antioxidative and pro-oxidative agent." *Nutrients* 12 (5): 1501. https://doi.org/10.3390/nu12051501

Kearney, Katherine J., Robert A. S. Ariëns, and Fraser L. Macrae. 2022. "The role of fibrin(Ogen) in wound healing and infection control." *Seminars in Thrombosis and Hemostasis* 48 (2): 174–87. https://doi.org/10.1055/s-0041-1732467

Khalili-Tanha, Ghazaleh, Evette S. Radisky, Derek C. Radisky, and Alireza Shoari. 2025. "Matrix metalloproteinase-driven epithelial-mesenchymal transition: Implications in health and disease." *Journal of Translational Medicine* 23 (April): 436. https://doi.org/10.1186/s12967-025-06447-w

Khatri, Mishti, Robert J. Naughton, Tom Clifford, Liam D. Harper, and Liam Corr. 2021. "The effects of collagen peptide supplementation on body composition, collagen synthesis, and recovery from joint injury and exercise: A systematic review." *Amino Acids* 53 (10): 1493–1506. https://doi.org/10.1007/s00726-021-03072-x

Kim, Do-Houn, Cesar A. Meza, Holly Clarke, Jeong-Su Kim, and Robert C. Hickner. 2020. "Vitamin D and endothelial function." *Nutrients* 12 (2): 575. https://doi.org/10.3390/nu12020575

Knight, Ashley, Eric W. Tan, and Ioanna W. Bolia-Kavouklis. 2024. "Critical literature review of bone grafts and orthobiologics for foot and ankle arthrodesis to facilitate optimal clinical decision making." *Foot & Ankle Orthopaedics*. https://doi.org/10.1177/2473011424S00227.

Korkmaz, Musa, and Meral Konca. 2020. "Comparison of effects of administration of Oral or topical boron on wound healing and oxidative stress in rats." *Kocatepe Veterinary Journal* 13 (1): 11–18. https://doi.org/10.30607/kvj.646939

Kremer, Michael, and Nicole Burkemper. 2024. "Aging skin and wound healing." *Clinics in Geriatric Medicine* 40 (1): 1–10. https://doi.org/10.1016/j.cger.2023.06.001

Kumar, Vinod, Jamal-Eddine Bouameur, Janina Bär, Robert H. Rice, Hue-Tran Hornig-Do, Dennis R. Roop, Nicole Schwarz, et al. 2015. "A keratin scaffold regulates epidermal barrier formation, mitochondrial lipid composition, and activity." *The Journal of Cell Biology* 211 (5): 1057–75. https://doi.org/10.1083/jcb.201404147

LaPelusa, Andrew, and Heeransh D. Dave. 2025. "Physiology, hemostasis." In *StatPearls*. Treasure Island (FL): StatPearls Publishing. http://www.ncbi.nlm.nih.gov/books/NBK545263/

Lebedeva, Svetlana A., Pavel A. Galenko-Yaroshevsky (Jr.), Mikhail Yu. Samsonov, Arkadiy B. Erlich, Anait V. Zelenskaya, Arus G. Margaryan, Maria Yu. Materenchuk, et al. 2023. "Molecular mechanisms of wound healing: The role of zinc as an essential microelement." *Research Results in Pharmacology* 9 (1): 25–39. https://doi.org/10.18413/rrpharmacology.9.10003

Lee, Jae-Seon, Su-Jin Oh, Hyun-Jung Choi, Joon Hee Kang, Seon-Hyeong Lee, Ji Sun Ha, Sang Myung Woo, Hyonchol Jang, Ho Lee, and Soo-Youl Kim. 2020. "ATP production relies on fatty acid oxidation rather than glycolysis in pancreatic ductal adenocarcinoma." *Cancers* 12 (9): 2477. https://doi.org/10.3390/cancers12092477

Lee, Joyce V., Corbett T. Berry, Karla Kim, Payel Sen, Taehyong Kim, Alessandro Carrer, Sophie Trefely, et al. 2018. "Acetyl-CoA promotes glioblastoma cell adhesion and migration through Ca^{2+}–NFAT signaling." *Genes & Development* 32 (7–8): 497–511. https://doi.org/10.1101/gad.311027.117

Li, Shiqi, Hui Zeng, Junli Fan, Fubing Wang, Chen Xu, Yirong Li, Jiancheng Tu, Kenneth P. Nephew, and Xinghua Long. 2023. "Glutamine metabolism in breast cancer and possible therapeutic targets." *Biochemical Pharmacology* 210 (April): 115464. https://doi.org/10.1016/j.bcp.2023.115464

Linju, M. C., and M. R. Rekha. 2023. "Proline Conjugated Chitosan as Wound Healing Material: In Vitro Studies on the Influence of the Scaffold on Collagen Production and Wound Healing." *International Journal of Biological Macromolecules* 242(July):124688. https://doi.org/10.1016/j.ijbiomac.2023.124688.

Liu, Huanzhu, Ruoyang Yu, Muwei Zhang, Xiaoyan Zheng, Lizi Zhong, Wanlin Yang, Yuqi Luo, Zifeng Huang, Jialing Zheng, and Hui Zhong. 2025. "Fibrinogen degradation products exacerbate alpha-synuclein aggregation by inhibiting autophagy via downregulation of Beclin1 in multiple system atrophy." *Neurotherapeutics*, e00538.

López-Sánchez, Laura M., Enrique Aranda, and Antonio Rodríguez-Ariza. 2020. "Nitric oxide and tumor metabolic reprogramming." *Biochemical Pharmacology*,176 (June): 113769. https://doi.org/10.1016/j.bcp.2019.113769

Lv, Bin, Naiting Shen, Zhangrong Cheng, Yuhang Chen, Hua Ding, Jishan Yuan, Kangchen Zhao, and Yukun Zhang. 2022. "Strategies for biomaterial-based spinal cord injury repair via the TLR4-NF-κB signaling pathway." *Frontiers in Bioengineering and Biotechnology* 9 (April): 813169. https://doi.org/10.3389/fbioe.2021.813169

Lyon, Peter, Victoria Strippoli, Byron Fang, and Luisa Cimmino. 2020. "B vitamins and one-carbon metabolism: Implications in human health and disease." *Nutrients* 12 (9): 2867. https://doi.org/10.3390/nu12092867

Fifi Mahamoud, Hassan Youssif Saleh, Esmat Sayed Abd El-Mageed, and Amna Desouky. 2021. "Effect of nutritional factor on burn wound healing." *Assiut Scientific Nursing Journal* 9 (27): 10–20. https://doi.org/10.21608/asnj.2021.105350.1263

Malone, Matthew, and Gregory Schultz. 2022. "Challenges in the diagnosis and management of wound infection." *British Journal of Dermatology* 187 (2): 159–66. https://doi.org/10.1111/bjd.21612

Martins, Kassya Lopes Epaminondas, Marcelo Martins Thomaz, Lais Nogueira Magno, Marina Clare Vinaud, Luciane Madureira Almeida, Pablo José Gonçalves, and Ruy De Souza Lino Junior. 2023. "Macroporous latex biomembrane from Hancornia speciosa modulates the inflammatory process and has a debridement effect on wound healing in rats." *Acta Cirúrgica Brasileira* 38: e385323. https://doi.org/10.1590/acb385323

Mathew-Steiner, Shomita S., Sashwati Roy, and Chandan K. Sen. 2021. "Collagen in wound healing." *Bioengineering* 8 (5): 63. https://doi.org/10.3390/bioengineering8050063

McDaniel, Jodi C., Martha Belury, Karen Ahijevych, and Wendy Blakely. 2008. "ω-3 fatty acids effect on wound healing." *Wound Repair and Regeneration : Official Publication of the Wound Healing Society [and] the European Tissue Repair Society* 16 (3): 337–45. https://doi.org/10.1111/j.1524-475X.2008.00388.x

Meikle, Thomas G., Kevin Huynh, Corey Giles, and Peter J. Meikle. 2021. "Clinical lipidomics: Realizing the potential of lipid profiling." *Journal of Lipid Research* 62 (January): 100127. https://doi.org/10.1016/j.jlr.2021.100127

Mierziak, Justyna, Marta Burgberger, and Wioleta Wojtasik. 2021. "3-hydroxybutyrate as a metabolite and a signal molecule regulating processes of living organisms." *Biomolecules* 11 (3): 402. https://doi.org/10.3390/biom11030402

Murai, Toshiyuki. 2015. "Lipid raft-mediated regulation of hyaluronan–CD44 interactions in inflammation and cancer." *Frontiers in Immunology* 6 (August): 420. https://doi.org/10.3389/fimmu.2015.00420

Muse, Mikel E., Karlie R. Shumway, and Jonathan S. Crane. 2023. "Physiology, epithelialization." In *StatPearls [Internet]*. StatPearls Publishing. https://www.ncbi.nlm.nih.gov/sites/books/NBK532977/

Mutlu, Nurshen, Liliana Liverani, Fatih Kurtuldu, Dušan Galusek, and Aldo R. Boccaccini. 2022. "Zinc improves antibacterial, anti-inflammatory and cell motility activity of chitosan for wound healing applications." *International Journal of Biological Macromolecules* 213 (July): 845–57. https://doi.org/10.1016/j.ijbiomac.2022.05.199

Nakakura, Asia, Beth A. Shields, and Leopoldo C. Cancio. 2025. "Carbohydrate and wound healing in critically Ill burn patients: A retrospective cohort study." *Journal of Burn Care & Research: Official Publication of the American Burn Association*, January, iraf006. https://doi.org/10.1093/jbcr/iraf006

Nuutila, Kristo, and Elof Eriksson. 2021. "Moist wound healing with commonly available dressings." *Advances in Wound Care* 10 (12): 685–98. https://doi.org/10.1089/wound.2020.1232

Oh, Hyun-Seok, Se Kwan Oh, Jum Seek Lee, Chunyan Wu, and Sung-Joon Lee. 2017. "Effects of L-arginine on growth hormone and insulin-like growth factor 1." *Food Science and Biotechnology* 26 (6): 1749–54. https://doi.org/10.1007/s10068-017-0236-6

Ouyang, Yu, and Itamar Willner. 2025. "Photomodulated transient catalytic constitutional dynamic networks and reaction circuits." *Angewandte Chemie International Edition* 64 (9): e202420787.

Panda, Venketesh K., and Sumant K. Mohanty. 2025. "Deciphering the Role of Tumor-Associ."

Parke, Milbrey A., Ariadna Perez-Sanchez, Dina H. Zamil, and Rajani Katta. 2021. "Diet and skin barrier: The role of dietary interventions on skin barrier function." *Dermatology Practical & Conceptual* 11 (1): e2021132.

Patten, Jennifer, and Karin Wang. 2021. "Fibronectin in development and wound healing." *Advanced Drug Delivery Reviews* 170 (March): 353–68. https://doi.org/10.1016/j.addr.2020.09.005

Pedreros-Rosales, Cristian, Aquiles Jara, Eduardo Lorca, Sergio Mezzano, Roberto Pecoits-Filho, and Patricia Herrera. 2023. "Unveiling the clinical benefits of high-volume hemodiafiltration: Optimizing the removal of medium-weight uremic toxins and beyond." *Toxins* 15 (9): 531. https://doi.org/10.3390/toxins15090531

Peng, Han, Qikui Xu, Ting Zhang, Jiakai Zhu, Jinheng Pan, Xiaoyu Guan, Shan Feng, Jianping Wu, and Qi Hu. 2025. "Molecular determinants for the association of human hormone-sensitive lipase with lipid droplets." *Nature Communications* 16 (1): 3497. https://doi.org/10.1038/s41467-025-58887-z

Penny, Harry, Regino Flores, Elisabeth Pennington, Anthony Pedersen, and Son Tran. 2022. "The role of macronutrients and micronutrients in wound healing: A narrative review." *Journal of Wound Care* 31 (Sup5): S14–22. https://doi.org/10.12968/jowc.2022.31.Sup5.S14

Peterson, Christine Tara, Dmitry A. Rodionov, Andrei L. Osterman, and Scott N. Peterson. 2020. "B vitamins and their role in immune regulation and cancer." *Nutrients* 12 (11): 3380. https://doi.org/10.3390/nu12113380

Phang, James M. 2023. "The regulatory mechanisms of proline and hydroxyproline metabolism: Recent advances in perspective." *Frontiers in Oncology* 12 (January): 1118675. https://doi.org/10.3389/fonc.2022.1118675

Phillipson, Mia, and Paul Kubes. 2019. "The healing power of neutrophils." *Trends in Immunology* 40 (7): 635–47. https://doi.org/10.1016/j.it.2019.05.001

Qin, Yuan, Xin Chen, Lili Bao, Lili Ren, Geng Dou, Jianing Lian, Shujuan Xing, et al. 2025. "Lipid metabolism of apoptotic vesicles accelerates cutaneous wound healing by modulating macrophage function." *Journal of Nanobiotechnology* 23 (February): 106. https://doi.org/10.1186/s12951-025-03194-4

Ren, Wenkai, Peng Bin, Yulong Yin, and Guoyao Wu. 2020. "Impacts of amino acids on the intestinal defensive system." In *Amino Acids in Nutrition and Health: Amino Acids in Systems Function and Health*, edited by Guoyao Wu, 133–51. Cham: Springer International Publishing. https://doi.org/10.1007/978-3-030-45328-2_8

Renne, Mike F., and Robert Ernst. 2023. "Membrane homeostasis beyond fluidity: Control of membrane compressibility." *Trends in Biochemical Sciences* 48 (11): 963–77. https://doi.org/10.1016/j.tibs.2023.08.004

Ri, Córdova-Chávez, Carrasco-Ruiz, M. F. , Rodríguez-Vera, D., Pérez-Capistran, T., Tamay-Cach, F., Scorei I. R., Abad-García, A., and Soriano-Ursúa M. A.. 2023. "Boron-containing compounds for prevention, diagnosis, and treatment of human metabolic disorders." *Biological Trace Element Research* 201 (5): 2222–2239. https://doi.org/10.1007/s12011-022-03346-9

Riehl, Terrence E., Srikanth Santhanam, Lynne Foster, Matthew Ciorba, and William F. Stenson. 2015. "CD44 and TLR4 mediate hyaluronic acid regulation of $Lgr5^+$ stem cell proliferation, crypt fission, and intestinal growth in postnatal and adult mice." *American Journal of Physiology – Gastrointestinal and Liver Physiology* 309 (11): G874–87. https://doi.org/10.1152/ajpgi.00123.2015

Sabbah, Asmaa Sayed, Manal Mahmoud Elkattan, Dalia Mohammed Labib, Mona Salah Eldin Hamdy, Nancy Samir Wahdan, and Al-shaimaa Mahmoud Aboulfotoh. 2024. "Role of von Willebrand factor level as a biomarker in acute ischemic stroke." *The Egyptian Journal of Neurology, Psychiatry and Neurosurgery* 60 (1): 63. https://doi.org/10.1186/s41983-024-00837-5

Saeg, Fouad, Rita Orazi, Gerald M. Bowers, and Jeffrey E. Janis. 2021. "Evidence-based nutritional interventions in wound care." *Plastic and Reconstructive Surgery* 148 (1): 226. https://doi.org/10.1097/PRS.0000000000008061

Saiki, Atsuhito, Yuka Takahashi, Shoko Nakamura, Shuhei Yamaoka, Kazuki Abe, Sho Tanaka, Yasuhiro Watanabe, et al. 2024. "Relationship between lipoprotein lipase derived from subcutaneous adipose tissue and cardio-ankle vascular index in Japanese patients with severe obesity." *Obesity Facts* 17 (3): 255–63. https://doi.org/10.1159/000537687

Samson, Faith Pwaniyibo, Tosin Esther Fabunmi, Ambrose Teru Patrick, Donghyun Jee, Diana R. Gutsaeva, and Wan Jin Jahng. 2021. "Fatty acid composition and stoichiometry determine the angiogenesis microenvironment." *ACS Omega* 6 (8): 5953–61. https://doi.org/10.1021/acsomega.1c00196

Santinho, Alexandre, Aymeric Chorlay, Lionel Foret, and Abdou Rachid Thiam. 2021. "Fat inclusions strongly alter membrane mechanics." *Biophysical Journal* 120 (4): 607–17. https://doi.org/10.1016/j.bpj.2021.01.009

Schoor, Natasja van, Renate de Jongh, and Paul Lips. 2024. "Chapter 54 – worldwide vitamin D status." In *Feldman and Pike's Vitamin D (Fifth Edition)*, edited by Martin Hewison, Roger Bouillon, Edward Giovannucci, David Goltzman, Mark Meyer, and JoEllen Welsh, 47–75. Academic Press. https://doi.org/10.1016/B978-0-323-91338-6.00004-5.

Schultz, Gregory S., Robert G. Sibbald, Vincent Falanga, Elizabeth A. Ayello, Craig Dowsett, Keith Harding, Marco Romanelli, Michael C. Stacey, and Wolfgang Vanscheidt. (2011). Wound bed preparation: a systematic approach to wound management. *Wound Repair and Regeneration*, 19 (Suppl 1), S1–S17. doi: 10.1111/j.1524-475X.2011.00783.x

Scopelliti, Fernanda, Caterina Cattani, Valentina Dimartino, Concetta Mirisola, and Andrea Cavani. 2022. "Platelet derivatives and the immunomodulation of wound healing." *International Journal of Molecular Sciences* 23 (15): 8370. https://doi.org/10.3390/ijms23158370

Sedighi-Pirsaraei, Nasrin, Amirhossein Tamimi, Faraz Sadeghi Khamaneh, Sana Dadras-Jeddi, and Naz Javaheri. 2024. "Boron in wound healing: A comprehensive investigation of its diverse mechanisms." *Frontiers in Bioengineering and Biotechnology* 12 (October): 1475584. https://doi.org/10.3389/fbioe.2024.1475584

Selmani, Atiđa, Ramona Jeitler, Christina Glader, Ivan Vidaković, Nataša Vujičić, Martin Sterrer, Nives Matijaković Mlinarić, Karin Kornmüller, and Eva Roblegg. 2024. "Collagen I-binding dynamics with and without hyaluronic acid-coated gold nanoparticles: Implications for enhanced wound healing applications." In *XVIII Christmas Biophysics Workshop: Book of Abstracts*, 8–9. Tarvisio, Italy, December 16–17, 2024.

Shams, Forough, Hamideh Moravvej, Simzar Hosseinzadeh, Ebrahim Mostafavi, Hadi Bayat, Bahram Kazemi, Mojgan Bandehpour, Elnaz Rostami, Azam Rahimpour, and Hamidreza Moosavian. 2022. "Overexpression of VEGF in dermal fibroblast cells accelerates the angiogenesis and wound healing function: In vitro and in vivo studies." *Scientific Reports* 12 (1): 18529. https://doi.org/10.1038/s41598-022-23304-8

Shenoy, Mahesh, Nishath Sayed Abdul, Zeeshan Qamar, Bader Musfer Al Bahri, Khalid Zuhair K. Al Ghalayini, and Ateet Kakti. 2022. "Collagen structure, synthesis, and its applications: A systematic review." *Cureus* (May). https://doi.org/10.7759/cureus.24856

Sherf-Dagan, Shiri, Tali Sinai, Ariela Goldenshluger, Inbal Globus, Yafit Kessler, Chaya Schweiger, and Tair Ben-Porat. 2021. "Nutritional assessment and preparation for adult bariatric surgery candidates: Clinical practice." *Advances in Nutrition* 12 (3): 1020–31. https://doi.org/10.1093/advances/nmaa121

Shields, Bridget. 2021. "Diet in wound care: Can nutrition impact healing?." *Cutis* 108 (6). https://doi.org/10.12788/cutis.0407

Siregar, Fitri Dona, and Wahyu Hidayat. 2023. "The role of vitamin D on the wound healing process: A case series." *International Medical Case Reports Journal* 16 (null): 227–32. https://doi.org/10.2147/IMCRJ.S402005

Soriano-Ursúa, Marvin A., Marlet Martínez-Archundia, Ahmet Kilic, Teresa Pérez-Capistran, Miriam A. Hernández-Zamora, Juan E. López-Ramos, and Eunice D. Farfán-García. 2025. "Formulations with boric acid or aryl-organoboron compounds for treating diabetic foot ulcers." *Scientia Pharmaceutica* 93 (1): 14. https://doi.org/10.3390/scipharm93010014

Starr, Nichola J., Mohammed H. Khan, Max K. Edney, Gustavo F. Trindade, Stefanie Kern, Alexander Pirkl, Matthias Kleine-Boymann et al. 2022. "Elucidating the molecular landscape of the stratum corneum." *Proceedings of the National Academy of Sciences* 119 (12): e2114380119. https://doi.org/10.1073/pnas.2114380119

Stegen, Steve, Gianmarco Rinaldi, Shauni Loopmans, Ingrid Stockmans, Karen Moermans, Bernard Thienpont, Sarah-Maria Fendt, Peter Carmeliet, and Geert Carmeliet. 2020. "Glutamine metabolism controls chondrocyte identity and function." *Developmental Cell* 53 (5): 530–544.e8. https://doi.org/10.1016/j.devcel.2020.05.001

Sumbria, Deepak, Engin Berber, Logan Miller, and Barry T. Rouse. 2021. "Modulating glutamine metabolism to control viral immuno-inflammatory lesions." *Cellular Immunology* 370 (December): 104450. https://doi.org/10.1016/j.cellimm.2021.104450

Tavukcuoglu, Zeynep, Umar Butt, Alessandra V. Sousa de Faria, Johannes Oesterreicher, Wolfgang Holnthoner, Saara Laitinen, Mari Palviainen, and Pia R. M. Siljander. 2024. "Platelet-derived extracellular vesicles induced through different activation pathways drive melanoma progression by functional and transcriptional changes." *Cell Communication and Signaling* 22 (1): 601.

Tenchov, Rumiana, Robert Bird, Allison E. Curtze, and Qiongqiong Zhou. 2021. "Lipid nanoparticles—From liposomes to mRNA vaccine delivery, a landscape of research diversity and advancement." *ACS Nano* 15 (11): 16982–15. https://doi.org/10.1021/acsnano.1c04996

Tombulturk, Fatma Kubra, Zeynep Gizem Todurga-Seven, Onder Huseyinbas, Sibel Ozyazgan, Turgut Ulutin, and Gonul Kanigur-Sultuybek. 2022. "Topical application of metformin accelerates cutaneous wound healing in streptozotocin-induced diabetic rats." *Molecular Biology Reports* 49 (1): 73–83. https://doi.org/10.1007/s11033-021-06843-7

Torres, Manuel, Sebastià Parets, Javier Fernández-Díaz, Roberto Beteta-Göbel, Raquel Rodríguez-Lorca, Ramón Román, Victoria Lladó, Catalina A. Rosselló, Paula Fernández-García, and Pablo V. Escribá. 2021. "Lipids in pathophysiology and development of the membrane lipid therapy: New bioactive lipids." *Membranes* 11 (12): 919.

Tsuruoka, Takuya, Akio Kodama, Shukuro Yamaguchi, Tomohiro Masutomi, Akio Koyama, Toyoaki Murohara, Kimihiro Komori, and Rei Shibata. 2022. "Zinc Deficiency impairs ischemia-induced angiogenesis." *JVS-Vascular Science* 3 (January): 30–40. https://doi.org/10.1016/j.jvssci.2021.09.023

Turkez, Hasan, Mehmet Enes Arslan, Abdulgani Tatar, and Adil Mardinoglu. 2021. "Promising potential of boron compounds against glioblastoma: In vitro antioxidant, anti-inflammatory and anticancer studies." *Neurochemistry International* 149 (October): 105137. https://doi.org/10.1016/j.neuint.2021.105137

VanBuren, Christine A., and Helen B. Everts. 2022. "Vitamin A in skin and hair: An update." *Nutrients* 14 (14): 2952. https://doi.org/10.3390/nu14142952

Wallert, Maria, Lisa Börmel, and Stefan Lorkowski. 2021. "Inflammatory diseases and vitamin E—What do we know and where do we go?." *Molecular Nutrition & Food Research* 65 (1): 2000097. https://doi.org/10.1002/mnfr.202000097

Wang, Chao, Becky K. Brisson, Masahiko Terajima, Qing Li, Kevt'her Hoxha, Biao Han, Abby M. Goldberg, et al. 2020. "Type III collagen is a key regulator of the collagen fibrillar structure and biomechanics of articular cartilage and meniscus." *Matrix Biology: Journal of the International Society for Matrix Biology* 85–86 (January): 47–67. https://doi.org/10.1016/j.matbio.2019.10.001

Wang, Xue, Zhangping Yu, Shengnan Zhou, Shiwei Shen, and Wei Chen. 2022. "The effect of a compound protein on wound healing and nutritional status." *Evidence-Based Complementary and Alternative Medicine* 2022 (1): 4231516. https://doi.org/10.1155/2022/4231516

Wang, Zitong, Feng Zhao, Chengcheng Xu, Qiqi Zhang, Haiyue Ren, Xing Huang, Cai He, Jiajie Ma, and Zhe Wang. 2024. "Metabolic reprogramming in skin wound healing." *Burns & Trauma* 12 (January):tkad047. https://doi.org/10.1093/burnst/tkad047

Weimann, Arved, Marco Braga, Franco Carli, Takashi Higashiguchi, Martin Hübner, Stanislaw Klek, Alessandro Laviano, et al. 2021. "ESPEN practical guideline: Clinical nutrition in surgery." *Clinical Nutrition* 40 (7): 4745–61. https://doi.org/10.1016/j.clnu.2021.03.031

Wiczew, Daniel, Natalia Szulc, and Mounir Tarek. 2021. "Molecular dynamics simulations of the effects of lipid oxidation on the permeability of cell membranes." *Bioelectrochemistry* 141 (October): 107869. https://doi.org/10.1016/j.bioelechem.2021.107869

Wijesinghe, Dayanjan S. 2020. "Chapter 10 – Role of lipid mediators in diabetic wound healing." In *Wound Healing, Tissue Repair, and Regeneration in Diabetes*, edited by Debasis Bagchi, Amitava Das, and Sashwati Roy, 181–95. Academic Press. https://doi.org/10.1016/B978-0-12-816413-6.00010-1

Wilgus, Traci A., Sashwati Roy, and Jodi C. McDaniel. 2013. "Neutrophils and wound repair: Positive actions and negative reactions." *Advances in Wound Care* 2 (7): 379–88. https://doi.org/10.1089/wound.2012.0383

Wilkinson, Holly N., and Matthew J. Hardman. 2020. "Wound healing: Cellular mechanisms and pathological outcomes." *Open Biology* 10 (9): 200223. https://doi.org/10.1098/rsob.200223

Williams, Paul T. 2020. "Gene-environment interactions due to quantile-specific heritability of triglyceride and VLDL concentrations." *Scientific Reports* 10 (March): 4486. https://doi.org/10.1038/s41598-020-60965-9

Wong, Rachel Si-Yin, Timothy Tan, Alexander Shao-Rong Pang, and Dinesh Kumar Srinivasan. 2025. "The role of cytokines in wound healing: From mechanistic insights to therapeutic applications." *Exploration of Immunology* 5: 1003183.

Wu, Guoyao, Cynthia J. Meininger, Catherine J. McNeal, Fuller W. Bazer, and J. Marc Rhoads. 2021. "Role of l-arginine in nitric oxide synthesis and health in humans." In *Amino Acids in Nutrition and Health: Amino Acids in Gene Expression, Metabolic Regulation, and Exercising Performance*, edited by Guoyao Wu, 167–87. Cham: Springer International Publishing. https://doi.org/10.1007/978-3-030-74180-8_10

Xiaoming, Luo, Zeng Xianmin, Cai. Liangren, and Zheng Xin. 2024. "Effects of celastrol on drug resistance of liver cancer cells through FAK/MEK/ERK signaling pathway." *China Pharmacy* 35 (20): 2477–2481.

Xu, Ming-shi, Xiao-fan Gu, Cong Li, Li-xuan Pan, Zi-xia Zhu, Meng Fan, Yun Zhao, Jian-fang Chen, Xuan Liu, and Xiong-wen Zhang. 2024. "A novel FAK-degrading PROTAC molecule exhibited both anti-tumor activities and efficient MDR reversal effects." *Acta Pharmacologica Sinica* 45 (10): 2174–85.

Yang, Yan, Yushan Yang, Jingsi Jiang, Zongzhou Wu, Jiuyuan Sun, Hui Zhi, ShiYu Chen, Le Kuai, Bin Li, and Haiqing Dong. 2024. "Arginine-nanoenzyme with timely angiogenesis for promoting diabetic wound healing." *ACS Applied Materials & Interfaces* 16 (8): 9640–55. https://doi.org/10.1021/acsami.3c13072

Yap, JiannWen, and Samantha Holloway. 2021. "Citation for: Evidence-based review of the effects of nutritional supplementation for pressure ulcer prevention." *Wiley Online Library*. Accessed April 21, 2025. https://onlinelibrary.wiley.com/action/showCitFormats?doi=10.1111%2Fiwj.13584

Ye, Zhenhong, Siyu Wang, Chunmei Zhang, and Yue Zhao. 2020. "Coordinated modulation of energy metabolism and inflammation by branched-chain amino acids and fatty acids." *Frontiers in Endocrinology* 11 (September): 617. https://doi.org/10.3389/fendo.2020.00617

Yiğit, Ümmühan, Arzu Özcan İlçe, Eylem Suveren, Cevher Altuğ, Çetin Boran, Serap Kolukısa, and Muhammet Büyükbayram. 2024. "The effect of novel boron ester derivatives on wound healing." *Sağlık Bakım ve Rehabilitasyon Dergisi* 3 (3): 33–46.

Yoo, Hee Chan, Ya Chun Yu, Yulseung Sung, and Jung Min Han. 2020. "Glutamine reliance in cell metabolism." *Experimental & Molecular Medicine* 52 (9): 1496–1516. https://doi.org/10.1038/s12276-020-00504-8

Younis, Ibby. 2020. "Role of oxygen in wound healing." *Journal of Wound Care* 29 (Sup5b): S4–10. https://doi.org/10.12968/jowc.2020.29.Sup5b.S4

Zhang, Hongyan, Wei-Xiong Li, Shipeng Tang, Yu Chen, Ling-Min Lan, Shanshan Li, Mingxin Xiong, et al. 2023a. "A boron-based probe driven theranostic hydrogel dressing for visual monitoring and matching chronic wound healing." *Advanced Functional Materials* 33 (51): 2305580. https://doi.org/10.1002/adfm.202305580

Zhang, Zhuangzhuang, Yajie Zhang, Yuanshan Liu, Penghui Zheng, Tong Gao, Bingqing Luo, Xingzhu Liu, Fanshu Ma, Jine Wang, and Renjun Pei. 2023b. "Water-retaining and separable adhesive hydrogel dressing for wound healing without secondary damage." *Science China Materials* 66 (8): 3337–46. https://doi.org/10.1007/s40843-022-2466-7

Zhao, Di, Yao-xing Dou, Ling-feng Zeng, Yan-hong Han, Fang-zheng Lin, Nan-jun Xu, Jun Liu, and Yu-ping Zeng. 2024. "The effects of extracellular matrix-degrading enzymes polymorphisms on intervertebral disc degeneration." *JOR Spine* 7 (4): e70012.

Zhaot, Qingsong, Xu Han, and Zheqi Zhang. 2025. "Induces angiogenesis to promote wound healing in rats with diabetic foot ulcers by activating TGF-B1/smad3 signaling pathway." *A Year in Review: Discussions in Molecular and Structural Endocrinology*, 36.

Zheng, ShengXing, JingXiang Yang, Xin Hu, Ming Li, Qian Wang, Rachel C. A. Dancer, Dhruv Parekh, Fang Gao-Smith, David R. Thickett, and ShengWei Jin. 2020. "Vitamin D attenuates lung injury via stimulating epithelial repair, reducing epithelial cell apoptosis and inhibits TGF-β induced epithelial to mesenchymal transition." *Biochemical Pharmacology* 177 (July): 113955. https://doi.org/10.1016/j.bcp.2020.113955

Zhu, Weiwen, Rui Wang, Zhijian Yang, Xuming Luo, Baoxi Yu, Jian Zhang, and Ming Fu. 2023. "GC-MS based comparative metabolomic analysis of human cancellous bone reveals the critical role of linoleic acid metabolism in femur head necrosis." *Metabolomics* 19 (10): 86. https://doi.org/10.1007/s11306-023-02053-3

11 Unveiling Artificial Intelligence for Prediction and Assessment of Wound Healing

Aarif Ali[1*], *Nahal Zahra*[2], *Javeed Iqbal Ahmad Bhat*[1,2], *Humayun Azad*[2], *Shameem Ahmad Shameem*[2], *and Faisal Amin*

[1] Division of Basic Sciences and Humanities, Faculty of Horticulture, Sher-e-Kashmir University of Agricultural Sciences and Technology of Kashmir, Shalimar, Srinagar, Jammu &d Kashmir, India

[2] Division of Environmental Sciences, Sher-e-Kashmir University of Agricultural Sciences and Technology of Kashmir, Shalimar, Srinagar, Jammu & Kashmir, India.

11.1 OVERVIEW OF ARTIFICIAL INTELLIGENCE (AI)

Artificial intelligence (AI) represents a domain within computer science that aims to create systems capable of undertaking tasks that generally necessitate human intelligence. These tasks involve learning, reasoning, problem-solving, perception, understanding language, and making decisions (Russell and Norvig, 2021). AI systems can be classified into narrow AI (which is designed for particular tasks) and general AI (which exhibits more extensive human-like cognitive functions). Recent developments in AI have been driven by the accessibility of large datasets, enhanced computing capabilities, and advanced algorithms, particularly deep learning. This progress has resulted in significant advancements in fields such as speech recognition, self-driving vehicles, medical diagnostics, and recommendation systems (LeCun et al., 2015). However, alongside its potential, AI presents ethical and societal issues concerning privacy, bias, accountability, and job displacement. The responsible development of AI necessitates a commitment to transparency, fairness, and alignment with human values (Floridi et al., 2018).

Wound healing is particularly well-suited for AI applications due to the diverse dataset it produces, which is ideal for analysis. Each wound provides visual indicators such as its size, texture, and color in images along with clinical data like glucose levels, signs of infection, and patient history (Anisuzzaman et al., 2022). In the past, clinicians have manually compiled this information, using rulers for measurements, visually assessing tissue condition, and reviewing charts, a method that is susceptible to mistakes and fatigue (Wang et al., 2018). Research indicates that variability among observers in wound measurement can be as high as 40%, and minor changes such as the onset of infection are frequently overlooked until complications develop (Sheehan et al., 2003). AI streamlines this process, enabling the analysis of hundreds of wounds in mere moments, identifying trends across different populations, and providing personalized insights.

* Corresponding author: buttaarif31@gmail.com

DOI: 10.1201/9781003644958-11

For instance, a deep learning model could integrate an image of a wound with a patient's glycated hemoglobin (A1c) history, forecasting infection risk with an accuracy of 92%, significantly surpassing human predictions (Anisuzzaman et al., 2022). The advancement of AI in the healthcare sector is largely attributed to the convergence of technologies. By the 2010s, the rapid increase in computing capabilities specifically, graphics processing units (GPUs) that can perform billions of calculations each second combined with big data and cloud computing, rendered AI feasible (LeCun et al., 2015). In the realm of wound care, this has resulted in innovations such as smartphone applications that evaluate wounds in real-time or predictive algorithms that estimate outcomes based on millions of historical cases. Historically, the management of wounds has been slower to evolve compared to other medical specialties cardiology adopted electrocardiograms (ECGs) in the 1930s, and oncology introduced chemotherapy in the 1940s often depending on visual assessments until the emergence of AI (Hess, 2019). Currently, the incorporation of AI into wound healing is rapidly progressing, fueled by the demand for accuracy in a domain where subjectivity has traditionally prevailed.

The strengths of AI are speed, scalability, and adaptability that positions it as a transformative force. A nurse equipped with an AI tool can evaluate a wound with the same precision as a specialist, a hospital can forecast resource requirements based on healing patterns, and a researcher can identify new correlations such as the relationship between wound pH and bacterial load (Mostafalu et al., 2018). However, AI is not a cure-all. Its dependence on high-quality data, computational power, and the trust of clinicians presents challenges that will be discussed later. For now, this summary highlights AI's ability to revolutionize wound care, shifting it from an art to a science and paving the way for its targeted applications in assessment and prediction.

11.2 PILLARS OF ARTIFICIAL INTELLIGENCE

AI is a multidisciplinary domain focused on developing machines that can replicate human intelligence, including abilities like learning, reasoning, perception, and decision-making. The extensive and complex nature of AI is underpinned by several fundamental "pillars", the essential areas and technologies that together form the field. Each pillar signifies a distinct area that provides vital functionality to AI systems. Wound assessment, an essential phase in the healing process, has traditionally depended on manual methods such as rulers, visual inspections, and clinician notes that are prone to inconsistency and subjectivity (Hess, 2019). AI brings forth a range of tools that revolutionize this process, utilizing sophisticated algorithms to improve accuracy, speed, and scalability. Three key AI technologies are particularly prominent in wound assessment: computer vision (CV), machine learning (ML), and natural language processing (NLP). At its essence, AI integrates knowledge from multiple disciplines, including mathematics, logic, linguistics, neuroscience, and cognitive psychology. It comprises subfields such as ML, where systems learn from data patterns; NLP which enables machines to interpret human language; and CV, which focuses on analyzing visual information (Goodfellow et al., 2016). Each technology possesses unique strengths, from image interpretation to data analysis and text insight extraction, collectively transforming the evaluation of wounds. The three pillars of AI are depicted in Figure 11.1.

11.2.1 Machine Learning

ML is a branch of AI dedicated to creating algorithms that enable computers to learn from data and make informed decisions or predictions (Mitchell, 1999). Instead of being directly programmed for specific tasks, ML systems are trained using data and employ statistical methods to recognize patterns and progressively enhance their performance. ML stands as the foremost pillar of AI, concentrating on algorithms that empower systems to learn from data and enhance their capabilities over time without the need for explicit programming (Shandilya et al., 2024). ML facilitates predictive

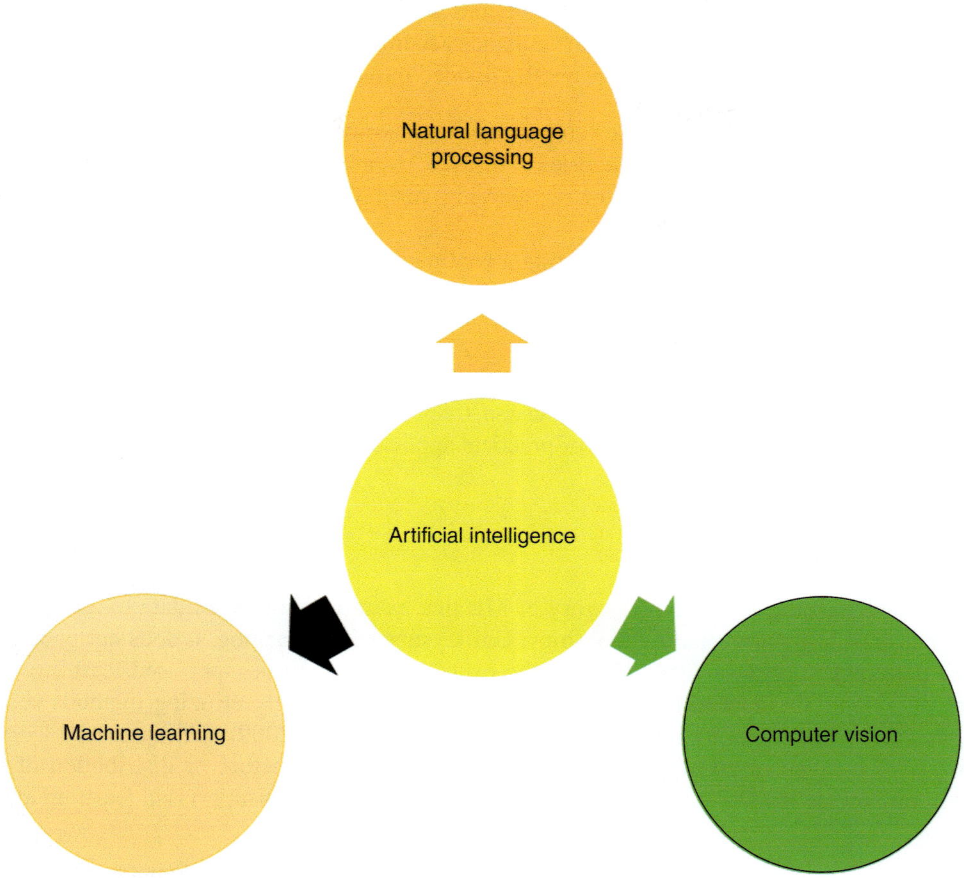

FIGURE 11.1 Pillars of artificial intelligence. The figure depicts the three pillars of AI.

analytics, pattern recognition, and adaptive behaviors. The domain is generally divided into four main categories as shown in Figure 11.2.

11.2.1.1 Supervised Learning

Supervised learning is a branch of ML in which an algorithm learns from data that is labeled (Shetty et al., 2022). This indicates that for every input in the training dataset, the associated correct output is also given. The objective of the algorithm is to understand the mapping or connection between the input features and the known output labels. During the training process, the supervised learning algorithm examines the labeled data to discover patterns and relationships. It subsequently applies this acquired knowledge to make predictions or classifications on new, unseen data. The model's

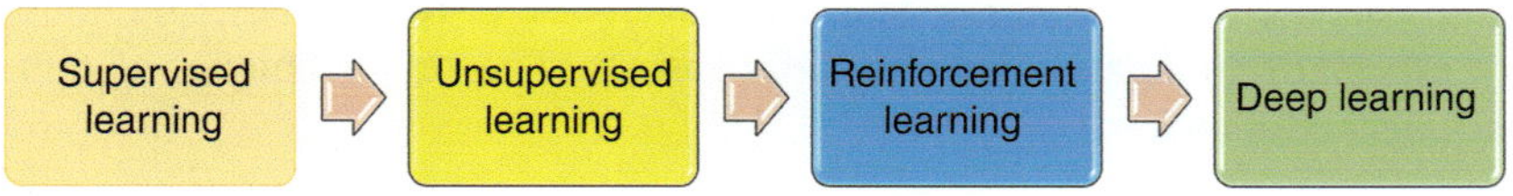

FIGURE 11.2 Categories of machine learning.

performance is assessed by comparing its predictions with the actual known outputs, and the model is continuously refined to reduce errors and enhance accuracy.

Supervised learning challenges are generally divided into two primary categories:

1. *Classification*: This pertains to forecasting a categorical result, where the model allocates input data to establish categories or classes. Instances include determining whether emails are spam or not spam, or recognizing various types of objects within images (Mitchell, 1997).
2. *Regression*: This pertains to forecasting a continuous numerical result. Instances include estimating house prices based on different characteristics or projecting sales revenue (James et al., 2013).

Algorithms frequently utilized in supervised learning include linear regression, logistic regression, support vector machines (SVMs), decision trees, and neural networks (NNs). Supervised learning is extensively used in a variety of practical applications, such as image recognition, fraud detection, NLP, and medical diagnosis.

11.2.1.2 Unsupervised Learning

Unsupervised learning refers to a category of ML that examines unlabeled data to identify concealed patterns, structures, and relationships. Unlike supervised learning, it does not necessitate labeled data or any prior understanding of the expected output. In unsupervised learning, algorithms investigate the data to uncover intrinsic structures, frequently utilizing methods such as clustering and dimensionality reduction. According Hastie et al. (2009) "Unsupervised learning encompasses a range of problems aimed at understanding the structure or distribution of data sets without any defined labels." The main approaches involved in unsupervised learning are the following.

11.2.1.2.1 Clustering

Clustering is a technique in unsupervised ML that groups data points into clusters, ensuring that points within the same cluster are more alike than those in different clusters (Maimon and Rokach, 2005). Similarity is typically measured using distance metrics or statistical characteristics. The algorithms commonly used in this method include the following:

- *K-Means Clustering*: This technique divides data into k clusters while reducing the variance within each cluster.
- *Hierarchical Clustering*: This method constructs a cluster tree through either agglomerative or divisive strategies.
- *Density-Based Spatial Clustering (DBSCAN)*: This algorithm identifies closely packed points and distinguishes outliers.

11.2.1.2.2 Dimensionality Reduction

Dimensionality reduction refers to an ML approach that focuses on minimizing the number of input variables (features) in a dataset while keeping as much relevant information intact (McInnes et al., 2018). Its main purpose is to streamline data, eliminate noise or redundancy, and boost algorithm performance especially in datasets with high dimensions. The key techniques of dimensionality reduction include the following:

- *Principal Component Analysis (PCA)*: PCA is a linear transformation method that projects data onto new axes (principal components) which capture the highest variance in the dataset (Jolliffe, 2011). The initial principal components typically account for most of the variance, enabling dimensionality reduction with minimal loss of information.

- *t-Distributed Stochastic Neighbor Embedding (t-SNE)*: t-SNE is a non-linear method that maintains local relationships in high-dimensional data by embedding it into a lower-dimensional space, typically 2D or 3D. This technique is particularly beneficial for visualizing complex datasets.
- *Linear Discriminant Analysis (LDA)*: LDA is a supervised technique for dimensionality reduction that enhances the distinction between various classes. In contrast to PCA, it utilizes class labels and emphasizes the direction that most effectively separates distinct classes.
- *Autoencoders*: Autoencoders are NNs designed to recreate their inputs. The bottleneck layer, which is the lowest dimensional hidden layer, captures a compressed version of the input data.

11.2.1.3 Reinforcement Learning

Reinforcement learning (RL) is a type of ML where an agent learns to make decisions through interaction with an environment to maximize a total reward (Sutton and Barto, 1998). This process resembles trial-and-error learning, akin to human learning, where the agent gets feedback (either rewards or penalties) based on its actions and modifies its behavior to enhance future results.

RL algorithms can be generally divided into two categories: model-free and model-based (Moerland et al., 2023; Szepesvari, 2022). Model-free algorithms do not create an explicit model of the environment, or more precisely, the Markov Decision Processes (MDP). They resemble trial-and-error methods that conduct experiments within the environment using actions and directly derive the optimal policy from these experiments. Model-free algorithms can be classified as either value-based or policy-based (Szepesvari, 2022). Value-based algorithms view the optimal policy as a direct outcome of accurately estimating the value function for each state (Polydoros and Nalpantidis, 2017). By utilizing a recursive relationship defined by the Bellman equation, the agent engages with the environment to sample trajectories of states and rewards. With a sufficient number of trajectories, the value function of the MDP can be approximated. Once the value function is established, finding the optimal policy becomes a straightforward task of acting greedily concerning the value function at every state in the process. Notable examples of value-based algorithms include SARSA and Q-learning. Conversely, policy-based algorithms directly estimate the optimal policy without the need to model the value function.

11.2.1.4 Deep Learning

Deep learning is a specific area within ML that makes use of artificial neural networks (ANNs) featuring multiple layers to examine data and uncover intricate patterns (Goodfellow et al., 2016). These networks, inspired by the human brain's structure, empower systems to identify complex details and forecast outcomes from large, often unstructured, datasets. Deep learning is based on ANNs, which consist of interconnected nodes (neurons) arranged in layers (Shukla and Choudhary, 2022). The term "deep" in deep learning signifies the numerous layers present in the NN, enabling the extraction of progressively complex features from data. Deep learning algorithms have the capability to automatically identify relevant features from data, minimizing the necessity for manual feature engineering, a practice often seen in traditional ML (Zhang et al., 2018). Just as humans learn from examples, deep learning models acquire knowledge from extensive datasets, enhancing their performance over time through iterative training. Deep learning has been widely applied across various domains, such as image recognition, NLP, speech recognition, and beyond.

11.2.2 Natural Language Processing

NLP is a field within AI dedicated to allowing computers to comprehend, interpret, generate, and interact with human languages (Jurafsky and Martin, 2020). It exists at the crossroads of linguistics, computer science, and AI and is essential for facilitating communication between machines and humans through natural language (Young et al., 2018). NLP enables computers to comprehend both spoken and written language, facilitating user interaction with machines through natural language, such as voice commands or text input (Silva and Gomez, 2025). It assists computers in analyzing text and speech to grasp the meaning, sentiment, and intent behind the language utilized. Additionally, NLP can generate text and speech: it allows machines to produce human-like text and speech, thereby enabling them to respond to users in a natural and coherent manner. NLP systems depend on established rules and patterns for language analysis. ML algorithms, such as deep learning, are employed to train models using extensive datasets of text and speech, allowing them to recognize patterns and relationships within language (Pouyanfar et al., 2018). Numerous NLP systems integrate both rule-based and ML approaches. Essentially, NLP is revolutionizing our interaction with technology, enhancing its accessibility and intuitiveness by connecting human language with machine comprehension.

NLP drives chatbots and virtual assistants such as Siri and Alexa, allowing them to comprehend and react to user inquiries (Shrivastava et al., 2025). NLP algorithms facilitate language translation, promoting smooth communication between various languages (Mohamed et al., 2024). NLP can evaluate text to ascertain the sentiment or opinion conveyed, which is beneficial for market research and monitoring social media. NLP is utilized in search engines to interpret user queries and deliver pertinent results. NLP can automatically condense extensive text, simplifying the understanding of key points. NLP algorithms are capable of detecting and filtering out unwanted emails and messages (Ismail et al., 2022).

11.2.3 Computer Vision

CV is an interdisciplinary area within AI that allows machines to perceive, interpret, and comprehend visual data from the environment like images or videos much like human vision operates (Szeliski, 2022). It integrates methods from AI, ML, image processing, and pattern recognition to facilitate the automation of visual perception tasks. The key techniques involved in CV are the following.

11.2.3.1 Image Processing

Image processing encompasses a range of computational methods aimed at improving, analyzing, and retrieving information from images. It serves as a fundamental aspect of CV, frequently acting as a preliminary step for more sophisticated AI and ML models. The objectives of image processing include enhancement, restoration, compression, analysis, segmentation, and transformation. Moreover, the key techniques in image processing are filtering, edge detection, histogram equalization, thresholding, morphological operations, color space conversion geometric transformation, and image registration (Gonzalez, 2009; Jain, 1989; Solomon and Breckon, 2011).

11.2.3.2 Convolutional Neural Networks (CNNs)

Convolutional neural networks (CNNs or ConvNets) represent a category of ANNs that are mainly utilized for the analysis of visual imagery, although their use cases also encompass other fields such as audio and text. They are particularly adept at autonomously learning features from input data, especially images, via convolution and pooling processes. This multi-layered architecture enables CNNs to recognize intricate patterns and objects within data, rendering them effective instruments for tasks such as image recognition, NLP, object detection, and beyond (Bacanin Dzakula, 2019; Koushik, 2016). It is more efficient than a standard network because it can autonomously recognize essential components from the input without requiring human involvement. A CNN generally consists of four distinct types of layers: convolutional; pooling; activation function; fully connected.

The convolutional layer plays a vital role in the overall architecture of CNNs. It consists of a collection of filters, also known as kernels, that are applied to the data prior to its utilization. Each kernel's dimensions include width, height, and weight is utilized to extract features from the input data. Initially, the weights in the kernel are assigned randomly, but they gradually become more refined through the training data. In essence, the feature map is created by merging the input image, represented by N-dimensional metrics, with these filters.

The pooling layer, commonly referred to as the down-sampling layer, serves to reduce the dimensionality of feature maps while preserving the most significant information. In this layer, a filter performs the pooling operation on the input data by sliding across it (max, min, avg). According to existing literature, maximum pooling is the most widely used method. The primary function of pooling, which helps to simplify the complexity of subsequent layers, is down-sampling. In the context of image processing, this can be likened to lowering the resolution. The number of filters remains unchanged during pooling. Max-pooling is among the most frequently employed pooling techniques. The image is segmented into rectangular subregions, and only the highest value found within each subregion is returned. A common size for max-pooling is 2×2.

The non-linearity layer comes after the convolution layer. This non-linearity enables the modification or cessation of the generated output (Alzubaidi et al., 2021; Du and Swamy, 2013; Zhang et al., 2019). It serves to either limit or excessively amplify the output. Each activation function across various NN types plays a crucial role in mapping inputs to outputs. The input value is derived from the weighted sum of the neuron's input along with its bias, if applicable (Alzubaidi et al., 2021; Du and Swamy, 2013; Zhang et al., 2019). This means that the activation function decides whether a neuron should activate in response to a specific input by producing the corresponding output. In the architecture of CNNs, non-linear activation layers are applied following all weight-bearing layers, which are also referred to as learnable layers, including fully connected layers and convolutional layers (Alzubaidi et al., 2021; Du and Swamy, 2013; Zhang et al., 2019).

Neurons are arranged into groups in the fully connected layer that resemble those observed in traditional NNs. Any node in a completely connected layer is thus directly linked to every node in the layers above and below it. Each node in the most recent frames of the pooling layer is connected as a vector from the fully connected layer to the top layer. These are the most commonly employed CNN parameters within these layers; however, they demand extensive training time (Alzubaidi et al., 2021; Du and Swamy, 2013; Zhang et al., 2019).

11.2.3.3 Transfer Learning

Transfer learning is a technique in ML that involves reusing a model created for one task as the foundation for a model on another, related task (Pan and Yang, 2009; Yosinski et al., 2014). Instead of building a model from the ground up, transfer learning leverages insights obtained from a large source dataset (such as ImageNet) and applies them to a smaller target dataset (Pan and Yang, 2009; Yosinski et al., 2014). Training deep CNNs from the beginning is costly and requires a lot of data. Numerous vision datasets (for instance, medical and satellite) have a scarcity of labeled data. The lower layers of CNNs capture universal visual features such as edges, corners, and textures. The widely used pre-trained CNN models include ResNet, EfficientNet, VGG, Inception, and MobileNet (Pan and Yang, 2009; Yosinski et al., 2014).

11.2.3.4 Object Detection

Object detection refers to a CV task that entails identifying and locating objects within images or videos. It seeks to answer: What objects are present? (classification) and where are they situated? (bounding box localization). Unlike image classification, which provides one label for each image, object detection deals with multiple objects of various types and positions. Object detection methods can generally be divided into two categories: traditional techniques and those based on deep learning. Traditional techniques, such as Viola–Jones and Histogram of Oriented Gradients (HOG), were widely used prior to the emergence of deep learning (Dalal and Triggs, 2005; Viola and Jones, 2001). The advent of deep learning has brought about remarkable progress, with models like YOLO, SSD, and Faster R-CNN taking the lead in this domain (Carion et al., 2020; Ren et al., 2015). These models provide enhanced accuracy and speed, particularly in real-time scenarios.

11.2.3.5 Image Segmentation

Image segmentation is a technique in CV that divides a digital image into various segments or pixel regions (Long et al., 2015; Ronneberger et al., 2015). These segments correspond to different objects or areas within the image, facilitating the analysis and comprehension of the image's content. In essence, it involves clustering pixels that share similar attributes (such as color, texture, or intensity) to streamline the image and emphasize key features (Long et al., 2015; Ronneberger et al., 2015). Well-known image segmentation models consist of U-Net, Fully Convolutional Networks (FCN), SegNet, Mask R-CNN, and the Segment Anything Model (SAM). These models are extensively utilized for tasks such as semantic segmentation, instance segmentation, and panoptic segmentation.

11.2.3.6 Vision Transformers (ViTs)

Vision transformers (ViTs) represent a groundbreaking deep learning architecture that modifies the transformer model, which was initially effective in NLP, for the purpose of image recognition (Dosovitskiy et al., 2020; Touvron et al., 2021). In contrast to conventional CNNs that depend on convolutions to identify local features, ViTs employ self-attention mechanisms to analyze relationships among image patches (Dosovitskiy et al., 2020; Touvron et al., 2021). This enables them to achieve a more thorough comprehension of images and their spatial relationships. ViTs mark a notable advancement in CV, providing a robust alternative to traditional CNNs, particularly when handling extensive datasets and necessitating a profound understanding of global image context (Dosovitskiy et al., 2020; Touvron et al., 2021). Although challenges such as data requirements and computational expenses persist, ongoing research is dedicated to enhancing their efficiency, adaptability, and interpretability. The outlook for ViTs in CV is bright, with anticipated advancements that will further broaden their capabilities and applications.

11.2.3.7 Generative Adversarial Networks (GANs)

Generative adversarial networks (GANs) represent a category of ML frameworks, specifically a deep learning architecture, designed to create new data that mimics a training dataset (Goodfellow et al., 2020; Ledig et al., 2017). GANs employ two NNs, known as a generator and a discriminator, which compete against one another in a "game" to generate increasingly realistic synthetic data (Goodfellow et al., 2020; Ledig et al., 2017). The generator takes random noise as input and attempts to create data (such as images, text, or audio) that resembles the training data. The discriminator, on the other hand, evaluates both the authentic data from the training set and the data produced by the generator, striving to differentiate between the two. These two networks are trained simultaneously in an adversarial fashion (Goodfellow et al., 2020; Ledig et al., 2017). The generator's goal is to deceive the discriminator, while the discriminator's objective is to accurately identify the counterfeit data. This "game" persists until the generator creates data that the discriminator cannot differentiate from real data. One prevalent application of GANs is in image generation. For instance, a GAN trained on a dataset of cat images can learn the features of cat images and subsequently generate new, realistic cat images that were not included in the original training dataset (Goodfellow et al., 2020; Ledig et al., 2017).

11.2.3.8 Optical Flow and Tracking

Optical flow and object tracking are interrelated methods in CV that examine motion within video sequences (Lucas and Kanade, 1981; Wojke et al., 2017). Optical flow assesses the perceived movement of objects or patterns across successive frames, whereas tracking utilizes this data to monitor the movement of particular objects over time (Lucas and Kanade, 1981; Wojke et al., 2017). In essence, optical flow supplies the essential information (pixel-level motion vectors) that tracking algorithms can leverage to establish object trajectories (Lucas and Kanade, 1981; Wojke et al., 2017). The common algorithms include Lucas–Kanade, Farneback, and DeepSORT.

11.3 USE OF AI TOOLS (MACHINE LEARNING (ML), NATURAL LANGUAGE PROCESSING (NLP), AND COMPUTER VISION (CV)) FOR WOUND ASSESSMENT

Wound assessment plays a vital role in clinical wound management, encompassing the evaluation of wound type, size, depth, tissue composition, and healing progress. Traditionally, this process is manual, subjective, and time-consuming. In clinical practice, skin wounds are commonly encountered (Subramaniam et al., 2021), which requires clinicians to base their treatment decisions and evaluations of wound healing progress on several factors, such as wound size, classification, and tissue composition (Masson-Meyers et al., 2020). These evaluations primarily rely on the subjective visual assessments made by physicians and clinical staff (Brem et al., 2006; Li et al., 2021), indicating a potential for the integration of emerging technologies (Sen et al., 2016). Inaccurate assessments can lead to severe consequences, including inappropriate dressing choices, missed non-healing wounds, and delays in specialist referrals (Wang et al., 2020). As the number of surgeries involving high-risk and multimorbid patients rises, effective perioperative wound management becomes essential. AI is valuable not only for surgical wounds but also plays a crucial role in assessing healing periods for chronic and burn wounds. AI is being utilized to enhance wound assessment by examining images and other data, resulting in more accurate, efficient, and personalized care. AI algorithms are capable of analyzing wound images to automatically detect wound boundaries, measure dimensions, and evaluate tissue types, which may lead to quicker and more precise diagnoses, optimized treatment strategies, and better healing outcomes. The use of these AI tools in wound assessment, measurement, classification, segmentation, and healing are described below.

11.3.1 Wound Tissue Classification

Accurate classification of wound tissue types such as granulation, necrotic (eschar/slough), and epithelial tissue is crucial in determining wound status and guiding clinical treatment decisions. Objective and consistent tissue classification enables appropriate interventions such as debridement or dressing selection and supports monitoring healing progress and identifying non-healing ulcers.

11.3.1.1 Clinical Significance of Tissue Types

11.3.1.1.1 Granulation Tissue

Characterized by new connective tissue and microscopic blood vessel formation, this tissue fills wound beds and appears moist and pink to beefy red depending on depth (Chairat et al., 2023).

11.3.1.1.2 Epithelial Tissue

Represents migrating keratinocytes covering the wound surface, visible as deep pink at wound edges advancing centrally (Chairat et al., 2023).

11.3.1.1.3 Necrotic Tissue

Comprises dead tissue needing removal to prevent infection. Eschar is the dry, dark, tough form, while slough is wet, soft, and variably colored (yellow, green, tan, or brown) (Chairat et al., 2023).

11.3.1.2 AI-Based Classification Techniques

Wound tissue classification tasks aim to identify various tissue types within wounds (e.g., necrotic, granulation) by employing ML and DL techniques, including CNNs and SVMs.

11.3.1.2.1 *Traditional Feature-Based ML Approaches*

Mukherjee and colleagues developed a computer-assisted tissue classification framework for chronic wounds by transforming RGB images into HSI color space, segmenting wound areas via fuzzy divergence thresholding and extracting color/textural features. A SVM with a third-order polynomial kernel achieved accuracies of 86.94% (granulation), 90.47% (slough), and 75.53% (necrosis), with an overall accuracy of 87.61% and Cohen's κ of 0.793 (Mukherjee et al., 2014).

Veredas and colleagues established an ML framework aimed at classifying wound tissue types such as granulation, slough, and necrosis from digital images (Veredas et al., 2015). The essential components of their strategy were the following:

- *Segmentation Method*: Used k-means clustering to isolate wound regions from the surrounding tissue,
- *Feature Selection*: Adopted a wrapper approach with recursive feature elimination, which allowed the model to emphasize the most significant predictors and decrease dimensionality without sacrificing classifier performance, and
- *Classification Algorithms Assessed*: SVM, NNs, and random forests (RF).

The best performance was achieved with SVM (around 88%), closely succeeded by RF (87.37%), with the NN (81.87%) coming in slightly lower underscoring the power of classical ML techniques in the realm of medical image analysis (Veredas et al., 2015).

Similarly, Chakraborty and his colleagues undertook a series of studies between 2016 and 2019 that focused on analyzing chronic wound tissue using both traditional image processing and ML methods (Chakraborty, 2017, 2019; Chakraborty et al., 2016). The insights and primary findings from these studies were as follows. Across the studies, segmentation accuracy consistently hovered around 99%, demonstrating the effectiveness of their preprocessing pipelines (Chakraborty, 2017, 2019; Chakraborty et al., 2016). The evolution from LDA (2016 and 2017) to more sophisticated ensemble methods like RF (2019) signifies a move towards more robust classifiers. The transition from smaller image sets (29–60 images) to a larger dataset (153 images) likely played a role in enhancing generalizability and accuracy (Chakraborty, 2017, 2019; Chakraborty et al., 2016). The highest classification accuracy of about 98% was achieved in 2017 through particle swarm optimization (PSO) enhanced processing (Chakraborty, 2017, 2019; Chakraborty et al., 2016). By 2019, RF provided a strong balance of performance (93.75%) and scalability across a larger image set (Chakraborty, 2017, 2019; Chakraborty et al., 2016).

Moreover, Nizam et al. (2018) concentrated solely on the task of differentiating tissue within the wound by bypassing segmentation, assuming that the wound boundaries had been predefined. The application of fuzzy C-means facilitated a probabilistic and nuanced method for classifying tissue regions, acknowledging that tissue boundaries in wound images are frequently ambiguous. The method's robustness is underscored by achieving nearly 90% accuracy for granulation and eschar, whereas the slightly lower accuracy for slough and epithelial tissue may indicate a visual similarity among specific tissue types or a class imbalance.

11.3.1.2.2 *Deep Learning-Based Methods*

DL has become a significant asset in automating and improving the classification of wound tissue by utilizing extensive datasets and directly learning distinguishing features from wound images. In the realm of wound tissue classification, DL models are capable of differentiating between various tissue types, including granulation tissue, necrotic (black) tissue, slough (yellow) tissue, epithelial tissue, infected or inflamed tissue. These classifications play a crucial role in tracking the healing process and informing clinical decisions.

11.3.1.2.2.1 EfficientNetB0 Encoder–Decoder Models EfficientNetB0, the foundational model of the EfficientNet series, is an exceptionally efficient CNN that strikes a remarkable balance between accuracy and computational efficiency. When employed as the encoder in an encoder–decoder framework, EfficientNetB0 captures rich hierarchical features from wound images, rendering it particularly well-suited for semantic segmentation tasks, including pixel-level tissue classification. The encoder–decoder framework is made up of two primary components:

- *Encoder: EfficientNetB0*: EfficientNetB0 employs compound scaling to uniformly adjust depth, width, and resolution. It consists of MBConv blocks (mobile inverted bottleneck convolutions) integrated with squeeze-and-excitation (SE) modules that assist the model in concentrating on significant features. In the context of wound tissue classification, the encoder captures textures, colors, and patterns specific to tissue at various scales.
- *Decoder: Custom or Predefined Segmentation Decoder*: The decoder is responsible for reconstructing the high-resolution spatial details that are lost during the downsampling process. It generally features upsampling layers, skip connections (such as those found in U-Net), and convolutional blocks to enhance the output. The result is a segmentation mask where each pixel is designated as a particular tissue type (for instance, granulation, slough, necrotic).

A model utilizing EfficientNetB0 as the encoder and a U-Net-like decoder was employed to categorize chronic wound tissues into four distinct classes, achieving a mean Intersection over Union (mIoU) exceeding 75%, thus surpassing conventional CNN-based models. A study applied this architecture to segment four tissue types (epithelial, granulation, slough, eschar) across 17,000 images (Reifs Jiménez et al., 2025). They reported a wound segmentation IoU of ~0.8644 and tissue segmentation IoU of ~0.7192. F1-scores were strong for slough (0.731) and eschar (0.802) but notably weaker for epithelial tissue (0.253) (Reifs Jiménez et al., 2025). A study employed EfficientNetB0 as the encoder paired with a Feature Pyramid Network (FPN)-based decoder to segment organs of the gastrointestinal tract from MRI scans (Scharp et al., 2024). The model achieved a Dice coefficient of 0.8975 and a Jaccard index of 0.8832, with a training time of roughly 2 hours and 25 minutes, outperforming both MobileNetV2 and ResNet50 in accuracy and speed (Scharp et al., 2024). Researchers investigated EfficientNet models B0–B7 as foundational architectures for U-Net in kidney tumor CT segmentation tasks (Abdelrahman and Viriri, 2023). The baseline EfficientNetB0 model showed impressive results, achieving mIoU and Dice scores of approximately 0.976 (IoU) and 0.987 (Dice) (Abdelrahman and Viriri, 2023). Enhanced versions such as B4 and B7 yielded even better segmentation outcomes, suggesting that deeper backbones contribute to improved performance (Abdelrahman and Viriri, 2023).

11.3.1.2.2.2 3D CNN-DeepMedic DeepMedic is an innovative architecture based on 3D convolutional neural networks (3D CNNs) that is specifically tailored for the segmentation of volumetric medical images, especially in the field of neuroimaging. Unlike conventional 2D CNNs, DeepMedic analyzes 3D image patches, allowing it to capture spatial context in all three dimensions (height, width, depth). This capability makes it particularly adept at detecting small or intricate structures, such as brain lesions, tumors, or other tissue irregularities in 3D medical imaging datasets. DeepMedic is extensively utilized for brain lesion segmentation (such as multiple sclerosis, stroke, and gliomas), liver tumor segmentation, 3D MRI prostate segmentation, cardiac segmentation, and 3D wound tissue segmentation.

A study conducted in 2024 compared DeepMedic with nnU-Net utilizing pediatric multi-institutional MRI data from 339 patients (Vossough et al., 2024). DeepMedic recorded mean Dice scores of 0.82 for the whole tumor, 0.66 for the enhancing tumor, and a notably lower score of 0.19 for peritumoral edema (Vossough et al., 2024). In contrast, nnU-Net demonstrated superior performance in nearly all subregions, achieving a Dice score of approximately 0.90 for the whole tumor (Vossough et al., 2024). The discussion emphasizes the multiscale architecture of DeepMedic, while also acknowledging its shortcomings in capturing the finer details of subregions when compared to U-Net-based models (Vossough et al., 2024). A study developed a multi-path 3D CNN for pressure ulcer tissue classification, achieving an area under the curve (AUC) of approximately 95%, a Dice similarity coefficient of 92%, and a percentage area difference (PAD) of around 13% (García-Zapirain et al., 2018). In the paired imaging of Computed Tomography Angiography-Digital Subtraction Angiography (CTA-DSA), DeepMedic demonstrated impressive segmentation accuracy for cerebral vessels and aneurysms with a Dice coefficient of approximately 0.868 and a connectivity-area-length (CAL) of around 0.971, surpassing expert raters in alignment with the ground truth (Patel et al., 2023). Additionally, another study focused on CTA-based aneurysm detection indicated a sensitivity of 92.3% and a positive predictive value (PPV) of 100%. However, the sensitivity decreased to 66.7% for aneurysms measuring 3 mm or less, while medium and large aneurysms were identified with a sensitivity of 100% (Liu et al., 2023). A study published on ArXiv in 2022 utilized DeepMedic for liver segmentation in abdominal CT scans, resulting in impressive metrics: 94.4% accuracy, 94.6% sensitivity, 91.9% specificity, and a Dice score of 93.1% (Humady et al., 2022). Additionally, a neuroscience research project that compared a new attention-guided 3D model with DeepMedic for segmenting amygdala subnuclei found that the new model outperformed DeepMedic in both Dice and average symmetric surface distance (ASSD), especially in the cortical–superficial subregions (Liu et al., 2020).

11.3.2 Wound Measurement

Precise wound measurement is foundational in clinical tracking, informing treatment plans, and evaluating healing trajectories. It enables quantification of temporal changes in size, depth, and overall healing rate which are key indicators in guiding therapeutic interventions.

11.3.2.1 Traditional Measurement Techniques

Conventional methods for measuring wounds play a crucial role in clinical settings, allowing for the monitoring of healing progress and assessment of treatment efficacy. Prior to the introduction of digital and 3D imaging technologies, various manual or semi-manual techniques were commonly employed.

11.3.2.1.1 Ruler-based (Length × Width)

The simplest, most accessible method involves measuring the greatest wound length and perpendicular width. However, this method presumes a rectangular shape and often overestimates wound area by roughly 41% particularly in irregular or large wounds and errors were found ranging between 29% and 43% in various clinical scenarios (Dumesnil et al., 2015; Jorgensen et al., 2016). Even with enhancements like elliptical modeling, accuracy diminishes as wound size increases (Jorgensen et al., 2016; Khoo and Jansen, 2016).

11.3.2.1.2 Manual Planimetry (Tracing)

Involves tracing wound boundaries on transparent film, then transferring onto graph paper to count grid squares. While more accurate for irregular shapes, it is time-consuming, prone to contamination, and less reliable near skin folds or on curved surfaces (Khoo and Jansen, 2016; Morris et al., 2018).

11.3.2.1.3 Digital Planimetry

Incorporates non-contact tracing onto tablets or imaging systems that calculate area digitally. Studies report excellent intraclass correlation (ICC) for wounds under 10 cm2 (ICC ≈ 0.95) but reduced reliability for larger wounds (ICC ≈ 0.70) (Khoo and Jansen, 2016; Jørgensen et al., 2016). Digital methods also minimize variability and speed up measurement relative to contact methods (Chang et al., 2011; Khoo and Jansen, 2016). Systems like Visitrak® provide fast, reliable measurements with correlation near 0.99, though they still involve contact with the wound (Chang et al., 2011).

11.3.2.2 AI-Enhanced and Imaging-Based Measurement

It involves the following approaches.

11.3.2.2.1 AI-Based Segmentation Frameworks

The automatic skin ulcer region assessment (ASURA) system utilizes a deep encoder–decoder model to detect wounds and a scale (e.g., a ruler) present in the image to estimate pixel density. It achieves a Dice score over 90% for segmentation accuracy and a 14% relative error in measuring wound area (in cm^2) when semi-automated (Chino et al., 2020).

11.3.2.2.2 Smartphone Application Validation

The tissue analytics (TA) smartphone app was clinically evaluated in patients with venous leg ulcers. Inter-rater reliability for length, width, and area measurements between TA and trained clinicians was good (0.799–0.919); intra-rater reliability on iOS and android devices was excellent (0.967–0.985) (Fong et al., 2023).

11.3.2.2.3 Deep Learning Image Segmentation

A systematic review underscored AI's strength in wound measurement, noting that DL models like U-Net and MASK R-CNN can achieve Dice coefficients up to ~0.95 for burn wound segmentation. U-Net models trained for diabetic foot ulcers (DFUs) and venous leg ulcers achieved AUC around 0.997, sensitivity of 0.993, and specificity of 0.943 (Liu et al., 2025a).

Table 11.1 summarizes the advantages and disadvantages of the methods.

11.3.3 Wound Segmentation

Segmentation refers to the process of delineating the wound area from surrounding healthy tissue in an image. This step is foundational for subsequent tasks such as wound measurement, tissue classification, and healing prediction. Without accurate segmentation, all downstream analytics risk compounding errors. In clinical wound care, segmentation is often challenging due to complex wound shapes, variable skin pigmentation, inconsistent lighting, and background artifacts like dressings or hair.

11.3.3.1 Traditional and Probability Map Approaches

Earlier computational methods relied heavily on color information and heuristic algorithms. A notable example is the Red, Yellow, Black, White (RYKW) probability map approach, which assigns likelihoods for each pixel belonging to specific wound or tissue categories based on color models (Fauzi et al., 2015). Segmentation is then achieved using optimal thresholding or region growing. This technique achieved an average segmentation accuracy of 75.1% across 80 wound images, with human inter-reader agreement ranging from 67% to 84% (Fauzi et al., 2015). While effective in some scenarios, it struggles with wounds exhibiting atypical colors or in poor lighting conditions.

TABLE 11.1
Comparative Summary of Methods with Their Advantages and Limitations

Method	Advantages	Limitations
Linear (ruler)	Easy, low cost	High overestimation, shape assumptions
Manual planimetry (tracing)	More accurate for irregular wounds	Labor intensive, contamination risk
Digital planimetry	Improved speed and reliability	Still contact-based; camera-angle sensitive
AI-based measurement	High accuracy, scalable, remote-capable	Dataset requirements, image quality dependency

11.3.3.2 Deep Learning-Based Segmentation Involves the Following Approaches

11.3.3.2.1 ASURA Framework

The ASURA system applies an encoder–decoder deep NN architecture to chronic skin ulcer images, integrating automatic detection of a visual scale (e.g., ruler) for pixel-to-centimeter conversion. This dual-task design allows both segmentation and accurate measurement in one pass. The system reports a Dice coefficient above 90% for segmentation accuracy, with only a 14% relative error in calculated wound area (Chino et al., 2020).

11.3.3.2.2 Detect-and-Segment (DS) Model

The detect-and-segment (DS) model is a type of CV architecture that initially identifies objects or areas of interest and subsequently divides them into significant subregions, such as boundaries, tissue types, or components. These models hold particular importance in fields such as medical imaging, autonomous driving, and scene comprehension. The DS method improves segmentation accuracy by first detecting and cropping the wound region before segmentation. This prevents the model from being distracted by irrelevant background features. On a DFU dataset, Matthews' correlation coefficient improved from 0.29 (full-image segmentation) to 0.85 with DS, and performance generalized well to four additional datasets. Remarkably, DS required up to 90% less training data without performance loss (Scebba et al., 2022).

11.3.3.2.3 LiDAR-Assisted Segmentation

Light detection and ranging (LiDAR)-assisted segmentation is a sophisticated imaging method that employs LiDAR to obtain depth data, enabling more accurate and automated 3D segmentation. This technique is particularly beneficial in areas such as medical imaging, autonomous vehicles, robotics, and remote sensing. Liu et al. (2023) combined traditional imaging with LiDAR depth mapping to provide accurate real-world scaling. A U-Net deep learning model achieved a Dice coefficient of 0.8448, significantly outperforming Mask R-CNN (0.5006). When tested in clinical settings, the system measured wound area with a mean relative error of 26.2% (Liu et al., 2023).

11.3.3.2.4 Lightweight MobileNetV2 Pipeline

A MobileNetV2 pipeline that is lightweight is perfect for environments with limited resources, including mobile applications, edge devices, or real-time medical diagnostics, particularly when speed, size, and efficiency are essential. To support mobile and telemedicine workflows, a MobileNetV2-based segmentation pipeline was developed for foot ulcer detection and boundary tracing. This lightweight architecture, trained on over 1000 annotated images, demonstrated performance comparable to heavier CNNs while being optimized for on-device execution.

11.3.3.2.5 Segmentation and Analysis Pipelines

A segmentation and analysis pipeline consists of a systematic series of steps aimed at inputting raw image data, segmenting significant structures (such as tissue, lesions, tumors), post-processing the segmentation outcomes, and analyzing them to derive valuable metrics or assist in decision-making. Some frameworks integrate segmentation with further analysis modules. For example, a deep CNN-based system can segment wounds, predict infection risk, and estimate healing progression in under 5 seconds per image, making it suitable for both point-of-care and high-throughput research environments (Wang et al., 2015).

Table 11.2 summarizes the wound segmentation approaches with their strengths and limitations.

11.3.4 Wound Healing

These techniques forecast the advancement of wound healing or pinpoint potential risks for delayed recovery by considering elements such as tissue condition and wound dimensions. ML

TABLE 11.2
Summary of Wound Segmentation Approaches

Approach	Strengths	Limitations
RYKW probability map	Simple, interpretable, works on standard images	Sensitive to lighting/color variations
ASURA (encoder–decoder)	High Dice score, integrates measurement scaling	Requires ruler or reference object in image
Detect and segment (DS)	Reduces background noise, data efficient	Slightly longer preprocessing pipeline
LiDAR-assisted U-Net	Adds depth info for better scaling accuracy	Requires specialized hardware
MobileNetV2 pipeline	Lightweight, mobile-friendly, telemedicine-ready	May underperform on very complex wound textures
Segmentation and analysis CNN	Combines multiple diagnostic tasks in one pipeline	Dependent on large annotated datasets

algorithms are employed to evaluate the probability of healing or complications, thereby supporting clinical decision-making. Importantly, the assessment of wound healing progress in chronic wounds frequently depends on the predictive scar factor, which is methodologically grounded in weekly evaluations conducted over a span of four consecutive weeks. This methodology is crucial for determining precise healing metrics and for offering comparative parameters that enable AI tools to enhance their learning and predictive abilities. In the domain of deep learning, numerous studies have investigated the use of CNNs for evaluating wounds. A reduced ResNet-18 model was utilized in (Lien et al., 2022) to identify granulation tissue in DFUs, achieving an Intersection over Union (IoU) rate exceeding 0.5 for tissue growth detection. Likewise, a DenseNet CNN framework incorporating patch-based attention was applied in (Liu et al., 2021) to evaluate various wound characteristics, attaining accuracy and F1 scores greater than 0.8. Another study presented the Semi-Supervised PMG EfficientNet architecture, which enhanced wound assessment accuracy to 90% by enriching the WoundNet dataset through a Progressive Multi-Granularity mechanism (Liu et al., 2023). In the realm of self-care, a smartphone application enabled patients to evaluate surgical wounds using different classifiers, including CART and Naive Bayes (NB), achieving over 90% accuracy in assessing wound conditions (Chen et al., 2018). In a different study, smartphone and tablet cameras were utilized to capture DFU images for predicting healing, using ResNet50 for feature extraction and RF and SVM for classification (Kim et al., 2020). The research revealed higher AUROC values (0.734) when integrating all features and demonstrated enhanced performance with handcrafted imaging features (0.760–0.794) compared to clinical features alone (Kim et al., 2020). Additionally, a large-scale AI model, AutoTrace, aimed at wound healing prognosis, was developed by Gupta et al. (2023) based on over 2 million wound evaluations, achieving an IoU score of 0.86 and an 8%–13% improvement over tools such as PUSH and BWAT. While PUSH is tailored for pressure injuries and BWAT is more general, covering venous, diabetic, and arterial ulcers, both have limitations in addressing various wound types (Gupta et al., 2023). In the realm of ML, various models were utilized to forecast wound healing outcomes and associated risk factors. The combination of NB classifiers with sequential wound characteristics notably enhanced the prediction of surgical site infections (SSI) when compared to baseline risk factors (Sanger et al., 2016). Furthermore, ML algorithms, including SVM, NB, and RF, were applied to predict challenging-to-heal DFUs, with the NB model yielding the most favorable results (AUC 0.864) (Wang et al., 2022).

11.4 AI FOR PRECISION WOUND HEALING

AI-driven methods are enabling a shift from subjective, episodic wound care to continuous, quantitative, and personalized wound management. By combining CV for image analysis, ML for prediction and decision support, and NLP to extract clinical context from notes and records, AI systems can (a) automate objective wound assessment, (b) predict individual healing trajectories and complications, and (c) support clinician decision-making and triage in real time (Griffa et al., 2024; Reifs Jiménez et al., 2025).

11.4.1 Machine Learning (ML): Prediction, Risk-Stratification, and Treatment Optimization

ML is essential for prediction, risk stratification, and optimizing treatment in numerous medical fields, such as wound healing, chronic disease management, and personalized medicine. Supervised ML models (RFs, gradient-boosted trees, NNs) trained on clinical features (wound area, depth, tissue composition, comorbidities, lab values, and prior healing rates) estimate likelihood of complete healing within defined timelines and identify wounds likely to stall or deteriorate (e.g., chronic and DFUs). These models have shown good discrimination (AUCs often >0.80) and practical utility for personalized follow-up (Berezo et al., 2022; Spinazzola et al., 2025). Ensemble models (e.g., XGBoost) using routine electronic health record (EHR) data can predict severe outcomes such as major amputation in DFU cohorts and help target early interventions to high-risk patients. Recent work demonstrates this approach yields clinically actionable risk stratification (Gao et al., 2025; Liu et al., 2025b). ML applied to -omics and transcriptomic datasets can identify molecules and pathways associated with poor healing (for instance, novel DFU targets reported via ML-driven analysis), opening pathways for adjunctive therapies (Yu et al., 2024). ML models provide individualized risk scores and can combine heterogeneous inputs (numeric labs, categorical comorbidities, time-series wound area). However, they require quality labeled data, careful handling of missingness, and external validation to avoid overfitting and bias (Berezo et al., 2022; Spinazzola et al., 2025).

11.4.2 Computer Vision (CV): Automated, Objective Wound Assessment

CV significantly transforms the automated and objective assessment of wounds, allowing for consistent, non-invasive, and real-time analysis of wound images. CV techniques lessen the burden on clinicians, remove subjective bias, and enhance smart wound-care tools particularly when utilized on mobile or edge devices. Deep architectures (U-Net, Mask R-CNN, EfficientNet encoders) segment wound boundaries and tissue types (granulation, slough, eschar) with high Dice scores in multiple studies; segmentation is the basis for accurate area/depth measurement and temporal comparison. Examples include ASURA, Swift Medical, and commercial tools such as TA that combine segmentation with calibrated scale detection for metric outputs (Griffa et al., 2024; Ramachandram et al., 2022). CV models quantify proportions of tissue types within the wound bed (e.g., percent granulation vs. slough), reducing inter-rater variability and improving objective documentation for clinical decisions such as debridement (Kabir et al., 2025; Ramachandram et al., 2022). Integration of thermal imaging or LiDAR depth sensors improves infection detection (local temperature rises) and real-world area estimation (3D surface mapping), respectively. These multimodal CV pipelines produce more robust measurements than RGB images alone (Morgado et al., 2025; Reifs et al., 2023). CV provides repeatable, fast, and scalable measures suitable for telemedicine and remote monitoring. Accuracy is affected by lighting, camera angle, skin tone, and presence/absence of a calibration object (ruler), so preprocessing, color normalization, and diverse training datasets are essential (Griffa et al., 2024; Ramachandram et al., 2022).

11.4.3 Natural Language Processing (NLP): Extracting Context from Clinical Text

NLP in the healthcare sector aims to derive significant insights from clinical documentation like physician notes, discharge summaries, pathology reports, and nursing records to enhance decision-making, facilitate research, and streamline automated processes. NLP pipelines (rule-based and transformer models like ClinicalBERT) extract wound descriptors, treatments used (dressings, topical agents), infection signs, and social determinants from free text nursing and physician notes, filling gaps where structured EHR fields are missing (Eguia et al., 2024; Scharp et al., 2024). NLP can identify cohorts (e.g., patients with non-healing wounds) for registry creation, retrospective research, and model training without exhaustive manual chart review (Kormilitzin et al., 2021). Automated summaries and trend extraction from narrative notes can flag documentation inconsistencies, prompt guideline-based orders, or generate structured wound reports for multidisciplinary teams (Eguia et al., 2024).

NLP unlocks longitudinal contextual data that images alone cannot provide (systemic disease control, antibiotics, prior interventions). Challenges include variable note quality, domain-specific vocabularies, and the need for de-identification and privacy compliance. Standardized terminologies (SNOMED CT, LOINC) and clinical validation are critical (Netscribes, 2025; Scharp et al., 2024).

Figure 11.3 shows the workflow of image acquisition and preprocessing in AI-driven wound analysis.

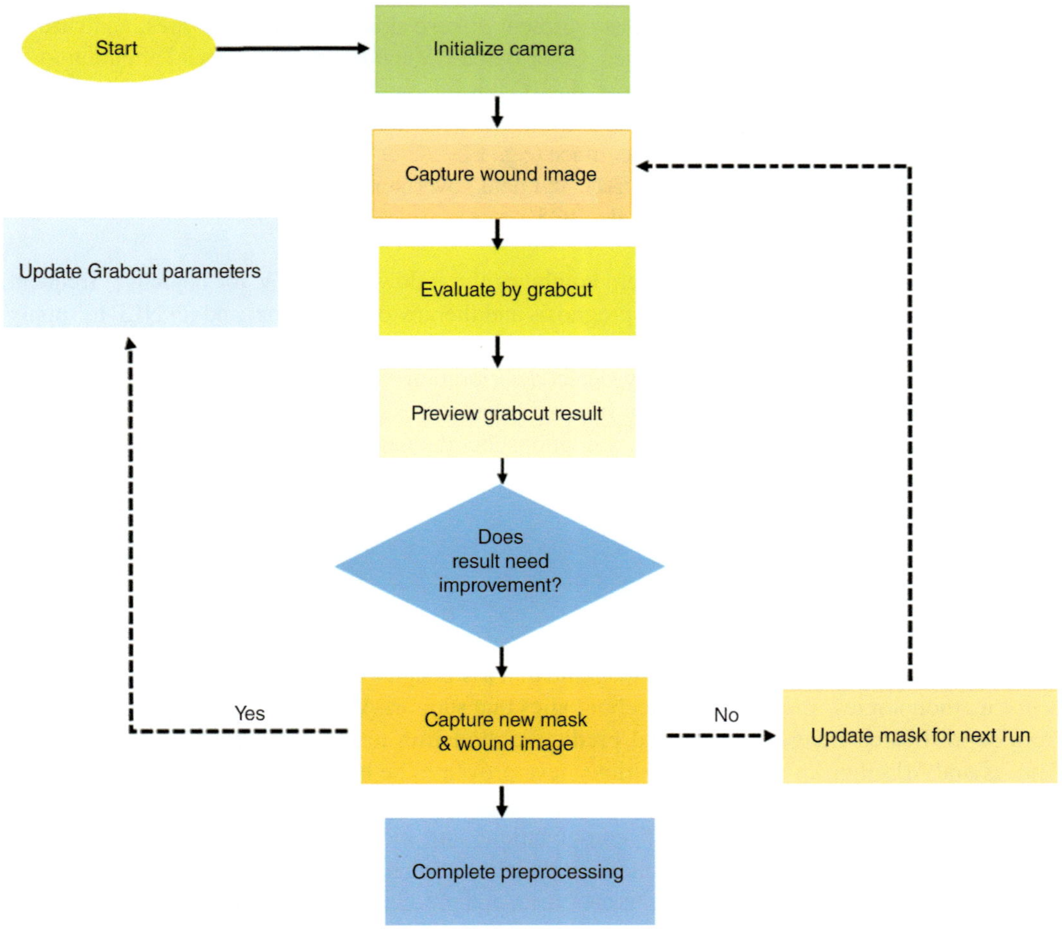

FIGURE 11.3 Workflow of image acquisition and preprocessing in AI-driven wound analysis, illustrating standardization steps before segmentation and measurement.

11.4.4 Multimodal AI and Closed-Loop Systems: the Path to Precision Wound Care

Multimodal systems combine CV (images/sensors), ML (predictive models), and NLP (clinical context) to create richer patient representations. Example frameworks (Deep Multimodal Wound Assessment Tool, DM-WAT) integrate wound images with clinical notes to support home-based triage and referral decisions, improving referral precision and lowering unnecessary clinic visits. Such systems are emblematic of precision wound care: personalized risk prediction, automated monitoring, and context-aware recommendations (Fard et al., 2025). Beyond software, closed-loop visions include smart dressings and sensors that monitor pH, temperature and exudate composition and feed data into ML models to prompt automated alerts or local therapeutic release (research in this area is emerging and multidisciplinary). These systems promise real-time interventions but require rigorous safety and regulatory proof (Al-Akayleh et al., 2025; Chen et al., 2024).

11.4.5 Implementation, Validation, and Regulatory Considerations

Clinical validation and generalizability: Models must be validated across different populations, imaging devices, and care settings; prospective trials are the gold standard for demonstrating clinical benefit (reduction in healing time, fewer complications) (Reifs Jiménez et al., 2025, Griffa et al., 2024).

- *Bias and Equity*: Training data must represent diverse skin tones, wound types, and care contexts to avoid biased performance. Color-based CV models require special attention to melanin-related color shifts (Griffa et al., 2024; Ramachandram et al., 2022).
- *Regulatory Pathway and Clinical Workflow*: AI tools that influence treatment decisions typically require medical device clearance (e.g., FDA/CE). Integration into clinician workflows, interpretability (explainable AI), and data governance (privacy, consent) are essential for adoption (Reifs Jiménez et al., 2025).

A mature precision wound-care system combines the following: (a) CV for objective, frequent measurements; (b) ML for personalized prognosis and therapy optimization; and (c) NLP to supply clinical context from notes and EHRs. When validated and integrated with clinician workflows, these components can reduce subjectivity, detect deterioration earlier, and guide resource allocation (e.g., early specialist referral), ultimately improving outcomes and efficiency. Current evidence is promising (multiple reviews and clinical validations) but the field needs more large prospective trials, standard datasets, and regulatorily robust deployments (Berezo et al., 2022; Griffa et al., 2024; Reifs Jiménez et al., 2025).

11.5 CONCLUSION

The incorporation of AI into the forecasting and evaluation of wound healing signifies a revolutionary change in clinical wound management. This chapter has examined the essential concepts, methodologies, and innovative technologies that support AI-enhanced wound care, ranging from sophisticated image analysis and predictive modeling to real-time monitoring via smart devices and ML algorithms. AI applications have proven to be highly effective in boosting diagnostic precision, enhancing patient outcomes, and alleviating the strain on healthcare systems by facilitating earlier identification of complications and more tailored treatment approaches. Methods such as DL, CNN, and multimodal data integration have exhibited potential in automating wound segmentation, categorizing tissue types, and predicting healing paths with greater reliability. Despite the growing use of AI in wound care, its widespread clinical implementation still encounters numerous obstacles, such as the need for data standardization, the interpretability of AI models, obtaining regulatory approval, and the integration into current healthcare workflows.

Ethical issues, including patient privacy, algorithmic bias, and the necessity for explainable AI, must also be thoroughly considered. Moving forward, it will be essential for clinicians, data scientists, engineers, and regulatory agencies to collaborate in order to convert AI innovations into safe, effective, and scalable clinical applications. As technology progresses, AI is set to become a vital element in the future of wound care, not as a substitute for clinical expertise but as a significant tool that enhances it.

GLOSSARY

Algorithm: A collection of procedures or computations employed by a computer to address issues or make choices, especially in the realm of machine learning.

Artificial intelligence (AI): A field within computer science dedicated to developing systems that can execute tasks usually necessitating human intelligence, including learning, reasoning, and problem-solving.

Artificial neural network: A computational framework inspired by the neural connections of the human brain, proficient in recognizing intricate patterns from information.

Computer vision: An area of artificial intelligence dedicated to empowering machines to analyze and make choices based on visual information (for instance, images of wounds). Chronic wound: a wound that does not progress through a typical, systematic, and timely healing process (such as diabetic foot ulcers or venous ulcers).

Convolutional neural network: A deep learning algorithm that excels in image recognition and classification, frequently utilized for analyzing wound images.

Deep learning: A branch of machine learning that employs neural networks with multiple layers (deep architectures), especially effective in recognizing images and speech.

Digital wound measurement: Use of digital technologies (such as mobile applications, cameras, and AI software) to assess the size, depth, and area of wounds.

Image segmentation: The method of partitioning a digital image into various regions or objects, essential for recognizing wound boundaries in AI applications.

IoMT (Internet of Medical Things): A system of interconnected medical devices and applications that gather, analyze, and transmit health information, including data from wound monitoring devices.

Machine learning: A branch of AI that enables systems to learn from data and enhance their performance through experience without the need for explicit programming.

Neural network: A collection of algorithms aimed at identifying patterns, modeled after the human brain's information processing.

Regression analysis: A statistical technique employed to assess relationships among variables, frequently used for predicting recovery duration.

Reinforcement learning: A category of machine learning in which an AI agent acquires decision-making skills through rewards or penalties.

ROC curve (receiver operating characteristic): A graphical representation that illustrates the effectiveness of a classification model, which is valuable for assessing how accurately the AI differentiates between healed and non-healed wounds.

Smart bandages: Sophisticated wound dressings that incorporate sensors and AI to track parameters such as pH, moisture, and temperature for immediate evaluation. Semantic segmentation: a deep learning technique that categorizes each pixel in an image into a specific class, such as identifying tissue types in a wound image.

Supervised learning: A machine learning approach in which the model is trained using a labeled dataset (where the outcomes are predetermined).

Wound evaluation: The procedure of assessing wound features (dimensions, depth, hue, discharge) to ascertain healing advancement or necessary treatment.

REFERENCES

Abdelrahman, Abubaker, and Serestina Viriri. "EfficientNet family U-Net models for deep learning semantic segmentation of kidney tumors on CT images." *Frontiers in Computer Science* 5 (2023): 1235622.

Al-Akayleh, Faisal, Rami A. Abdel-Rahem, Mayyas Al-Remawi, Faisal Aburub, Ibrahim S. I. Al-Adham, and Ahmed S. A. Ali Agha. "AI-driven tools and methods for wound healing: towards precision wound care and optimized outcomes." In 2025 1st International Conference on Computational Intelligence Approaches and Applications (ICCIAA), pp. 1–5. IEEE, 2025.

Alzubaidi, Laith, Jinglan Zhang, Amjad J. Humaidi, Ayad Al-Dujaili, Ye Duan, Omran Al-Shamma, José Santamaría, Mohammed A. Fadhel, Muthana Al-Amidie, and Laith Farhan. "Review of deep learning: concepts, CNN architectures, challenges, applications, future directions." *Journal of Big Data* 8, no. 1 (2021): 53.

Anisuzzaman, D. M., Chuanbo Wang, Behrouz Rostami, Sandeep Gopalakrishnan, Jeffrey Niezgoda, and Zeyun Yu. "Image-based artificial intelligence in wound assessment: a systematic review." *Advances in Wound Care* 11, no. 12 (2022): 687–709.

Bacanin Dzakula, Nebojsa. "Convolutional neural network layers and architectures." In *Sinteza 2019-international scientific conference on information technology and data related research*, 445–451. Singidunum University, 2019.

Berezo, Matthew, Joshua Budman, Daniel Deutscher, Cathy Thomas Hess, Kyle Smith, and Deanna Hayes. "Predicting chronic wound healing time using machine learning." *Advances in Wound Care* 11, no. 6 (2022): 281–296.

Brem, Harold, Peter Sheehan, Harvey J. Rosenberg, Jillian S. Schneider, and Andrew J. M. Boulton. "Evidence-based protocol for diabetic foot ulcers." *Plastic and Reconstructive Surgery* 117, no. 7S (2006): 193S–209S.

Carion, Nicolas, Francisco Massa, Gabriel Synnaeve, Nicolas Usunier, Alexander Kirillov, and Sergey Zagoruyko. "End-to-end object detection with transformers." In *European conference on computer vision*, 213–229. Cham: Springer International Publishing, 2020.

Chairat, Sawrawit, Sitthichok Chaichulee, Tulaya Dissaneewate, Piyanun Wangkulangkul, and Laliphat Kongpanichakul. "AI-assisted assessment of wound tissue with automatic color and measurement calibration on images taken with a smartphone." *Healthcare* 11, no. 2 (2023): 273.

Chakraborty, Chinmay. "Chronic wound image analysis by particle swarm optimization technique for Tele-wound network." *Wireless Personal Communications* 96, no. 3 (2017): 3655–3671.

Chakraborty, Chinmay. "Computational approach for chronic wound tissue characterization." *Informatics in Medicine Unlocked* 17 (2019): 100162.

Chakraborty, Chinmay, Bharat Gupta, Soumya K. Ghosh, Dev K. Das, and Chandan Chakraborty. "Telemedicine supported chronic wound tissue prediction using classification approaches." *Journal of Medical Systems* 40, no. 3 (2016): 68.

Chang, Angela Christine, Bronwyn Dearman, and John Edward Greenwood. "A comparison of wound area measurement techniques: visitrak versus photography." *Eplasty* 11 (2011): e18.

Chen, Ming-Yao, Ming-Qi Cao, and Tian-Ying Xu. "Progress in the application of artificial intelligence in skin wound assessment and prediction of healing time." *American Journal of Translational Research* 16, no. 7 (2024): 2765.

Chen, Yung-Wei, Jui-Tse Hsu, Chih-Chieh Hung, Jin-Ming Wu, Feipei Lai, and Sy-Yen Kuo. "Surgical wounds assessment system for self-care." *IEEE Transactions on Systems, Man, and Cybernetics: Systems* 50, no. 12 (2018): 5076–5091.

Chino, Daniel Y. T., Lucas C. Scabora, Mirela T. Cazzolato, Ana E. S. Jorge, Caetano Traina-Jr, and Agma J. M. Traina. "Segmenting skin ulcers and measuring the wound area using deep convolutional networks." *Computer Methods and Programs in Biomedicine* 191 (2020): 105376.

Dalal, Navneet, and Bill Triggs. "Histograms of oriented gradients for human detection." In 2005 IEEE Computer Society Conference on Computer Vision and Pattern Recognition (CVPR'05), vol. 1, pp. 886–893. IEEE, 2005.

Dosovitskiy, Alexey, Thomas Unterthiner, Mostafa Dehghani, et al. "An image is worth 16 × 16 words: transformers for image recognition at scale." arXiv preprint arXiv:2010.11929, 2020.

Du, Ke-Lin, and Madisetti N. S. Swamy. *Neural networks and statistical learning.* London: Springer Science & Business Media, 2013.

Dumesnil, Etienne, Divya Konda, Gabriel Aldaz, Hnin Ookhin, David Pickham, and Lauren Aquino Shluzas. "Design and validation of a dynamic digital ruler for hands-free chronic wound assessment." In International Design Engineering Technical Conferences and Computers and Information in Engineering Conference, vol. 57106, p. V003T14A009. American Society of Mechanical Engineers, 2015.

Eguia, Hans, Carlos Luis Sánchez-Bocanegra, Franco Vinciarelli, Fernando Alvarez-Lopez, and Francesc Saigí-Rubió. "Clinical decision support and natural language processing in medicine: systematic literature review." *Journal of Medical Internet Research* 26 (2024): e55315. https://www.jmir.org/2024/1/e55315/

Fard, Reza Saadati, Emmanuel Agu, Palawat Busaranuvong, Deepak Kumar, Shefalika Gautam, Bengisu Tulu, and Diane Strong. "Multimodal AI for home wound patient referral decisions from images with specialist annotations." *IEEE Journal of Translational Engineering in Health and Medicine* (2025). https://ieeexplore.ieee.org/abstract/document/11078373

Fauzi, Mohammad Faizal Ahmad, Ibrahim Khansa, Karen Catignani, Gayle Gordillo, Chandan K. Sen, and Metin N. Gurcan. "Computerized segmentation and measurement of chronic wound images." *Computers in Biology and Medicine* 60 (2015): 74–85.

Floridi, Luciano, Josh Cowls, Monica Beltrametti, Raja Chatila, Patrice Chazerand, Virginia Dignum, Christoph Luetge, et al. "AI4People—an ethical framework for a good AI society: opportunities, risks, principles, and recommendations." *Minds and Machines* 28, no. 4 (2018): 689–707.

Fong, Khi Yung, Tina Peiting Lai, Kai Siang Chan, Isabel Jia Le See, Cheng Cheng Goh, Sivakami Muthuveerappa, Audrey Huimin Tan, Shanying Liang, and Zhiwen Joseph Lo. "Clinical validation of a smartphone application for automated wound measurement in patients with venous leg ulcers." *International Wound Journal* 20, no. 3 (2023): 751–760.

Gao, Lei, Zi-Xuan Liu, and Jiang-Ning Wang. "Predictive model and risk analysis for outcomes in diabetic foot ulcer using eXtreme Gradient Boosting algorithm and SHapley Additive exPlanation." *World Journal of Diabetes* 16, no. 7 (2025): 104789.

García-Zapirain, Begona, Mohammed Elmogy, Ayman El-Baz, and Adel S. Elmaghraby. "Classification of pressure ulcer tissues with 3D convolutional neural network." *Medical & Biological Engineering & Computing* 56, no. 12 (2018): 2245–2258. https://link.springer.com/article/10.1007/s11517-018-1835-y

Gonzalez, Rafael C, and Richard E. Woods. Digital image processing (3rd ed.). New Delhi: Pearson Education India. (2009) ISBN: 978-81-317-1313-9.

Goodfellow, Ian, Yoshua Bengio, Aaron Courville, and Yoshua Bengio. *Deep learning.* Vol. 1, no. 2. Cambridge: MIT Press, 2016.

Goodfellow, Ian., Jean Pouget-Abadie, Mehdi Mirza, Bing Xu, David Warde-Farley, Sherjil Ozair, Aaron Courville, and Yoshua Bengio. "Generative adversarial networks." *Communications of the ACM* 63, no. 11 (2020): 139–144.

Griffa, Davide, Alessio Natale, Yuri Merli, Michela Starace, Nico Curti, Martina Mussi, Gastone Castellani, Davide Melandri, Bianca Maria Piraccini, and Corrado Zengarini. "Artificial intelligence in wound care: a narrative review of the currently available mobile apps for automatic ulcer segmentation." *BioMedInformatics* 4, no. 4 (2024): 2321–2337.

Gupta, Rishabh, Lucas Goldstone, Shira Eisen, Dhanesh Ramachandram, Amy Cassata, Robert D. J. Fraser, Jose L. Ramirez-GarciaLuna, Robert Bartlett, and Justin Allport. "Towards an AI-based objective prognostic model for quantifying wound healing." *IEEE Journal of Biomedical and Health Informatics* 28, no. 2 (2023): 666–677.

Hastie, Trevor, Robert Tibshirani, Jerome H. Friedman, and Jerome H. Friedman. *The elements of statistical learning: data mining, inference, and prediction.* Vol. 2. New York: Springer, 2009.

Hess, Cathy Thomas. "Misdiagnosis of wounds." *Advances in Skin & Wound Care* 32, no. 3 (2019): 144.

Humady, Khaled, Yasmeen Al-Saeed, Nabila Eladawi, Ahmed Elgarayhi, Mohammed Elmogy, and Mohammed Sallah. "Efficient liver segmentation with 3D CNN using computed tomography scans." arXiv preprint arXiv:2208.13271, 2022.

Ismail, Safaa S. I., Romany F. Mansour, Rasha M. Abd El-Aziz, and Ahmed I. Taloba. "Efficient e-mail spam detection strategy using genetic decision tree processing with NLP features." *Computational Intelligence and Neuroscience* 2022, no. 1 (2022): 7710005.

Jain, Anil K. Fundamentals of digital image processing. 1989.

James, Gareth, Daniela Witten, Trevor Hastie, and Robert Tibshirani. *An introduction to statistical learning: with applications in R.* Vol. 103. New York: Springer, 2013.

Jolliffe, Ian. "Principal component analysis." In *International encyclopedia of statistical science*, 1094–1096. Berlin, Heidelberg: Springer, 2011.

Jurafsky, Dan, and James H. Martin. Speech and language processing. Draft of 3rd edition. 2020.

Kabir, Muhammad Ashad, Nidita Roy, Md Ekramul Hossain, Jill Featherston, and Sayed Ahmed. "Deep learning for wound tissue segmentation: a comprehensive evaluation using a novel dataset." arXiv preprint arXiv:2502.10652, 2025. https://arxiv.org/abs/2502.10652

Khoo, Rachel, and Shirley Jansen. "The evolving field of wound measurement techniques: a literature review." *Wounds: a Compendium of Clinical Research and Practice* 28, no. 6 (2016): 175–181.

Kim, Renaid B., Jonathan Gryak, Abinash Mishra, Can Cui, SM Reza Soroushmehr, Kayvan Najarian, and James S. Wrobel. "Utilization of smartphone and tablet camera photographs to predict healing of diabetes-related foot ulcers." *Computers in Biology and Medicine* 126 (2020): 104042.

Kormilitzin, Andrey, Nemanja Vaci, Qiang Liu, and Alejo Nevado-Holgado. "Med7: a transferable clinical natural language processing model for electronic health records." *Artificial Intelligence in Medicine* 118 (2021): 102086.

Koushik, Jayanth. "Understanding convolutional neural networks." arXiv preprint arXiv:1605.09081, 2016.

LeCun, Yann, Yoshua Bengio, and Geoffrey Hinton. "Deep learning." *Nature* 521, no. 7553 (2015): 436–444.

Ledig, Christian, Lucas Theis, Ferenc Huszár, Jose Caballero, Andrew Cunningham, Alejandro Acosta, Andrew Aitken, et al. "Photo-realistic single image super-resolution using a generative adversarial network." In Proceedings of the IEEE conference on computer vision and pattern recognition, pp. 4681–4690. 2017.

Li, Shuxin, Paul Renick, Jon Senkowsky, Ashwin Nair, and Liping Tang. "Diagnostics for wound infections." *Advances in Wound Care* 10, no. 6 (2021): 317–327.

Lien, Angela Shin-Yu, Chen-Yao Lai, Jyh-Da Wei, Hui-Mei Yang, Jiun-Ting Yeh, and Hao-Chih Tai. "A granulation tissue detection model to track chronic wound healing in DM foot ulcers." *Electronics* 11, no. 16 (2022): 2617.

Liu, Huazhen, Wenbin Sun, Weihuang Cai, Kaidi Luo, Chunxiang Lu, Aoxiang Jin, Jiantao Zhang, and Yuanyuan Liu. "Current status, challenges, and prospects of artificial intelligence applications in wound repair theranostics." *Theranostics* 15, no. 5 (2025a): 1662.

Liu, Tom J., Hanwei Wang, Mesakh Christian, Che-Wei Chang, Feipei Lai, and Hao-Chih Tai. "Automatic segmentation and measurement of pressure injuries using deep learning models and a LiDAR camera." *Scientific Reports* 13, no. 1 (2023): 680. https://www.nature.com/articles/s41598-022-26812-9

Liu, Xiujuan, Jun Mao, Ning Sun, Xiangrong Yu, Lei Chai, Ye Tian, Jianming Wang, et al. "Deep learning for detection of intracranial aneurysms from computed tomography angiography images." *Journal of Digital Imaging* 36, no. 1 (2023): 114–123.

Liu, Yilin, Brendon M. Nacewicz, Gengyan Zhao, Nagesh Adluru, Gregory R. Kirk, Peter A. Ferrazzano, Martin A. Styner, and Andrew L. Alexander. "A 3D fully convolutional neural network with top-down attention-guided refinement for accurate and robust automatic segmentation of amygdala and its subnuclei." *Frontiers in Neuroscience* 14 (2020): 260.

Liu, Zixuan, Dehua Wei, Jiangning Wang, and Lei Gao. "Predicting major amputation risk in diabetic foot ulcers using comparative machine learning models for enhanced clinical decision-making." *Scientific Reports* 15, no. 1 (2025b): 28103.

Liu, Ziyang, Emmanuel Agu, Peder Pedersen, Clifford Lindsay, Bengisu Tulu, and Diane Strong. "Comprehensive assessment of fine-grained wound images using a patch-based CNN with context-preserving attention." *IEEE Open Journal of Engineering in Medicine and Biology* 2 (2021): 224–234.

Liu, Ziyang, Emmanuel Agu, Peder Pedersen, Clifford Lindsay, Bengisu Tulu, and Diane Strong. "Chronic wound image augmentation and assessment using semi-supervised progressive multi-granularity efficientnet." *IEEE Open Journal of Engineering in Medicine and Biology* 5 (2023): 404–420.

Long, Jonathan, Evan Shelhamer, and Trevor Darrell. "Fully convolutional networks for semantic segmentation." In Proceedings of the IEEE conference on computer vision and pattern recognition, pp. 3431–3440. 2015.

Lucas, Bruce D., and Takeo Kanade. "An iterative image registration technique with an application to stereo vision." In IJCAI'81: 7th International Joint Conference on Artificial Intelligence, vol. 2, pp. 674–679. 1981.

Maimon, Oded, and Lior Rokach, eds. *Data mining and knowledge discovery handbook*. Vol. 2, no. 2005. New York: Springer, 2005.

Masson-Meyers, Daniela S., Thiago A. M. Andrade, Guilherme F. Caetano, Francielle R. Guimaraes, Marcel N. Leite, Saulo N. Leite, and Marco Andrey C. Frade. "Experimental models and methods for cutaneous wound healing assessment." *International Journal of Experimental Pathology* 101, no. 1–2 (2020): 21–37.

McInnes, Leland, John Healy, and James Melville. "Umap: uniform manifold approximation and projection for dimension reduction." arXiv preprint arXiv:1802.03426, 2018.

Mitchell, Tom M. "Does machine learning really work?" *AI Magazine* 18, no. 3 (1997): 11.

Mitchell, Tom M. "Machine learning and data mining." *Communications of the ACM* 42, no. 11 (1999): 30–36.

Moerland, Thomas M., Joost Broekens, Aske Plaat, and Catholijn M. Jonker. "Model-based reinforcement learning: a survey." *Foundations and Trends® in Machine Learning* 16, no. 1 (2023): 1–118.

Mohamed, Yasir Abdelgadir, Akbar Khanan, Abdul Hakim H. M., Mohamed, Mohamed Bashir, Mousab A. E. Adiel, and Muawia A. Elsadig. "The impact of artificial intelligence on language translation: a review." *IEEE Access* 12 (2024): 25553–25579.

Morgado, Ana C., Rafaela Carvalho, Ana Filipa Sampaio, and Maria JM Vasconcelos. "Enhancing chronic wound assessment through agreement analysis and tissue segmentation." *Scientific Reports* 15, no. 1 (2025): 22244. https://www.nature.com/articles/s41598-025-06703-5

Morris, John N., Katherine Berg, Eva Topinkova, Leonard C. Gray, and Erez Schachter. "Developing quality indicators for in-patient post-acute care." *BMC Geriatrics* 18, no. 1 (2018): 161.

Mostafalu, Pooria, Ali Tamayol, Rahim Rahimi, Manuel Ochoa, Akbar Khalilpour, Gita Kiaee, Iman K. Yazdi, et al. "Smart bandage for monitoring and treatment of chronic wounds." *Small* 14, no. 33 (2018): 1703509.

Mukherjee, Rashmi, Dhiraj Dhane Manohar, Dev Kumar Das, Arun Achar, Analava Mitra, and Chandan Chakraborty. "Automated tissue classification framework for reproducible chronic wound assessment." *BioMed Research International* 2014, no. 1 (2014): 851582.

Netscribes. "NLP in EHR: a game-changer for unstructured data." *Netscribes Blog* (June 2025). https://dev.netscribes.com/nlp-in-ehr/

Nizam, Khajista, Mohammad Faizal Ahmad Fauzi, Nurul Nadia Ahmad, and H. K. Nair. "Characterization of tissues in chronic wound images." In 2018 IEEE Student Conference on Research and Development (SCOReD), pp. 1–5. IEEE, 2018.

Pan, Sinno Jialin, and Qiang Yang. "A survey on transfer learning." *IEEE Transactions on Knowledge and Data Engineering* 22, no. 10 (2009): 1345–1359.

Patel, Tatsat R., Aakash Patel, Sricharan S. Veeturi, Munjal Shah, Muhammad Waqas, Andre Monteiro, Ammad A. Baig, et al. "Evaluating a 3D deep learning pipeline for cerebral vessel and intracranial aneurysm segmentation from computed tomography angiography–digital subtraction angiography image pairs." *Neurosurgical Focus* 54, no. 6 (2023): E13.

Polydoros, Athanasios S., and Lazaros Nalpantidis. "Survey of model-based reinforcement learning: applications on robotics." *Journal of Intelligent & Robotic Systems* 86, no. 2 (2017): 153–173.

Ramachandram, Dhanesh, Jose Luis Ramirez-GarciaLuna, Robert D. J. Fraser, Mario Aurelio Martínez-Jiménez, Jesus E. Arriaga-Caballero, and Justin Allport. "Fully automated wound tissue segmentation using deep learning on mobile devices: cohort study." *JMIR mHealth and uHealth* 10, no. 4 (2022): e36977.

Reifs Jiménez, David, Lorena Casanova-Lozano, Sergi Grau-Carrión, and Ramon Reig-Bolaño. "Artificial intelligence methods for diagnostic and decision-making assistance in chronic wounds: a systematic review." *Journal of Medical Systems* 49, no. 1 (2025): 1–39.

Reifs, David, Lorena Casanova-Lozano, Ramon Reig-Bolano, and Sergi Grau-Carrion. "Clinical validation of computer vision and artificial intelligence algorithms for wound measurement and tissue classification in wound care." *Informatics in Medicine Unlocked* 37 (2023): 101185. https://www.sciencedirect.com/science/article/pii/S2352914823000278

Ren, Shaoqing, Kaiming He, Ross Girshick, and Jian Sun. "Faster R-CNN: towards real-time object detection with region proposal networks." *Advances in Neural Information Processing Systems* 28 (2015).

Ronneberger, Olaf, Philipp Fischer, and Thomas Brox. "U-net: convolutional networks for biomedical image segmentation." In *International conference on Medical image computing and computer-assisted intervention*, 234–241. Cham: Springer International Publishing, 2015.

Russell, Stuart J., and Peter Norvig. Artificial intelligence: a modern approach, global edition 4e. 2021.

Sanger, Patrick C., Gabrielle H. van Ramshorst, Ezgi Mercan, Shuai Huang, Andrea L. Hartzler, Cheryl A. L. Armstrong, Ross J. Lordon, William B. Lober, and Heather L. Evans. "A prognostic model of surgical site infection using daily clinical wound assessment." *Journal of the American College of Surgeons* 223, no. 2 (2016): 259–270.

Scebba, Gaetano, Jia Zhang, Sabrina Catanzaro, Carina Mihai, Oliver Distler, Martin Berli, and Walter Karlen. "Detect-and-segment: a deep learning approach to automate wound image segmentation." *Informatics in Medicine Unlocked* 29 (2022): 100884.

Scharp, Danielle, Mollie Hobensack, Anahita Davoudi, and Maxim Topaz. "Natural language processing applied to clinical documentation in post-acute care settings: a scoping review." *Journal of the American Medical Directors Association* 25, no. 1 (2024): 69–83.

Sen, Chandan K., Subhadip Ghatak, Surya C. Gnyawali, Sashwati Roy, and Gayle M. Gordillo. "Cutaneous imaging technologies in acute burn and chronic wound care." *Plastic and Reconstructive Surgery* 138, no. 3S (2016): 119S–128S.

Shandilya, Shishir Kumar, Agni Datta, Yash Kartik, and Atulya Nagar. "Role of artificial intelligence and machine learning." In *Digital resilience: navigating disruption and safeguarding data privacy*, 313–399. Cham: Springer Nature Switzerland, 2024.

Shetty, Shruthi H., Sumiksha Shetty, Chandra Singh, and Ashwath Rao. "Supervised Machine Learning: Algorithms and Applications." In Fundamentals and Methods of Machine and Deep Learning: Algorithms, Tools and Applications, edited by Pradeep Singh, 1–16. Hoboken, NJ, USA: Wiley & Sons, 2022.

Shrivastava, Neeraj, Pushpa Tewari, S. Sujatha, Srinivasa Rao Bogireddy, Neeraj Varshney, and Vinod Sharma. "Natural Language Processing for Conversational AI: Chatbots and Virtual Assistants." In *Proceedings of the 2025 IEEE International Conference on Interdisciplinary Approaches in Technology and Management for Social Innovation (IATMSI)*, vol. 3, 1–6. Gwalior, India: IEEE, 2025.

Shukla, Vidit, and Shilpa Choudhary. "Deep Learning in Neural Networks: An Overview." In *Deep Learning in Visual Computing and Signal Processing*, 29–53. Singapore: Springer, 2022.

Silva, Jorge, and Maria Gomez. "Advancements in human-computer interaction through natural language processing and voice recognition." *Journal of Computing Innovations and Applications* 3, no. 1 (2025): 8–12.

Solomon, Chris, and Toby Breckon. *Fundamentals of Digital Image Processing: A Practical Approach with Examples in MATLAB*. Chichester, UK: John Wiley & Sons, 2011.

Spinazzola, Elisabetta, Guillaume Picaud, Sara Becchi, Monica Pittarello, Elia Ricci, Marc Chaumont, Gérard Subsol, Fabio Pareschi, Luc Teot, and Jacopo Secco. "Chronic ulcers healing prediction through machine learning approaches: preliminary results on diabetic foot ulcers case study." *Journal of Clinical Medicine* 14, no. 9 (2025): 2943. https://www.mdpi.com/2077-0383/14/9/2943

Subramaniam, Thayaalini, Mh Busra Fauzi, Yogeswaran Lokanathan, and Jia Xian Law. "The role of calcium in wound healing." *International Journal of Molecular Sciences* 22, no. 12 (2021): 6486.

Sutton, Richard S., and Andrew G. Barto. *Reinforcement learning: an introduction*. Vol. 1, no. 1. Cambridge: MIT Press, 1998.

Szeliski, Richard. *Computer Vision: Algorithms and Applications*. 2nd ed. Cham, Switzerland: Springer Nature, 2022.

Szepesvari, Csaba. *Algorithms for Reinforcement Learning*. Cham, Switzerland: Springer Nature, 2022.

Touvron, Hugo, Matthieu Cord, Matthijs Douze, Francisco Massa, Alexandre Sablayrolles, and Hervé Jégou. "Training data-efficient image transformers & distillation through attention." In International conference on machine learning, pp. 10347–10357. PMLR, 2021.

Veredas, Francisco J., Rafael M. Luque-Baena, Francisco J. Martín-Santos, Juan C. Morilla-Herrera, and Laura Morente. "Wound image evaluation with machine learning." *Neurocomputing* 164 (2015): 112–122.

Viola, Paul, and Michael Jones. "Rapid Object Detection Using a Boosted Cascade of Simple Features." In *Proceedings of the 2001 IEEE Computer Society Conference on Computer Vision and Pattern Recognition (CVPR 2001)*, vol. 1, I–511–I–518. Kauai, HI: IEEE, 2001.

Vossough, Arastoo, Nastaran Khalili, Ariana M. Familiar, Deep Gandhi, Karthik Viswanathan, Wenxin Tu, Debanjan Haldar, et al. "Training and comparison of nnU-Net and DeepMedic methods for autosegmentation of pediatric brain tumors." *American Journal of Neuroradiology* 45, no. 8 (2024): 1081–1089.

Wang, Changhan, Xinchen Yan, Max Smith, Kanika Kochhar, Marcie Rubin, Stephen M. Warren, James Wrobel, and Honglak Lee. "A unified framework for automatic wound segmentation and analysis with deep convolutional neural networks." In 2015 37th Annual International Conference of the IEEE Engineering in Medicine and Biology Society (EMBC), pp. 2415–2418. IEEE, 2015. https://pubmed.ncbi.nlm.nih.gov/26736781/

Wang, Chuanbo, D. M. Anisuzzaman, Victor Williamson, Mrinal Kanti Dhar, Behrouz Rostami, Jeffrey Niezgoda, Sandeep Gopalakrishnan, and Zeyun Yu. "Fully automatic wound segmentation with deep convolutional neural networks." *Scientific Reports* 10, no. 1 (2020): 21897. https://www.nature.com/articles/s41598-020-78799-w

Wang, Shiqi, Chao Xia, Qirui Zheng, Aiping Wang, and Qian Tan. "Machine Learning Models for Predicting the Risk of Hard-to-Heal Diabetic Foot Ulcers in a Chinese Population." *Diabetes, Metabolic Syndrome and Obesity: Targets and Therapy* 15 (2022): 3347–3359.

Wang, Yichuan, LeeAnn Kung, and Terry Anthony Byrd. "Big data analytics: understanding its capabilities and potential benefits for healthcare organizations." *Technological Forecasting and Social Change* 126 (2018): 3–13.

Wojke, Nicolai, Alex Bewley, and Dietrich Paulus. "Simple online and realtime tracking with a deep association metric." In 2017 IEEE International Conference on Image Processing (ICIP), pp. 3645–3649. IEEE, 2017.

Yosinski, Jason, Jeff Clune, Yoshua Bengio, and Hod Lipson. "How transferable are features in deep neural networks?" *Advances in Neural Information Processing Systems* 27 (2014).

Young, Tom, Devamanyu Hazarika, Soujanya Poria, and Erik Cambria. "Recent trends in deep learning based natural language processing." *IEEE Computational Intelligence Magazine* 13, no. 3 (2018): 55–75.

Yu, Xin, Zhuo Wu, and Nan Zhang. "Machine learning-driven discovery of novel therapeutic targets in diabetic foot ulcers." *Molecular Medicine* 30, no. 1 (2024): 215.

Zhang, Qianru, Meng Zhang, Tinghuan Chen, Zhifei Sun, Yuzhe Ma, and Bei Yu. "Recent advances in convolutional neural network acceleration." *Neurocomputing* 323 (2019): 37–51.

Zhang, Qingchen, Laurence T. Yang, Zhikui Chen, and Peng Li. "A survey on deep learning for big data." *Information Fusion* 42 (2018): 146–157.

Index

H

I

J

K

L

M

N

O

P

Q

R